T0334519

Power Generation Technologies

Power Generation Technologies

Foundations, Design and Advances

Masood Ebrahimi

Associate Professor, Mechanical Engineering Department, University of Kurdistan, Iran

ACADEMIC PRESS

An imprint of Elsevier

ELSEVIER

Academic Press is an imprint of Elsevier
125 London Wall, London EC2Y 5AS, United Kingdom
525 B Street, Suite 1650, San Diego, CA 92101, United States
50 Hampshire Street, 5th Floor, Cambridge, MA 02139, United States
The Boulevard, Langford Lane, Kidlington, Oxford OX5 1GB, United Kingdom

Copyright © 2023 Elsevier Inc. All rights reserved.

No part of this publication may be reproduced or transmitted in any form or by any means, electronic or mechanical, including photocopying, recording, or any information storage and retrieval system, without permission in writing from the publisher. Details on how to seek permission, further information about the Publisher's permissions policies and our arrangements with organizations such as the Copyright Clearance Center and the Copyright Licensing Agency, can be found at our website: www.elsevier.com/permissions.

This book and the individual contributions contained in it are protected under copyright by the Publisher (other than as may be noted herein).

Notices
Knowledge and best practice in this field are constantly changing. As new research and experience broaden our understanding, changes in research methods, professional practices, or medical treatment may become necessary.

Practitioners and researchers must always rely on their own experience and knowledge in evaluating and using any information, methods, compounds, or experiments described herein. In using such information or methods they should be mindful of their own safety and the safety of others, including parties for whom they have a professional responsibility.

To the fullest extent of the law, neither the Publisher nor the authors, contributors, or editors, assume any liability for any injury and/or damage to persons or property as a matter of products liability, negligence or otherwise, or from any use or operation of any methods, products, instructions, or ideas contained in the material herein.

ISBN: 978-0-323-95370-2

For information on all Academic Press publications visit our
website at https://www.elsevier.com/books-and-journals

Publisher: Charlotte Cockle
Acquisitions Editor: Graham Nisbet
Editorial Project Manager: Andrae Akeh
Production Project Manager: Prasanna Kalyanaraman
Cover Designer: Matthew Limbert

Typeset by TNQ Technologies

Working together
to grow libraries in
developing countries

www.elsevier.com • www.bookaid.org

Dedicated to the scientists who,
They protect the planet from global warming and greenhouse gases.

Contents

About the author

Masood Ebrahimi received his PhD in Mechanical Engineering from the K. N. Toosi University of technology. Dr. Ebrahimi has taught university and industrial courses in power plants, thermodynamics, fluid mechanics, turbomachinery, heat exchangers, engines, operation and maintenance of industrial equipment, and troubleshooting procedures. Dr. Ebrahimi published a book in 2015 with Elsevier titled "*Combined Cooling Heating And Power, Decision-Making, Design, And Optimization*"; in addition, he has translated a book from English to Persian titled "*Gas Pipeline Hydraulics.*" Furthermore, Dr. Ebrahimi has published numerous high-quality journal articles mostly in the Elsevier journals such as *Energy, Energy and Buildings, Applied Thermal Engineering, Energy Conversion and Management,* and *Journal Of Cleaner Production.* He has completed several industrial projects and presented many articles in the conferences worldwide. He is currently associate Prof. of mechanical engineering at the University of Kurdistan.

Preface

Electricity has improved the quantity and quality of human life and increased public welfare. It is enough to look around us. We will realize that if there is no electricity, almost nothing can be done. The clothes we wear, the food we cook, the car we travel with, the lighting of houses, streets and parks, the movies we watch in cinemas, the programs we follow, the radio listen to, the mobile phones we use to call our families, Internet, computers, 2D and 3D printers, elevators we use in High-rise towers and buildings, fire alarm, and extinguishing systems, the water we drink at home, traffic lights at junctions, airport control towers, airplanes flying, launching satellites, trains, and thousands of other things all depend on electricity. We humans have become too dependent on electricity. Our dependence on electricity and power is becoming a threat. The buildings in which we live and work, the transportation system that moves us, and the industries that produce our needs alone account for more than 73% of the total energy consumed in the world. The production of electricity, heat, and transportation also has the greatest impact on the production of greenhouse gases. Greenhouse gases are a serious threat to human existence on the planet earth by causing global warming. If the global warming is not controlled, the increase of the average temperature reaches more than 1.5°C. It means that the efforts made in the Paris Agreement and the Net Zero Emissions by 2050 Scenario (NZE) have failed, and we must be ready to face its disastrous consequences. What many people especially, the politicians, do not know is that the global warming crisis will lead to other crises about drinking water and land, which will be enough reasons for international conflicts and even world wars.

While we are highly dependent on electricity, we must do something to limit its effects on the environment. This planet should be handed over to the future and our children in a better way than it is now. With their tireless efforts, scientists have always come to the aid of their fellows in difficult situations. The last example was the COVID-19 pandemic, if it was not for their efforts, we would have lost more friends and relatives. In the upcoming discussion of this book, which deals with power generation technologies, the efforts of scientists and industrialists who have made many efforts to reduce environmental pollutants by trying to develop, diversify, and improve power generation systems will be discussed.

The contents of this book have been prepared and organized so that students of mechanical engineering, electrical engineering, employees of power plants, and companies active in the field of power generation technologies can use it for learning or retraining. Almost all the chapters have been enriched with practical examples, and most of the examples have been analyzed parametrically or optimized by using coding in Engineering Equation Solver (EES) software. The purpose of the examples is to deepen the reader's understanding of the content, and the purpose of the codes presented after the examples is to give the reader the opportunity to learn coding in the EES and to solve more complex or different problems by developing the given codes. Correspondingly, these codes provide the possibility of investigating the effect of different design parameters on the evaluation criteria of power generation cycles. In addition, at the end of each chapter, some problems are designed, in which most of them can be solved by developing the codes given in the chapter.

In the end, the author wishes that this book can provide acceptable knowledge and experience to the readers and be used by professors, students, and industries as a reliable source along with other sources in the field of power generation technologies. Correspondingly, the author is willing to accept any kind of criticism, suggestions, comments, and corrections provided by readers and specialists.

Masood Ebrahimi

July 25, 2022

Introduction to power generation

1

Chapter outline

1. Power importance

Life without power looks like a king without a throne. Education, health, water, food, internet, telephone, refrigerator, and any other essential things for today's life are founded on power or its most useful form, electricity. Without these things, life quality will flash back to hundreds years ago. Table 1.1 shows the 20 greatest engineering achievements in 20th century identified by the National Academy of Engineering at the United States [1]. It shows while "electrification" is on the top of list, approximately 18 out of 19 other achievements rely on electricity and power.

In an analysis published by Power Magazine [2], it was revealed that there is a statistical connection between the Human Development Index (HDI) and electricity consumption. In that analysis, they considered 38 countries and divided them into two nineteen-country groups of high electricity consumers with annual per capita electricity usage of at or above 2000 kWh and low electricity consumers with annual

Table 1.1 20 greatest engineering achievements in 20th century [1].

1. Electrification	**11.** Highway
2. Automobile	**12.** Spacecraft
3. Airplane	**13.** Internet
4. Water supply and distribution	**14.** Imaging
5. Electronics	**15.** Household appliances
6. Radio and TV	**16.** Health technologies
7. Agricultural mechanization	**17.** Petroleum and petrochemical technologies
8. Computers	**18.** Laser and fiber optics
9. Telephone	**19.** Nuclear technologies
10. Air conditioning and refrigeration	**20.** High performance materials

Power Generation Technologies. https://doi.org/10.1016/B978-0-323-95370-2.00005-3
Copyright © 2023 Elsevier Inc. All rights reserved.

per capita electricity usage of below 2000 kWh. They found out countries with higher electricity consumption, have higher mean HDI.

Fig. 1.1 shows the share of population with access to electricity. Here, the electricity access means, having an electricity source that can provide basic lighting, phone charge, or power a radio for 4 h a day.

Unfortunately, there are still some countries in the world, mostly in Africa, with electricity access of less than 50% and in some cases even less than 10%. It is clear that HDI and also the GDP[1] per capita in these countries would be very low (Fig. 1.2) in comparison with the world GDP average.

2. Power in statistics

Now that the importance of power in human life is undeniable, we need to know how it is generated? What does it need to be generated? and What sources of energy can be used to generate power?

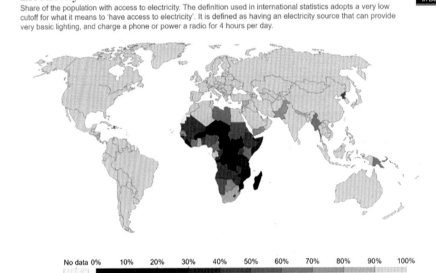

Figure 1.1 The share of population with access to electricity (ourworldindata.org) [3].

---

[1] Gross domestic product (GDP) per capita measures a country's economic output per person and is calculated by dividing the GDP of a country by its population.

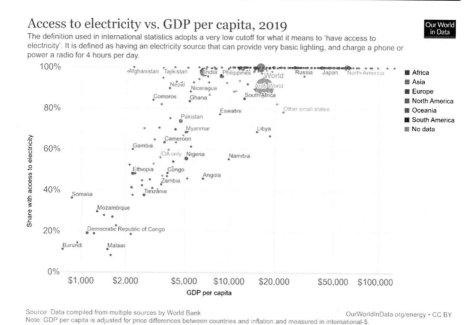

Figure 1.2 Share of population access to electricity versus gross domestic product (GDP) per capita (ourworldindata.org) [3].

Fig. 1.3 shows the electricity production shares in the world by the various energy source. It shows that coal, oil, gas, hydro, wind, solar, and other renewable energies such as bio-fuel, biogas, geothermal, etc. are used today to generate electricity. In addition, it shows the need for electricity has increased more than 150% in comparison with 1985. However, as it can be seen, the share of different energy sources is different, and the share of some energies is too small. Special arrangements must be made to ensure that reliable electricity with reasonable price is available to consumers everywhere, anytime. To reach this point, electricity generation technologies and energy resources must improve from different points of views.

First of all, various technologies must be used to generate electricity. Dependency on only one technology or energy source is fatal, but combining of different power generators using divers energy resources can improve the reliability, availability, and price of the electricity generation. Fig. 1.4 shows although various energy sources are used for electricity generation, but still fossil fuels are the dominant energy source. In 2020, fossil fuels, nuclear, and renewables produce 60.92%, 10.12%, and 28.97% of the total electricity, while 10 years ago, in 2010, they produced 67.04%, 12.9%, and 20.06% of the total electricity. As it can be seen, the share of fossil fuels and nuclear are reduced during the 10 years, but the share of renewable electricity has increases by 8.91%. Although the share of renewables is still low, but if the current trend continues in the following decades, the share of renewables will increase significantly. This increase of the renewable energy share is very welcome to the world. Because it reduces

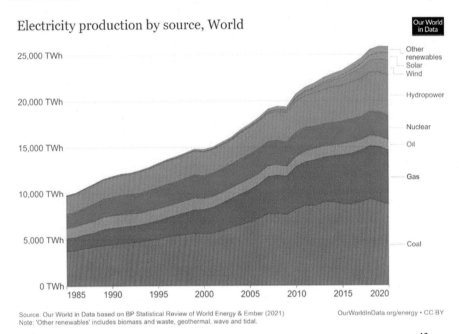

Figure 1.3 Electricity production by source in the world in terawatt hour (1TWh $= 10^{12}$Wh) (ourworldindata.org) [3].

air pollution and greenhouse gases (GHGs), and improves passive defense systems for the time of crises, such as earth quakes, wars, etc. Fig. 1.5 shows the rapid growth of solar energy generation in the world and some pioneer countries during the last decade. This sharp growth of solar energy generation is due to significant reduction of solar PV (photovoltaic) module cost from 106.09 $\frac{US\$}{W}$ in 1976 to 0.38 $\frac{US\$}{W}$ in 2019. The lower the solar PV price index, the greater the popularity of using this system, especially if the technology can compete with fossil fuel—based technologies in terms of initial cost (Fig. 1.6). Developing renewable energies reduces the death toll due to energy production as well. Fig. 1.7 shows the death rates due to energy production from different energy sources per terawatt hour (TWh). It shows that brown coal kills about 32.72 people per 1 TWh, while it is only 0.02 people for 1 TWh of solar and hydropower energies.

According to the discussions and the data, renewable energies have many more advantages over the fossil fuels. They need more developments and more price reduction. In addition, they need more efficiency improvements. To achieve this, special attention must be paid to the researches and investigations on these energies. Fig. 1.8 shows the installed global renewable energy capacity in MW by hydropower, wind, solar, bioenergy, and geothermal. It shows that hydropower plants are at the front head of renewables, while the rapid growth of wind and solar shares proves that there are still great potentials that need to be paid more attention.

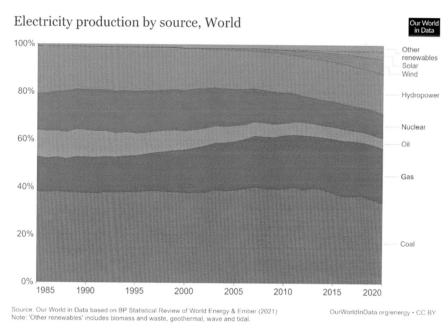

Figure 1.4 Relative electricity production by source in the world (ourworldindata.org) [3].

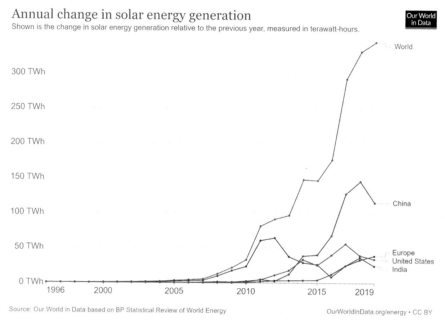

Figure 1.5 Annual change in solar energy generation (ourworldindata.org) [3].

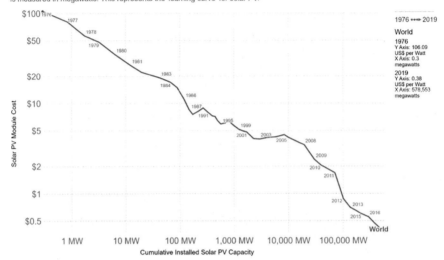

Figure 1.6 Solar photovoltaic module cost during the time versus cumulative capacity (ourworldindata.org) [3].

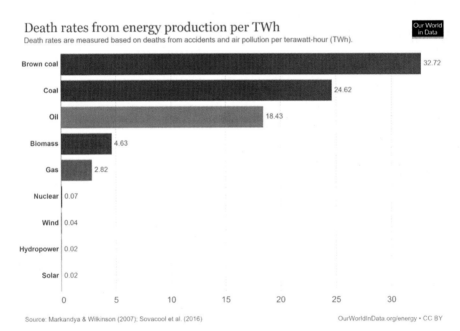

Figure 1.7 The number of death toll per TWh of different energy sources (ourworldindata.org) [3].

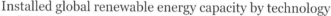

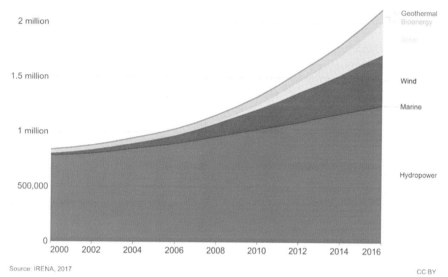

Source: IRENA, 2017   CC BY

Figure 1.8 The installed global renewable energy capacity in MW by technology (ourworldindata.org) [3].

3. Power in history

The idea of power generation in human mind began when they thought of making tools, so that they could use them to do things easier, faster, with higher quality, and at a lower cost. Our ancestors have always needed to generate power to transfer water from rivers to higher places, plowing lands, harvesting agricultural products, carrying large stones in constructions, turning water and windmills, and so on.

The Industrial Revolution shifted to new manufacturing processes in Britain, continental Europe, and the United States from about 1760 to the period between 1820 and 1840. This transfer included the transfer from manual production to machinery, new chemical production, and iron making, increasing use of steam and hydropower, development of machinery, and the emergence of mechanized factory systems [4].

During the Industrial Revolution, when power took on a new position and its impact on the mass production of products in demand, storage, transmission, and distribution was determined, scientists, and craftsmen made many efforts to generate more efficient power in various ways by using different energy sources. The Industrial Revolution began around 1830 with limited and important innovations in textiles, steam power, iron production, and machine tools [5].

The development of the steam engine was one of the most important features of the Industrial Revolution. However, at the beginning of Industrial Revolution, most of the

Table 1.2 The timeline of power generation and power technologies (briefed, and arranged from Power Magazine) [6].

1740s	Electric conduction by Englishman Stephen Gray, which led to the invention of glass friction generators in Germany.
1830s	The first electromagnetic generator, known as the *Faraday disk,* was discovered by British scientist Michael Faraday.
	Discovery of the photovoltaic (PV) effect by French scientist Edmond Becquerel.
1860s	The industrial use of modern dynamo autonomously by Sir Charles Wheatstone, Werner von Siemens and Samuel Alfred Varley.
1880s	DC hydropower electricity at the Wolverine Chair Co. by Michigan's Grand Rapids Electric Light and Power Co.
	The world's first central DC hydroelectric station powered a paper mill in Appleton, Wisconsin.
	The Willamette Falls Station in Oregon City, Oregon. The nation's first AC hydroelectric plant came online.
	Thomas Edison achieved his vision of a full-scale central power station with a system of conductors to distribute electricity to end-users in New York City.
	The first solar cell was created by New York inventor Charles Fritts.
	Sir Charles Parsons, built the first steam turbine generator, 2 years later improve by introducing the first condensing turbine, which drove an AC generator.
	Charles Brush from Ohio invented a 60-foot wind turbine in his backyard.
1890s	Charles Curtis offered an invention of a different turbine to the General Electric.
	Coal-fired power plants with capacity in the 1−10 MW range, prepared with a steam generator, an economizer, evaporator, and a superheater section.
	The world's first pump storage hydro power plant in Switzerland was built.
	GE developed a 500-kW Curtis turbine generator, which admitted high pressure steam to drive rapid rotation of a shaft-mounted disk
	GE delivered the world's first 5-MW steam turbine to the Commonwealth Edison Co. of Chicago.
1910s	The coal-fired power plant efficiency improved by using feedwater heaters on the turbines, and air preheaters on the steam generators to 15%.
1920s	Introducing of once-through boilers and reheat for power plants, along with the Benson steam generator.
1940s	GE installed its first commercial heavy-duty gas turbine for power generation of 3.5 MW.
	The world's first wind turbine with capacity of 1.25 MW connected to the grid on a hill in Castleton, Vermont, called Grandpa's Knob.
1950s	The first supercritical once-through steam generator introduced for steam power plants.
	The firing temperatures of gas turbines reached 1500 °F.
	The first nuclear reactor to produce electricity in Idaho.
	Silicon solar cells were produced commercially.
1960s	The first fully fired boiler combined cycle gas turbine (CCGT) power plant came online, and the first CCGT was outfitted with a heat recovery steam generator (HRSG).
1970s	The world's first integrated coal gasification combined cycle power plant.
	The firing temperatures of gas turbines reached 2000°F
1980s	The first utility-scale wind farms constructed in California.
	The first megawatt-scale PV power station, developed by ARCO Solar, came online in Hesperia, California.
2000s	Residential solar power systems were available for sale.

industrial power was produced by water and wind. By 1800, in Britain, about 10,000 hp was provided by steam and increased to 210,000 hp by 1815.

The Second Industrial Revolution, also known as the Technological Revolution, was the stage of rapid standardization and industrialization from the late 19th century to the early 20th century.

A new revolution began with electricity and electrification. By the 1890s, industrialization in these areas created the first giant industrial corporation companies such as US Steel, General Electric, and Standard Oil. In the following, a timeline of power generation is presented from the Power Magazine. This time line has focused on the first presentation of power technologies including thermal, hydro, wind, and solar systems (Table 1.2).

4. Problems

1. Find out about the beginning of biogas-fueled power plants.
2. What is the electricity consumption share of different sectors such as buildings, agriculture, industry, transportation, etc.? draw or find an appropriate graph.
3. What is the current average electricity price in the world?
4. What is the maximum inlet steam temperature to the steam turbines at the moment?
5. What is the maximum inlet temperature to the gas turbines at the moment?
6. Draw a timeline graph to show the efficiency improvement in steam power plants during the years.
7. Draw a timeline graph to show the efficiency trend in gas turbines during the time.
8. Draw a graph to show the efficiency variations in solar PV panels during the history.
9. Draw a graph to show the efficiency change in wind turbines and hydro turbines during the years.
10. When the first dry cooling tower was invented? By whom?
11. When the first intercooler was used in the gas turbines? by which company?
12. Find the GDP and electricity access share of your own country and compare it with the world average and your neighbor countries.
13. Determine how much of the electricity generation in your country is provided by the renewable energies
14. How much do you buy electricity and what is the mean price in the world?
15. What kind of renewable energies are available in your country? Do you use them for electricity generation in your country?
16. Does your homes are equipped with solar PV panels?

References

[1] http://www.greatachievements.org/.
[2] J. Clemente, The statistical connection between electricity and human development, Powertech. Mag. (2010).
[3] https://ourworldindata.org/.

[4] D.S. Landes, The Unbound Prometheus, Press Syndicate of the University of Cambridge, 1969, ISBN 978-0-521-09418-4, p. 104.

[5] S. Muntone, Second Industrial Revolution, Education.com. The McGraw-Hill Companies, 2013. Retrieved 14 October 2013.

[6] A. Harvey, A. Larson, S. Patel, History of Power: The Evolution of the Electric Generation Industry, Power Magazine, 2020.

Thermodynamics of power plant

2

Chapter outline

Whether or not to construct a power plant in a particular region depends on various analyses. The power plant type (thermal, hydro, solar, etc.) dictates some specific analyses that may not be required for other types of power plants.

For instance, if we are supposed to build a combined cycle gas turbine[1] (CCGT) power plant, it is vital to have access to a steady source of water for the whole lifetime of the power plant. In addition, consuming water for power production must not endanger the water resources needed for agriculture, drinking, and other essential demands. Utilizing a wet cooling tower for the CCGT means the water consumption[2] of about $1 m^3 / MW_e h$. However, using a dry cooling tower would reduce the water consumption to a minimum of about $10^{-3} m^3 / MW_e h$ [1]. It means that a 1000 MW_e power plant equipped with a wet cooling tower would consume $1 (m^3 / MW_e h) \times 1000 (MW_e) \times 6000 \text{ h} = 6 \times 10^6 m^3$ of water for 6000 h of operation per year.

The water consumption of a single person per year is estimated to be 110 to 140 m^3 [2]; this means that the CCGT power plant with capacity of 1000 MW_e consumes the required domestic water for a city with population of about 43,000 to 54,000. According to the discussions given, raising a CCGT with wet cooling tower may endanger the water sources of a city with a considerable population of about 50,000. Hence, available and renewable water source is a prerequisite for a CCGT.

As other examples, the wind and solar irradiation studies are prerequisites for installing wind turbines and solar photovoltaic panels.

---

[1] In a CCGT, the hot exhaust gas of the gas turbine is reused in a heat recovery steam generator to produce steam for a bottoming steam turbine cycle. This technique which is also called *cogeneration* is used to increase the cycle electrical efficiency by producing more electricity.

[2] The volume of water removed from a water source such as well, lake or river is called *"withdrawal water"*. Some portion of withdrawal water may return to its source but the portion which is used and not returned to the main source is called *"water consumption"*. Hence withdrawal water equals water consumption plus returned water.

Power Generation Technologies. https://doi.org/10.1016/B978-0-323-95370-2.00006-5
Copyright © 2023 Elsevier Inc. All rights reserved.

However, beside the analyses that should be carried out regarding the prerequisites of a power plant, economic, environmental, and thermodynamical evaluations are the most essential required analyses before constructing a power plant. In his chapter, attention is given to the thermodynamic concepts, which are needed to evaluate the performance of a power plant from energy and exergy point of views. For this purpose, the mass conservation law, first law of thermodynamics, and second law of thermodynamics are presented, and the text is enriched with practical examples related to the power plant equipment to make a deep understanding about the main concepts.

In addition, at the end of this chapter, some exercises are also provided that not only help to understand the thermodynamic concepts much better but also give some new technical data about other types of power plants and related equipment. This helps the reader to become more familiar with the power related technologies.

1. Conservation of matter

In the absence of nuclear reaction, the amount of mass contained in an arbitrary volume of V remains constant, while its volume may change during time. This statement is called *conservation of matter, continuity equation,* or *mass conservation law* and can be formulated as below [3]:

$$\frac{D}{Dt} \int_V \rho dV = 0 \qquad (2.1)$$

In which D/Dt is the material derivative and $\int_V \rho dV$ is the mass of matter contained in the arbitrary volume of V.

However, in power plant design, usually, the deferential form of equations is not used, instead what is most needed is the fluid flow behavior in steady operation.[3] The transient behavior mostly occurs in the start-up and shut down processes.

The mass conservation law for a control volume (CV) such as that shown in Fig. 2.1 can be written as below:

$$\sum \dot{m}_{in} - \sum \dot{m}_{out} = \left(\frac{dm}{dt}\right)_{C.V.} \quad (\text{kg}/\text{s}) \qquad (2.1a)$$

in which m is the mass contained in the control volume. It is obvious that if there are no inputs and outputs to the CV or the inlet and outlet mass flows are equal, the changes in the mass of the control volume would be zero (in this case the process is called steady state).

---

[3] Steady operation means that the operational parameters such as temperature, pressure, speed, mass flow rate etc do not change with time anymore. For most of industrial equipment there is a steady operation that occurs after a specific time after start up. Most of power plant processes are transient only in the start up or shut down.

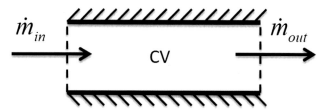

Figure 2.1 A general control volume (CV) with general inputs and outputs of mass flow.

$$\left(\frac{dm}{dt}\right)_{CV} = 0$$

or:

$$\sum \dot{m}_{in} = \sum \dot{m}_{out} \tag{2.2}$$

The following example shows the importance of mass conservation law in the compressor of a gas turbine.

Example 2.1. Stall is a phenomenon that may occur in axial or centrifugal compressors due to flow separation over the blades or the vanes of impellers.[4] In the start-up process of a gas turbine, the inlet guide vanes (IGVs) of the compressor are not well adjusted and do not action correctly according to the control system commands. As a result, the IGVs become less open as the rotor speed increases. Due to this problem, the rotor speed increases more than it was expected, and consequently, the air flow regime becomes sonic[5] as it passes over the compressor blades and causing flow separation or stall. The wakes generated due to stall block the air flow through the compressor partially and a portion of air cannot leave the compressor (Fig. 2.2). If this phenomenon is not well controlled, the trapped air inside the compressor casing would increase very fast and causes vibration a severe damage to the compressor. If we are supposed to control the stall, the trapped air must be removed from the compressor by opening bleed-valves and also increasing the inlet air flow to the compressor simultaneously.

It is assumed that stall has started at time $t = t0$ and mass has been trapped with a time dependent rate of:

$$\left(\frac{dm}{dt}\right)_{trapped} = -\frac{b}{2T}(t - t0)^2 + b(t - t0), \quad b > 0$$

---

[4] The rotating component of an axial or centrifugal compressor that increases the velocity and pressure of flowing gas through the compressor is called blade and impeller respectively. They are installed on a shaft. The shaft with other components installed on it, is called rotor assembly.

[5] If the Mach number (Ma) of air stream equals 1 the flow regime is called sonic. If Ma $\cong$ 1, Ma<1, Ma>1 and Ma>>1 the flow regime is called transonic, subsonic, supersonic and hypersonic respectively. In addition, Ma = V/c in which V is the air speed and c is the speed of sound in the air.

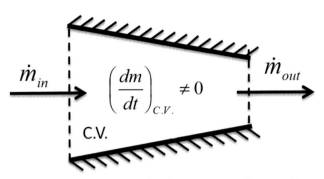

Figure 2.2 The schematic of mass balance for the compressor of a gas turbine under stall.

At $t = t0 + T$, the control system has been activated and bleed-valves open for removing the trapped air until $t = t0 + 2T$ when stall is fully controlled, and all of the trapped air is removed. If we have only T seconds to control the stall, determine:

A: The inlet and outlet mass flow rates of the compressor for the following two scenarios:

Scenario 1: The trapped air is extracted from the compressor casing and fed into the compressor inlet.

Scenario 2: The trapped air is extracted from the compressor casing and released into the exhaust of the gas turbine.

B: If the bleed-valves remove the trapped air with a constant rate, what would be its flow rate?

C: According to your engineering feeling, which scenario would be faster in controlling the stall? Which scenario do you recommend to be used on a compressor of a gas turbine?

Solution. Before activation of the control system and opening the bleed-valve, the mass conservation law can be written as below:

$$\left(\dot{m}_{in} - \dot{m}_{out}\right) = \left(\frac{dm}{dt}\right)_{C.V.}$$

By applying the scenarios as demonstrated in Figs. 2.3 and 2.4, the mass conservation law at $t = t0 + 2T$ will be:

$$\left(\dot{m}_{in} - \dot{m}_{out}\right)|_{t=t0+2T} = 0$$

A: Scenario 1 If the first scenario is applied, the inlet and outlet air flow rate of the compressor during the control process time can be formulated as below:

Inlet flow rate to the compressor $= \dot{m}_{in}|_{t=t1} + \left(\frac{dm}{dt}\right)_{bv}$

Outlet flow rate from the compressor exit $= \dot{m}_{in}|_{t=t1} + \left(\frac{dm}{dt}\right)_{bv} - \left(\frac{dm}{dt}\right)_{CV}$

$$t0 + T < t1 < t0 + 2T$$

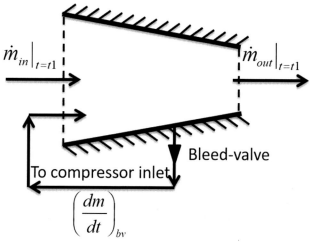

Figure 2.3 The schematic of mass balance for the compressor by applying scenario 1.

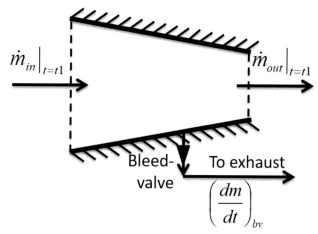

Figure 2.4 The schematic of mass balance for the compressor by applying scenario 2.

In which $\dot{m}_{in}|_{t=t1} = \alpha_1 \dot{m}_{in}|_{t=t0}$ that α_1 is the flow adjustment coefficient for the first scenario and is a function of rotor speed and time. In addition, $\left(\frac{dm}{dt}\right)_{bv}$ is the bleed-valve mass flow rate.

A: Scenario 2 In the second scenario, the bleed valves evacuate the trapped air into the turbine exhaust. Therefore, the inlet air to the compressor should be increased with a different flow adjustment coefficient of α_2.

Air flow rate at the compressor inlet $= \dot{m}_{in}|_{t=t1}$

Air flow rate at the compressor exit $= \dot{m}_{in}|_{t=t1} - \left(\frac{dm}{dt}\right)_{bv} - \left(\frac{dm}{dt}\right)_{CV}$

$$t0 + T < t1 < t0 + 2T$$

In which $\dot{m}_{in}|_{t=t1} = \alpha_2 \dot{m}_{in}|_{t=t0}$ that α_2 is the flow adjustment coefficient for the second scenario and is a function of rotor speed and time.

B: It is assumed that the bleed-valve flow rate is constant, and it must remove all of the accumulated air from the compressor. Hence:

$$\int_{t0}^{t0+2T} \left(\frac{dm}{dt}\right)_{CV} dt = \int_{t0+T}^{t0+2T} \left(\frac{dm}{dt}\right)_{bv} dt$$

$$\rightarrow T\left(\frac{dm}{dt}\right)_{bv} = \int_{t0}^{t0+2T} \left(-\frac{b}{2T}(t-t0)^2 + b(t-t0)\right) dt$$

$$= \left(-\frac{b}{6T}(t-t0)^3 + \frac{b}{2}(t-t0)^2\right)\Big|_{t0}^{t0+2T}$$

$$\rightarrow \left(\frac{dm}{dt}\right)_{bv} = \frac{2bT}{3}$$

C: In order to compare the scenarios, the input mass flow rate at the compressor inlet must be equal in the two scenarios, hence:

$$\alpha_1 \dot{m}_{in}|_{t=t0} + \left(\frac{dm}{dt}\right)_{bv} = \alpha_2 \dot{m}_{in}|_{t=t0} \rightarrow \alpha_2 = \alpha_1 + \frac{2bT}{3\dot{m}_{in}|_{t=t0}}$$

This means that flow adjustment coefficient for the second scenario, α_2 is bigger than α_1, in the other words, the second scenario needs more suction power, while in the first scenario, air is extracted from a high-pressure point inside the compressor and injected into the lowest pressure point. This is a positive point for the first scenario.

However, in the first scenario, hot gas is bypassed into the compressor inlet that increases the compression work. This is a negative point for the first scenario.

If decision is to be made based on the energy consumption only, extra suction power in the second scenario should be compared with the extra compression power in the first scenario. The one with smaller energy requirement is the answer.

If decision is to be made based on the response time to control the stall, the first scenario would act faster because opening a valve can increase the inlet flow rate considerably very fast and reduces the rotor speed but in the second scenario air flow must increase by opening IGVs and increasing rotor speed simultaneously. Therefore, according to the response time of the control system, the first scenario would be chosen.

More discussions

The system that controls air flow extraction from the compressor casing is called the antisurge system. The antisurge is a system to protect compressor from backflow to the compressor exit due to low flow rate at the compressor exit. More information about surge and stall can be found in Chapter 6.

Example 2.2. It is common to extract steam from different stages of the steam turbine casing for preheating the condensate leaving the condenser[6] before entering the steam generator. In steam turbine cycle power plants, this process is done by utilizing open and closed feedwater heaters (FWH).[7] Fig. 2.5 shows a steam turbine operating in steady condition with an open-FWH, condenser, and pumps. Derive the mass balance equation for the steam turbine, open-FWH, and condenser. Assume that no leakage occurs from different components.

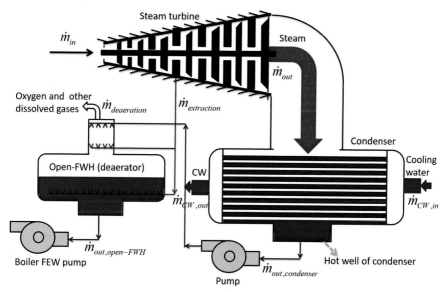

Figure 2.5 The extraction of steam and preheating of boiler feedwater.

---

[6] Condenser is an indirect contact shell and tube heat exchanger. Cooling water is flowing inside the tubes while steam pours down over the tubes from the turbine exit manifold.

[7] Open feedwater heater is a direct contact heat exchanger but closed feedwater heater is an indirect contact heat exchanger. The open feedwater heater is also used for steam deaeration, injecting make-up water to the cycle, producing net positive suction head for the boiler feedwater pump and also injecting chemical additives for water treatment purposes.

Solution. Steam entering the turbine leaves it from two points. One from the extraction for preheating purpose in the open-FWH and the other from the turbine exit to be condensed in the condenser. Hence, the mass balance for the steam turbine can be written according to Eq. (2.1) as below:

$$\sum \dot{m}_{in} - \sum \dot{m}_{out} = \left(\frac{dm}{dt}\right)_{CV} = 0$$

$$\sum \dot{m}_{in} = \sum \dot{m}_{out} \xrightarrow{yields} \dot{m}_{in} = \dot{m}_{extraction} + \dot{m}_{out}$$

By dividing both sides of the mass balance equation to $\dot{m}_{in}$ and assuming that $y = \dot{m}_{open-fwh}/\dot{m}_{in}$ it can be concluded that:

$$\dot{m}_{extraction} = y\dot{m}_{in} \text{ and } \dot{m}_{out} = (1-y)\dot{m}_{in}$$

The mass balance for the condenser includes two equations, one for the steam and the other for the cooling water.

For the cooling water (CW):

$$\dot{m}_{CW, out} = \dot{m}_{CW,in}$$

and for the steam turbine:

$$\dot{m}_{out,condenser} = (1-y)\dot{m}_{in}$$

The mass balance for the open-FWH includes 4 main streams of an inlet steam from the extraction, outlet dissolved gasses due to deaeration, inlet and outlet liquid water streams.

$$\dot{m}_{out,condenser} + \dot{m}_{extraction} = \dot{m}_{deaeration} + \dot{m}_{out, open-FWH}$$

$$\xrightarrow{yields} \dot{m}_{in} = \dot{m}_{deaeration} + \dot{m}_{out, open-FWH}$$

2. First law of thermodynamics

In different parts of power plants, energy is transforming from one form to another. Many examples of energy conversion can be found. For example, solar heat is transformed to electricity when it is absorbed by the solar photovoltaic cells. Fossil fuel energy is released in the combustion chamber of a gas turbine and is transformed to rotational motion of the gas turbine rotor or generator to produce power or electricity. Electricity is consumed in the feed water pump of a steam power plant to produce hydraulic power and water flow. Wind kinetic energy is absorbed by a wind turbine to rotate the turbine blades and produce electricity. A thermoelectric generator converts temperature difference to electricity, etc. As it can be seen always an input energy is required to achieve another form of energy. During the energy transformation process, some part of energy is always wasted to receive a desired output. It is obvious that the input energy equals the desired output plus the wasted energy. Simply, this is the energy balance or the first law of thermodynamics. Hence:

First law of thermodynamics is actually another statement of the *energy conservation principal* and also called *energy balance*. It says that *the total energy of an isolated system is constant, energy is neither created nor destroyed but it can be transformed from one form to another.*

The general form of the first law of thermodynamics for a system such as that shown in Fig. 2.6 can be written as below [4]:

$$\dot{E}_{in} - \dot{E}_{out} = \frac{dE_{system}}{dt} \, (\mathrm{W}) \tag{2.3}$$

In which $\dot{E}_{in}$ and $\dot{E}_{out}$ are the input and output rates of energy transfer by heat, work, and mass, and $\frac{dE_{system}}{dt}$ is the rate of change in the internal, kinetic, potential, etc energies of the system.

It is evident that in a steady state process, the change in the energy of the system would be zero:

$$\frac{dE_{system}}{dt} = 0$$

Hence:

$$\dot{E}_{in} = \dot{E}_{out} \tag{2.4}$$

The rate of energy transfer by mass flow can be written as below:

$$\dot{E} = \dot{m}\theta \tag{2.5}$$

In which, θ is the total energy per unit of mass and is determined by summation of enthalpy, kinetic, and potential energies.

$$\theta = h + ke + pe$$

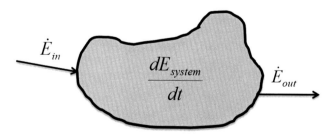

Figure 2.6 Schematic of an arbitrary system and first law of thermodynamics.

$$ke = \frac{V^2}{2}$$

$$pe = gz \tag{2.6}$$

Hence, the energy transfer by the mass flow rate of $\dot{m}$ is determined by Eq. (2.7).

$$\dot{E} = \dot{m}\left(h + \frac{V^2}{2} + gz\right) (\text{W}) \tag{2.7}$$

The rate of energy transfer by work and heat are shown by $\dot{W}$ and $\dot{Q}$, respectively. Most of power plant analyses should be done under steady-state condition. In such cases, Eq. (2.4) can be rewritten as Eq. (2.8):

$$\dot{Q}_{in} + \dot{W}_{in} + \sum_{in} \dot{m}\left(h + \frac{V^2}{2} + gz\right) = \dot{Q}_{out} + \dot{W}_{out} + \sum_{out} \dot{m}\left(h + \frac{V^2}{2} + gz\right) (\text{W}) \tag{2.8}$$

In most of equipment used in power plants (except for nozzles and diffusers that change velocity significantly), the potential and kinetic energy differences between the inlet and outlet flows are negligible. As a result, the energy equation for such systems can be written as Eq. (2.9).

$$\dot{Q}_{in} + \dot{W}_{in} + \sum_{in} \dot{m}h = \dot{Q}_{out} + \dot{W}_{out} + \sum_{out} \dot{m}h (\text{W}) \tag{2.9}$$

Eq. (2.9) can also be written in a more squeezed form as Eq. (2.10):

$$\dot{Q}_{net} - \dot{W}_{net} = \sum_{out} \dot{m}h - \sum_{in} \dot{m}h (\text{W}) \tag{2.10}$$

In which

$$\dot{Q}_{net} = \dot{Q}_{in} - \dot{Q}_{out} \cdot \dot{W}_{net} = \dot{W}_{out} - \dot{W}_{in} \tag{2.11}$$

Example 2.3. A high-pressure steam turbine (HPT) receives 1 kg.s^{-1} of superheated steam from a steam generator at 15 MPa and 540°C (Fig. 2.7). The steam pressure drops to 3 MPa after passing through the HPT. To increase the cycle thermal efficiency, it is recommended to reheat the HPT exit steam to its inlet temperature before entering the intermediate pressure turbine (IPT).

Determine the power produced by the HPT and the fuel energy consumed by the reheater. In addition, show the process in a temperature-entropy (T-s) diagram of water. The adiabatic efficiency of the HP turbine is 0.87.

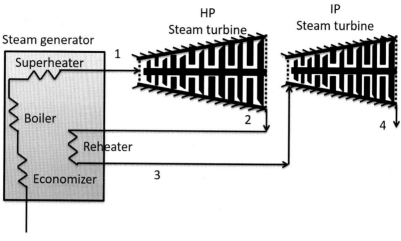

Figure 2.7 Demonstration of the process related to Example 2.3.

Solution. It is assumed that the turbine and all pipes containing steam are well insulated, and no heat loss occurs through the equipment. In addition, the pressure drop due to friction and fittings for the reheater tubes is ignored. Under these assumptions, the energy balance for the HP turbine is as below:

$$\dot{Q}_{net,HPT} = 0$$

$$\dot{W}_{in,HPT} = 0$$

Using Eq. (2.10):

$$\dot{W}_{out,HPT} = \dot{m}_1(h_1 - h_2)$$

To calculate the power produced by the HPT, the enthalpy at state points of 1 and 2 should be determined:

State 1: $P_1 = 15$ MPa and $T_1 = 540°C \rightarrow h_1 = 3422$ kJ.kg^{-1}, $s_1 = 6.487$ kJ.kg^{-1}.K^{-1}

The adiabatic efficiency of the HPT is given. The adiabatic efficiency of a turbine is defined as the ratio of rate of work production in real process to the rate of work production in isentropic process. Hence:

$$\eta_{turbine,ad} = \frac{h_1 - h_2}{h_1 - h_{2s}} \tag{2.12}$$

In which h_{2s} is the enthalpy at state 2 under isentropic condition that can be determined as below:

$$s_{2s} = s_1, \ P_2 = 3\text{MPa} \rightarrow h_{2S} = 2964 \text{ kJ.kg}^{-1}$$

Therefore:

$$h_2 = \eta_{turbine, \ ad}(h_{2s} - h_1) + h_1$$

$$\rightarrow h_2 = 0.87(2964 - 3422) + 3422$$

$$\rightarrow h_2 = 3023 \text{ kJ.kg}^{-1}$$

Hence:

$$\dot{W}_{out,HPT} = 1(3422 - 3023)$$

$$\dot{W}_{out,HPT} = 399 \text{ kW}$$

The energy balance of the reheater can also be written as below:

$$\dot{W}_{net, \ reheater} = 0$$

$$\dot{Q}_{out,reheater} = 0$$

$$\dot{Q}_{in,reheater} = \dot{m}_2(h_3 - h_2)$$

Enthalpy at state 3 can be determined according to the data give:
State 3: $P_1 = 3$ MPa and $T_3 = T_1 \rightarrow h_3 = 3547 \text{ kJ.kg}^{-1}$
Hence:

$$\dot{Q}_{in,reheater} = 1(3547 - 3023)$$

$$\dot{Q}_{in,reheater} = 524 \text{ kW}$$

The real and isentropic process through the HPT and reheater are shown in Fig. 2.8. It shows that the exit temperature in the real process is higher due to internal irreversibilities such friction.

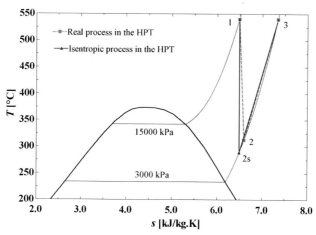

Figure 2.8 The real and isentropic process through the high-pressure steam turbine (HPT) and reheater.

The codes written for the Example 2.3 in the EES is given below:

Code 2.1. The code written for the Example 2.3 in the EES.

```
"Example 2.3:"
m_dot[1]=1 [kg/s]
m_dot[2]=m_dot[1]
m_dot[3]=m_dot[2]
T[1]=540  [C]
P[1]=150*100 [kPa]
P[2]=0.2*P[1]
T[3]=T[1]
P[3]=P[2]
eta_adiabatic_HP=0.87
H[1]=Enthalpy(Water,T=T[1],P=P[1])
s[1]=Entropy(Water,T=T[1],P=P[1])
s_isentropic2=s[1]
h_s_2=Enthalpy(Water,s=s_isentropic2,P=P[2])
HS[2]=h_s_2
SS[1]=s[1]
SS[2]=s[1]
SS[3]=S[3]
TS[2]=Temperature(Water,P=P[2],h=HS[2])
TS[1]=T[1]
TS[3]=T[3]
eta_adiabatic_HP=(H[1]-H[2])/(H[1]-h_s_2)
W_HPT=m_dot[1]*(H[1]-H[2])
T[2]=Temperature(Water,P=P[2],h=h[2])
s[2]=Entropy(Water,T=T[2],P=P[2])
s[3]=Entropy(Water,T=T[3],P=P[3])
H[3]=Enthalpy(Water,T=T[3],P=P[3])
Q_reheat=m_dot[2]*(H[3]-H[2])
```

Example 2.4. Consider the cycle presented in Example 2.3. If the IPT is supposed to be fed with the steam leaving the reheater, and its exit pressure drops to 1 MP. The adiabatic efficiency of the IPT is 0.85. Calculate:

a. How much power would be produced by the IP turbine? Calculate the total power production by HPT and IPT.

b. If no reheat was done, how much power could be generated by the IPT?

Solution. Part a

It is assumed that the turbine and all pipes containing steam are well insulated, and no heat loss occurs through the equipment. In addition, the pressure drop due to friction and fittings for the reheater tubes is ignored. Under these assumptions, the energy balance for the IP turbine is as below:

$$\dot{Q}_{net,IPT} = 0$$

$$\dot{W}_{in,IPT} = 0$$

Using Eq. (2.10):

$$\dot{W}_{out,IPT} = \dot{m}_3(h_3 - h_4)$$

The adiabatic efficiency of the IPT is given. Hence:

$$\eta_{ad} = \frac{h_3 - h_4}{h_3 - h_{4s}}$$

In which h_{4s} is the enthalpy at state 4 under isentropic condition that can be determined as below:

$$s_{4s} = s_3, \ P_4 = 1 \ \text{MPa} \rightarrow h_{4S} = 3187 \text{kJ.kg}^{-1}$$

Therefore:

$$h_4 = \eta_{ad}(h_{4s} - h_3) + h_3$$

$$\rightarrow h_4 = 0.85(3187 - 3547) + 3547$$

$$\rightarrow h_4 = 3241 \ \text{kJ.kg}^{-1}$$

Hence:

$$\dot{W}_{out,IPT} = 1(3547 - 3241)$$

$$\dot{W}_{out,IPT} = 306 \ \text{kW}$$

The total power generated by IPT and HPT is also calculated as below, and the T-s diagram of the process in the HPT,reheater, and IPT are shown in Figs. 2.9 and 2.10. As the digram shows the teperature at the turbines outlets in real conditions is higher than the temperature in isentropic coditions. This is mainly due to friction and other irreversibilities.

$$\dot{W}_{HPT+IPT} = \dot{W}_{out,IPT} + \dot{W}_{out,IPT}$$

$$\rightarrow \dot{W}_{HPT+IPT} = 306 + 399 = 705 \text{ kW}$$

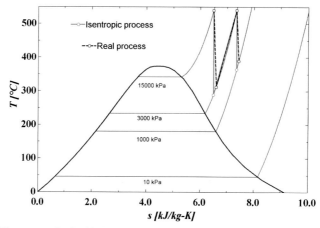

Figure 2.9 The process in the high-pressure steam turbine (HPT), first reheat, and intermediate pressure turbine (IPT).

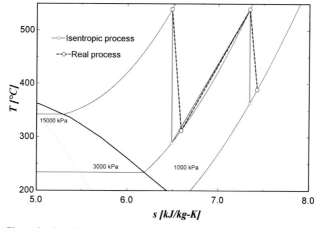

Figure 2.10 Closer look at the process presented in Fig. 2.9.

Part b

If no reheat was done, the schematic cycle and state points would be as Fig. 2.11: According to the cycle the energy balance for the IPT would change as below:

$$\dot{Q}_{net,IPT} = 0$$

$$\dot{W}_{in,IPT} = 0$$

Using Eq. (2.10):

$$\dot{W}_{out,IPT} = \dot{m}_2(h_2 - h_3)$$

The adiabatic efficiency of the IPT is given. Hence:

$$\eta_{ad} = \frac{h_2 - h_3}{h_2 - h_{3s}}$$

In which h_{4s} is the enthalpy at state 4 under isentropic condition that can be determined as below:

$$s_{3s} = s_2, \ P_3 = 1 \text{ MPa} \rightarrow h_{3S} = 2780 \text{ kJ.kg}^{-1}$$

Therefore:

$$h_3 = \eta_{ad}(h_{3s} - h_2) + h_2$$

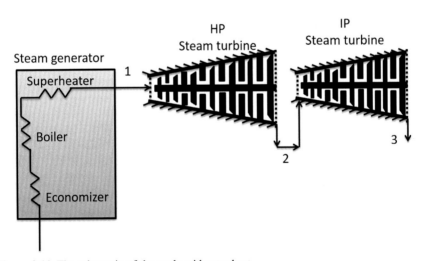

Figure 2.11 The schematic of the cycle without reheat.

$$\rightarrow h_3 = 0.85(2780 - 3023) + 3023$$

$$\rightarrow h_3 = 2816 \text{ kJ.kg}^{-1}$$

The T-s diagram of the process for the part b is shown in Fig. 2.12 and the power output by IPT would be:

$$\dot{W}_{out,IPT} = 1(3023 - 2816)$$

$$\dot{W}_{out,IPT} = 207 \text{ kW}$$

Conclusions. According to the results, reheating would increase the power generation by the IPT from 207 kW to 306 kW. In addition at part a, the temperature outlet from IPT is 389.2°C that is significantly higher with respect part b that is 195.4 °C. Having higher temperature means more power generation potential for the low-pressure turbines (LPT) that can be installed downstream of the HPT. It should be mentioned that using reheat is one of the most effective methods to increase the steam cycle electrical efficiency. This topic will be discussed in details in the following chapters.

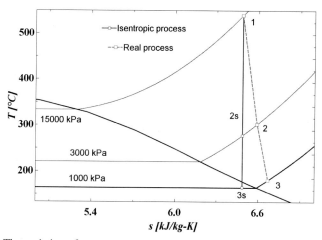

Figure 2.12 The variation of temperature versus entropy for part b of Example 2.4.

The codes written for the Example 2.4 in the EES is given below:

Code 2.2. The code written for the Example 2.4 in the EES.

```
"Example 2.4: part a"

m_dot[1]=1 [kg/s]
m_dot[2]=m_dot[1]
m_dot[3]=m_dot[2]
T[1]=540  [C]
P[1]=150*100 [kPa]
P[2]=0.2*P[1]
T[3]=T[1]
P[3]=P[2]
eta_adiabatic_HP=0.87
H[1]=Enthalpy(Water,T=T[1],P=P[1])
s[1]=Entropy(Water,T=T[1],P=P[1])
s_isentropic2=s[1]
h_s_2=Enthalpy(Water,s=s_isentropic2,P=P[2]
)
HS[2]=h_s_2
SS[1]=s[1]
SS[2]=s[1]
SS[3]=S[3]
TS[2]=Temperature(Water,P=P[2],h=HS[2])
TS[1]=T[1]
TS[3]=T[3]
eta_adiabatic_HP=(H[1]-H[2])/(H[1]-h_s_2)
W_HPT=m_dot[1]*(H[1]-H[2])
T[2]=Temperature(Water,P=P[2],h=h[2])
s[2]=Entropy(Water,T=T[2],P=P[2])
s[3]=Entropy(Water,T=T[3],P=P[3])
H[3]=Enthalpy(Water,T=T[3],P=P[3])
Q_reheat=m_dot[2]*(H[3]-H[2])
P[4]=1000
s_isentropic4=s[3]
eta_adiabatic_IP=0.85
m_dot[4]=m_dot[3]
h_s_4=Enthalpy(Water,s=s_isentropic4,P=P[4]
)
HS[4]=h_s_4
eta_adiabatic_IP=(H[3]-H[4])/(H[3]-h_s_4)
W_IPT=m_dot[3]*(H[3]-H[4])
T[4]=Temperature(Water,P=P[4],h=h[4])
s[4]=Entropy(Water,T=T[4],P=P[4])
TS[4]=Temperature(Water,P=P[4],h=HS[4])
SS[4]=s[3]
SS[3]=s[1]
xs[3]=quality(Water,s=ss[3],h=Hs[3])
```

```
"Example 2.4: part b"
m_dot[1]=1 [kg/s]
m_dot[2]=m_dot[1]
T[1]=540  [C]
P[1]=150*100 [kPa]
P[2]=0.2*P[1]
eta_adiabatic_HP=0.87
H[1]=Enthalpy(Water,T=T[1],P=P[1])
s[1]=Entropy(Water,T=T[1],P=P[1])
s_isentropic2=s[1]
h_s_2=Enthalpy(Water,s=s_isentropic2,P=P[2])
HS[2]=h_s_2
SS[1]=s[1]
SS[2]=s[1]
TS[2]=Temperature(Water,P=P[2],h=HS[2])
TS[1]=T[1]
eta_adiabatic_HP=(H[1]-H[2])/(H[1]-h_s_2)
W_HPT=m_dot[1]*(H[1]-H[2])
T[2]=Temperature(Water,P=P[2],h=h[2])
s[2]=Entropy(Water,T=T[2],P=P[2])
P[3]=1000
s_isentropic3=s[2]
eta_adiabatic_IP=0.85
m_dot[3]=m_dot[2]
h_s_3=Enthalpy(Water,s=s_isentropic3,P=P[3])
HS[3]=h_s_3
eta_adiabatic_IP=(H[2]-H[3])/(H[2]-h_s_3)
W_IPT=m_dot[2]*(H[2]-H[3])
T[3]=Temperature(Water,s=s[3],h=h[3])
s[3]=Entropy(Water,P=P[3],h=h[3])
TS[3]=Temperature(Water,s=s_isentropic3,h=HS[3]
)
```

3. Second law of thermodynamics

By experience, we know that heat can transfer from a high-temperature space to the low-temperature one without using any equipment. Fluid can flow from a high-pressure spot to the low-pressure one by utilizing no equipment. Consider we are supposed to do these two processes inversely. In the first case, a refrigeration cycle or a heat pump must go between the high- and low-temperature spaces to make the heat transfer possible. In the second case, depending on the fluid state, a pump or a compressor is required to transport fluid from the low-pressure spot to the high-pressure spot. In both cases, work must be consumed to make the reverse processes possible.

Another example is the conversion of work to heat and vice versa. Work can be converted to heat directly, but for the reverse process, a heat engine is required. Heat engines receive high-quality heat from a heat source and reject a portion of the inlet heat as the waste heat to a heat sink and produce a useful net power. A heat source or a heat sink that can transfer heat isothermally is called *heat reservoir*.

According to the examples and discussions, the second law of thermodynamics concerns about the direction of processes. It states that *processes occur in a certain direction not in any direction*. A process to happen requires satisfying both the first and the second laws of thermodynamics. There are two important statements of the second law, namely, *Kelvin−Planck statement* and *Clausius statement*.

According to the *Kelvin−Planck statement, no heat engine can produce a net amount of power by exchanging heat with one heat reservoir only*. In other words, heat engines must operate between a heat source and a heat sink. They receive heat from the heat source, produce some net power, and reject the remaining low-quality heat to the heat sink. It means wasting some heat is inevitable if we are supposed to convert heat to work.

Clausius statement of the second law states that *no device can transfer heat from a low-temperature space to a high-temperature space without leaving an impact on the surroundings*.

If a process is *reversible* the system and its surroundings can return to their original conditions. If only the system can restore to its original condition, the process is called *internally reversible* but *externally irreversible*. A system to be reversible needs to be internally and externally reversible. Some main reasons of irreversibility include friction, heat transfer due to finite temperature difference, combustion, mixing, etc.

A *Carnot cycle* is an internally reversible heat engine that includes two isentropic and two isothermal processes (Fig. 2.13). It receives heat of Q_H at constant temperature of T_H and rejects waste heat of Q_L to the heat sink at constant temperature of T_L. It also includes isentropic compression (1−2) and expansion (3−4) processes. When the

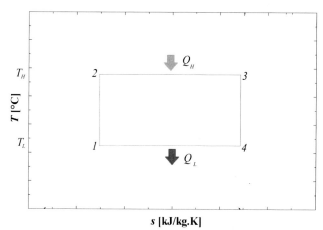

Figure 2.13 The Carnot cycle, an internally reversible cycle.

Carnot cycle is internally reversible, it means that all pressure drops and temperature increase due to friction is ignored. In addition, all of the compression/expansion equipment (such as pump, compressor, turbine, etc.), piping, and heat exchangers (such as boiler, condenser, combustion chamber, etc.) are well insulated.

The *Carnot principal* states that the highest thermal efficiency of a heat engine operating between T_H and T_L is limited by the thermal efficiency of the Carnot cycle operating between the same temperature limits. In addition, it says that the thermal efficiency of all internally reversible heat engines operating between the same temperature limits is the same and is equal to:

$$\eta_{Carnot} = 1 - \frac{T_L}{T_H} \tag{2.13}$$

In which T is in Kelvin.

Important conclusion

According to the definition of the Carnot cycle efficiency, the higher the heat addition temperature (T_H), the higher the Carnot cycle efficiency. Likewise, the lower the heat rejection temperature (T_L), the higher the Carnot cycle efficiency. According to these two important conclusions, if we are supposed to increase the thermal efficiency of an arbitrary heat engine, what ever we do should result in higher average temperature of heat addition or lower average temperature of heat rejection.

In the second law of thermodynamics, *entropy* is the magnitude of disorder in the systems. Examples of disorder are friction, mixing, combustion, heat addition, etc. Entropy enables us to quantify the part of heat that can be converted to power or mechanical work in a heat engine. The higher the *entropy generation* in a process, the less the power generation. Hence, the entropy generation minimization is a powerful approach to increase the power cycles efficiency. If no entropy change occurs during a process, the process is called an *isentropic process*. In the second law of thermodynamics, *Clausius inequality* states that the *cyclic integral of the Q/T is always negative or zero*.

$$\oint \frac{Q}{T} \leq 0 \tag{2.14}$$

The Clausius inequality is valid for all processes. However, the equality holds for internally reversible or totally reversible cycles. The inequality is valid for irreversible cycles.

For the especial case of internal reversible processes, it means that the cyclic integral is a property that does not depend on the path of integration (since friction, and pressure drop are ignored, and external surfaces are well insulated) and depend only on the state parameters. Clausius found out that this integral is a thermodynamic property and called it entropy and used S (kJ /K) to show entropy. Therefore:

$$dS = \oint \left(\frac{Q}{T}\right)_{int.rev} \tag{2.15}$$

Entropy is an extensive property and can be given in the unit of mass as an intensive property s (kJ /K.kg). If we are supposed to determine the entropy change from state 1 to state 2 in an internally reversible cycle, we have:

$$S_2 - S_1 = \int_1^2 \left(\frac{Q}{T}\right)_{int,\ rev} \tag{2.16}$$

Entropy can be transferred by heat or mass and can be generated (S_{gen}) by irreversibilities such as friction, mixing, flow separation, explosion, combustion, etc. The entropy generation is a nonnegative parameter. It is zero for reversible processes.

$$S_{gen} \begin{cases} > 0 & \text{for irreversible process} \\ = 0 & \text{for reversible process} \\ < 0 & \text{impossible process} \end{cases} \tag{2.17}$$

For an arbitrary system transferring heat, work, and mass with its surrounding the entropy balance can be written as below (Fig. 2.14):

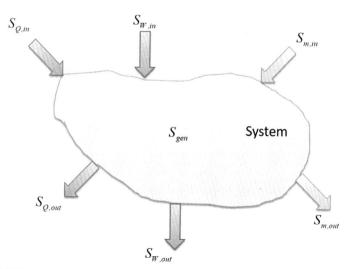

Figure 2.14 The entropy generation and entropy transfer to/from a system.

$$(S_{in} - S_{out})_{net\ entropy\ transfer} + S_{gen} = \Delta S_{system} \qquad (2.18)$$

The entropy balance equation can be stated as, *the total entropy change of a system is due to net entropy transfer and entropy generation within the system.*

The entropy transfer by heat, work, and mass is determined as below:

$$S_Q = \frac{Q}{T}, \ S_W = 0, \ S_m = ms \qquad (2.19)$$

In the rate form, the entropy balance is as below:

$$\left(\dot{S}_{in} - \dot{S}_{out}\right)_{net\ rate\ of\ entropy\ transfer} + \dot{S}_{gen} = \frac{dS_{system}}{dt} \qquad (2.20)$$

Another important concept that has emerged from the second law of thermodynamics is the *exergy. Exergy is the potential of energy to do work.* By maximizing the *exergy efficiency,* the conditions to produce highest amount of power would be determined.

For an arbitrary system transferring heat, work, and mass with its surrounding the exergy (*X*) balance can be written as below (Fig. 2.15):

$$(X_{in} - X_{out})_{net\ entropy\ transfer} - X_{destroyed} = \Delta X_{system} \qquad (2.21)$$

The exergy balance equation can be stated as; *the total exergy change of a system is due to net exergy transfer and exergy destruction within the system.* The exergy destruction is due to entropy generation inside the system (irreversibilities).

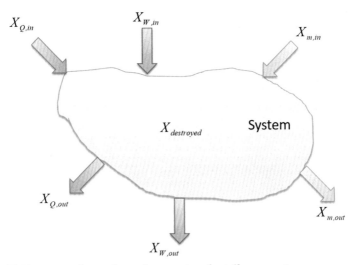

Figure 2.15 The exergy destruction and exergy transfer to/from a system.

The exergy transfer by heat, work, and mass are determined as below:

$$X_Q = Q\left(1 - \frac{T_0}{T}\right), \ X_W = W, \ X_m = m\psi \tag{2.22}$$

The exergy per unit of mass for a thermodynamic state with respect to the *dead state* can be calculated as below:

$$\psi = (h - h_0) - T_0(s - s_0) + \frac{V^2}{2} + gz \tag{2.23}$$

In the rate form, the exergy balance is as below:

$$\underbrace{\left(\dot{X}_{in} - \dot{X}_{out}\right)}_{net \ rate \ of \ entropy \ transfer} - \dot{X}_{destroyed} = \frac{dX_{system}}{dt} \tag{2.24}$$

A system is in *dead state* if it is thermodynamically in equilibrium with its surrounding. In thermodynamic, most of time, it is $T_0 = 25\,°C$ and $P_0 = 1$ atm. However, in some cases, the dead state is the temperature and pressure of the heat sink that the power cycle is rejecting heat to it.

The exergy destruction of a process or a system can be evaluated as below:

$$X_{destroyed} = T_0 S_{gen} \tag{2.25}$$

In the rate form, the exergy destruction rate would be:

$$\dot{X}_{destroyed} = T_0 \dot{S}_{gen} \tag{2.26}$$

The exergy efficiency or the *second law efficiency* can be used for every component of power generation cycles, and the cycle as well.

The second law efficiency of heat engines can also be determined by dividing the actual thermal efficiency of the heat engine to the reversible thermal efficiency of the heat engine.

$$\eta_{II} = \frac{\eta_{th}}{\eta_{rev}} \tag{2.27}$$

It can also be defined as ratio of the recovered exergy to the supplied exergy to the system or process as below:

$$\eta_{II} = \frac{X_{recovered}}{X_{supplied}} = 1 - \frac{X_{destroyed}}{X_{supplied}} \tag{2.28}$$

This definition can be extended for power producer or power consumer equipment. For example, the second law efficiency of power producer equipment such as gas turbines, steam turbines, thermoelectric generators, fuel cells, etc. can be defined as below:

$$\eta_{II} = \frac{\dot{W}_{actual}}{\dot{W}_{rev}} \qquad (2.29)$$

For power consumer equipment such as pumps, fans and compressors, the second law efficiency would be defined as below:

$$\eta_{II} = \frac{\dot{W}_{rev}}{\dot{W}_{actual}} \qquad (2.30)$$

Example 2.5. Saturated steam at pressure of 15 MPa is supposed to be superheated to 600°C by using the superheater shown in Fig. 2.16. The steam flow rate is 1 kg/s, and the superheaters' tubes approximately contain 7 kg of steam in steady operation. Determine the entropy change, entropy generation, required heat rate for the process, and heat source temperature for:

a. The process is internally reversible
b. The process experiences 10% pressure drop due to friction loss, bends, and fittings. Draw the T-s diagram for both of processes.

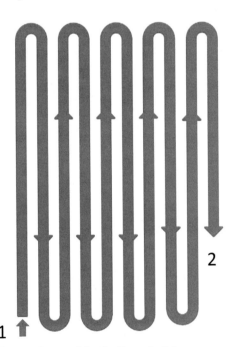

Figure 2.16 The superheater tubes used for the Example 2.5.

Solution. Part a

It is assumed that the process is internally reversible in part a. Care must be taken that if a process is internally reversible it does not mean that the entropy change is zero because entropy can change due to entropy transfer by heat or mass. Here, in this case, entropy change occurs only due to heat transfer to the saturated steam. The steam properties at points 1 and 2 would be as below:

$$P_1 = 15 \text{ MPa}, \ x = 1 \rightarrow T_1 = 342.2°C, \ h_1 = 2610 \frac{\text{kJ}}{\text{kg}}, \ s_1 = 5.309 \frac{\text{kJ}}{\text{kg.K}}$$

$$P_2 = 15 \text{ MPa}, \ T_2 = 600°C \rightarrow h_2 = 3581 \frac{\text{kJ}}{\text{kg}}, \ s_2 = 6.677 \frac{\text{kJ}}{\text{kg.K}}$$

For an internally reversible process, entropy generation is zero, hence:

$$s_{gen} = 0$$

And the entropy change:

$$\Delta s_a = s_2 - s_1 = 6.677 - 5.309 = 1.368 \frac{\text{kJ}}{\text{kg.K}}$$

And the required heat rate for the process:

$$\dot{Q}_a = \dot{m}_1(h_2 - h_1) = 1(3581 - 2610)$$

$$\rightarrow \dot{Q}_a = 971.5 \text{ kW}$$

Since the process is internally reversible, the entropy generation is zero and the total entropy change is because of heat transfer. To determine the temperature of heat source for an internally reversible process:

$$\Delta s_a = \frac{Q_a}{T_a}$$

$$\rightarrow T_a = \frac{\dot{Q}_a \times t}{\Delta s_a}$$

In which t is the time that it takes for the steam flow to be superheated from 342.2°C to the temperature of 600°C. For this purpose, the time-dependent energy balance for the superheater is written as below:

$$M\frac{dh}{dt} = \dot{m}_1(h - h_1) \rightarrow \int_0^t dt = \frac{M}{\dot{m}_1} \int_{h_1}^{h_2} \frac{dh}{h - h_1}$$

in which M is the steam mass in the superheater's tubes, therefore

$$t = \frac{M}{\dot{m}_1} \ln\left(\frac{h_2}{h_1}\right) = \frac{7}{1} \ln\left(\frac{3581}{2610}\right)$$

$$\rightarrow t = 2.215\text{s}$$

$$\rightarrow T_a = \frac{\dot{Q}_a \times \frac{M}{\dot{m}_1} \ln\left(\frac{T_2}{T_1}\right)}{\Delta s_a} = \frac{971.5 \times 2.215}{1.368}$$

$$\rightarrow T_a = 1573.15\text{K} = 1300°\text{C}$$

Part b
The entropy change in part b is due to heat transfer and friction as well.

$$P_1 = 15\text{MPa}, \ x = 1 \rightarrow T_1 = 342.2°\text{C}, \ h_1 = 2610\frac{\text{kJ}}{\text{kg}}, \ s_1 = 5.309\frac{\text{kJ}}{\text{kg.K}}$$

$$P_2 = 0.9 \times 15 = 13.5 \text{ MPa}, \ T_2 = 600°\text{C} \rightarrow h_2 = 3595\frac{\text{kJ}}{\text{kg}}, \ s_2 = 6.737\frac{\text{kJ}}{\text{kg.K}}$$

Hence:

$$\Delta s_b = s_2 - s_1 = 6.737 - 5.309 = 1.428\frac{\text{kJ}}{\text{kg.K}}$$

The entropy generation is the difference of the entropy change between part a and part b.

$$s_{gen} = \Delta s_b - \Delta s_a = 1.428 - 1.368$$

$$\rightarrow s_{gen} = 0.06\frac{\text{kJ}}{\text{kg.K}}$$

$$\dot{Q}_b = \dot{m}_1 (h_2 - h_1) = 1(3595 - 2610)$$

$$\rightarrow \dot{Q}_b = 984.6\text{kW}$$

Since the process is irreversible the entropy generation is not zero and the total entropy change is because of heat transfer and friction. To determine the temperature of heat source for an internally reversible process:

$$\Delta s_b = \frac{Q_b}{T_b} + s_{gen}$$

$$\rightarrow T_b = \frac{\dot{Q}_b \times t}{\Delta s_b - s_{gen}}$$

From the part a, we have:

$$t = \frac{M}{\dot{m}_1} \ln\left(\frac{h_2}{h_1}\right) = \frac{7}{1} \ln\left(\frac{3595}{2610}\right)$$

$$\rightarrow t = 2.241 \text{ s}$$

Hence:

$$\rightarrow T_b = \frac{\dot{Q}_b \times \frac{M}{\dot{m}_1} \ln\left(\frac{h_2}{h_1}\right)}{\Delta s_b - s_{gen}} = \frac{984.6 \times 2.241}{1.43 - 0.06}$$

$$\rightarrow T_b = 1482.15\text{K} = 1209°\text{C}$$

And the T-s diagram for the reversible and irreversible superheating processes is depicted in Fig. 2.17:

Conclusions. The second law efficiency for the superheating process in part b can be determined as below:

$$\dot{S}_{gen} = \dot{m}_1 s_{gen}$$

$$\eta_{II} = 1 - \frac{X_{destroyed}}{X_{supplied}} = 1 - \frac{T_0 \dot{S}_{gen}}{X_{Q_b}}$$

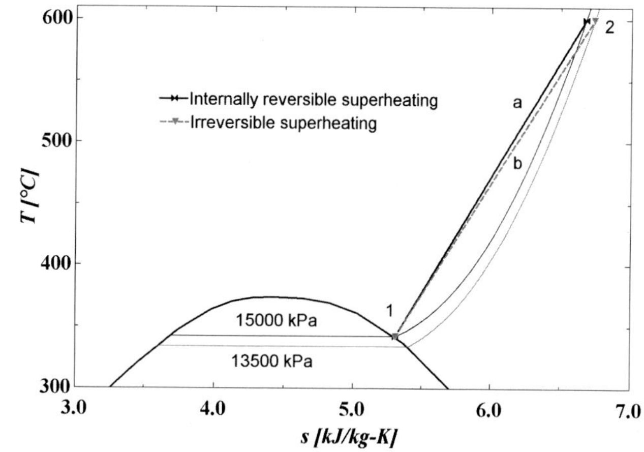

Figure 2.17 Internally reversible and irreversible superheating processes.

$$= 1 - \frac{(25 + 273.15) \times 1 \times 0.06}{986.6 \times \left(1 - \frac{25 + 273.15}{1482.15}\right)} = 0.9771 = 97.71\%$$

While the initial and final temperatures of the superheating process are kept the same in parts a and b, but due to having 10% pressure loss (irreversibility) in part b, the heat required for superheating is more, but its quality is lower (having smaller heat source temperature). As a result, internal irreversibility has caused more fuel consumption in the heating process.

The code written for the Example 2.5 is given below:

Code 2.3. The code written for the Example 2.5 in the EES.

```
"Example 2.5:"
K=273.15
m_dot[1]=1 [kg/s]
m_dot[2]=m_dot[1]

Pa[1]=150*100 [kPa]
Ta[1]=T_sat(Water,P=Pa[1])
Ha[1]=Enthalpy(Water,T=Ta[1],x=1)
Sa[1]=entropy(Water,T=Ta[1],x=1)

Ta[2]=600 [C]
Pa[2]=Pa[1]
Ha[2]=Enthalpy(Water,T=Ta[2],P=Pa[2])
Sa[2]=entropy(Water,T=Ta[2],P=Pa[2])
Delta_Sa=Sa[2]-Sa[1]
Qa=m_dot[1]*(Ha[2]-Ha[1])
T_source_a+K=DELTA_t_a*Qa/Delta_Sa

Pb[1]=150*100 [kPa]
Tb[1]=T_sat(Water,P=Pb[1])
Hb[1]=Enthalpy(Water,T=Tb[1],x=1)
Sb[1]=entropy(Water,T=Tb[1],x=1)

Tb[2]=600 [C]
Pb[2]=0.9*Pb[1]
Hb[2]=Enthalpy(Water,T=Tb[2],P=Pb[2])
Sb[2]=entropy(Water,T=Tb[2],P=Pb[2])
Delta_Sb=Sb[2]-Sb[1]
Qb=m_dot[1]*(Hb[2]-Hb[1])
Sgen=Delta_Sb-Delta_Sa
Delta_Sb=DELTA_t_b*Qb/(T_source_b+K)-Sgen
M=7*m_dot[1]
DELTA_t_a=M/m_dot[1]*ln((ha[2])/(ha[1]))
DELTA_t_b=M/m_dot[1]*ln((hb[2])/(hb[1]))
S_dot_gen=m_dot[1]*Sgen
X_Q_b=Qb*(1-(25+K)/(T_source_b+K))
eta_II=1-(25+K)*S_dot_gen/X_Q_b
```

Example 2.6. A thermoelectric generator (TEG) uses Seebeck effect to convert a temperature difference to electricity. It receives heat of Q_H from the heat source at temperature of T_H and rejects heat of Q_L to the heat sink at temperature of T_L while producing the electricity of E_{TEG}. For a TEG module, the heat rate to the hot side at temperature of 65°C is 70 W, and it rejects 67 W of heat to the cold side at temperature of 20°C. Determine the difference between the thermal efficiency of the TEG and the corresponding Carnot cycle efficiency. What is the reason for the difference? What is your recommendation to increase the TEG efficiency? (Fig. 2.18)

Solution. The Carnot cycle efficiency can be calculated as below:

$$\eta_{Carnot,\ TEG} = \left(1 - \frac{T_L}{T_H}\right) \times 100 = \left(1 - \frac{20 + 273.15}{65 + 273.15}\right) \times 100$$

$$\eta_{Carnot,\ TEG} = 13.3\%$$

The thermal efficiency can also be calculated as below:

$$\eta_{Thermal,\ TEG} = \left(1 - \frac{\dot{Q}_L}{\dot{Q}_H}\right) \times 100 = \left(1 - \frac{67}{70}\right) \times 100$$

$$\eta_{Thermal,\ TEG} = 4.3\%$$

As it can be seen the difference between reversible and real thermal efficiencies is very considerable. It shows that the recovered exergy from the TEG is very low. The exergy potential of the TEG can be determined as below:

$$\dot{X}_Q = \dot{Q}_H\left(1 - \frac{T_L}{T_H}\right) = 70\left(1 - \frac{20 + 273.15}{65 + 273.15}\right)$$

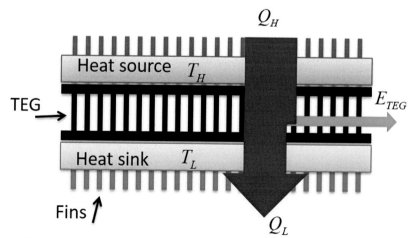

Figure 2.18 The thermoelectric generator construction and energy flow.

$$\dot{X}_Q = 9.3\text{W}$$

The exergy recovered in the TEG is actually the electricity generated by the TEG and is:

$$\dot{X}_E = \dot{Q}_H - \dot{Q}_L = 70 - 67$$

$$\dot{X}_E = 3\text{W}$$

Hence, the exergy destruction would be:

$$\dot{X}_{destroyed} = \dot{X}_Q - \dot{X}_E = 9.3 - 3$$

$$\dot{X}_{destroyed} = 6.3\text{W}$$

The exergy destruction can also be calculated as below:

$$\dot{X}_{destroyed} = T_0 \dot{S}_{gen} = T_0 \left(\frac{\dot{Q}_L}{T_L} - \frac{\dot{Q}_H}{T_H} \right)$$

$$= (20 + 273.15) \left(\frac{67}{20 + 273.15} - \frac{70}{65 + 273.15} \right)$$

$$\dot{X}_{destroyed} = 6.3\text{W}$$

The exergy efficiency of the TEG can be calculated as below:

$$\eta_{II,TEG} = 1 - \frac{X_{destroyed}}{X_{supplied}} = 1 - \frac{6.3}{9.3}$$

$$\eta_{II,TEG} = 32.2\%$$

The second law efficiency can also be calculated by the following equation:

$$\eta_{II,TEG} = \frac{\eta_{th}}{\eta_{rev}} = \frac{\eta_{Thermal,\ TEG}}{\eta_{Carnot,\ TEG}} = \frac{4.3}{13.3}$$

$$\eta_{II,TEG} = 32.2\%$$

According to the exergy definition, if the magnitudes of the $\dot{Q}_H$ and $\dot{Q}_L$ are kept the same, the best way to decrease exergy destruction (or increasing exergy efficiency) is to have higher T_H and lower T_L. It should be noted that the material used in the TEG should withstand the higher temperature at the source side as well.

The code written for the Example 2.6 is given below:

Code 2.4. The code written for the Example 2.6 in the EES.

```
"Example 2.6:"
K=273.15
T_L=20
T_H=65
Q_L=67
Q_H=70
T0=20
E_TEG=Q_H-Q_L
eta_carnot=1-(T_L+K)/(T_H+K)
eta_TEG=1-Q_L/Q_H
S_gen=-(Q_H/(T_H+K)-Q_L/(T_L+K))
X_Q=Q_H*(1-(T0+K)/(T_H+K))
X_detruction1=-(Q_H/(T_H+K)-Q_L/(T_L+K))*(T0+K)
X_detruction2=X_Q-E_TEG
eta_II_1=1-X_detruction2/X_Q
eta_II_2=eta_TEG/eta_carnot
```

4. Problems

1. A wet cooling tower (Fig. 2.19) is used for heat removal from the saturated steam and converting it to the saturated water in the condenser of a steam power plant. The mass flow rate of

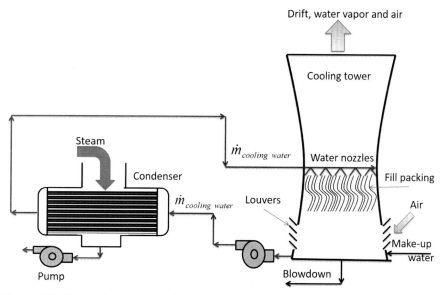

Figure 2.19 The condenser and cooling tower related to problem 1.

cooling water entering the condenser is $\dot{m}_{cooling\ water} = 8500\ (kg.s^{-1})$. Water loss occurs due to drift, evaporation, and blowdown.

Drift happens when the water droplets leaving the tower from its top with the air flow and the water loss due to drift is estimated to be about $\dot{m}_{drift} = 0.002\dot{m}_{cooling\ water}$.

Evaporation happens because of evaporative cooling process that is the most important phenomenon in the cooling tower. Water vapor leaves the tower from its top with the air stream. The water loss due to evaporation is approximately $\dot{m}_{evaporation} = 0.01\dot{m}_{cooling\ water}$.

Due to water evaporation, the concentration of the cooling water increases that this can cause more fouling on the condenser tube bundles and, as a result, narrowing the air and water passages and less heat transfer. To avoid this problem, a portion of the cooling water should be extracted from the tower basin as the blowdown. The magnitude of the blowdown is measured by $\dot{m}_{blowdown} = \frac{\dot{m}_{evaporation}}{Cycles-1}$, in which $Cycles$ is the cycles of concentration that is usually between 5 and 10.

To avoid cooling water flow rate reduction, a make-up water line is designed to compensate the water losses. Plot the makeup water required for the cooling tower at different cycles of concentrations.

2. A thermal photovoltaic (PVT) system (Fig. 2.20) is used to generate electricity and warm water. The warm water is supposed to be stored in a well-insulated storage tank. If the water inlet temperature and mass flow rate to the PVT are $T_{in} = 15\ °C$ and $0.02\ kg.s^{-1}$, respectively, and the water outlet temperature from the PVT is estimated as

$$T_{out,PVT} = 55\exp\left[-\frac{(t-13)^2}{32}\right],$$ in which t is the hour from 6:00 to 20:00 in a summer day. Determine the followings:

   A: The instantaneous mass of water inside the storage tank.
   B: The instantaneous temperature of water inside the storage tank.
   C: Plot the temperature of water at the outlet of PVT and inside of the storage tank on a single graph. Explain the trends.

3. If the storage tank in the exercise 2 initially contained M = 1008 kg of water and all pipes, the collector and pump are primed with water. The pump circulates water through the PVT during the sunshine time (Fig. 2.21). The heat absorbed by the water after passing through the PVT is $Q_{abs} = 550\exp\left[-\frac{(t-13)^2}{32}\right]$, in which t is the hour from 6:00 to 20:00 in a summer day. Determine the requests in parts A, B, and C in the previous exercise. Compare the results of both cases with and without circulation. Determine the impact of changing M from 500 to 1000 kg on the storage tank temperature for every increase of 100 kg for M.

4. A centrifugal pump (Fig. 2.22) is primed[8] with water, and the operator starts the pump without opening the discharge valve by mistake. The pump impeller with mass of m_i rotates with rotational velocity of n(rpm) by using an electromotor that consumes E (W) of

---

[8] When it is said that a pump is primed it means that gas molecules inside the pump is vented to the atmosphere and is filled with liquid.

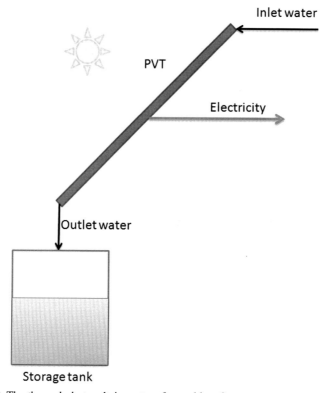

Inlet water

PVT

Electricity

Outlet water

Storage tank

Figure 2.20 The thermal photovoltaic system for problem 2.

electricity. The electromotor and pump efficiencies are η_{elec} and η_{pump}, respectively. The pump casing has a mass of m_c. The mass of water inside the pump casing is m_w, and its initial temperature is $T_1(^\circ C)$. It takes θ seconds for the operator to recognize that the discharge valve is closed. Determine the water temperature increase due to this mistake. The pump casing is well insulated and no heat transfer occurs through the casing.

5. In the exercise 4, if it is supposed to design a line which is called minimum flow line (MFL) to avoid the temperature rise due to wrong start up with closed discharge valve, determine the relation between the water temperature inside the pump and mass flow rate of this line (m_{MFL}). Assume that the minimum flow line is connected to a sink (Fig. 2.22).

6. In the exercise 4, if the pump was operated correctly, determine the temperature variation of water from inlet to the outlet of pump.

7. In the steam generator unit of a steam power plant an attemperator (Fig. 2.23) is used to control the superheated steam temperature by injecting water into the steam line. The inlet steam temperature and pressure are 580°C and 12 MPa. To avoid pressure drop, the inlet water pressure should be the same as 12 MPa. If we are not allowed to change the steam mass flow rate at the outlet more than $\alpha = 10\%$, determine a control function for the attemperating water temperature and its mass flow rate to control the outlet steam temperature at 540°C. What is the minimum value of α for the temperature control? Attention should be paid

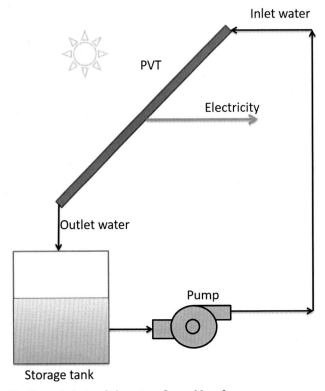

Figure 2.21 The thermal photovoltaic system for problem 3.

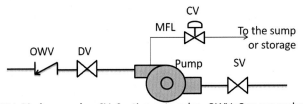

DV: Discharge valve, SV: Suction gate valve, OWV: One way valve,
MFL: Minimum flow line

Figure 2.22 Pump and suction line and discharge line for problem 4.

that you are not allowed to use two-phase flow in the attemperator system. Assume the mass flow rate of the inlet superheated steam to be $1\,\frac{kg}{s}$.

8. Drinking water of a village is provided from natural sources coming from the mountains and is stored in a storage that is constructed on a hill with height of $H = 200$ m at point A (Fig. 2.24). The water is pumped due to the hydrostatic head to the consumers using a piping system. The farthest consumer is located on a hillside at point B.

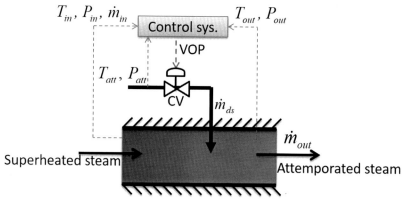

att: attemporated, in: input, out: output, CV: Control valve
VOP: Valve opening percentage

Figure 2.23 The schematic of the attemperator system for problem 7.

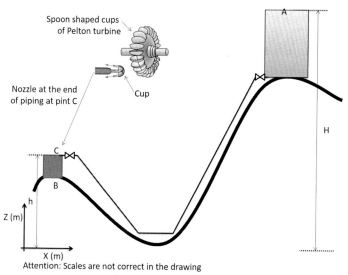

Figure 2.24 The schematic of the water storage system, piping, and hydro-turbine for problem 8.

A mechanical engineering student, who lives at the village, notices that the head produced by the water storage is much more than that, is required by the consumer at point B. He proposes installing a hydrokinetic turbine at point C to produce some electricity by using the excess head. If the head loss due to the friction and fittings between points A and C is h_f, calculate the magnitude of electricity per unit of water mass flow rate that can be produced by the hydrokinetic turbine with hydraulic efficiency of $\eta_{turbine}$.

9. A power generation cycle (Fig. 2.25), base on a steam turbine receives 1 kg.s^{-1} of super-heated steam from a steam generator at 10 MPa and 540°C. The hot gas temperature inside the steam generator due to combustion reaches 1000°C, and after passing through the steam generator, its temperature decreases to 200 °C at the stack exhaust. The steam pressure drops to 50 kPa after passing through the turbine and then enters a condenser. Assume that the output of the condenser is saturated liquid. A feedwater pump pressurizes the saturated liquid to 10 MPa before entering the steam generator unit. Neglect all pressure drops in all components and piping. The adiabatic efficiency of the turbine and pump are 0.87 and 0.85, respectively.

Determine the thermal efficiency of the cycle, exergy efficiency of the cycle, heat loss through the condenser, inlet heat to the steam generator, mass flow rate of air through the steam generator, and steam quality at the turbine exit.

In addition, to understand the cycle behavior, determine the impact of inlet temperature to the steam turbine if it could be changed from 500°C to 600°C on heat loss, heat input, thermal efficiency, and exergy efficiency of the cycle. Assume the combustion gases inside the furnace of the steam generator as pure air. In addition, the air pressure inside the steam generator is 105 kPa. The ambient temperature and pressure are 27 °C and 101 kPa.

10. Consider the cycle introduced in exercise 9. Determine the impact of maximum and minimum operating pressure of the cycle if they could change as 15 kPa $\leq P_{max} \leq$ 100 kPa and 5 MPa $\leq P_{min} \leq$ 15 MPa, respectively on the exergy efficiency of the cycle, heat loss through the condenser, inlet heat to the steam generator, and steam quality at the turbine exit.

11. Determine the minimum temperature difference between the combustion gases and steam/water in exercise 9.

12. What would be the impact of changing the exhaust temperature from 100°C to 250°C on the minimum temperature difference between the combustion gases (hot stream) and steam/water (cold stream) in the cycle?

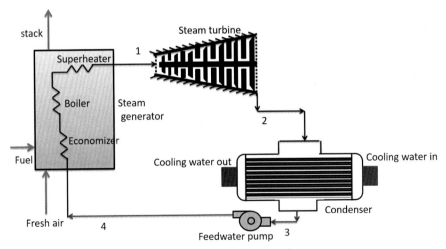

Figure 2.25 The simple cycle of steam turbine for problem 9.

References

[1] IEA, Water for Energy, is energy becoming a thirstierre source? OECD/IEA, USA, 2012.
[2] USGS, The USGS Water Science School, 2016. https://water.usgs.gov.
[3] I.G. Currie, Fundamental Mechanics of Fluids, Marcel Dekker Inc., New York. Basel, 2003.
[4] Y.A. Cengel, M.A. Boles, Thermodynamics, an engineering approach, eighth ed., McGraw-Hill, New York, 2015.

Economics of power generation

Chapter outline

1.   Importance of the economic analyses

Beside the thermodynamic analysis presented in Chapter 2, the economical assessment of power plants is vital. Imagine you are graduated from university and you are employed in a company that consults, designs, and establishes different kinds of power plants for the investors in this field.

A building owner comes to your company and asks to provide electricity for his residential building with maximum electrical demand of 100 kW. Besides, he wants to use a system that not only pay his initial investment back but also make him some profit during the lifetime of the electricity generator unit.

What would you recommend for the building owner?

Providing 100 kW of electricity may be easy but being profitable during the lifetime of the project is not simple. It means that you need to carry out economic analyses for many scenarios and electricity provision strategies to find the profitable cases. These evaluations may include considering many power generation units such as buying electricity from grid, installing micro steam turbines, micro gas turbines, different types of fuel cells, reciprocating internal/external combustion engines, solar photovoltaic, wind turbines, etc. or a combination of these systems.

Some of the power generators such as micro gas turbines, reciprocating engines, and fuel cells have high quality waste heat that can be recovered for providing the

Power Generation Technologies. https://doi.org/10.1016/B978-0-323-95370-2.00004-1
Copyright © 2023 Elsevier Inc. All rights reserved.

heating demand in the building and also reducing heating bills and increasing profitability of the project. In this case, you need to investigate the heating and cooling systems of the building and determine if it is possible to use the waste heat in the existing heating/cooling systems.

Some of the electricity generators such as the solar photovoltaic and wind turbine may only be used for certain conditions. They need to do a feasibility study about the solar heat gain and wind condition in the region. Installing a photovoltaic system in a region with very low solar irradiation may oblige you to install extra solar panels that increase the initial cost. In addition, for using solar panels or wind turbines in a residential building, you may need to determine the area needed for installation of the units and make sure about the strength of the building structure against extra load that will be applied to the building structure due to installing solar panels or wind turbine.

When you are designing a *distributed generation*[1] system of electricity system being *on-grid, off-grid,* or *hybrid*[2] have impact on the economic evaluations. Purchasing or selling electricity from or to the public grid also change the economic criteria of the project. During the low-load time, you can sell electricity to the grid and buy it back during the peak-load time. In addition, the time- or load-dependent tariff of electricity should be considered in the analyses. Another economic parameter that should be considered is the yearly operation and maintenance costs of different power generators.

As you can see for a small-scale 100 kW DG power plant, you need to consider many parameters that have impact on the economic criteria. This shows the importance of the topic that is going to be discussed in this chapter. To discuss the importance of the economic analysis more deeply, another example is discussed in the following.

Now another investor from the United States comes to your company and asks for consultation about installing a power plant for a small city with population of about 5000 inhabitants and 1500 households.

What would you recommend for the building owner?

There are many differences between providing electricity for a building or a city. Providing electricity for the small city clearly needs a bigger power plant, which means higher equipment cost.

---

[1] Distributed generation (DG) of electricity refers to the generation of electricity at the place of consumption that is usually in small to micro scales; it approximately omits the transmission and distribution electrical losses and provides the opportunity to make use of the waste heat of the power generators if there was any. In addition, a DG can be on-grid/grid-tie (connected to the public grid) or off-grid/islanding (not connected to the public electricity grid).

[2] An on-grid system can produce electricity only when the public electricity grid is available. They must connect to the grid to operate, and in case of existing excess electricity it can be sold to the grid. The reason that an on-grid system does not produce electricity during the blackout or outage is for the safety, because during the outage some people are troubleshooting and repairing the possible faults of the grid and exporting electricity to the grid may cause electrocuting. An off-grid system utilizes battery bank that can store extra electricity for the time of outage or peak load. A hybrid system is an off-grid system that is also connected to grid and during the outage it does not export electricity to the grid but it stores the extra electricity in the battery bank. In general, an on-grid system is the cheapest for investors and households.

The yearly average electricity consumption by an American household was 10766 kWh

in 2016 [1]. According to this statistics $10766\frac{kWh}{year} = \dfrac{10766\dfrac{kWh}{year}}{365\dfrac{day}{year}\times24\dfrac{h}{day}} = 1.23$ kW, that is,

the average electrical capacity needed per household. Hence, the average capacity of the power plant for the small city would be $1500 \times 1.23 = 1845$ kW. It is clear that at the peak-load time, the electricity demand would be higher. In order to have an estimation of the peak-load capacity of the power plant, you need to know the *annual load factor*.[3] The load factor has a considerable impact on the economics of the power plant because having higher load factor means that the plant capacity is used for more power generation during a longer time that means smaller $\frac{cost}{kWh}$. Briefly installing a power plant for a consumer with higher annual load factor is more economic because it results in smaller $\frac{cost}{kWh}$, that the *cost* refers to the fixed costs of the power plant installation. For example, a load factor of 0.8 results in smaller $\frac{cost}{kWh}$ in comparison with a load factor of 0.7.

Since the equipment cost is the main part of the initial investment costs, the type of the power generator would be very important, because the price of different power generators with the same capacity is different. However, providing electricity for a city obliges you to care about the power *availability*,[4] *reliability*[5], and *quality*.[6] It is clear that using highly available, reliable power generator with high-quality power generation costs more [2].

Installing, operating, and maintenance of a bigger power plant need engineers, operators, technician, etc. as personals and employees that must be paid monthly.

A big power plant like this one needs a land for its installation, buying this land means cost.

Transmission and distribution losses in the electricity grid are important in this case because electricity should be provided in a power plant and then transmitted and distributed among the consumers. This loss will reduce the overall efficiency of power delivery and increase the electricity cost for the supplier and customer as well.

Big power plants like this one should be on-grid because in case of overhaul maintenance or forced shutdown due to unexpected failures, the electricity demand of the city should be provided by the other power plants operating in other places. This means that the local grid of the small city should be connected to the other grids. This also will cause some extra costs.

---

[3] Annual load factor is defined as the ratio of the average load to the peak-load during a year.

[4] "The ability of equipment to successfully perform its required function at a stated instant of time or over stated period of time is called availability" [2].

[5] "Reliability is the probability that a device will satisfactorily perform a specified function for a specified period of time under given operating conditions" Reliability can be defined as $R = \left(1 - \frac{FOH}{PH}\right) \times 100$ in which FOH is the forced outage hours, and PH is the operational period in hours [2].

[6] The quality of electricity received from a power plant can be evaluated according to the outage of electricity, the harmonic content, variation in the voltage level, and the transient voltage and current. The smaller these parameters the higher the power quality will be [2].

As it can be seen, the economics of power plant and power generation depends on many parameters that should be identified by the designer. The economic criterion that is supposed to be used for the evaluation should consider all the impacting parameters. According to the examples and the discussions given, economic analysis is essential for establishing every power plant system from micro scale to mega scale. For this purpose, many economic criteria can be used to compare different scenarios from different point of views. This chapter is devoted to the economic criteria, which are vital for economic analyses of power plants. Among the criteria the payback period (PB), net present value (NPV), internal rate of return (IRR), etc. will be presented, and practical examples and exercises are also provided to understand the economic concepts more deeply.

2. Economical evaluation criteria

The economics of power plants starts from the time when the idea of power generation comes to the mind of the investor.

The most important questions that would be asked by the investor are:

1 How much do I invest initially?
2 How much do I earn yearly?

The answer of these questions can roughly tell the investor about the future of the investment and getting some more details makes it clear what will happen for the investment after the lifetime of the project.

Hence, to start the economic evaluations, first of all, we should be able to estimate the initial investment costs.

2.1 Initial investment costs (I)

The first economic parameter that an investor needs to know about the investment is the initial capital that should be invested in the project just before electricity generation. The initial investment may include but are not limited to the following parameters.

a. Equipment costs include the power generator, and its subsystems
b. Footprint or the land that the power plant should be installed there
c. Labor payments
d. Materials and machines needed for the construction of the buildings, roads and foundations inside the power plant site
e. Project and construction management
f. Engineering and consultation fees
g. Project contingency
h. Project financing (interest during construction)
i. Energy bills payments during power plant construction
j. Required permissions and licenses needed to be taken from different organizations

The investment cost (I) and investment index (i) are defined as below:

$$I = \sum_{i=1}^{n} I_i \tag{3.1}$$

$$i = \frac{I}{C} \tag{3.2}$$

In which C is the power plant capacity.

Example 3.1. An investor has asked a company to evaluate the initial investment cost of a power plant in $\frac{\$}{kW}$ for different capacities of 100, 500, 1000, 3000, and 5000 kW using reciprocating internal combustion engines. An engineer from the company after several hours of studies and price inquiry from equipment manufacturers provides the data presented in Table 3.1. You are asked to calculate total investment money in \$ and the investment index of $\frac{\$}{kW}$ for each power plant capacity. According to the results, what is your conclusion?

Solution. Since all of the costs are given in $\frac{\$}{kW}$, it would be very easy to calculate the investment index of i for each capacity as below:

$$i_{100} = 1000 + 10 + +260 + 340 + 200 + 200 + 70 + 30 = 2110\,\frac{\$}{kW}$$

$$i_{500} = 880 + 9 + 60 + 300 + 180 + 180 + 60 + 40 = 1709\,\frac{\$}{kW}$$

$$i_{1000} = 760 + 8 + 40 + 250 + 150 + 150 + 50 + 50 = 1458\,\frac{\$}{kW}$$

$$i_{3000} = 520 + 7 + 30 + 249 + 90 + 90 + 30 + 50 = 1057\,\frac{\$}{kW}$$

Table 3.1 Capital costs of a power plant based on the reciprocating internal combustion engines [3].

Capacity (kW)	100	500	1000	3000	5000
Costs all in $/kW					
Generator set package	1000	880	760	520	520
Land	10	9	8	7	6
Interconnect/electrical	260	60	40	30	20
Labor/material	340	300	250	240	250
Project and construction management	200	180	150	90	70
Engineering and fees	200	180	150	90	70
Project contingency	70	60	50	30	30
Project financing (interest during construction)	30	40	50	50	50

$$i_{5000} = 520 + 6 + 20 + 250 + 70 + 70 + 30 + 50 = 1016 \frac{\$}{kW}$$

And the total capital cost in dollar can be calculated as below:

$$I_{100} = C_{100} \times i_{100} = 100 \times 2110 = 211000\$$$

$$I_{500} = C_{500} \times i_{500} = 500 \times 1709 = 854500\$$$

$$I_{1000} = C_{1000} \times i_{1000} = 1000 \times 1458 = 1458000\$$$

$$I_{3000} = C_{3000} \times i_{3000} = 3000 \times 1057 = 3171000\$$$

$$I_{5000} = C_{5000} \times i_{5000} = 5000 \times 1016 = 5080000\$$$

According to the results, as the size of power plant increases, the investment index decreases. The results show that the investment index of 100 kW power plant is more than twice the 5000 kW power plant.

2.2 Net cash flow (CF)

Net cash flow is the annual net profit of the power plant. It is the summation of the positive cash flows (incomes) and negative cash flows (expenses). The higher the annual net cash flow, the more profitable the power plant.

In order to calculate *CF*, the possible yearly incomes and expenses should be recognized. The income sources (*In*) may include but are not limited to the followings.

a. The electricity sold to the grid.
b. Subsides that may be received from government and international organizations
c. United Nation rewards due to reduction of carbon production[77]
d. The possible recovered heat from the waste heat sources that can be sold for production of power, cooling, heating, etc.

The expense sources (*Ex*) of power generation include but are not limited to the followings.

a. Fuel costs, if nonrenewable energy sources are used (such as fossil fuels)
b. Operation and maintenance costs
c. *Carbone tax*[88] that should be paid for extra carbone emission production
d. Operational loan installments that should be paid regularly. Care should be taken; if a loan is used for construction of the power plant, it should be considered in the initial investment costs only.
e. Payments and salaries of the operators, engineers, etc.

The *CF* can be calculated as below:

$$CF_y = \sum In - \sum Ex \qquad (3.3)$$

In which, *In* and *Ex* are the incomes and expenses of power plant during the year of *y*.

Example 3.2. A gas turbine power plant with capacity of 2 MW of electricity production works 7500 h of year at full load and sells electricity to the residential customers at the United States. The average electricity price for residential sector in the United States is 13.31 $\frac{\text{cents}}{\text{kWh}}$. The operation and maintenance costs of the power plant is 1 $\frac{\text{cent}}{\text{kWh}}$ [4]. The gas turbine is fueled with natural gas with an average price of 3.67 $\frac{\$}{1000\text{ft}^3}$ in 2018. The power plant electrical efficiency is 35%, and five operators are working at this power plant that each one receives about 6000 $\frac{\$}{\text{month}}$. Each cubic feet of natural gas has 0.29 kWh of energy. Determine the yearly net cash flow of the power plant.

Solution. The data given in the problem can be listed as below:

$$E = 2000 \text{ kW}$$

$$T_{\text{operating}} = 7500 \text{ h}$$

$$E_{\text{tariff}} = 0.1331 \frac{\$}{\text{kWh}}$$

$$NG_{\text{tariff}} = 0.00367 \frac{\$}{\text{ft}^3}$$

$$\eta_{\text{electrical}} = 0.35$$

$$1 \text{ft}^3 \text{ of NG} = 0.29 \text{ kWh}$$

$$i_{O\&M} = 0.01 \frac{\$}{\text{kWh}}$$

$$N_{\text{personel}} = 5$$

$$\text{Salary} = 6000 \frac{\$}{\text{month.person}}$$

The only positive cash flow is due to selling electricity, but the negative cash flows include the fuel cost, operation and maintenance cost, and salary, which is paid monthly. Hence, the annual incomes and expenses are as below:

$$In_E = E(\text{kW}) \times T_{\text{operating}}\left(\frac{\text{h}}{\text{year}}\right) \times E_{\text{tariff}}\left(\frac{\$}{\text{kWh}}\right)$$

$$= 2000(\text{kW}) \times 7500(\text{h}) \times 0.1331\left(\frac{\$}{\text{kWh}}\right) = 1.997 \times 10^6 \frac{\$}{\text{year}}$$

$$Ex_{NG} = V_{NG}\left(\text{ft}^3\right) \times NG_{\text{tariff}}\left(\frac{\$}{\text{ft}^3}\right)$$

$$\eta_{\text{electrical}} = \frac{E(\text{kW})}{NG_{\text{energy}}(\text{kW})} \to NG_{\text{energy}}(\text{kW}) = \frac{E(\text{kW})}{\eta_{\text{electrical}}} \to$$

$$V_{NG}\left(\text{ft}^3\right) = \frac{NG_{\text{energy}}(kW) \times T_{\text{operating}}\left(\frac{\text{h}}{\text{year}}\right)}{0.29\dfrac{\text{kWh}}{\text{ft}^3}}$$

$$Ex_{NG} = \frac{E(\text{kW}) \times T_{\text{operating}}\left(\frac{\text{h}}{\text{year}}\right)}{\eta_{\text{electrical}} \times 0.29\dfrac{\text{kWh}}{\text{ft}^3}} \times NG_{\text{tariff}}\left(\frac{\$}{\text{ft}^3}\right)$$

$$= \frac{2000(\text{kW}) \times 7500(\text{h})}{0.35 \times 0.29\dfrac{\text{kWh}}{\text{ft}^3}} \times 0.00367\left(\frac{\$}{\text{ft}^3}\right) = 542365\frac{\$}{\text{year}}$$

$$Ex_{O\&M} = E(\text{kW}) \times T_{\text{operating}}\left(\frac{\text{h}}{\text{year}}\right) \times i_{O\&M}\left(\frac{\$}{\text{kWh}}\right)$$

$$= 2000(\text{kW}) \times 7500(\text{h}) \times 0.01\left(\frac{\$}{\text{kWh}}\right) = 150000\frac{\$}{\text{year}}$$

$$Ex_{\text{personel}} = N_{\text{personel}}(\text{person}) \times Salary\left(\frac{\$}{\text{month.person}}\right) \times 12\left(\frac{\text{months}}{\text{year}}\right)$$

$$= 5(\text{person}) \times 6000\left(\frac{\$}{\text{month.person}}\right) \times 12\left(\frac{\text{months}}{\text{year}}\right) = 360000\frac{\$}{\text{year}}$$

And now the annual net cash flow can be calculated.

$$CF = In_E - Ex_{NG} - Ex_{O\&M} - Ex_{\text{personel}}$$

$$= 1.997 \times 10^6 - 542365 - 150000 - 360000 = 944135\ \$/year$$

$$\to CF = 944135\frac{\$}{\text{year}}$$

It means that this power plant can produce net profit of 944135 $ per year. The calculations are coded in the EES as demontrated in Code 3.1.

Code 3.1. The code written for Example 3.2.

```
E=2000                    "kW"
Operating_time=7500       "hours"
E_tariff=0.1331               "$/kWh"
NG_tariff=3.67*10^(-3)      "$/ft^3"
eta_electrical=0.35
c_conversion=0.29         " 1ft^3=0.29 kWh"
i_O&M=0.01                 "$/kWh"
N_personel=5
salary=6000
eta_electrical=E/F_energy
V_NG=F_energy*Operating_time/c_conversion
In_e=E*Operating_time*E_tariff
Ex_NG=V_NG*NG_tariff
Ex_O&M=E*Operating_time*i_O&M
Ex_personel=N_personel*salary*12
CF=In_e-Ex_NG-Ex_O&M-Ex_personel
```

According to the investment cost and annual net cash flow of the project, an important criterion can be calculated that makes the decision-making much easier for the investor. It is called payback period that is introduced in the following.

2.3 Payback period

Payback period is the time period that it takes for the projects to pay the initial investment costs back to investor. This criterion can provide an initial tool for the investor to make the right decision. Hence, the *PB* is defined as below:

$$PB = \frac{I}{CF} \tag{3.4}$$

It is important to explain that this criterion may cause some misunderstanding in some special cases.

a. The magnitude of *CF* may change from year to year. This change can be due to load change, salary change, changing the fuel and electricity tariff, etc.

b. The *CF* may stay the same, but its value due to increasing or decreasing of inflation rate can change.

c. Less electricity generation due to efficiency reduction that usually happens because of equipment aging.

According to the points mentioned above, it is recommended to include some of those impacting parameters in the *CF* calculations for different years.

Example 3.3. Consider a power plant with a constant *CF* for the *n* years of its life time of operation. The initial investment is *I*, and the inflation rate is *r*. If you are supposed to calculate a modified payback period for this project, how do you calculate it? Compare the results for the conventional payback period and modified payback period for these data $I = 10^7\$$, $CF = 2 \times 10^6\ \$$, $n = 15$ years and $r = 0.05$, which one gives a more realistic data for decision-making.

Solution. Since the project has faced a constant inflation rate, the real value of the CF in the future years should be calculated, and then an average CF for the n years be calculated.

The present value of the CF at the end of the first year:

$$CF_1 = \frac{CF}{1+r}$$

The present value of the CF at the end of the second year:

$$CF_2 = \frac{CF}{(1+r)^2}$$

The value of the CF at the end of the third year:

$$CF_3 = \frac{CF}{(1+r)^3}$$

And the value of the CF at the end of the nth year:

$$CF_n = \frac{CF}{(1+r)^n}$$

To calculate the average CF during the n years of operation of the power plant, we have:

$$\overline{CF} = \frac{\sum_{i=1}^{n} CF_i}{n} = \frac{CF}{n} \sum_{i=1}^{n} \frac{1}{(1+r)^i}$$

Then, the modified payback period can be calculated as below:

$$PB_{modified} = \frac{I}{\overline{CF}} = \frac{I}{\dfrac{CF}{n}\sum_{i=1}^{n}\dfrac{1}{(1+r)^i}} = \frac{I}{CF} \times \frac{n}{\sum_{i=1}^{n}\dfrac{1}{(1+r)^i}}$$

$$= PB \left(\frac{n}{\sum_{i=1}^{n}\dfrac{1}{(1+r)^i}} \right)$$

$$PB_{\text{modified}} = PB \left(\frac{n}{\sum_{i=1}^{n} \frac{1}{(1+r)^i}} \right) \qquad (3.5)$$

According to the above equation:

$$\frac{PB}{PB_{\text{modified}}} = \frac{\sum_{i=1}^{n} \frac{1}{(1+r)^i}}{n} \leq 1$$

Hence:

$$PB \leq PB_{\text{modified}}$$

By substituting the data given in the example, we get:

$$PB = \frac{I}{CF} = \frac{10^7}{2 \times 10^6} = 5 \text{ years}$$

$$PB_{\text{modified}} = \frac{I}{\overline{\overline{CF}}}$$

$$\overline{CF} = \frac{CF}{n} \sum_{i=1}^{n} \frac{1}{(1+r)^i} = \frac{2 \times 10^6}{15} \sum_{i=1}^{15} \frac{1}{(1+0.05)^i} = 1.384 \times 10^6 \text{ \$}$$

Hence:

$$PB_{\text{modified}} = \frac{10^7}{1.384 \times 10^6} = 7.2 \text{ years}$$

The results show that if the real value of money would be considered, the modified payback period will be 7.2 years, while the conventional method payback period is 5 years. As a result, considering the time value of money gives a better understanding about the investment in the project.

The code written in EES for the example three is given below:

Code 3.2. The code written for Example 3.3.

```
Function CF_COEFFICIENT(n)
    PV:=0
    i:=0
10: i:=i+1
    PV:=1/(1+0.05)^i+PV
    If (i<N) Then GoTo 10
    CF_COEFFICIENT:=PV
End

n=15
Y= CF_COEFFICIENT(n)
I_investment=10^7
CF=2*10^6
CF_BAR=CF/n*Y
PB_modified=I_investment/CF_BAR
PB=I_investment/CF
```

An investor needs to know if a project is profitable or unprofitable. In addition, he/she needs to know about the magnitude of the profit that can be achieved in the project. The criterion that can help the investors about the profitability and magnitude of profit is net present value (*NPV*), which is introduce in the following.

2.4   Net present value (NPV)

NPV is the summation of all cash flows of the project from the zero time (the time when it was decided to build a power plant) to the death time (the time when the power plant cannot work anymore). In this summation, the present value of each cash flow should be considered. In addition, the salvage value of the project should be considered as well. Hence:

$$NPV = -I + S + \sum_{y=1}^{n} \frac{CF_y}{(1+r)^y} \tag{3.6}$$

In which, S is the present value of the project salvage, n is the project lifetime in year, and r is the interest rate.

$$NPV \begin{cases} > 0 & \text{profitable} \\ < 0 & \text{unprofitable} \\ = 0 & \text{neutral} \end{cases} \tag{3.7}$$

The magnitude of *NPV* shows the real profit that has been achieved during the project lifetime.

Example 3.4. A 5 MWe gas turbine power plant works 7500 h of year at full load and sells electricity to the residential customers at the United States. The complete installation cost of the power plant is 2200 $\frac{\$}{kW}$. The average electricity price for residential sector in the United States is 13.31 $\frac{cents}{kWh}$. The operation and maintenance costs of the power plant are 0.0074 $\frac{\$}{kWh}$. The gas turbine is fueled with natural gas with an average price of 3.67 $\frac{\$}{1000ft^3}$ in 2018. The power plant electrical efficiency is 37%, and 8 operators are working at this power plant that each one receives about 6500 $\frac{\$}{month}$. Each cubic feet of natural gas has 0.29 kWh of energy. The annual interest rate in the United States can be assumed as 2%, and the power plant can operate for 15 years. The salvage present value of the power plant will be about 10% of its initial investment cost. Calculate the *NPV* and *PB* of the project.

Solution. The data given in the problem can be listed as below:

$$E = 5000 \text{ kW}$$

$$T_{operating} = 7500 \frac{h}{year}$$

$$E_{tariff} = 0.1331 \frac{\$}{kWh}$$

$$NG_{tariff} = 0.00367 \frac{\$}{ft^3}$$

$$\eta_{electrical} = 0.37$$

$$1ft^3 \text{ of } NG = 0.29 \text{ kWh}$$

$$i_{O\&M} = 0.0074 \frac{\$}{kWh}$$

$$N_{personel} = 8$$

$$Salary = 6500 \frac{\$}{month.person}$$

The only positive cash flow is due to selling electricity, but the negative cash flows include the fuel cost, operation and maintenance cost, and salary, which is paid monthly. Hence, the annual incomes and expenses are as below:

$$In_E = E(\text{kW}) \times T_{\text{operating}} \left(\frac{\text{h}}{\text{year}}\right) \times E_{\text{tariff}} \left(\frac{\$}{\text{kWh}}\right)$$

$$= 5000(\text{kW}) \times 7500 \left(\frac{\text{h}}{\text{year}}\right) \times 0.1331 \left(\frac{\$}{\text{kWh}}\right) = 4.991 \times 10^6 \frac{\$}{\text{year}}$$

$$Ex_{NG} = V_{NG}\left(\text{ft}^3\right) \times NG_{\text{tariff}} \left(\frac{\$}{\text{ft}^3}\right)$$

$$\eta_{\text{electrical}} = \frac{E(\text{kW})}{NG_{\text{energy}}(\text{kW})} \rightarrow NG_{\text{energy}}(\text{kW}) = \frac{E(\text{kW})}{\eta_{\text{electrical}}} \rightarrow$$

$$V_{NG}\left(\text{ft}^3\right) = \frac{NG_{\text{energy}}(\text{kW}) \times T_{\text{operating}} \left(\frac{\text{h}}{\text{year}}\right)}{0.29 \frac{\text{kWh}}{\text{ft}^3}}$$

$$Ex_{NG} = \frac{E(\text{kW}) \times T_{\text{operating}} \left(\frac{\text{h}}{\text{year}}\right)}{\eta_{\text{electrical}} \times 0.29 \frac{\text{kWh}}{\text{ft}^3}} \times NG_{\text{tariff}} \left(\frac{\$}{\text{ft}^3}\right)$$

$$= \frac{5000(\text{kW}) \times 7500 \left(\frac{\text{h}}{\text{year}}\right)}{0.37 \times 0.29 \frac{\text{kWh}}{\text{ft}^3}} \times 0.00367 \left(\frac{\$}{\text{ft}^3}\right) = 1.283 \times 10^6 \frac{\$}{\text{year}}$$

$$Ex_{O\&M} = E(\text{kW}) \times T_{\text{operating}} \left(\frac{\text{h}}{\text{year}}\right) \times i_{O\&M} \left(\frac{\$}{\text{kWh}}\right) =$$

$$5000(\text{kW}) \times 7500 \left(\frac{\text{h}}{\text{year}}\right) \times 0.0074 \left(\frac{\$}{\text{kWh}}\right) = 277500 \frac{\$}{\text{year}}$$

$$Ex_{\text{personel}} = N_{\text{personel}}(\text{person}) \times \text{Salary} \left(\frac{\$}{\text{month} \cdot \text{person}}\right) \times 12(\text{months})$$

$$= 8(\text{person}) \times 6500 \left(\frac{\$}{\text{month} \cdot \text{person}}\right) \times 12 \frac{\text{month}}{\text{year}} = 624000 \frac{\$}{\text{year}}$$

And now the annual net cash flow can be calculated.

$$CF = In_E - Ex_{NG} - Ex_{O\&M} - Ex_{\text{personel}}$$

$$= 4.991 \times 10^6 - 1.283 \times 10^6 - 277500 - 624000 = 2.807 \times 10^6 \frac{\$}{\text{year}}$$

$$\rightarrow CF = 2.807 \times 10^6 \frac{\$}{\text{year}}$$

It means that this power plant can produce net profit of $2.807 \times 10^6 \$$ per year. The NPV of the project can be determined as below:

$$NPV = -I + S + \sum_{y=1}^{n} \frac{CF_y}{(1+r)^y} = -i \times C + 0.1 \times i \times C + \sum_{y=1}^{15} \frac{2.807 \times 10^6}{(1+0.02)^y}$$

$$= -2200 \left(\frac{\$}{\text{kW}} \right) \times 5000(\text{kW}) + 0.1 \times 2200 \left(\frac{\$}{\text{kW}} \right) \times 5000(\text{kW})$$

$$+ \sum_{y=1}^{15} \frac{2.807 \times 10^6 \$}{(1+0.02)^y} = 2.617 \times 10^7 \$$$

This means that the project will pay the initial investment back during the 15 years and makes 2.617×10^7 $ profit for the investor as well.

The conventional and modified payback periods are as below:

$$PB = \frac{I}{CF} = \frac{2200 \left(\frac{\$}{\text{kW}} \right) \times 5000(\text{kW})}{2.807 \times 10^6 \frac{\$}{\text{year}}} = 3.9 \text{ years} \cong 4 \text{ years}$$

$$PB_{\text{modified}} = PB \left(\frac{n}{\sum_{i=1}^{n} \frac{1}{(1+r)^i}} \right) = 3.9 \left(\frac{15}{\sum_{i=1}^{15} \frac{1}{(1+0.02)^i}} \right) = 4.57 \text{ years}$$

According to the PB_{modified}, the project will pay the initial investment back after 4.57 years and produce profits for the remaining life time, which is 10.43 years $(15 - 4.57 = 10.43)$.

The code written for this example is given below:

Code 3.3. The code written for Example 3.4.

```
Function CF_COEFFICIENT(n)
    PV:=0
    i:=0
10: i:=i+1
    PV:=1/(1+0.02)^i+PV
    If (i<N) Then GoTo 10
    CF_COEFFICIENT:=PV
End


n=15
Y= CF_COEFFICIENT(n)
E=5000                    "kW"
Operating_time=7500       "hours"
i_GT=2200                 "$/kW"
E_tariff=0.1331                    "$/kWh"
NG_tariff=3.67*10^(-3)    "$/ft^3"
eta_electrical=0.37
c_conversion=0.29         " 1ft^3=0.29 kWh"
i_O&M=0.0074              "$/kWh"
N_personel=8
salary=6500
eta_electrical=E/F_energy
V_NG=F_energy*Operating_time/c_conversion
In_e=E*Operating_time*E_tariff
Ex_NG=V_NG*NG_tariff
Ex_O&M=E*Operating_time*i_O&M
Ex_personel=N_personel*salary*12
CF=In_e-Ex_NG-Ex_O&M-Ex_personel
I_investment=i_GT*E
CF_BAR=CF/n*Y
PB_modified=I_investment/CF_BAR
PB=I_investment/CF
Salvage=0.1*I_investment
NPV=-I_investment+Salvage+CF*Y
```

2.5 Internal rate of return (IRR)

Internal rate of return is a criterion for investigating the possible profitability of an investment. The term internal refers that this criterion neglects the external parameters such as economical imbalance, risky conditions, inflation, war, etc. It measures a discount rate of *IRR* that sets the net present value of the project equal to zero. Hence:

$$-I + S + \sum_{y=1}^{n} \frac{CF_y}{(1 + IRR)^y} = 0 \tag{3.8}$$

If the interest rate for the project is r, the profitability condition would be as below:

$$IRR \begin{cases} > r & \rightarrow \text{profitable} \\ < r & \rightarrow \text{unprofitable} \\ = r & \rightarrow \text{neutral} \end{cases} \tag{3.9}$$

Example 3.5. Determine the internal rate of return for the Example 3.4 and discuss the profitability condition of this power plant.

Solution. According to the results of the Example 3.4 and Eq. (3.8) we have:

$$-I + S + \sum_{y=1}^{n} \frac{CF_y}{(1 + IRR)^y} = 0 \rightarrow i \times C = 0.1 \times i \times C + \sum_{y=1}^{15} \frac{2.807 \times 10^6 \$}{(1 + IRR)^y} \rightarrow$$

$$= 2200 \left(\frac{\$}{kW} \right) \times 5000(kW)$$

$$= 0.1 \times 2200 \left(\frac{\$}{kW} \right) \times 5000(kW)$$

$$+ \sum_{y=1}^{15} \frac{2.807 \times 10^6 \$}{(1 + IRR)^y} \rightarrow IRR \cong 0.28$$

According to the Example 2.4, the interest rate was 0.02, but the IRR is 0.28. It means that the project is profitable. The margin between 0.02 and 0.28 tells that investing in this project is highly safe and profitable.

The code written for this example is given below:

Code 3.4. The code written for Example 3.5.

```
E=5000            "kW"
Operating_time=7500    "hours"
i_GT=2200       "$/kW"
E_tariff=0.1331            "$/kWh"
NG_tariff=3.67*10^(-3)      "$/ft^3"
eta_electrical=0.37
c_conversion=0.29      " 1ft^3=0.29 kWh"
i_O&M=0.0074            "$/kWh"
N_personel=8
salary=6500
eta_electrical=E/F_energy
V_NG=F_energy*Operating_time/c_conversion
In_e=E*Operating_time*E_tariff
Ex_NG=V_NG*NG_tariff
Ex_O&M=E*Operating_time*i_O&M
Ex_personel=N_personel*salary*12
CF=In_e-Ex_NG-Ex_O&M-Ex_personel
I_investment=i_GT*E
Salvage=0.1*I_investment

I_investment=Salvage+CF*(1/(1+IRR)^1+1/(1+IRR)^2+1/(1+IRR)^3+1/(1+IRR)^4+1/(1+IRR)^5+1/(1+I
RR)^6+1/(1+IRR)^7+1/(1+IRR)^8+1/(1+IRR)^9+1/(1+IRR)^10+1/(1+IRR)^11+1/(1+IRR)^12+1/(1+IRR
)^13+1/(1+IRR)^14+1/(1+IRR)^15)
```

2.6 Depreciation

While the electricity demand has increased due to population growth, automation, industrialization, etc., the existing power plants are getting older and their equipment are aging. Erosion, corrosion, friction, vibration, fouling, thermal stress, etc. are among the phenomena that have limited the power plants lifespan. To continue electricity generation, the investors can think of *repowering*[9] the old power plants with up to date technologies [5−7].

Repowering needs money; therefore, the investor can have a plan to annually deposit a quantity of money to give them the opportunity to repower the power plant with a new one when the old power plant is retired. The amount of money that should be set aside annually is called depreciation. The depreciation should be taken from the income. The depreciation rate can be estimated by different methods such as straight line method, percentage method, sinking fund method, and unit method. The formulation of the sinking fund method is given below:

The money deposited at the end of the first year:	d
The money deposited during the second year:	$d + d \times r = d(1 + r)$
The money deposited during the third year:	$d(1 + r) + d(1 + r)r = d(1 + r)^2$
The money deposited during the nth year:	$d(1 + r)^n - 1$
Summation of deposits during n years:	$\left(\frac{(1+r)^n - 1}{r}\right)d$

In which, r is the annual rate of compound interest on the invested capital, and n is the lifetime of the equipment. The summation of the deposits should balance the investment cost of the new equipment minus the salvage value of the retired equipment. Hence:

$$\left(\frac{(1+r)^n - 1}{r}\right)d = I - S$$

Therefore:

$$d = (I - S)\left(\frac{r}{(1+r)^n - 1}\right) \tag{3.10}$$

---

[9] Repowering of a power plant may include adding or replacing the old equipment and out-of-date technologies with the new equipment that are up-to-date technologically. The repowering may involve changing the whole power plant or only some equipment. The purpose of repowering can be improvement in the economics, reducing environmental pollution, extending the lifespan, increasing availability reliability, operability and maintainability of the existing power plant.

In other words, d is the amount of money that should be set aside each year from the income.

Care must be taken about the cost of land and construction of buildings, foundations, etc. If the repowered power plant is supposed to be installed in the previous site, the cost of land, construction, foundations, etc. should be subtracted from the initial investment costs required for the repowering. Under these conditions, the depreciation rate would be:

$$d = \left(I - I_{\text{land, cons}} - S\right)\left(\frac{r}{(1+r)^n - 1}\right) \tag{3.11}$$

Example 3.6. Consider the power plant discussed in Example 3.4. The investor is supposed to replace the whole plant after 15 years in the existing land and constructions. If the cost for the land, construction, etc. is estimated to be 200 $\frac{\$}{\text{kW}}$, what would be the depreciation rate for this project. What percent of the annual net cash flow should be set aside for repowering?

Solution. According to the data given, the total installation index cost for the power plant in Example 3.4 was 2200 $\frac{\$}{\text{kW}}$, hence:

$$d = \left(I - I_{\text{land, cons}} - S\right)\left(\frac{r}{(1+r)^n - 1}\right) = \left((i - i_{\text{land, cons}}) \times C - 0.1 \times i \times C\right)$$

$$\left(\frac{r}{(1+r)^n - 1}\right) = \left((2200 - 200)\left(\frac{\$}{\text{kW}}\right) \times 5000(\text{kW}) - 0.1\right.$$

$$\times (2200 - 200)\left(\frac{\$}{\text{kW}}\right) \times 5000(\text{kW})\left(\frac{0.02}{(1 + 0.02)^{15} - 1}\right) = 514648 \frac{\$}{\text{year}}$$

According to the amount of the annual net cash flow, which was calculated in Example 3.4, $CF = 2.807 \times 10^6 \frac{\$}{\text{year}}$, about $\dfrac{514648 \frac{\$}{\text{year}}}{2.807 \times 10^6 \frac{\$}{\text{year}}} \times 100 = 18.33\%$ of the annual net cash flow should be set aside as depreciation rate for repowering purpose.

Code 3.5. The code written for Example 3.6.

```
E=5000              "kW"
i_GT=2200           "$/kW"
i_landcons=200        "$/kW"
I_investment=i_GT*E
I_land_cons=i_landcons*E
CF=2.807*10^6
Salvage=0.1*I_investment
n=15
r=0.02
d=(r/((1+r)^n-1))*(I_investment-I_land_cons-Salvage)
PERCENTAGE=d/CF*100
```

3. Problems

1. Look for the lifetime, maintenance index cost, and investment index cost of a 500 MWe power plant based on reciprocating internal combustion engines, gas turbines, steam turbine, solid oxide fuel cells, solar photovoltaic, and wind turbine.
2. Find the electricity and fuel tariffs in your own country. Is the tariff uniform or consumption/time dependent? Draw the tariffs on a plot.
3. Determine the investment index price in $\frac{\$}{kW}$, and total cost in United States Dollar (USD) for installing 1 MW solar photovoltaic system in your own living area. Consider equipment costs, interconnect/electrical, labor/material, project and construction management, engineering and fees, project contingency, and project financing (interest during construction) in your evaluations.
4. If you could sell the electricity produced by the solar photovoltaic in problem 1 to the electricity grid, determine the annual net cash flow for the project, the conventional and modified payback period for the project
5. Determine the net present value, depreciation rate, and internal rate of return for the project discussed in Example 3.1 in your own country.
6. A steam turbine power plant with capacity of 500 MWe works 7500 h of year at full load and sells electricity to the residential customers at the United States. The complete installation cost of the power plant is 1200 $\frac{\$}{kW}$. The average electricity price for residential sector in the United States is 13.31 $\frac{cents}{kWh}$. The operation and maintenance costs of the power plant are 0.004 $\frac{\$}{kWh}$. The boiler is fueled with natural gas with an average price of 3.67 $\frac{\$}{1000ft^3}$ in 2018. The power plant electrical efficiency is 40%, and 200 operators are working at this power plant that each one receives about 6500 $\frac{\$}{month}$. Each cubic feet of natural gas has 0.29 kWh of energy. The annual interest rate in the US can be assumed as 2%, and the power plant can operate for 30 years. The salvage present value of the power plant will about 10% of its initial investment cost. Determine:
   A. The depreciation rate, *NPV*, *IRR*, conventional, and modified *PB* of the project.
   B. Plot the impact of changing the interest rate from 0 to 0.05 on the *NPV*, *d*, and *PB*.
   C. Plot the impact of operational hours of the power plant on the *NPV*, *d*, and *PB* when it changes from 6000 h to 8000 h.
   D. Plot the impact of natural gas tariff for $\overline{+}$30% change in its price on the *NPV*, *d*, and *PB*
   E. Plot the impact of electricity tariff for $\overline{+}$30% change in its price on the *NPV*, *d*, and *PB*

7. A solid oxide fuel cell power plant with capacity of 5 MWe works 7500 h of year at full load and sells electricity to the residential customers at the United States. The complete installation cost of the power plant is 3500 $\frac{\$}{kW}$. The average electricity price for residential sector in the United States is 13.31 $\frac{cents}{kWh}$. The operation and maintenance costs of the power plant are 0.023 $\frac{\$}{kWh}$. The fuel cell is fueled with natural gas with an average price of 3.67 $\frac{\$}{1000\ ft^3}$ in 2018. The fuel cell electrical efficiency is 55%, and 20 operators are working at this power plant that each one receives about 8000 $\frac{\$}{month}$. Each cubic feet of natural gas has 0.29 kWh of energy. The annual interest rate in the United States can be assumed as 2%, and the power plant can operate for 17 years. The salvage present value of the power plant will about 10% of its initial investment cost. Determine:

A. The depreciation rate, *NPV*, *IRR*, conventional, and modified *PB* of the project. Compare the results with exercise 6. Is the project profitable? If not, for what magnitude of the investment index cost the project will be profitable?

B. Plot the impact of changing the interest rate from 0 to 0.05 on the *NPV*, *d* and *PB*.

C. Plot the impact of operational hours of the power plant on the *NPV*, *d* and *PB* when it changes from 6000 h to 8000 h.

D. Plot the impact of natural gas tariff for $\overline{+}$ 30% change in its price on the *NPV*, *d* , and *PB*.

E. Plot the impact of electricity tariff for $\overline{+}$ 30% change in its price on the *NPV*, *d* and *PB*.

References

[1] U.S. Energy Information Administration, Annual Energy Outlook, EIA, United States, 2017.

[2] de Souza, G. Francisco Martha, Thermal Power Plant, Performance Analysis, Springer, London Dordrecht Heidelberg New York, 2012.

[3] Energy and Environmental Analysis, Inc. an ICF Company, Technology Characterization: Reciprocating Engines. Washington DC: Combined heat and power partnership program, 2008.

[4] D. Balevic, R. Burger, D. Forry, Heavy-duty Gas Turbine Operating and Maintenance, GE Energy, Atlanta, 2009.

[5] A.K. Raja, A.P. Srivastava, M. Dwivedi, Power Plant Engineering, New Age International, New Delhi, 2006.

[6] S. Duffuaa, A. Raouf, J. Dixon, Planning and Control of Maintenance Systems: Modeling, Wiley, New York, 1999.

[7] A. Smith, G. Hinchcliffe, RCM Gateway to World Class Maintenance, Butterworth-Heinemann, Elsevier, Oxford, 2004.

Energy sources for power generation

Chapter outline

Our ancestors relied more on the strength of their muscles than anything else to do their jobs. In addition, they burned wood and dried animal waste for cooking and heating. After a while, they tamed animals and used their muscular strength for various applications such as transportation, war, and agriculture. With the beginning of the industrial revolution and the use of fossil fuels (coal, oil, and gas), human life changed a lot in all aspects, including the level of welfare, economy, health, production, education, etc., and the effects of this revolution are still visible.

Scientists know that the amount of fossil fuels is limited. Despite increasing discoveries, as well as improvements in fuel extraction methods, the time will come when the recyclable rate of fossil fuels will begin to decline until it reaches its end point.

Hubbert's Peak Theory [1], also known as the *Hubbert curve* was published in 1956 for any given region, a fossil fuel production curve would follow a bell-shaped curve, with production first increasing following discovery of new resources and improved extraction methods, peaking, then ultimately declining as resources became depleted.

Power Generation Technologies. https://doi.org/10.1016/B978-0-323-95370-2.00002-8
Copyright © 2023 Elsevier Inc. All rights reserved.

Many researchers have tried to validate the Hubert curve with the real fossil energy production curve in some regions of the world, but since the discovery of new energy sources and the improvement of extraction methods may reverse the trend in the short term, in many cases, these validations have not been successful (Fig. 4.1).

However, in the long run, we need to look for more reliable energy sources than fossil fuels. The new energy resources must be nonpolluting, unlimited, renewable, and as far as possible distributed around the world to prevent energy monopolies in some regions. Today, we are using divers energy resources including fossil, nuclear, hydro, solar, wind, bio, and geothermal. However, their share in our energy consumption is different and, in some cases, very small in comparison with the existing potentials. Extreme use of fossil fuels has polluted atmosphere, and greenhouse gases have caused global warming. To stabilize global temperature, all man-made pollution and greenhouse gases must be removed. For this purpose, the **net zero emission (NZE) by 2050** scenario under the **Paris Agreement** is triggered to control the global warming. In this program, the impact and share of different energy sources on climate change are evaluated, and steps toward achieving the goal are stated. In the following, the energy sources for power generation are presented, their share in the current global power production is reported, and steps to reach the NZE scenario are stated.

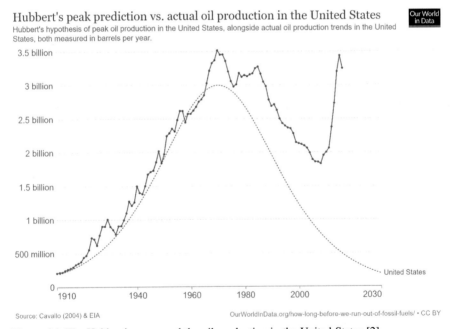

Figure 4.1 The Hubbert's curve and the oil production in the United States [2].

1. Fossil fuels

Fossil fuels including coal, oil, and gas are the mostly known fossil fuels, which provide about 60% of the total electricity production in the world. Fig. 4.2 shows the trend of using fossil fuels for electricity production in the world and some countries or continents. It shows that China and Africa have been using fossil fuels more than the world average for electricity production during 1985—2020; however, china has started decreasing the share of fossil fuels for electricity production. The United States also started to decline the share of fossil fuels from 2007, and in 2020, it is below the world average. But Europe has always used less fuel for electricity production in comparison with the world average. Europe is providing 38.03% of its total electricity by using fossil fuels.

What matters now is that how much fossil fuel is reserved in our planet and how long it can last? According to the available data, the oil, gas, and coal reserves in the world is increasing while we still have no data about fossil fuel reserves in great regions of the world. Figs. 4.3 and 4.4 show the distribution of coal and gas reserves around the world in billion tons and trillion cubic meter.

Due to the new discoveries of the coal, gas, and oil reserves around the world, the overall trend of fossil fuel reserves is increasing according to Fig. 4.5.

The units of reserves for coal, gas, and oil are ton, million cubic feet, and barrel. And G in the vertical axis is Giga, which means 10^9. For example, 1000G tons of

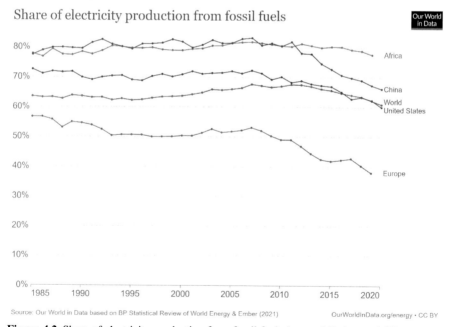

Figure 4.2 Share of electricity production from fossil fuels (ourworldindata.org) [2].

Coal reserves, 2020

Proved reserves is generally taken to be those quantities that geological and engineering information indicates with reasonable certainty can be recovered in the future from known reservoirs under existing economic and operating conditions.

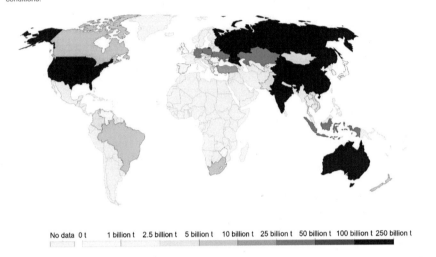

No data 0 t   1 billion t   2.5 billion t   5 billion t   10 billion t   25 billion t   50 billion t   100 billion t   250 billion t

Source: BP Statistical Review of World Energy   OurWorldInData.org/fossil-fuels/ • CC BY

Figure 4.3 The coal reserves distribution around the world in 2020 (ourworldindata.org) [2].

Gas Reserves, 2019

Proved reserves is generally taken to be those quantities that geological and engineering information indicates with reasonable certainty can be recovered in the future from known reservoirs under existing economic and operating conditions.

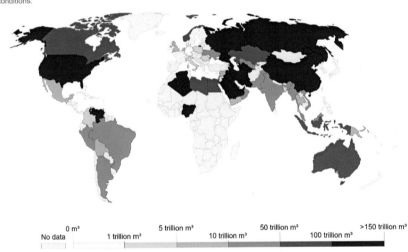

0 m³   5 trillion m³   50 trillion m³   >150 trillion m³
No data   1 trillion m³   10 trillion m³   100 trillion m³

Source: BP Statistical Review of World Energy   OurWorldInData.org/fossil-fuels/ • CC BY

Figure 4.4 The gas reserves distribution around the world in 2020 (ourworldindata.org) [2].

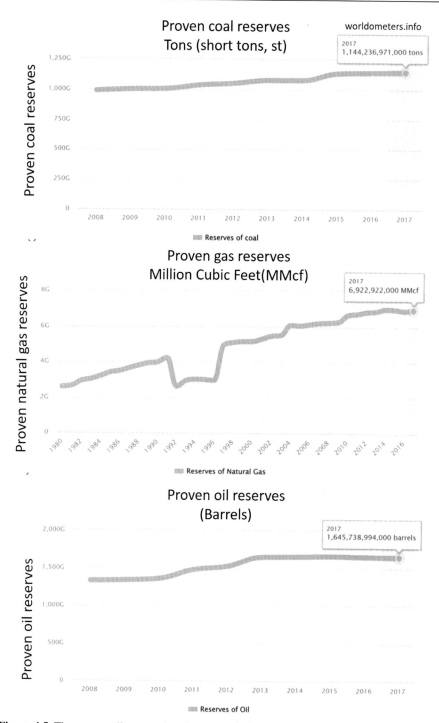

Figure 4.5 The proven oil, gas, and coal reserves during years of extractions until 2017 (worldometers.info) [3].

coal is 10^{12} tons of coal. According to the curves, the increasing rate of gas reserves is much higher than oil and coal.

The most critical question is that how long does it take to consume the remaining fossil fuel reserves? To determine this, a *reserve to production ratio (R/ P)* must be calculated. In this ratio, the reserves value is divided by the current production (consumption) rate, and the result will tell us how many years later the existing fossil fuels will be finished. It is clear that this ratio may change according to the changes in the production rate, new discoveries, and promotions in the extraction techniques. According to the production and consumption rates of coal, oil, and gas reserves in 2017, they will be finished 133, 47, and 52 years after 2017 (Fig. 4.6).

Even if we double the remaining years of the life of fossil fuels, it is still negligible in comparison with the human life on Earth before. It is very unpleasant to imagine that suddenly, the inhabitants of the earth lose most of their sources of energy and electricity, but it is truth, and we must prepare ourselves for such day. Man must take other ways of providing energy very seriously, especially renewable energy sources. In the following, the main fossil fuels are discussed in more details.

1.1   Coal

Coal is a shining piece of black stone that burns and produces heat and light. Cavemen used coal for heating as well as cooking. Coal generates more heat per unit volume and weight than wood and can burn longer. In the modern era, people and various industries from the 1800s began to use coal to heat houses, move locomotives and ships, or metal smelting furnaces to make steel and copper, etc. Although human is consuming and finishing coal reserves very fast, but its production has taken million years.

Fig. 4.7 shows the peatification and coalification processes. Peatification is the accumulation and breaking down of the plant materials in the swamps and wetlands. The buried peat under pressure and heat during very long times will change to different ranks of coal such as lignite (brown coal), sub-bituminous, bituminous, and anthracite.

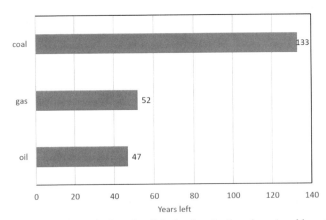

Figure 4.6 Years left for the main three fossil fuels of coal, oil, and gas (worldometers.info) [3].

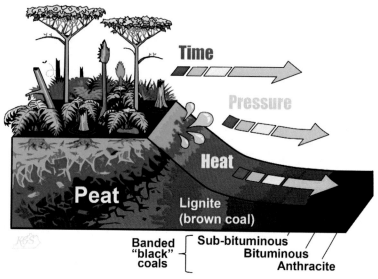

Figure 4.7 The peatification and coalification processes.
From Greb (2021). Copyright Kentucky Geological Survey, 2021. https://www.uky.edu/KGS/. (Accessed on 4 August 2022). Used with permission.

The rank of a coal is specified by the amount and time duration of applying pressure and heat on the peats. The process of changing peat to coal due to heat and pressure is called coalification. Coal can be mined from underground by using different methods such as drift mine, slope mine, shaft mine, area mine, highwall (auger) mine, contour mine, and mountaintop mine. Fig. 4.8 shows the mining methods of coal.

The dominant material in different ranks of coal is carbon with variable amounts of other elements, mainly hydrogen, sulfur, oxygen, and nitrogen. The colas can be classified according to their carbon content and heating value. Anthracite comprises of 86%−97% carbon and has the highest heating value among all coal types. Bituminous coal contains 45%−86% carbon; it is used to generate electricity and is an important fuel for using in the iron and steel industry. Bituminous heating value is less than anthracite but more than sub-bituminous. Sub-bituminous coal usually contains 35%−45% carbon. Lignite contains about 25%−35% of carbon; since it has not been under severe pressure and heat for a long time, it has high moisture content and has the lowest heating value among the four types of coal. The energy content of four types of coal is shown in Table 4.1 [5].

Coal is the major fuel in electricity generation. It can be used in solid state and pulverized form in steam power plants. It can also be gasified or liquefied, which are very easy to transport through pipelines or storage in tanks. Gasification is the conversion of coal into a mixture of carbon monoxide, hydrogen, carbon dioxide, methane, and other hydrocarbons. Gasification is done through incomplete and controlled burning of coal with injection of air and water steam. Since the average ratio of hydrogen to carbon (HCR) in coal is about 0.8, which is lower than in liquid (HCR = 2) and gaseous

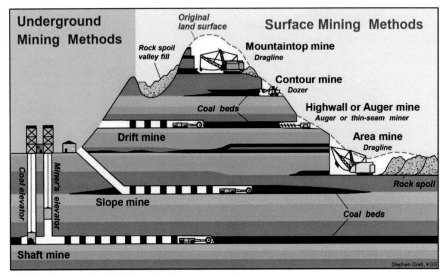

Figure 4.8 Mining methods for extraction of underground coal.
From Greb (2021). Copyright Kentucky Geological Survey, 2021. https://www.uky.edu/KGS/.
(Accessed on 4 August 2022). Used with permission.

Table 4.1 The energy content and carbon content of coal types.

Type of coal	Energy content (MJ/kg)	Carbon content
Anthracite	31–36	86%–97%
Bituminous	25–35	45%–86%
Sub-bituminous	19–30	35%–45%
Lignite	12–20	25%–35%

(HCR = 4) fossil fuels, any process of converting coal to gaseous or liquid fuels must be done by injecting some hydrogen directly or accompanied by water vapor. The gas produced in the gasification process can be used as fuel for electricity production in thermal power plants or as synthesis gas for liquefaction.

Any conversion of coal to a liquid fuel is called liquefaction and can be done directly or indirectly. There are four common ways to convert coal to liquid fuel including 1-pyrolysis and hydrocarbonization, 2-solvent extraction, 3-catalytic liquefaction, and 4-indirect liquefaction.

1.2 Fuel oil

Fuel oil, or furnace oil, mostly consists of residues from crude-oil distillation. Furnace oil is used in the steam generator of steam turbine power plants and different industries such as steal.

Fuel oils can be divided into two main categories: residual fuel oils and distilled fuel oils. Distilled fuel oils evaporate and condense during the distillation process and therefore have a certain boiling range and do not contain compounds with a high boiling point. Residual fuel oils are obtained from the residues of crude thermal cracking distillation, and their compositions are more complex than distilled fuel oils.

The American Society for Testing and Materials (ASTM) D396 standard classifies fuel oils according to their boiling range, composition, and other physical properties in six classes from fuel oil No. 1 to 6.

The heating value of the fuel oils is in the range of 42−47 MJ/kg, which in comparison with coal has a higher heating value than all coal types [6].

1.3 Fuel gas

Fuel gas is any type of fuel that is in the form of gas in environmental conditions. Fuel gas may be a combination of hydrocarbons such as methane, propane, hydrogen, carbon dioxide, carbon monoxide, and so on. Fuel gas can be easily delivered to the consumers using transmission and distribution pipelines. Fuel gas can also be liquefied for storage purposes. The most common type of gas fuel currently used in the world is natural gas. Fuel gas is divided into two general categories depending on their source of production. Manufactured fuel gas are extracted from gas wells. The manufactured fuel gas can have different sources, for the production of which we usually need a station or a factory. This type of fuel gas is obtained from materials such as coal, biomass, water, syngas, wood gas, acetylene, etc. One of the most important types of fuel gas that is extracted from underground wells is natural gas. Natural gas, if mixed with some hydrogen (about 4%−9% of the total energy), is called H_2CNG. Natural gas is odorless and colorless, so its leakage is not detectable and may cause fire and explosion in case of buildup. Harmless chemical odors such as Mercaptan are commonly used to solve the problem and detect leaks [6]. It mixes with natural gas and smells like rotten eggs. Table 4.2 shows the heating value of some gas fuels:

2. Nuclear energy

Nuclear energy is actually the heat energy that releases during the nuclear reactions of *fission* or *fusion*. To discuss these reactions, first we need to review the structure of atoms. Gases, liquids, and solids are made of molecules, and molecules are made up of atoms. Each atom comprises of protons, electrons, and neutrons. The protons and neutrons are at the core or nucleus of atoms, while electrons surround the nucleus and move around it. Protons and electrons carry positive and negative electrical charges, while neutron has no electrical charge. Great amount of energy exists in the bonds that hold the nucleus together. By breaking these bond through the fission, nuclear energy will be released that can be used for different purposes, mostly power and electricity generation. Uranium is a nonrenewable fuel that is mostly used for fission in nuclear power plants. In nuclear fission, neutron collides with a uranium

Table 4.2 The net heating value of some gas fuels [7].

Gas	MJ/ m^3	MJ/ kg	Gas	MJ/ m^3	MJ/ kg
Blast furnace gas	3.42	2.74	Hydrogen sulfite	25.03	17.39
Butane - C$_4$H$_{10}$	110.92	46.46	Landfill gas	17.73	
Butylene (butene)	107.15	45.17	Methane—CH$_4$	33.90	49.85
Carbon monoxide—CO	12.03	10.16	Natural gas (typical)	31.67	40.70
Carburetted water gas	18.92	24.57	Octane saturated with water	118.11	24.29
Coal gas	5.55	38.37	Pentane	137.07	44.94
Coke oven gas	19.15	35.51	Propane - C$_3$H$_8$	88.34	46.13
Digester gas (sewage or biogas)	23.13	23.68	Propene (propylene) - C$_3$H$_6$	81.26	45.65
Ethane -C$_2$H$_6$	60.73	47.20	Propylene	81.41	45.78
Ethylene	57.01	47.74	Water gas (bituminous)	8.90	10.39
Hydrogen (H$_2$)	10.24	120.08			

atom and splits it into some neutrons and lighter elements while releasing huge amount of energy. When fission starts, the produced neutrons keep colliding with other atoms and fission repeats itself as illustrated in Fig. 4.9. This process is called *nuclear chain reaction* that must be controlled in nuclear power plants to produce the heat demand by the nuclear steam generator. In fact, in a nuclear power plant, nuclear reactors are responsible for providing energy demand by the steam generation [8,9].

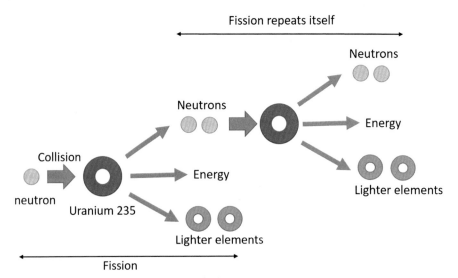

Figure 4.9 Fission and self-repeating of fission.

Uranium fuel is molded into ceramic pellets; each ceramic pellet generates energy approximately as 150 gallons of oil. These pellets are stacked in 12-foot metal fuel rods Fig. 4.10, hundreds of fuel rods together produce a fuel bundle, which is also called fuel assembly. A reactor core is made of many fuel assemblies.

Uranium is a common metal, such as tin, or zinc, which is found almost everywhere in the earth's crust as well as in seas with different concentrations. Table 4.3 shows typical concentrations of the uranium.

The uranium resources by country is listed in Table 4.4.

Uranium-235, which can be used to make nuclear fuel for use in nuclear power plants, is so rare that it accounts for about 0.7% of all uranium. U-235 must be enriched to the level of 3%–5% for commercial nuclear reactors; then, it can be used for producing fuel pellets, fuel rods, and fuel bundles.

Fusion is generating larger atoms by combining or fusing atoms. During the fusing process, heat will be released. Fusion is also the source of energy in sun and stars; fusion is still under development and is not still economical because of the difficulties in controlling the reaction. Fusion process is schematically illustrated in Fig. 4.11.

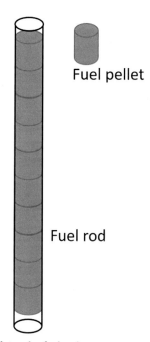

Figure 4.10 Uranium fuel pellet and a fuel rod.

Table 4.3 Typical natural uranium concentrations [10,11].

Very high-grade ore (Canada)—20% U	200,000 ppm U
High-grade ore—2% U	20,000 ppm U
Low-grade ore—0.1% U	1000 ppm U
Very-low-grade ore (Namibia)—0.01% U	100 ppm U
Granite	3—5 ppm U
Sedimentary rock	2—3 ppm U
Earth's continental crust (average)	2.8 ppm U
Seawater	0.003 ppm U

Table 4.4 Uranium resources by country in 2019 [10,11].

Country	Tons U	Percentage of world (%)	Country	tons U	Percentage of world (%)
Australia	1,692,700	28	Mongolia	143,500	2
Kazakhstan	906,800	15	Uzbekistan	132,300	2
Canada	564,900	9	Ukraine	108,700	2
Russia	486,000	8	Botswana	87,200	1
Namibia	448,300	7	Tanzania	58,200	1
South Africa	320,900	5	Jordan	52,500	1
Brazil	276,800	5	USA	47,900	1
Niger	276,400	4	Other	295,800	5
China	248,900	4			
World total: 6,147,800 tons of uranium					

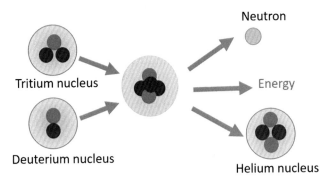

Figure 4.11 The fusion of tritium and deuterium.

3. Hydropower

Hydropower can be achieved from falling or moving water. The higher the height or velocity, the more the potential for power generation. More or less, there is the potential of hydropower around the world, and this energy source can be used to improve human well-being. The most important challenge facing human and energy use in the coming century is global warming due to greenhouse gases. In the **Net Zero Emission by 2050** report released in 2021, International Energy Agency (IEA) [12] called for the global hydropower capacity to double by 2050 to keep global temperature rising below 1.5°C. In other words, in the next 30 years, hydropower plants should be launched as much as in the last 100 years, which is equivalent to 1300 GW. The total hydropower installed until 2020 is shown in Fig. 4.12. To keep the global temperature rise below 1.5°C, the yearly increase of hydropower capacity in the world must be at least 2.3%. According to the 2020 report of IEA, the current average age of hydropower plants is 31.3 years. While installing new hydropower plants, care must be taken about the previously installed hydropower plants, they are getting old and most of them need repair, and repowering.

Fig. 4.13 shows the age of the installed hydro power fleets at 2020 around the world. It shows that China has the youngest hydropower fleet with age of 15 years, and North America has the oldest fleet with the age of 50 years.

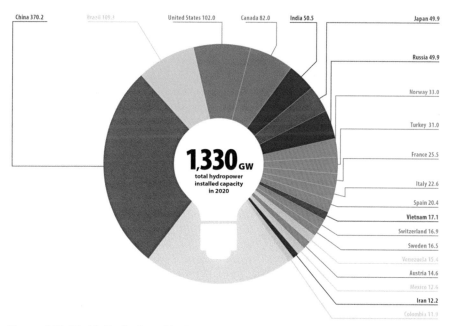

Figure 4.12 World distribution of hydropower generation (*IHA*, International Hydropower Association) [13].

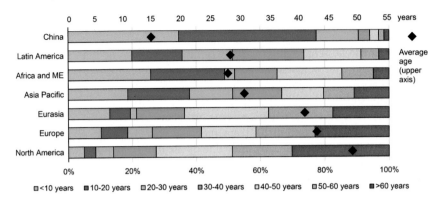

IEA. All rights reserved.

Note: ME = Middle East.

Figure 4.13 Age of installed hydropower capacity, 2020.
Courtesy of IEA.

In a research published by Hoes OAC et al. (2017) [13], it was estimated that the gross theoretical hydropower potential in the world is approximately 52×10^6 GWh/year divided over 11.8 million locations. In addition, they presented the distribution of different hydropower scales around the world, which is shown in Fig. 4.14. According to the data, the number of locations (N) with the potential of different hydropower scales is as below:

Total locations: $N_{total} = 11.8 \times 10^6$, micro scale locations: $N_{micro} = 8 \times 10^6$, mini scale locations: $N_{mini} = 2.7 \times 10^6$, small-scale locations: $N_{small} = 0.88 \times 10^6$, and large-scale locations: $N_{large} = 0.16 \times 10^6$.

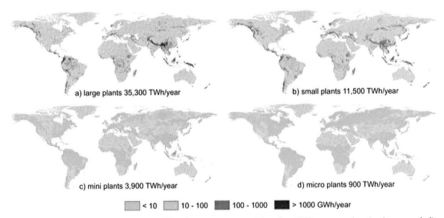

Figure 4.14 Global distribution of hydropower capacities for different scales (a, b, c, and d) around the world, TWh = 10^{12} Wh (open access) [14].

Hydropower plants can work for about 80 years, which is the highest lifetime among the power generators. The lifetime of other technologies is as below:

Battery storage (10 years), solar PV (25 years), onshore and offshore wind (25 years), gas-fired power plants (30 years), coal-fired and geothermal steam power plants (40 years), and nuclear power plants (60 years).

The main challenge for hydropower plants is their dependency on climate, hydrological conditions, water runoff patterns, and water reservoir volumes. These parameters can be strongly affected by the rainfall, snow accumulation, and melting. Small power plants such as those installed on rivers can be easily affected by the seasonal hydrological fluctuations, and their power generation may vary or even stopped during a year. On the contrary, big water storages such as dams can be used for decorrelation of the hydropower plants (partially or totally) from the hydrological fluctuations.

In addition to the hydropower potentials on the ground, a huge potential of hydropower exists in the seas and oceans. Waves, tides, salinity gradient, and ocean thermal energy conversion (OTEC) are great sources of power generation. The salinity gradient refers to the salinity difference between the see water and fresh water coming from the river at the place where it enters the sea. The OTEC refers to the temperature difference between the cooler deep and warmer surface seawater in the ocean that can be used to run a heat engine such as an organic Rankine cycle working with a fluid with low boiling temperature such as ammonia. In this case, the surface warm layer of seawater is used as the heat source, and the deep cold layers are used in the condenser. Seawater temperature decreases with depth; it may vary from 30°C at the sea surface to −1°C at the seabed.

In areas with low latitudes,[1] the temperature difference between sea surface and the seabed is high, but in the arctic regions, due to extreme cold, the water at the sea surface becomes denser and heavier, it constantly changes its place with deeper layers that have higher temperatures. Slowly, the temperature difference between sea surface and the seabed decreases and approaches the isothermal state. Hence, the potential for OTEC in these regions is low.

A report published by International Renewable Energy Agency (IRENA) [15] says that the cumulative value of ocean energy technologies is between 45,000 TWh and potentially well above 130,000 TWh of electricity per year. To know the importance of the ocean potential, the global electricity demand was 25,814 TWh in 2019. It means that oceans can generate more than four times of the electricity demand of world.

4. Solar energy

Sun distance from earth is about 150 Mkm, and speed of light is 0.3 Mkm/s, that means sun radiation takes about 500 s to reach earth. The amount of solar radiation received

---

[1] Low latitude regions are placed between the Equator and 30°N/S. The mid-latitude regions are placed between 30°N/S and 60°N/S. And the high latitude regions are placed between 60°N/S and the north and south poles.

on earth varies with cloud cover, mist and air pollution, latitude of the location, and the time of the year. It is interesting to know that earth is continuously receiving 172 TW of solar energy that is more than 10,000 times the world's total energy demand. The solar energy can be used for producing heat or electricity. Different solar systems are designed and manufactured for solar heating such as air heater collectors, liquid heater collectors, heat pipes, concentrated solar heaters, etc.

In addition, photovoltaic panels with different types are also manufactured for converting solar radiation to electricity. Figs. 4.15 and 4.16 show the global horizontal irradiation on earth in terms of $\frac{kWh}{m^2}$, and global photovoltaic power potential in terms of kWh/kWp on daily and yearly basis. The term kWh/kWp shows that how many hours in day or a year the peak power of the PV panels can be achieved [16,17].

The term kWp is the kilowatt peak and is the peak power of a photovoltaic system under Standard Test Conditions (STC): irradiance of 1000 W/m^2, a module temperature at 25°C, and a solar spectrum of AM 1.5 (ASTM G173-03(2020)) [18].

The International Energy Agency reports that to reach the **Net Zero Emissions by 2050** Scenario, 6970 TWh of PV power must be installed until 2030. This means that PV farms must increase with an average annual rate of 24% during 2020−2030 [19

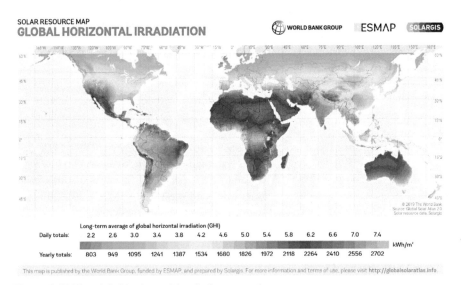

Figure 4.15 The global horizontal irradiation on earth.
Global Solar Atlas 2.0, a free, web-based application is developed and operated by the company Solargis s.r.o. on behalf of the World Bank Group, utilizing Solargis data, with funding provided by the Energy Sector Management Assistance Program (ESMAP). https://solargis.com/maps-and-gis-data/download/world. (Accessed on 4 August 2022). Used with permission.

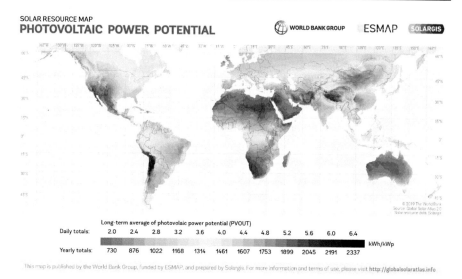

Figure 4.16 The global photovoltaic power potential on earth.
Global Solar Atlas 2.0, a free, web-based application is developed and operated by the company Solargis s.r.o. on behalf of the World Bank Group, utilizing Solargis data, with funding provided by the Energy Sector Management Assistance Program (ESMAP). https://solargis.com/maps-and-gis-data/download/world. (Accessed on 4 August 2022). Used with permission.

Example 4.1. If daily kWh/kWp of a location is 5, by installing a PV panel with a specified peak power of $P(\text{kWp})$, determine the daily and yearly potential for photovoltaic power generation in this site:

Solution.

$$E_{PV}\left(\frac{\text{kWh}}{\text{day}}\right) = P(\text{kWp}) \times \left(\frac{\text{kWh}}{\text{kWp}}\right)_{daily} = P(\text{kWp}) \times 5\frac{\left(\frac{\text{kWh}}{\text{kWp}}\right)}{\text{day}} = 5P\left(\frac{\text{kWh}}{\text{day}}\right)$$

$$E_{PV}\left(\frac{\text{kWh}}{\text{year}}\right) = 365\left(\frac{\text{day}}{\text{year}}\right)E_{PV}\left(\frac{\text{kWh}}{\text{day}}\right) = 1825P\left(\frac{\text{kWh}}{\text{year}}\right)$$

Example 4.2. You are supposed to erect a solar photovoltaic farm with capacity of 10^8 $\frac{\text{kWh}}{\text{year}}$ by using a kind of PV panel with peak power of 500 W. If the daily kWh/ kWp of the farm site is 5.5, determine the number of PV panels required for the farm.

Solution.

$$E_{PV}\left(\frac{\text{kWh}}{\text{year}}\right) = n_{PV} \times 365\left(\frac{\text{day}}{\text{year}}\right) \times P(\text{kWp}) \times \left(\frac{\text{kWh}}{\text{kWp}}\right)_{\text{daily}}$$

$$= n_{PV} \times 365\left(\frac{\text{day}}{\text{year}}\right) \times 0.500(\text{kWp}) \times 5.5\frac{\left(\frac{\text{kWh}}{\text{kWp}}\right)}{\text{day}} = 10^8 \left(\frac{\text{kWh}}{\text{year}}\right) \rightarrow n_{PV}$$

$$= 99626.4 \cong 99627 \text{ Panels}$$

Hence, the number of PV panels with peak power of 500 W for the solar PV farm with capacity of $10^8 \frac{\text{kWh}}{\text{year}}$ is 99,627 panels. They must be installed in the optimum direction and orientation in the farm.].

5. Wind energy

Wind energy is classified as onshore and offshore. The wind turbines installed on land are called onshore wind turbines, while the turbines installed in the seas and oceans and use the wind over open sea surface for electricity generation are called offshore turbines.

The International Energy Agency, as a part of its offshore wind outlook 2019, has assessed the technical potential of offshore wind by geospatial analysis, and the results showed that the best offshore wind sites close to shore could generate about 36,000 $\frac{\text{TWh}}{\text{year}}$, which is more than the global electricity demand in 2019 [20]. Figs. 4.17 and

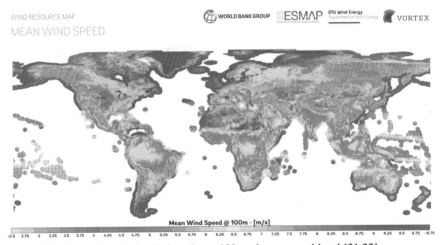

Figure 4.17 The global wind velocity atlas at 100 m above ground level [21,22]. Used with permission.

WIND RESOURCE MAP

WIND POWER DENSITY POTENTIAL

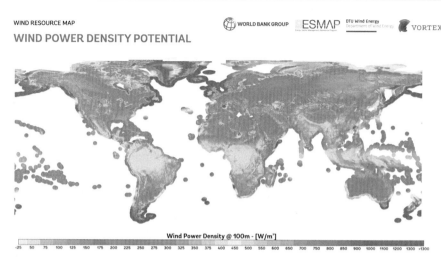

Figure 4.18 The global wind power density atlas at 100 m above ground level [21,22]. Used with permission.

4.18 show the global mean wind speed, and wind power density potential atlases at 100 m above ground level.[2] The power density is plotted in terms of $\frac{W}{m^2}$ that is the potential of power generation per unit of swept area by the turbine blades [21,22].

The International Energy Agency reports that to reach the Net Zero Emissions by 2050 Scenario, 8000 TWh of wind power must be generated at 2030. This means that wind farms must increase with an average yearly rate of 18% during 2021−2030. That means it is necessary to add onshore and offshore capacities annually by 310 and 80 GW [20].

Example 4.3. The wind power density at a location is 1000 $\frac{W}{m^2}$. If we are supposed to install a wind turbine with a tower height of 100 m and blade size of 75 m, how much power can be extracted from wind by installing this turbine at the location?

---

[2] This wind resource map provides an estimate of mean wind power density at 100 m above surface level. Power density indicates wind power potential, part of which can be extracted by wind turbines. The map is derived from high-resolution wind speed distributions based on a chain of models, which downscale winds from global models (∼30 km), to mesoscale (3 km) to microscale (250 m). The Weather Research & Forecasting (WRF) mesoscale model uses ECMWF ERA-5 reanalysis data for atmospheric forcing, sampling from the period 1998−2017. The WRF output at 3 km resolution is generalized and downscaled further using the WAsP software, plus terrain elevation data at 150 m resolution, and roughness data at 300 m resolution. The microscale wind climate is sampled on calculation nodes every 250 m. For the microscale modeling, the terrain data is derived from the digital elevation models from Viewfinder Panoramas. The WAsP microscale modeling uses a linear flow model. For steep terrain, this modeling becomes more uncertain, most likely leading to an overestimation of mean wind speeds on ridges and hilltops. Users are recommended to inspect the terrain complexity of their region of interest [21,22].

Solution.

$$\text{Power Potential} = A \times \text{WPD} = \pi R^2 \times \text{WPD} = 3.14 \times 75^2 \times 1000$$
$$= 17.66 \text{ MW}$$

Discussion

The power that can be extracted from wind depends on the wind turbine design and its efficiency. The commercialized wind turbines evolution is shown in Fig. 4.19. Comparing the above result (17.66 MW with that in Fig. 4.19 (6 MW)) for the same wind turbine size shows that efficiency improvement to extract more power from wind needs more efforts and investigations.

Example 4.4. If you are supposed to erect two offshore wind farms with the same ca-pacities of 1000 MW, how many turbine unites with tower height of 150 m and blade size of 120 m should be installed, if the wind power density at the site location is (a) 1100 $\frac{W}{m^2}$ (b) 600 $\frac{W}{m^2}$.

Solution.

a)

$$\text{Power Potential} = n_a A \times WPD_a = n_a \pi R^2 \times WPD_a = n_a 3.14 \times 120^2 \times 1100$$
$$= 1000 \text{ } MW$$

$$\rightarrow n_a = \frac{1000 \times 10^6}{3.14 \times 120^2 \times 1100} = 20.11 \cong 21 \text{ units}$$

That means we should consider 21 wind turbines for the wind turbine farm.

b)

$$\frac{n_a}{n_b} = \frac{WPD_b}{WPD_a} \rightarrow n_b = n_a \frac{WPD_a}{WPD_b} = 20.11 \frac{1100}{600} = 36.86 \cong 37 \text{ units}$$

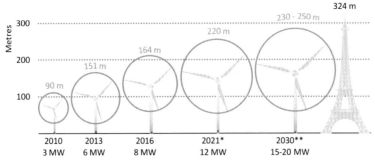

Figure 4.19 Development of largest commercially available wind turbines until 2021 [20].

Discussion

The number of wind turbines from 21 units in the first site increased to 37 units in the second farm. It shows that choosing the right site for a wind turbine farm can significantly reduce the cost per installed kW of electricity. In the second case, we must pay extra money for 16 extra units of wind turbines, more land, and extra operation and maintenance costs.

6. Bioenergy

Bioenergy as one of the energy sources identified today can meet a large part of human energy needs. Bioenergy is obtained from biomass, and it is in fact provided from organic materials such as plants. Biomass can be used for production of solid, liquid, or gaseous biofuels (for heating and transportation), bioelectricity, and bioproducts [23].

Biomass is a renewable energy that can be derived from crop wastes, forest residues, purpose-grown grasses, woody energy crops, microalgae, urban wood waste, and food waste.

Biomass sources can be used directly as fuel or converted to fuel after certain processes. These sources are called biomass *feedstocks*. The biomass feedstocks can be classified as dedicated energy crops, agricultural crop residues, forestry residues, algae, wood processing residues, municipal waste, and wet waste.

Dedicated energy crops are nonfood crops that can be grown in low-fertility soils specifically for biomass supply. These are divided into two general categories: herbaceous and woody. Herbaceous plants are perennial grasses that are harvested annually after 2−3 years to achieve full productivity. These include switch grass, bamboo, sweet sorghum, tall fescue, wheat grass, and more. Wood products are fast-growing trees that are harvested within 5−8 years after planting. These include hybrid spruce, hybrid willow, green ash, black walnut, sweet gum, and sycamore.

Some plants used to produce bioenergy can grow in less fertile soils. The growth of these plants in such lands, in addition to producing bioenergy, improves soil health and creates habitat for some wildlife.

Agricultural crop residues are a part of agricultural products that remain on the ground after harvesting the main products and can be used to produce bioenergy without any disruption in the process of food production. For example, we can refer to the residues of agricultural products such as wheat straw, rice straw, sugarcane, corn, leaves and stems of sunflower seeds, etc [23].

Forest products used to produce bioenergy are divided into two categories. The first kind is trees that are completely cut down and used to produce bioenergy, and the second one is the residue of trees that cannot be sold as wood and are left in the forest after harvesting the marketable parts. Also, some diseased trees that are not suitable for sale due to their unsuitable timber shape can also be collected and used to produce bioenergy Fig. 4.20.

Products and processes that use wood, such as paper industry, furniture, etc., always have wastes such as wood chips and sawdust, which are rich in energy to produce bioenergy. These wastes are called wood processing residues.

Algae as a feedstock for bioenergy refers to a diverse group of organisms, including microalgae, macro-algae, and cyanobacteria. Many algae use sunlight and nutrients to build biomass, which contains essential components including lipids, proteins, and carbohydrates that can be converted into biofuels. Algae can grow in freshwater, brackish water, groundwater, seawater, treated industrial wastewater, municipal wastewater, etc.

Another source for bioenergy production is commercial and residential waste, such as paper, cardboard, plastics, rubber, leather, textiles, and food waste. They must be separated and sorted at the origin of waste generation. This source of biofuel is called municipal waste.

Fig. 4.21, shows the global and some selected countries and continents share in biofuel production. It shows that global biofuel production has reached 1110 TWh in 2019, but due to the COVID-19 pandemic, it has declined to 1043 TWh in 2020. According to the reports by IEA, 718 TWh electricity is generated by the biofuel in 2020,

Figure 4.20 Trees are a great source for earth habitants, we need to make a balance, deforesting is not acceptable at all.
Photo by author, taken from Kamyaran, Akasheh village.

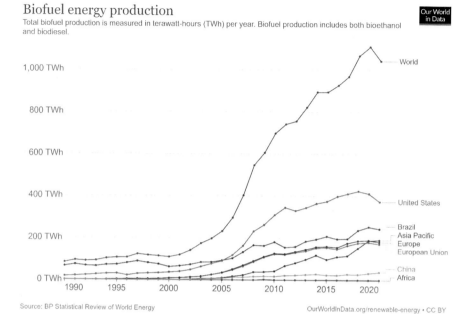

Figure 4.21 Biofuel production in the world and some selected countries and continents (https://ourworldindata.org/) [2].

which it must reach more than 1400 TWh in 2030 to meet the **Net Zero Emission by 2050**. It means that annually 15 GW of electricity must be added to reach the goal [24].

7. Geothermal

Geothermal energy comes from the interior layers of earth (Fig. 4.22). It is renewable because it is continuously produced due to slow decay of radioactive particles in the earth's core.

The temperature at the iron core is as hot as the sun surface and is about 10,800°F; the pressure in this core is about 3.6 Million atm. The temperature in the mantel ranges from 392°F at the interior surface of earth's crust to about 7230°F at the mantle internal boundary.

The energy of the hot rocks can be extracted to produce steam for electricity production. Reservoirs of steam or hot water with temperatures higher than about 108°C can be utilized for electricity generation, while the lower quality steam or hot water can be used for heating, cooling (by using thermally activated chillers), drying, or other purposes.

Geologists find the right places to build a geothermal power plant. They estimate the amount of energy in a geothermal source and examine ways to extract it. They

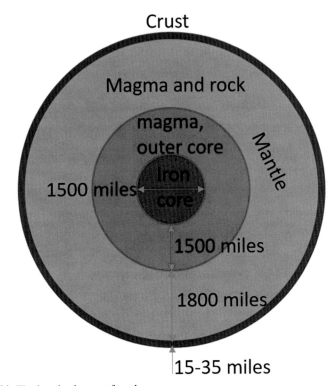

Figure 4.22 The interior layers of earth.

also investigate the environmental effects of energy extraction and erecting a geothermal power plant [25,26].

Fig. 4.23 shows the global and some selected countries and continents share in geothermal energy capacities. It shows that global geothermal capacity has reached 14,000 MW in 2020. According to the reports by IEA, 94 TWh power is generated by the geothermal in 2020, which it must reach more than 330 TWh in 2030 to meet the **Net Zero Emission by 2050**. It means that annually 3.6 GW of capacity must be added to reach the goal [27].

8. Hydrogen energy

Hydrogen is a clean fuel that produces heat and water when reacts with oxygen. Hydrogen is found very sparingly naturally and must be produced by different methods from various sources containing hydrogen molecules. The most common methods for hydrogen production are reforming (steam and dry methods) and hydrolysis. Reforming uses oil, coal, and gaseous fuels, such as natural gas and biogas, for hydrogen production. The electrolysis method is used to convert water molecules into

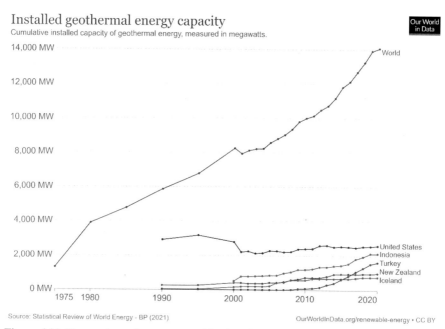

Figure 4.23 The geothermal energy capacities installed in the world and some selected countries (https://ourworldindata.org/) [2].

hydrogen and oxygen and requires electricity and heat to perform this process. The reforming process requires heat, and this heat may be supplied using renewable energy, and heat recovery from processes that have heat loss or burning fossil fuels. The electricity required for the electrolysis can also be supplied from various power plants in off-peak times or from electricity generated by renewable sources such as solar, wind, hydro, and bio. The hydrogen produced can be stored and used in a variety of ways to generate power, electricity, or heat. For example, hydrogen produced from surplus solar electricity can be stored, and when there is no sunlight, we can use the hydrogen to generate electricity in fuel cells. Hydrogen generated from excess wind electricity can also be used to generate electricity in the fuel cell for the times when there is no wind.

In addition to the well-matured and commercialized methods such as electrolysis and reforming, some other methods can be used for hydrogen production. In general, the hydrogen production methods can be classified as thermochemical processes, electrolytic processes (electrolysis), direct solar water splitting processes, and biological processes.

Examples of the thermochemical processes are natural gas reforming, biomass gasification, biomass-derived liquid reforming, and solar thermochemical hydrogen (STCH). The electrolytic processes include alkaline, polymer electrolyte membrane (PEM), and solid oxide electrolyzer cell (SOEC). Direct solar water splitting processes

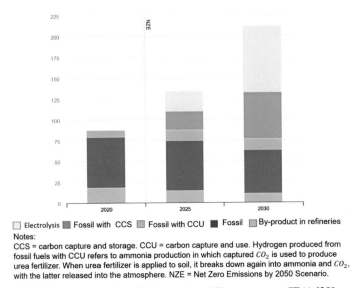

Notes:
CCS = carbon capture and storage. CCU = carbon capture and use. Hydrogen produced from fossil fuels with CCU refers to ammonia production in which captured CO_2 is used to produce urea fertilizer. When urea fertilizer is applied to soil, it breaks down again into ammonia and CO_2, with the latter released into the atmosphere. NZE = Net Zero Emissions by 2050 Scenario.

Figure 4.24 The global production of hydrogen by different sources (IEA) [28].

include photoelectrochemical (PEC) and photobiological, and finally examples of biological processes are microbial biomass conversion and photobiological.

According to the reports by IEA, 2020, hydrogen production costs by natural gas, coal, and renewable energies are 0.9–3.2 USD/kg, 1.2–2.2 USD/kg, and 3–7.5 USD/kg. The main demand for hydrogen in the world is for refining and ammonia production. However, leading companies such as General Electric and Siemens are proposing power turbines working with hydrogen.

Fig. 4.24 shows the global production of hydrogen by different sources. It shows that global capacity has reached 90 Mt in 2020. According to the reports by IEA, hydrogen production must reach more than 212 Mt in 2030 to meet the **Net Zero Emission by 2050**. It means that annually 12.2 Mt of capacity must be added to reach the goal [28,29].

9. Problems

1 Why *Hubbert's Peak Theory* is not valid for most of oil producer countries?
2 What kind of coal has the highest energy content? How much is it?
3 How coal can be converted to gas fuel?
4 Determine the total energy consumption of your country by looking into the references introduced in this chapter.
5 What portion of energy is provided by renewable energies in your country?
6 Provide a list of specifications of general electric gas turbines that use hydrogen as fuel.
7 Can you produce hydropower by placing very tiny turbine in the water pipes in your home?
8 Determine the kWh/kWp of solar PV power in your region according to the solar atlas.

9 Determine the wind power density of your region according to the wind atlas.

10 Does your country produce power by bioenergy? How much?

11 Does your country produce power by using geothermal energy? How much?

12 Do you believe that world will achieve the net zero emission by 2050? What are the obstacles?

13 What are *negative emissions* when talking about net zero emission by 2050?

14 Draw the schematic diagram for the hydropower plants based on the ocean thermal energy conversion (OTEC) and salinity gradient.

15 What regions in the world are more proper for hydropower plant by using OTEC?

16 If daily kWh/kWp of a location is 4.4, for a PV panel with a specified peak power of P(kWp) determine the daily and yearly potential for photovoltaic power generation in this site.

17 You are supposed to erect a solar photovoltaic farm with capacity of $10^6 \frac{kWh}{year}$ by using a kind of PV panel with peak power of 280 W. If the daily kWh/kWp of the farm site is 5, determine the number of PV panels required for the farm.

18 The wind power density at a location is 700 $\frac{W}{m^2}$. If we are supposed to install a wind turbine with a tower height of 100 m and blade size of 75 m, how much power can be extracted from wind by installing this turbine at the location?

19 If you are supposed to erect two offshore wind farms with the same capacities of 500 MW how many turbine unites with tower height of 150 m and blade size of 110 m should be installed if the wind power density at the site location is a) 400 $\frac{W}{m^2}$b) 800 $\frac{W}{m^2}$.

References

[1] M.K. Hubbert, Nuclear energy and the fossil fuels, in: Presented before the Spring Meeting of the Southern District, American Petroleum Institute, Plaza Hotel, San Antonio, Texas, March 7−8−9, 1956.

[2] https://ourworldindata.org/. (Accessed on 4 August 2022).

[3] https://www.worldometers.info/. (Accessed on 4 August 2022).

[4] Greb, Copyright Kentucky Geological Survey, 2021, 2021. https://www.uky.edu/KGS/ (Accessed 4 August 2022).

[5] https://www.eia.gov/energyexplained/coal/. (Accessed on 4 August 2022).

[6] https://www.britannica.com/.

[7] https://www.engineeringtoolbox.com/. (Accessed on 4 August 2022).

[8] https://world-nuclear.org/. (Accessed on 4 August 2022).

[9] P. Wexler, Encyclopedia of Toxicology, third ed., 2014, ISBN 978-0-12-386455-0.

[10] https://www.eia.gov/energyexplained/nuclear/. (Accessed on 4 August 2022).

[11] https://world-nuclear.org/information-library/nuclear-fuel-cycle/nuclear-resources/ supply-of-uranium.aspx. (Accessed on 4 August 2022).

[12] https://www.eia.gov/energyexplained/hydropower/. (Accessed on 4 August 2022).

[13] www.hydropower.org. (Accessed on 4 August 2022).

[14] O.A.C. Hoes, L.J.J. Meijer, R.J. van der Ent, N.C. van de Giesen, Systematic high-resolution assessment of global hydropower potential, PLoS One 12 (2) (2017) e0171844, https://doi.org/10.1371/journal.pone.0171844.t001.

[15] https://www.irena.org/publications/2020/Dec/Innovation-Outlook-Ocean-Energy-Technologies. (Accessed on 4 August 2022).

[16] Global Solar Atlas 2.0, a free, web-based application is developed and operated by the company Solargis s.r.o. on behalf of the World Bank Group, utilizing Solargis data, with funding provided by the Energy Sector Management Assistance Program (ESMAP). https://solargis.com/maps-and-gis-data/download/world. (Accessed on 4 August 2022).

[17] https://globalsolaratlas.info. (Accessed on 4 August 2022).

[18] https://www.astm.org/g0173-03r20.html. (Accessed on 4 August 2022).

[19] Solar PV − analysis - IEA. (Accessed on 4 August 2022).

[20] World Energy Outlook Special Report, International Energy Agency (IEA), Offshore wind outlook, 2019. www.iea.org (Accessed on 4 August 2022).

[21] Global Wind Atlas 3.0, A free, web-based application developed, owned and operated by the technical University of Denmark (DTU). The global wind atlas 3.0 is released in partnership with the world bank group, utilizing data provided by Vortex, using funding provided by the energy Sector Management Assistance program (ESMAP).

[22] https://globalwindatlas.info. (Accessed on 4 August 2022).

[23] Bioenergy Technologies Office. https://www.energy.gov/eere/bioenergy/bio-benefits-basics. (Accessed on 4 August 2022).

[24] https://www.iea.org/reports/bioenergy-power-generation. (Accessed on 4 August 2022).

[25] https://www.americangeosciences.org/critical-issues/geothermal-energy-basics. (Accessed on 4 August 2022).

[26] Geothermal electricity generation, DOE https://www.energy.gov/eere/geothermal/electricity-generation. (Accessed on 4 August 2022).

[27] https://www.iea.org/reports/geothermal-power. (Accessed on 4 August 2022).

[28] https://www.iea.org/data-and-statistics/charts/global-hydrogen-demand-by-production-technology-in-the-net-zero-scenario-2020-2030. (Accessed on 4 August 2022).

[29] https://www.iea.org/reports/the-future-of-hydrogen. (Accessed on 4 August 2022).

Steam power plant, design

Chapter outline

Most of the world electricity demand is supplied by the steam power plants (SPPs) that are widely installed all over the world. They were mostly designed to be fueled with coal, but excessive consumption of coal with old combustion techniques has emitted so much air pollutants into the atmosphere and has caused significant negative impact on the environment. Since coal is still a vital source [1] [1] for electricity production, significant attempts are carried out to invent combustion improvement techniques[2] to extract more thermal energy from the same amount of coal. According to the World Coal Association reports, 1% increase of efficiency in a conventional pulverized coal combustion SPP results in 2%–3% reduction of CO_2 [2]. Nowadays, SPPs are fueled with different fossil fuels, nuclear energy, geothermal energy, biomass, biogas, etc. Since SPPs are highly flexi fuel, wherever inexpensive and high-quality heat sources such as coal, biomass, biogas, solid wastes, gas turbine exhaust gas, refinery residual oil, and refinery off gases are available, they can be considered as a reliable power

---

[1] According to the Energy Information Administration (EIA) report in 2018 U.S. have used diverse energy sources of natural gas (35%), coal (27%), nuclear (19%), renewable (including hydro (7%), wind (6%), solar (1.6%), biomass (1.5%), geothermal (0.4%)), and petroleum (1%)) for electricity production. According to this report natural gas is used in both gas turbines and steam turbine power plants but approximately all of the coal is used in SPPs for electricity generation [1].

[2] Different techniques such as 1- combustion control optimization, 2- cooling system heat loss recovery, 3- flue gas heat recovery, 4- low-rank coal drying, and 5- sootblower optimization can be used to increase combustion efficiency of coal [2].

Power Generation Technologies. https://doi.org/10.1016/B978-0-323-95370-2.00007-7
Copyright © 2023 Elsevier Inc. All rights reserved.

generation system. The steam turbines can be used for different applications such as combined heat and power, mechanical derive to start a pump, a compressor a generator, or for district cooling and heating system.

The start-up time of large SPPs require producing high-quality and high-pressure steam from the liquid working fluid that may take even 24 h. Due to long start-up time, when they are started, they should be working for a long time. They are usually utilized to provide the base load of demand.

In comparison with other power generator units such as gas turbines, micro-turbines, fuel cells, etc. the lifetime of SPP is very long, and it may operate for more than 35 years in normal operation. In reality, there are steam turbines being in service over 50 years. In general, the lifetime of an SPP mostly depends on the operating temperature, operating pressure, material used in the components, water treatment quality, steam generation control, operating and maintenance programs of the power plant. The lifespan of the SPP can be further extended by repowering. Repowering of a power plant may include adding or replacing the old equipment and out-of-date technologies with the new equipment that are up-to-date technologically. The repowering may involve changing the whole power plant or only some equipment. The purpose of repowering can be improvement in the economics, reducing environmental pollution, extending the lifespan, increasing availability, reliability, operability and maintainability of the existing power plant.

The steam turbines exist in broad capacities from 50 kW to 500 MW; they are less available to some extent in comparison with other technologies such fuel cells, gas turbines, or micro-gas turbines. Steam turbines are extremely reliable with long overhaul intervals. They have less maintenance cost per kW with respect to other technologies such fuel cells, gas turbines, or microgas turbines.

SPPs have a steam generation unit, condenser, cooling tower, water treatment unit, fuel treatment system, etc.; therefore, they need more land per kW for installation of the equipment.

In case of using water-cooled condenser for the SPP, availability of cooling water for the entire lifetime of the power plant is extremely important. The water cooled condensers can utilize dry or wet cooling towers, they both need water but with significantly different rates. Utilizing a wet cooling tower, means the *water consumption*[3] of about $1 m^3 / MW_e h$. However, using a dry cooling tower would reduce the water consumption to a minimum of about $10^{-3} m^3 / MW_e h$ [3]. In addition, consuming water for power production must not endanger the water resources needed for agriculture, drinking, and other essential demands.

In the present chapter, the SPP cycle operation and processes are discussed; it is analyzed thermodynamically, different efficiency improvement techniques are applied, optimum reheater, optimum placement, and number of feedwater heaters (FWH) are presented. Many practical examples are presented. The EES code for each example

---

[3] The volume of water removed from a water source such as well, lake or river is called "withdrawal water". Some portion of withdrawal water may return to its source but the portion which is used and not returned to the main source is called "water consumption". Hence withdrawal water equals water consumption plus returned water.

is also attached to the solution of examples for the students to extend the code and do more analyses on the cycles. Supercritical SPP is also discussed, and single, double, and triple reheaters are optimized for this cycle.

1. SPP technology description

According to the thermodynamic principals, an SPP cycle like every other heat engines comprises of four essential processes of *compression, heat addition, expansion,* and *heat rejection* (Figs. 5.1 and 5.2).

Compression: In an SPP, the process 1-2 is done by a pump, which is called feed-pump, and if the working fluid is water, it is called feedwater pump. The pump duty is to provide a specified flow rate of liquid at a required head (or pressure). A pump must be coupled with a *driver*[4] to operate. The driver can be an electromotor for example. The electromotor converts electricity to rotation, and the driver rotation is transmitted to the pump's rotor through the *coupling*.[5] And, finally pump converts the rotor rotation to hydraulic power by using impellers that are mounted on the pump's rotor. Liquids can be considered as incompressible approximately; hence, the power required for their pressurizing is very small in comparison with gas compression for the same mass flow rate and pressure. If the internal irreversibilities such as friction could be neglected for the pump, the process could be assumed as isentropic (1'-2').

Heat addition: The heat addition process 2-3 is done in a *steam generator* unit by combusting different kinds of gaseous, liquid, or solid fuels such as natural gas, diesel, or coal, respectively. In addition, renewable energy sources such as solar, geothermal, biomass, biogas, etc. can be used as the energy resource for steam generation. Depending on the design and application, the steam may be saturated, superheated, or supercritical. In most of the modern SPPs, the working fluid, which absorbs heat, is *demineralized water*.[6] However, depending on the heat source quality (temperature), operating temperature, operating pressure, material used in the components, national and international standards and regulation, different kind of fluids such as mercury, organic fluids, etc. can be used as well. The process (2'-3') in Fig. 5.1 is internally

---

[4] A driver is a device that is used to rotate a pump, compressor a fan etc. It can be an electromotor, steam turbine, gas turbine, different reciprocating engines, hydro turbines etc. In SPP cycles the Feed-pump is usually driven by an electromotor or steam turbine.

[5] Coupling is a key element that connects the driver shaft to the pump shaft. It can be rigid or adjustable. The rigid coupling is used when very small vibration is allowed and clearances between components is too small, hence the driver and driven shafts must be perfectly aligned. However, the flexible couplings are used when misalignment capacity is needed. This misalignment capacity may be due to bigger clearances between internal and external components, or thermal growth of different components with different materials. The flexible couplings can be mechanically flexible (such as gear, chain, and grid spring), metallic-material flexible (such as disc, and diaphragm), or elastomeric-material flexible (such as jaw, and bonded tire).

[6] Demineralized water is water which its dissolved minerals are removed by distillation, deionization or membrane filtration. Water should be demineralized to avoid fouling on the SPP components which are in direct contact with water.

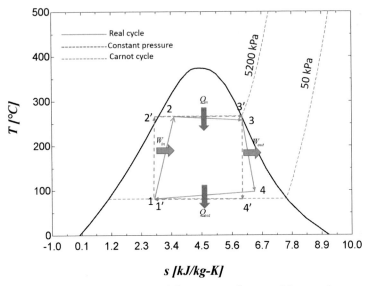

Figure 5.1 T–s diagram of the four essential processes of a general heat engine.

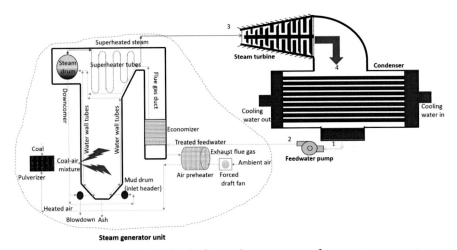

Figure 5.2 Steam turbine cycle with the four main components of pump, steam generator, turbine, and condenser (cycle No. 1).

reversible because the pressure loss due to friction, static head, and fitting losses (static and dynamic losses) are neglected. The process (2-3) is the real process.

Expansion: In an SPP, the steam expansion occurs in the turbine due to pressure reduction as it approaches the turbine exit. Steam turbine is a rotary equipment that receives hot and pressurized steam from the steam generator. As the steam is passing through the turbine's stages, its energy (in the forms of heat and pressure) converts to rotation of the turbine rotor. Every stage of turbine includes a stationary row of blades (or nozzles) on the casing inside, and a rotary row of blades mounted on the rotor. The stationary blades convert pressure and temperature of steam to velocity and guide it to attack the rotary blades. Due to steam attack, its kinematic energy is transmitted to the rotor and impels it to rotate the turbine shaft and produces mechanical power. This mechanical power can be used to rotate an electric generator for electricity production.

Condensation: In the SPP, the outlet pressure of steam turbine is not allowed to drop below a specified minimum value due to some technical reasons. Firstly, to avoid erosion due to the two-phase flow in the last stages of turbine, the steam quality at the turbine outlet must be higher than 88% for most of the steam turbines. In addition, very low pressure at the turbine outlet may disrupt the heat rejection process. Because of these technical problems, the outlet of steam turbine is mostly steam and must be condensed before being pressurized. The condensation of steam is carried out by a heat exchanger, which is called *condenser*. It is in a shell-and-tube heat exchanger where cooling water flows through the tube bundles, while steam pours down on the tubes in a cross-flow pattern. The condensed water is collected in a chamber just below the tubes. This chamber is called *hot well*. The condensed water which is saturated liquid is ready to be re-pumped into the steam generator again for continuing the cycle. However, the saturated liquid before entering the steam generator is usually preheated by the *feedwater heaters*.[7] This preheating improves the cycle efficiency. Feedwater heaters are discussed in details later in this chapter.

2. The Rankine cycle

An SPP with its four most vital components including turbine, condenser, pump, and steam generator is depicted in Fig. 5.2. The average temperature of the combustion gases inside the steam generator is T_H, and the ambient temperature in which heat loss is transferred to it is T_L.

As discussed in Chapter 2, the *Carnot principal* states that the highest thermal efficiency of a heat engine operating between a heat source and a heat sink with average temperatures of T_H and T_L is limited by the thermal efficiency of the Carnot cycle operating between the same temperature limits. In addition, it states that the thermal efficiency of all internally reversible heat engines operating between the same temperature limits is the same and is equal to:

$$\eta_{Carnot} = 1 - \frac{T_L}{T_H} \tag{5.1}$$

In which T is in Kelvin.

---

[7] Feedwater heaters are heat exchangers used to transfer heat between steam that is extracted from turbine and the sub-cooled feed water. The feedwater heaters can be close or open. In the open type the hot and cold streams are in direct contact and get mixed together but in the close one the hot and cold streams do not touch each other.

Fig. 5.3A shows the real and ideal (Carnot) steam engine cycle operating between two temperature limits of $T_H = 266.4°C$ and $T_L = 81.3°C$. The Carnot cycle efficiency of this heat engine would be:

$$\eta_{Carnot} = 1 - \frac{81.3 + 273.15}{266.4 + 273.15} = 0.34$$

It means that the thermal efficiency of the heat engine when all of the four processes could be done internally reversible could reach 34%. However, in addition to the existing irreversibilities, there are some technical problems in the Carnot cycle that make it unsuitable in its present form for a real or even ideal steam turbine cycle.

Contrary to the gas turbines that the working fluid remains in the gaseous form during the cycle, in an SPP (Fig. 5.2), the working fluid converts from sub-cooled liquid (at point 2) to two-phase flow (in the steam drum), dry steam (point 3), then two-phase flow (point 4), and saturated liquid (at point 1). This characteristic of the SPP creates some technical limitations.

1. The steam turbine inlet (state point 3) must be dry steam to avoid erosion and corrosion of the turbine nozzles and rotor. If liquid droplets of the working fluid could enter the steam turbine, they will imping the nozzles and rotating parts like bullets due to very high speed of rotation and cause vibration, erosion, corrosion, and finally decreasing the lifespan of the steam turbine. To solve this problem, the state point 3 in Fig. 5.3A must be located in the superheated region but on the same constant pressure line (Fig. 5.3B).

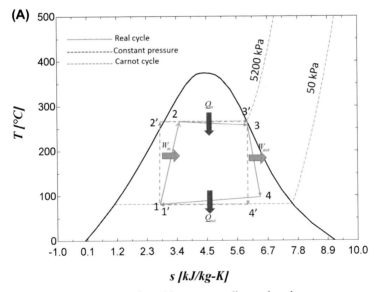

Figure 5.3A The Carnot steam cycle and its corresponding real cycle.

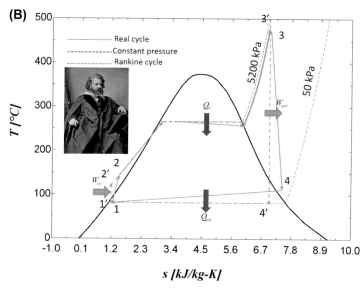

Figure 5.3B The Rankine and real steam power plant (SPP) cycle. The photo is the Scottish engineer William J.M. Rankine.

2. Another problem of the cycle in Fig. 5.3A is the quality of the working fluid at point 1 (Fig. 5.3A). In an SPP, liquid must enter a pump for pressurizing. If two-phase flow enters, a pump it will be damaged very soon due to cavitation, vibration, and erosion. Hence, the point 1 in Fig. 5.3A must be moved toward the saturated liquid line and point 1 in Fig. 5.3B. At this point, saturated liquid enters the pump, and it will be pressurized to sub-cooled liquid with the required pressure for the steam to enter the steam turbine while overcoming the pressure losses inside the steam generator tubes.

By applying the two points mentioned above to the Carnot cycle in Fig. 5.3A, the new cycle will be Fig. 5.3B. The internally reversible cycle in Fig. 5.3B is called *Rankine cycle,* which was first described by the Scottish engineer William J.M. Rankine in 1859.

The Rankine cycle $(1'-2'-3'-4'-1')$ helps to understand the behavior of the real cycle more simply. For this purpose, in the following, the cycle is analyzed from energy and exergy point of views.

Starting from the state point of $1'$ the saturated liquid at this point enters the pump and leaves it at state point of $2'$. Hence, the inlet power, $\dot{W}_{in}$, and inlet specific work, w_{in}, consumed by the pump can be calculated as Eq. (5.2).

$$\dot{W}_{in} = \dot{m}_{1'}(h_{2'} - h_{1'}) \tag{5.2}$$

$$w_{in} = \frac{\dot{W}_{in}}{\dot{m}_{1'}}$$

Since liquids can be considered as incompressible, the specific volume remains constant ($v_{2'} = v_{1'}$). In addition, for the isentropic compression process entropy remains constant. Hence, the specific work can be calculated as Eq. (5.3) for the isentropic compression.

$$dh = Tds + vdP$$

$$s_{2'} = s_{1'}$$

$$\rightarrow h_{2'} - h_{1'} = v_{1'}(P_{2'} - P_{1'})$$

$$w_{in} = v_{1'}(P_{2'} - P_{1'}) \tag{5.3}$$

In which $s, v, h, P, \dot{W}, w$, and $\dot{m}$ are specific entropy, specific volume, specific enthalpy, pressure, power, specific work, and mass flow rate, respectively.

Due to internal irreversibilities in the real process (1–2), the entropy will change, and as a result, the power consumed by the pump will increase. The extra power is consumed to overcome the internal friction of the pump and other possible irreversibilities (losses). The ratio of the power consumption in the isentropic process to the real process is called the pump *adiabatic efficiency* or *pump exergy efficiency* and can be written as Eq. (5.4):

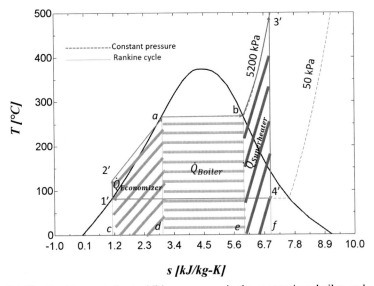

Figure 5.4 The Rankine cycle heat addition processes in the economizer, boiler, and superheater.

$$\eta_{ad,pump} = \frac{\dot{W}_{isentropic}}{\dot{W}_{actual}} = \frac{h_{s,out} - h_{in}}{h_{a,out} - h_{in}} = \frac{\int_{P_{in}}^{P_{s,out}} vdP}{\int_{S_{in}}^{S_{a,out}} Tds + \int_{P_{in}}^{P_{a,out}} vdP} \tag{5.4}$$

In which subscripts of a, s, in, and out stand for actual process, isentropic process, inlet, and outlet of the pump. It is clear that $\int_{S_{a,out}}^{S_{in}} Tds$ would be positive for the pump. This parameter shows the power consumed (or exergy destruction) to overcome the internal irreversibilities such as friction.

After the pump, the steam generator produces superheated steam from the sub-cooled liquid. The phase change from the sub-cooled liquid to the superheated steam occurs in the steam generator, and it comprises the following three processes (Fig. 5.4) for a *sub-critical*[8] Rankine cycle:

$2' - a$: Changing from sub-cooled liquid to saturated liquid that occurs in the *economizer*. The economizer makes use of flue gas heat before leaving the steam generator. Hence, it can be considered as a heat recovery heat exchanger.

The heat rate received by the working fluid in the economizer section is calculated by Eq. (5.5):

$$\dot{Q}_{Economizer} = \dot{m}_{1'}(h_a - h_{2'}) \tag{5.5}$$

$a - b$: Changing from the saturated liquid to the saturated steam that occurs in the *boiler* or the *evaporator* section. This section includes *steam drum, mud drums* (or *inlet headers*), *water wall tubes, risers,* and *downcomers*. The saturated liquid first enters the steam drum, then it goes down to the mud drums through the downcomers. From the mud drums, liquid water flows into the water wall tubes where it receives heat and starts boiling and going up to the risers. The risers are connected to the steam drum; therefore, the two-phase flow enters the steam drum. Steam drum is used to separate the saturated steam from its liquid. The saturated steam goes out from the top to the superheaters while the saturated liquid flows into the downcomers again (Fig. 5.2).

The heat rate received by the working fluid in the boiler section is:

$$\dot{Q}_{Boiler} = \dot{m}_{1'}(h_b - h_a) \tag{5.6}$$

$b - 3'$: Changing from the saturated steam to the superheated steam occurs in the superheater tubes. Superheating may be carried out in single or double steps. Superheaters are located at the hottest zone of a steam generator. The heat rate received by the working fluid in the superheater section is:

---

[8] If the steam generator operating pressure is below the critical pressure of the working fluid the cycle is called sub-critical. However, if the operating pressure is higher than the critical pressure of the working fluid it is called super-critical cycle. In a super-critical cycle, the two-phase flow does not exist because at the critical state point liquid flashes into superheated steam suddenly.

$$\dot{Q}_{Superheater} = \dot{m}_{1'}(h_{3'} - h_b) \tag{5.7}$$

In addition to the thermodynamic equations given to calculate the heat addition in the economizer, boiler, and superheater, they can be calculated according to the surface area (A) under the process $2' - a - b - 3'$ as below:

$$\dot{Q}_{Economizer} = A_{2'-a-d-c-1'-2'},$$

$$\dot{Q}_{Boiler} = A_{a-b-e-d-a}, \dot{Q}_{Superheater} = A_{b-3'-4'-f-e-b} \tag{5.8}$$

The total heat received by the working fluid is then:

$$\dot{Q}_{in} = \dot{Q}_{Economizer} + \dot{Q}_{Boiler} + \dot{Q}_{Superheater} = \dot{m}_{1'}(h_{3'} - h_{2'})$$

$$= A_{1'-2'-a-b-3'-4'-f-e-d-c-1'} \tag{5.9}$$

The turbine receives high-pressure superheated steam from the steam generator to produce power. During an isentropic expansion, the output power, $\dot{W}_{out}$, and specific work, w_{out}, produced by the turbine would be:

$$\dot{W}_{out} = \dot{m}_{1'}(h_{3'} - h_{4'}) \tag{5.10}$$

$$w_{out} = \frac{\dot{W}_{out}}{\dot{m}_{1'}}$$

In addition, according to Eq. (5.3) for the isentropic process of $3' - 4'$ we have:

$$dh = vdP \rightarrow h_{3'} - h_{4'} = \int_{P_{4'}}^{P_{3'}} vdP \tag{5.11}$$

In which $v = v(P, T)$.

The ratio of the power generation in the real process to the isentropic one is called the turbine *adiabatic efficiency* or *turbine exergy efficiency* and can be written as Eq. (5.12):

$$\eta_{ad,turbine} = \frac{\dot{W}_{actual}}{\dot{W}_{isentropic}} = \frac{h_{in} - h_{a,out}}{h_{in} - h_{s,out}} = \frac{\int_{S_{a,out}}^{S_{in}} Tds + \int_{P_{a,out}}^{P_{in}} vdP}{\int_{P_{s,out}}^{P_{in}} vdP} \tag{5.12}$$

The adiabatic efficiency shows the deviation of the actual turbine from the isentropic, insulated turbine. It is clear that $\int_{S_{a,out}}^{S_{in}} Tds$ would be negative for the turbine because $s_{a,out} > s_{in}$. This term shows the exergy destruction due to internal irreversibilities.

The process of heat rejection occurs in the condenser. The amount of heat rate that must be wasted through the condenser in the Rankine cycle is calculated by Eq. (5.13):

$$\dot{Q}_{out} = \dot{m}_{1'}(h_{4'} - h_{1'}) \tag{5.13}$$

According to the first law of thermodynamics for a closed cycle such as Rankine cycle the net heat rate and net power of the Rankine cycle are equal (Eq. 5.14):

$$\dot{Q}_{in} + \dot{W}_{in} - \dot{Q}_{out} - \dot{W}_{out} = 0$$

$$\left(\dot{Q}_{in} - \dot{Q}_{out}\right) - \left(\dot{W}_{out} - \dot{W}_{in}\right) = 0$$

$$\dot{Q}_{net} = \dot{Q}_{in} - \dot{Q}_{out}$$

$$\dot{W}_{net} = \dot{W}_{out} - \dot{W}_{in}$$

$$\rightarrow \dot{Q}_{net} = \dot{W}_{net} \tag{5.14}$$

The thermal efficiency of the Rankine cycle can also be determined by Eq. (5.15):

$$\eta_{th} = \frac{\text{desired output}}{\text{supplied energy}} = \frac{\dot{W}_{net}}{\dot{Q}_{in}} = \frac{\dot{Q}_{in} - \dot{Q}_{out}}{\dot{Q}_{in}} = 1 - \frac{\dot{Q}_{out}}{\dot{Q}_{in}} \tag{5.15}$$

The exergy analysis of the Rankine cycle can be carried out as a guideline for calculating the second law efficiency of the SPP cycles. The exergy rate balance for steady flow devices can be written as Eq. (5.16):

$$\underbrace{\left(\dot{X}_{in} - \dot{X}_{out}\right)}_{\text{net rate of entropy transfer}} - \dot{X}_{destroyed} = 0 \tag{5.16}$$

Starting from the pump for steady state conditions, it can be written that:

$$\dot{X}_{in,pump} = \dot{X}_{\dot{W}_{in}} + \dot{X}_{1'} = \dot{W}_{in} + \dot{m}_{1'}((h_{1'} - h_0) - T_0(s_{1'} - s_0))$$

$$\dot{X}_{out,pump} = \dot{X}_{2'} = \dot{m}_{1'}((h_{2'} - h_0) - T_0(s_{2'} - s_0))$$

$$\dot{X}_{destroyed,pump} = \dot{W}_{in} - \dot{m}_{1'}((h_{2'} - h_{1'}) - T_0(s_{2'} - s_{1'}))$$

Therefore, the exergy destruction by the pump would be as Eq. (5.17):

$$\dot{X}_{destroyed,pump} = \dot{m}_{1'}T_0(s_{2'} - s_{1'}) \tag{5.17}$$

It is evident if the pressurizing process is isentropic, then $s_{2'} = s_{1'}$ and $\dot{X}_{destroyed, \ pump} = 0$. In a steady state, internally reversible process, $\dot{X}_{destroyed, pump} = 0$ and the reversible power consumption would be as below:

$$\dot{X}_{in,pump} = \dot{X}_{out,pump}$$

$$\dot{X}_{W_{in,rev}} + \dot{X}_{1'} = \dot{X}_{2'}$$

$$\dot{W}_{in,rev} = \dot{X}_{2'} - \dot{X}_{1'}$$

Hence, the second law efficiency of the pump would be as Eq. (5.18):

$$\eta_{II,pump} = \frac{\dot{W}_{in,rev}}{\dot{W}_{in,actual}} = \frac{\dot{X}_{2'} - \dot{X}_{1'}}{\dot{m}_{1'}(h_{2'} - h_{1'})} = \frac{\psi_{2'} - \psi_{1'}}{h_{2'} - h_{1'}}$$

$$\eta_{II,pump} = \frac{\psi_{2'} - \psi_{1'}}{h_{2'} - h_{1'}} \tag{5.18}$$

The above equation can also be achieved as below

$$\eta_{II,pump} = 1 - \frac{\dot{X}_{destroyed,pump}}{\dot{X}_{supplied,pump}} = 1 - \frac{\dot{m}_{1'} T_0(s_{2'} - s_{1'})}{\dot{W}_{in,actual}}$$

$$= \frac{\dot{m}_{1'}(h_{2'} - h_{1'}) - \dot{m}_{1'} T_0(s_{2'} - s_{1'})}{\dot{m}_{1'}(h_{2'} - h_{1'})} = \frac{\psi_{2'} - \psi_{1'}}{h_{2'} - h_{1'}}$$

For the steam generator (SG), the exergy balance for a steady-state condition can be formulated as below:

$$\dot{X}_{in,SG} = \dot{X}_{\dot{Q}_{in}} + \dot{X}_{2'} = \dot{Q}_{in}\left(1 - \frac{T_0}{T_{H-mean}}\right) + \dot{m}_{1'}((h_{2'} - h_0) - T_0(s_{2'} - s_0))$$

$$\dot{X}_{out,SG} = \dot{X}_{3'} = \dot{m}_{1'}((h_{3'} - h_0) - T_0(s_{3'} - s_0))$$

$$\dot{X}_{destroyed,SG} = -\dot{m}_{1'}((h_{3'} - h_{2'}) - T_0(s_{3'} - s_{2'})) + \dot{Q}_{in}\left(1 - \frac{T_0}{T_{H-mean}}\right)$$

$$= \dot{m}_{1'} T_0(s_{3'} - s_{2'}) - \dot{Q}_{in}\frac{T_0}{T_{H-mean}}$$

Hence, the exergy destruction by the steam generator would be as Eq. (5.19):

$$\dot{X}_{destroyed,SG} = \dot{m}_{1'} T_0(s_{3'} - s_{2'}) - \dot{Q}_{in}\frac{T_0}{T_{H-mean}} \tag{5.19}$$

The exergy efficiency of the steam generator would be as Eq. (5.20):

$$\eta_{II,SG} = 1 - \frac{\dot{m}_{1'} T_0 (s_{3'} - s_{2'}) - \dot{Q}_{in} \dfrac{T_0}{T_{H-mean}}}{\dot{Q}_{in} \left(1 - \dfrac{T_0}{T_{H-mean}}\right)} = \frac{\dot{Q}_{in} - \dot{m}_{1'} T_0 (s_{3'} - s_{2'})}{\dot{Q}_{in} \left(1 - \dfrac{T_0}{T_{H-mean}}\right)} = \frac{\dot{X}_{3'} - \dot{X}_{2'}}{\dot{X}_{\dot{Q}_{in}}}$$

$$\eta_{II,SG} = \frac{\dot{X}_{3'} - \dot{X}_{2'}}{\dot{X}_{\dot{Q}_{in}}} \tag{5.20}$$

In the above equation, T_{H-mean} is the heat source mean temperature. In this case, the heat source is the combustion products inside the steam generator.

The exergy balance for the steam turbine can be written as below:

$$\dot{X}_{in,turbine} = \dot{X}_{3'} = \dot{m}_{1'}((h_{3'} - h_0) - T_0(s_{3'} - s_0))$$

$$\dot{X}_{out,turbine} = \dot{X}_{\dot{W}_{out}} + \dot{X}_{4'} = \dot{W}_{out} + \dot{m}_{1'}((h_{4'} - h_0) - T_0(s_{4'} - s_0))$$

$$\dot{X}_{destroyed,turbine} = -\dot{W}_{out} - \dot{m}_{1'}((h_{4'} - h_{3'}) - T_0(s_{4'} - s_{3'}))$$

Therefore, the exergy destruction by the turbine would be as Eq. (5.21):

$$\dot{X}_{destroyed,turbine} = \dot{m}_{1'} T_0 (s_{4'} - s_{3'}) \tag{5.21}$$

The internally reversible power produced by the turbine happens when $\dot{X}_{destroyed,turbine} = 0$, hence, for the steady-state condition:

$$\dot{X}_{in,turbine} = \dot{X}_{out,turbine}$$

$$\dot{X}_{3'} = \dot{X}_{\dot{W}_{out,rev}} + \dot{X}_{4'}$$

$$\dot{X}_{\dot{W}_{out,rev}} = \dot{X}_{3'} - \dot{X}_{4'}$$

$$\dot{W}_{out,rev} = \dot{X}_{3'} - \dot{X}_{4'}$$

The second law efficiency of the turbine would be as Eq. (5.22):

$$\eta_{II,turbine} = \frac{\dot{W}_{out,actual}}{\dot{W}_{out,rev}} = \frac{\dot{m}_{1'}(h_{3'} - h_{4'})}{\dot{X}_{3'} - \dot{X}_{4'}} = \frac{h_{3'} - h_{4'}}{\psi_{3'} - \psi_{4'}}$$

$$\eta_{II,turbine} = \frac{h_{3'} - h_{4'}}{\psi_{3'} - \psi_{4'}} \tag{5.22}$$

Eq. (5.22) can also be achieved as below:

$$\eta_{II,turbine} = 1 - \frac{\dot{X}_{destroyed,turbine}}{\dot{X}_{supplied,turbine}} = 1 - \frac{\dot{m}_{1'}T_0(s_{4'} - s_{3'})}{\dot{X}_{3'} - \dot{X}_{4'}}$$

$$= \frac{\dot{m}_{1'}((h_{3'} - h_{4'}) - T_0(s_{3'} - s_{4'})) - \dot{m}_{1'}T_0(s_{4'} - s_{3'})}{\dot{X}_{3'} - \dot{X}_{4'}} = \frac{h_{3'} - h_{4'}}{\psi_{3'} - \psi_{4'}}$$

The condenser exergy balance can also be written as below:

$$\dot{X}_{in,condenser} = \dot{X}_{4'} = \dot{m}_{1'}((h_{4'} - h_0) - T_0(s_{4'} - s_0))$$

$$\dot{X}_{out,condenser} = \dot{X}_{\dot{Q}_{out}} + \dot{X}_{1'} = \dot{Q}_{out}\left(1 - \frac{T_0}{T_{L-mean}}\right)$$

$$+ \dot{m}_{1'}((h_{1'} - h_0) - T_0(s_{1'} - s_0))$$

$$\dot{X}_{destroyed,condenser} = -\dot{m}_{1'}((h_{1'} - h_{4'}) - T_0(s_{1'} - s_{4'})) - \dot{Q}_{out}\left(1 - \frac{T_0}{T_{L-mean}}\right)$$

$$= \dot{m}_{1'}T_0(s_{1'} - s_{4'}) + \dot{Q}_{out}\frac{T_0}{T_{L-mean}}$$

Hence, the exergy destruction by the condenser would be as Eq. (5.23):

$$\dot{X}_{destroyed,condenser} = \dot{m}_{1'}T_0(s_{1'} - s_{4'}) + \dot{Q}_{out}\frac{T_0}{T_{L-mean}} \qquad (5.23)$$

In the above equation, T_{L-mean} is the heat sink mean temperature. In this case, the heat sink is the ambient where the heat loss is finally rejected into it.

The exergy efficiency of the condenser would be calculated as below:

$$\eta_{II,condenser} = 1 - \frac{\dot{m}_{1'}T_0(s_{1'} - s_{4'}) + \dot{Q}_{out}\frac{T_0}{T_{L-mean}}}{\dot{X}_{4'} - \dot{X}_{1'}}$$

$$= \frac{\dot{m}_{1'}((h_{4'} - h_{1'}) - T_0(s_{4'} - s_{1'})) - \dot{m}_{1'}T_0(s_{1'} - s_{4'}) - \dot{Q}_{out}\frac{T_0}{T_{L-mean}}}{\dot{X}_{4'} - \dot{X}_{1'}}$$

$$= \frac{\dot{m}_{1'}(h_{4'} - h_{1'}) - \dot{Q}_{out}\frac{T_0}{T_{L-mean}}}{\dot{X}_{4'} - \dot{X}_{1'}} = \frac{\dot{Q}_{out} - \dot{Q}_{out}\frac{T_0}{T_{L-mean}}}{\dot{X}_{4'} - \dot{X}_{1'}}$$

$$= \frac{\dot{Q}_{out}\left(1 - \frac{T_0}{T_{L-mean}}\right)}{\dot{X}_{4'} - \dot{X}_{1'}} = \frac{\dot{X}_{\dot{Q}_{out}}}{\dot{X}_{4'} - \dot{X}_{1'}}$$

Hence, the condenser exergy efficiency would be as Eq. (5.24):

$$\eta_{II,condenser} = \frac{\dot{X}_{\dot{Q}_{out}}}{\dot{X}_{4'} - \dot{X}_{1'}} \tag{5.24}$$

The exergy balance for the cycle can also be written as below:

$$\dot{X}_{in,Rankine} = \dot{X}_{\dot{W}_{in}} + \dot{X}_{\dot{Q}_{in}} = \dot{W}_{in} + \dot{Q}_{in}\left(1 - \frac{T_0}{T_{H-mean}}\right)$$

$$\dot{X}_{out,Rankine} = \dot{X}_{\dot{W}_{out}} + \dot{X}_{\dot{Q}_{out}} = \dot{W}_{out} + \dot{Q}_{out}\left(1 - \frac{T_0}{T_{L-mean}}\right)$$

$$\dot{X}_{destroyed,Rankine} = -\dot{W}_{out} - \dot{Q}_{out}\left(1 - \frac{T_0}{T_{L-mean}}\right) + \dot{W}_{in} + \dot{Q}_{in}\left(1 - \frac{T_0}{T_{H-mean}}\right)$$

$$= -\left(\dot{W}_{out} - \dot{W}_{in}\right) + \left(\dot{Q}_{in} - \dot{Q}_{out}\right) - T_0\left(\frac{\dot{Q}_{in}}{T_{H-mean}} - \frac{\dot{Q}_{out}}{T_{L-mean}}\right)$$

$$= -\dot{W}_{net} + \dot{Q}_{net} - T_0\left(\frac{\dot{Q}_{in}}{T_{H-mean}} - \frac{\dot{Q}_{out}}{T_{L-mean}}\right)$$

Hence, the total exergy destruction by the SPP cycle would be as Eq. (5.25):

$$\dot{X}_{destroyed,Rankine} = T_0\left(\frac{\dot{Q}_{out}}{T_{L-mean}} - \frac{\dot{Q}_{in}}{T_{H-mean}}\right) \tag{5.25}$$

The reversible net power produced by the Rankine cycle would happen when $\dot{X}_{destroyed,Rankine} = 0$, hence:

$$\dot{X}_{destroyed,Rankine} = T_0\left(\frac{\dot{Q}_{out}}{T_{L-mean}} - \frac{\dot{Q}_{in}}{T_{H-mean}}\right)_{rev} = 0$$

$$\frac{\dot{Q}_{out}}{T_{L-mean}} = \frac{\dot{Q}_{in}}{T_{H-mean}} \rightarrow \frac{\dot{Q}_{out}}{\dot{Q}_{in}} = \frac{T_{L-mean}}{T_{H-mean}}$$

As a result, the reversible thermal efficiency of the Rankine cycle would be:

$$\eta_{th,rev} = 1 - \left(\frac{\dot{Q}_{out}}{\dot{Q}_{in}}\right)_{rev} = 1 - \frac{T_{L-mean}}{T_{H-mean}}$$

Therefore, the exergy efficiency of the Rankine cycle would be as Eq. (5.26):

$$\eta_{II,Rankine} = \frac{\eta_{th}}{\eta_{th,rev}} = \frac{1 - \dfrac{\dot{Q}_{out}}{\dot{Q}_{in}}}{1 - \dfrac{T_{L-mean}}{T_{H-mean}}} = \frac{\dot{Q}_{in} - \dot{Q}_{out}}{\dot{Q}_{in}\left(1 - \dfrac{T_{L-mean}}{T_{H-mean}}\right)} = \frac{\dot{W}_{net}}{\dot{Q}_{in}\left(1 - \dfrac{T_{L-mean}}{T_{H-mean}}\right)}$$

$$\eta_{II,Rankine} = \left(\frac{1}{\eta_{th,rev}}\right)\frac{\dot{W}_{net}}{\dot{Q}_{in}} \tag{5.26}$$

Special care must be taken about T_{L-mean} and T_{H-mean}, which are the mean temperature of the heat sink and heat source, respectively. For an SPP, the heat source temperature is the combustion gas (CG) mean temperature, while the heat sink is the ambient because the heat rejected in the condenser will be released into ambient by the cooling tower. Therefore, to calculate the T_{H-mean}, we have the following energy balance between hot stream (combustion gases) and cold stream (cycle working fluid) by assuming that the steam generator is well insulated and the process in both sides of combustion gas and working fluid is internally reversible.

$$\dot{m}_{1'}(h_{3'} - h_{2'}) = \dot{m}_{CG} T_{H-mean}(s_{CG,in} - s_{CG,out})$$

In addition:

$$\dot{m}_{CG}(h_{CG,in} - h_{CG,out}) = \dot{m}_{1'}(h_{3'} - h_{2'})$$

Hence, the heat source mean temperature would be as Eq. (5.27):

$$T_{H-mean} = \frac{h_{CG,in} - h_{CG,out}}{s_{CG,in} - s_{CG,out}} \tag{5.27}$$

Since most of the combustion gases is air, considering air instead of the combustion gases usually result in a good approximation.

Since the heat which is absorbed by the cooling water in the condenser must be rejected to the environment, therefore the heat sink mean temperature is the ambient temperature

$$T_{L-mean} = T_{ambient} \tag{5.28a}$$

In some cases, the cooling water system may be a once through system and the cooling water is taken from the sea and then returned to the sea directly. In such cases:

$$T_{L-mean} = T_{sea\ water} \hspace{6cm} (5.28b)$$

3.   The working fluid of SPP

The working fluid that can be used in the SPPs can be divided into three categories of dry, isentropic and wet. This classification is done according to the shape of the steam/ liquid saturation-line of the working fluid in the $T - s$ diagram. Examples of the three types of organic fluids are provided in Fig. 5.5.

According to the shape of the T-s diagram of the wet fluids such as water, wet fluids must be superheated before entering the turbine; otherwise, saturated steam condenses just after entering the turbine and the whole stages of the turbine will operate under two-phase flow. This causes the turbine internal components to erode and corrode very fast due to bombardment by the liquid droplets. However, if SPP is supposed to operate based on dry or isentropic fluid, superheating is not required, and after entering the saturated steam to the turbine, it gets superheated while expanding through the turbine; this can be easily seen for cyclohexane in Fig. 5.5.

In addition, due to small area under the T-s diagram of dry and isentropic fluids such as R13 and R123, they are more talented for small-scale power production; on the contrary, the wet fluids such as water have a very wide T-s diagram; they create a vast area under the T-s diagram that makes them very suitable for large scale SPPs.

Furthermore, in comparison with the wet fluids, dry, and isentropic fluids have smaller critical temperature that makes them more proper for operating at low temperature; hence, they are best suited to be used with low-temperature waste heat sources.

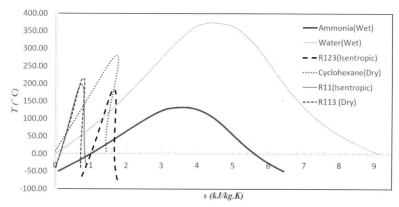

Figure 5.5 Mollier diagrams for wet, isentropic, and dry organic fluids.

Using water, as the working fluid in the low pressure, low-temperature SPP is not a good idea, because, electrical efficiency decreases and exergy destruction increases.

According to the discussions small-scale SPPs need a working fluid with low critical temperature and pressure. Small-scale SPPs working with organic fluids are called organic rankine cycle (ORC). On the contrary, the wet fluids are more proper for the large-scale SPPs operating at high temperature and pressure.

Table 5.1 presents some sample fluids, their critical properties, and type. Dry and isentropic organic fluids with positive steam saturation-line slope are good choices for small-scale low-temperature/pressure ORCs.

A good organic fluid for using in an ORC has small specific volume, lower critical temperature and pressure, higher cycle efficiency for the same operating conditions in comparison with other fluids, lower cost, less toxicity, smaller surface tension, higher thermal conductivity, less viscosity, greater molecular weight, thermal stability, compatibility with the turbine material and lubrication oil, nonflammable, noncorrosive, and an isentropic Mollier diagram.

Example 5.1.   Consider a basic internally reversible Rankine cycle. The steam generator operating pressure is 30 bar. The steam temperature at the turbine inlet is 250°C, and the saturated liquid temperature at the pump inlet is 36°C. If the combustion gases inside the steam generator could be estimated as pure air, determine the cycle thermal efficiency, exergy efficiency, and steam rate of the cycle. The cycle is supposed to operate with different working fluids of water, R123, ammonia, cyclohexane, R11, and R113. The air temperature and pressure at the inlet and outlet of the steam generator are 300°C, 1.1 bar and 110°C, 1 bar, respectively. The mass flow rate of steam in the cycle is $1 \frac{\text{kg}}{\text{s}}$ and ambient temperature is 20°C. In addition, the cooling water with temperature of 20°C is available for the condenser.

Assumptions.   The combustion gas is estimated as pure air; all of the equipment are well insulated, and every pressure loss due to friction and internal irreversibilities are neglected.

Table 5.1 Characteristics of different organic fluid used in the simulation.

Working fluid	T_{cr} (K)	P_{cr} (barg)	Type
Water	674.15	220.64	Wet
R134a	374.2	40.59	Wet
Ammonia	405.4	113.3	Wet
Propane	369.7	42.47	Wet
Cyclohexane	535.5	40.75	Dry
HFE7100	468.5	22.29	Dry
R113	487.3	34.39	Dry
R123	456.7	36.68	Isentropic
n-Butane	425.1	37.96	Isentropic
Isobutene	497.7	36.40	Isentropic
R11	471.2	44.08	Isentropic

Solution. When water is considered as the working fluid:
The working fluid properties (for water) at the pump inlet are:

$$T_1 = 36\,°C, x_1 = 0\% \rightarrow P_1 = 0.06 \text{ bar}, h_1 = 150.8\,\frac{kJ}{kg}, s_1 = 0.5185\,\frac{kJ}{kg.K}$$

The working fluid properties at the turbine inlet are:

$$P_3 = 30 \text{ bar}, T_3 = 250\,°C \rightarrow s_3 = 6.286\,\frac{kJ}{kg.K}, h_3 = 2855\,\frac{kJ}{kg}$$

The working fluid properties at the condenser inlet are:

$$s_4 = s_3, P_4 = P_1 \rightarrow h_4 = 1934\,\frac{kJ}{kg}$$

The working fluid properties at the steam generator inlet (or pump outlet) are:

$$P_2 = P_3, s_2 = s_1 \rightarrow h_2 = 153.8\,\frac{kJ}{kg}$$

The amount of heat addition and heat rejection to/from the working fluid would be:

$$\dot{Q}_{in} = \dot{m}(h_3 - h_2) = 1 \times (2855 - 153.8) = 2701 \text{ kW}$$

$$\dot{Q}_{out} = \dot{m}(h_4 - h_1) = 1 \times (1934 - 150.8) = 1783 \text{ kW}$$

The net power produced by the Rankine cycle is:

$$\dot{W}_{net} = \dot{Q}_{in} - \dot{Q}_{out} = 2701 - 1783 = 918.1 \text{ kW}$$

The Rankine cycle thermal efficiency is:

$$\eta_{Rankine} = 1 - \frac{\dot{Q}_{out}}{\dot{Q}_{in}} = 1 - \frac{1783}{2701} = 0.3399$$

The steam rate of the power plant is:

$$S.R. = \frac{\dot{m}}{\dot{W}_{net}} = \frac{1}{918.1} = 0.001089\,\frac{kg}{kW.s}$$

Combustion gas properties at the inlet of superheater would be:

$$P_{air,in} = 1.1 \text{ bar}, T_{air,in} = 300\,°C \rightarrow h_{air,in} = 578.9\,\frac{kJ}{kg}, s_{air,in} = 6.337\,\frac{kJ}{kg.K}$$

Combustion gas properties at the out let of economizer would be:

$$P_{air,out} = 1 \text{ bar}, T_{air,out} = 110\,°C \rightarrow h_{air,out} = 384.2\,\frac{kJ}{kg}, s_{air,out} = 5.952\,\frac{kJ}{kg.K}$$

The combustion gas is estimated as pure air; therefore, the mean temperature of the combustion gas as the heat source would be:

$$T_{H-mean} = \frac{h_{air,in} - h_{air,out}}{s_{air,in} - s_{air,out}} = \frac{578.9 - 384.2}{6.337 - 5.952} = 505.9 \text{ K}$$

Mean temperature of the heat sink that is the atmosphere temperature would be:

$$T_{L-mean} = 20 + 273.15 = 298.15 \text{ K}$$

The combustion gas mass flow is:

$$\dot{m}_{air} = \frac{\dot{m}_{steam}(h_3 - h_2)}{h_{air,in} - h_{air,out}} = \frac{1 \times (2855 - 153.8)}{578.9 - 384.2} = 13.87\,\frac{kg}{s}$$

The cycle exergy efficiency would be:

$$\eta_{II,Rankine} = \frac{\dot{W}_{net}}{\dot{Q}_{in}\left(1 - \dfrac{T_{L-mean}}{T_{H-mean}}\right)} = \frac{918.1}{2701\left(1 - \dfrac{298.15}{505.9}\right)} = 0.80$$

Following the same procedure, for the other 5 working fluids the results are presented in Table 5.2.

Discussions. According to the results, for the same initial conditions, water is the best choice as the working fluid. To find out why, we should pay attention to the water saturation pressure for the given saturated liquid temperature at the pump inlet. In this example, the inlet temperature of pump is assumed to be 36°C. At this temperature, water has the lowest saturation pressure of 0.06 bar, while the condensation goes on without any problem. One may ask why not considering the condenser pressure of 0.06 bar for other working fluids? To answer this question, we should know the corresponding saturation temperature for this pressure. At this pressure, the saturation temperature of water, R123, ammonia, cyclohexane, R11, and R113 is 36°C, −31.97°C, not applied (N.A.), 8.89 °C, −37.41°C, and −17.57°C, respectively. It is evident if the cooling water with temperature of 20°C is available, except for water, not only the condensation will not happen but also due to higher temperature of cooling water, the steam will be superheated in the condenser. As a result, the pressure of 0.06 bar except for water is not feasible. Finally, the limit for the operating pressure of the condenser depends on the available cooling water temperature, the lower the cooling water temperature is the lower the condenser operating pressure will be.

Table 5.2A The thermodynamic properties for the six working fluids used in the Rankine cycle in Example 5.1.

	T_1	P_1	h_1	s_1	P_2	h_2	s_2	T_3	P_3	h_3	s_3	P_4	h_4	s_4
Water	36	0.06	150.8	0.5185	30	153.8	0.5185	250	30	2855	6.286	0.05945	1934	6.286
R123	36	1.35	237.7	1.129	30	239.7	1.129	250	30	567	1.91	1.352	492.7	1.91
Ammonia	36	13.9	371	1.582	30	373.7	1.582	250	30	2031	6.189	13.9	1856	6.189
Cyclohexane	36	0.21	52.7	0.1789	30	56.6	0.1789	250	30	590.3	1.46	0.2093	448.9	1.46
R11	36	1.53	65.2	0.2415	30	68.15	0.2415	250	30	361.1	0.9431	1.527	286.9	0.9431
R113	36	0.68	66.65	0.2467	30	70.08	0.2467	250	30	355.9	0.9244	0.6768	285.6	0.9244

Table 5.2B The thermodynamic properties of air for the six working fluids used in the Rankine cycle in Example 5.1.

	$T_{air,in}$	$P_{air,in}$	$h_{air,in}$	$s_{air,in}$	$T_{air,out}$	$P_{air,out}$	$h_{air,out}$	$s_{air,out}$	$\dot{m}_{air}$
Water	300	1.1	578.9	6.337	110	1	384.2	5.952	13.87
R123	300	1.1	578.9	6.337	110	1	384.2	5.952	1.681
Ammonia	300	1.1	578.9	6.337	110	1	384.2	5.952	8.514
Cyclohexane	300	1.1	578.9	6.337	110	1	384.2	5.952	2.741
R11	300	1.1	578.9	6.337	110	1	384.2	5.952	1.504
R113	300	1.1	578.9	6.337	110	1	384.2	5.952	1.468

Table 5.2C The evaluation criteria for the six working fluids used in the Rankine cycle in Example 5.1.

	$\dot{Q}_{in}$	$\dot{Q}_{out}$	$\dot{W}_{net}$	$\eta_{Rankine}$	S.R.	T_{H-mean}	T_{L-mean}	$\eta_{II,Rankine}$	$T_{sat@0.06bar}$
Water	2701	1783	918.1	0.3399	0.001089	505.9	298.15	0.8083	36.17
R123	327.4	255.1	72.31	0.2209	0.01383	505.9	298.15	0.5252	−31.97
Ammonia	1658	1485	172.6	0.1041	0.005793	505.9	298.15	0.2476	N.A.
Cyclohexane	533.7	396.2	137.6	0.2577	0.007269	505.9	298.15	0.6129	8.892
R11	292.9	221.7	71.22	0.2431	0.01404	505.9	298.15	0.5781	−37.41
R113	285.8	219	66.83	0.2338	0.01496	505.9	298.15	0.5561	−17.57

In Table 5.2, the units of the parameters are the same as the units given for the calculations presented for water as the working fluid.

Another important criterion, which was calculated is steam ratio (S.R.). It shows the required steam mass flow for generation of 1 kW of power. It is clear that the smaller the S.R., the better the cycle. From this point of view, as well, water is the best working fluid. The exergy efficiency of the cycle working with water is also highest.

According to the results, from the thermodynamic point of view, water is the best among the six working fluids. In addition, water is cheap, available, nontoxic, and nonflammable. Considering all of these advantages have made water as the number one choice working fluid for the SPPs. The code developed for the Example 5.1 is shown in Code 5.1 in the following.

Code 5.1. The code developed for Example 5.1 when water is the working fluid (Cycle No. 1).

```
R$='WATER'
T_L_air=20
T_air_in=300
T_air_out=110
P_air_in=1.1
P_air_out=1
P[3]=30
T[1]=36
T[3]=250
P[2]=P[3]
P[4]=P_SAT(R$, T=T[1])
P[1]=P[4]
h[3]=enthalpy(R$, T=T[3], P=P[3])
s[3]=entropy(R$, T=T[3], P=P[3])
s[4]=s[3]
h[4]=enthalpy(R$, P=P[4], s=s[4])
h[1]=enthalpy(R$, P=P[1], X=0)
s[1]=entropy(R$, P=P[1], X=0)
s[2]=s[1]
h[2]=enthalpy(R$, P=P[2], s=s[2])
Q_in=h[3]-h[2]
Q_out=h[4]-h[1]
eta=1-Q_out/Q_in
W_NET=Q_in-Q_out
SR=1/W_NET
T_L_m=Q_out/(s[4]-s[1])
T_H_m=Q_in/(s[3]-s[2])
X_Q_IN=Q_in*(1-(T_L_air+273.15)/T_H_air)
ETA_EXERGY=W_NET/X_Q_IN
h_air_in=ENTHALPY(Air,T=T_air_in)
h_air_out=ENTHALPY(Air,T=T_air_out)
s_air_in=ENTROPY(Air,T=T_air_in,P=P_air_in)
s_air_out=ENTROPY(Air,T=T_air_out,P=P_air_out)
m_dot_air*(h_air_in-h_air_out)=h[3]-h[2]
h[3]-h[2]=m_dot_air*T_H_air*(s_air_in-s_air_out)
```

Example 5.2. Consider a basic SPP cycle. The steam generator inlet pressure is 30 bar. The steam temperature at the turbine inlet is 250 °C and the saturated liquid temperature at the pump inlet is 36°C. The pressure loss due to friction in the steam generator is 10% of the inlet pressure, while in the condenser, it is 3%. The adiabatic efficiency of the turbine and pump are 0.87 and 0.85, respectively. If the combustion gases inside the steam generator could be estimated as pure air. Determine the cycle thermal efficiency, exergy efficiency, and steam rate of the cycle. The cycle is supposed to operate with different working fluids of water, R123, ammonia, cyclohexane, R11, and R113. The air temperature and pressure at inlet and outlet of the steam generator are 300 °C, 1.1 bar and 110 °C, 1 bar, respectively. The mass flow rate of steam in the cycle is $1\frac{kg}{s}$ and ambient temperature is 20°C. In addition, the cooling water with temperature of 20°C is available for the condenser.

Assumptions. The combustion gas is estimated as pure air, and all of the equipment are well insulated.

Solution. When water is considered as the working fluid:
The working fluid properties at the pump inlet are:

$$T_1 = 36°C, x_1 = 0\% \rightarrow P_1 = 0.06 \text{ bar}, h_1 = 150.8\frac{kJ}{kg}, s_1 = 0.5185\frac{kJ}{kg.K}$$

The working fluid properties at the turbine inlet are:

$$P_3 = 0.9\,P_2 = 0.9 \times 30 = 27 \text{ bar}, T_3 = 250°C \rightarrow s_3 = 6.357\frac{kJ}{kg.K}, h_3 = 2870\frac{kJ}{kg}$$

The working fluid properties at the condenser inlet are:

$$s_{s-4} = s_3, P_4 = 1.03P_1 = 0.06124 \text{ bar} \rightarrow h_{s-4} = 1959\frac{kJ}{kg}$$

$$h_4 = h_3 - \eta_{ad,turbine} \times (h_3 - h_{s-4}) = 2870 - 0.87(2870 - 1959) = 2077\frac{kJ}{kg}$$

$$s_4 = 6.357\frac{kJ}{kg.K}$$

The working fluid properties at the steam generator inlet are:

$$P_2 = 30, s_{s-2} = s_1 \rightarrow h_{s-2} = 153.8\frac{kJ}{kg}$$

$$h_2 = h_1 + \frac{h_{s-2} - h_1}{\eta_{ad,pump}} = 148.5 + \frac{153.8 - 148.5}{0.85} = 154.3\frac{kJ}{kg}$$

The amount of heat addition and heat rejection to/from the working fluid would be:

$$\dot{Q}_{in} = \dot{m}(h_3 - h_2) = 1 \times (2870 - 154.3) = 2715 \text{ kW}$$

$$\dot{Q}_{out} = \dot{m}(h_4 - h_1) = 1 \times (2077 - 148.5) = 1927 \text{ kW}$$

The net power produced by the Rankine cycle is:

$$\dot{W}_{net} = \dot{Q}_{in} - \dot{Q}_{out} = 2715 - 1927 = 788.8 \text{ kW}$$

The Rankine cycle thermal efficiency is:

$$\eta_{thermal} = 1 - \frac{\dot{Q}_{out}}{\dot{Q}_{in}} = 1 - \frac{1927}{2715} = 0.2905$$

The steam rate of the power plant is:

$$S.R. = \frac{\dot{m}}{\dot{W}_{net}} = \frac{1}{788.8} = 0.001268 \frac{\text{kg}}{\text{kW.s}}$$

Combustion gas properties at the inlet of superheater would be:

$$P_{air,in} = 1.1 \text{ bar}, T_{air,in} = 300°C \rightarrow h_{air,in} = 578.9 \frac{\text{kJ}}{\text{kg}}, s_{air,in} = 6.337 \frac{\text{kJ}}{\text{kg.K}}$$

Combustion gas properties at the out let of economizer would be:

$$P_{air,out} = 1 \text{ bar}, T_{air,out} = 110°C \rightarrow h_{air,out} = 384.2 \frac{\text{kJ}}{\text{kg}}, s_{air,out} = 5.952 \frac{\text{kJ}}{\text{kg.K}}$$

The combustion gas is estimated as pure air therefore, the mean temperature of the combustion gas as the heat source would be:

$$T_{H-mean} = \frac{h_{air,in} - h_{air,out}}{s_{air,in} - s_{air,out}} = \frac{578.9 - 384.2}{6.337 - 5.952} = 505.9 \text{ K}$$

Mean temperature of the heat sink, that is, the atmosphere temperature would be:

$$T_{L-mean} = 20 + 273.15 = 298.15 \text{ K}$$

The combustion gas mass flow is:

$$\dot{m}_{air} = \frac{\dot{m}_{steam}(h_3 - h_2)}{h_{air,in} - h_{air,out}} = \frac{1 \times (2870 - 154.3)}{578.9 - 384.2} = 13.95 \frac{\text{kg}}{\text{s}}$$

The cycle exergy efficiency would be:

$$\eta_{II,SPP} = \frac{\dot{W}_{net}}{\dot{Q}_{in}\left(1 - \dfrac{T_{L-mean}}{T_{H-mean}}\right)} = \frac{788.8}{2715\left(1 - \dfrac{298.15}{505.9}\right)} = 0.69$$

Following the same procedure, for the other 5 working fluids the results are presented in Table 5.3.

Discussions. In this example, the impact of internal irreversibilities in the turbine, pump, steam generator, and condenser on the evaluation criteria was investigated. The results show that internal irreversibilities reduce the cycle thermal and exergy efficiencies. Considering the internal irreversibilities do not change, the main conclusion about the best working fluid. The code written in the EES for this example is given below.

Code 5.2. The code developed for Example 5.2 when water is the working fluid cycle.

```
R$='WATER'
P_air_in=1.1
P_air_out=1
T_L_air=20
eta_adiabatic_turbine=0.87
eta_adiabatic_pump=0.85
T_air_in=300
T_air_out=110
P[2]=30
T[1]=36
T[3]=250
P[3]=P[2]-0.1*P[2]
P[1]=P_SAT(R$, T=T[1])
P[4]=P[1]+0.03*P[1]
h[3]=enthalpy(R$, T=T[3], P=P[3])
s[3]=entropy(R$, T=T[3], P=P[3])
eta_adiabatic_turbine=(h[3]-h[4])/(h[3]-h_4_s[1])
s_4_s[1]=s[3]
h_4_s[1]=enthalpy(R$, P=P[4], s=s_4_s[1])
s[4]=entropy(R$, h=h[4], P=P[4])
h[1]=enthalpy(R$, P=P[1], X=0)
s[1]=entropy(R$, P=P[1], X=0)
s_2_s[1]=s[1]
h_2_s[1]=enthalpy(R$, P=P[2], s=s_2_s[1])
eta_adiabatic_pump=(h_2_s[1]-h[1])/(h[2]-h[1])
s[2]=entropy(R$, h=h[2], P=P[2])
Q_in=h[3]-h[2]
Q_out=h[4]-h[1]
eta=1-Q_out/Q_in
W_NET=Q_in-Q_out
SR=1/W_NET
h_air_in=ENTHALPY(Air,T=T_air_in)
h_air_out=ENTHALPY(Air,T=T_air_out)
s_air_in=ENTROPY(Air,T=T_air_in,P=P_air_in)
s_air_out=ENTROPY(Air,T=T_air_out,P=P_air_out)
m_dot_air*(h_air_in-h_air_out)=h[3]-h[2]
h[3]-h[2]=m_dot_air*T_H_air*(s_air_in-s_air_out)
X_Q_in=Q_in*(1-(T_L_air+273.15)/T_H_air)
ETA_EXERGY=W_NET/X_Q_in
```

Table 5.3A The thermodynamic properties for the six working fluids used in the Rankine cycle in Example 5.1.

	T_1	P_1	h_1	s_1	P_2	h_2	s_2	T_3	P_3	h_3	s_3	P_4	h_4	s_4
Water	36	0.06	150.8	0.5185	30	154.3	0.5202	250	27	2870	6.357	0.06124	2077	6.739
R123	36	1.35	237.7	1.129	30	240	1.13	250	27	569.3	1.918	1.393	506.6	1.94
Ammonia	36	13.9	371	1.582	30	374.2	1.584	250	27	2035	6.245	14.32	1907	6.287
Cyclohexane	36	0.21	52.7	0.1789	30	57.29	0.1812	250	27	753.3	1.774	0.2156	585.7	1.837
R11	36	1.53	65.2	0.2415	30	68.67	0.2526	250	27	363.4	0.9528	1.573	300.7	0.9764
R113	36	0.68	66.65	0.2467	30	70.68	0.2596	250	27	359	0.9339	0.6971	299	0.9553

Table 5.3B The thermodynamic properties of air for the six working fluids used in the Rankine cycle in Example 5.1.

	$T_{air,in}$	$P_{air,in}$	$h_{air,in}$	$s_{air,in}$	$T_{air,out}$	$P_{air,out}$	$h_{air,out}$	$s_{air,out}$	$\dot{m}_{air}$
Water	300	1.1	578.9	6.337	110	1	384.2	5.952	13.95
R123	300	1.1	578.9	6.337	110	1	384.2	5.952	1.691
Ammonia	300	1.1	578.9	6.337	110	1	384.2	5.952	8.531
Cyclohexane	300	1.1	578.9	6.337	110	1	384.2	5.952	3.574
R11	300	1.1	578.9	6.337	110	1	384.2	5.952	1.514
R113	300	1.1	578.9	6.337	110	1	384.2	5.952	1.481

Table 5.3C The evaluation criteria for the six working fluids used in the Rankine cycle in Example 5.1.

	$\dot{Q}_{in}$	$\dot{Q}_{out}$	$\dot{W}_{net}$	$\eta_{thermal}$	S.R.	T_{H-mean}	T_{L-mean}	$\eta_{II,Rankine}$
Water	2715	1927	788.8	0.2905	0.001268	505.9	298.15	0.6908
R123	329.3	268.9	60.33	0.1832	0.01658	505.9	298.15	0.4357
Ammonia	1661	1536	124.8	0.07511	0.008016	505.9	298.15	0.1786
Cyclohexane	696	533	163	0.2341	0.006137	505.9	298.15	0.5568
R11	294.7	235.5	59.21	0.2009	0.01689	505.9	298.15	0.4777
R113	288.3	232.3	56	0.1942	0.01786	505.9	298.15	0.4618

In Table 5.3A–C, the units of the parameters are the same as the units given for the calculations presented for water as the working fluid.

4. Efficiency improvement techniques in SPPs

Efficiency improvement is one of the most important concerns of power plant owners and electricity producer/seller companies. The reason behind this concern is the money that can be earned more or be expensed less annually. For a big power plant such as an 1000 MW one, 1% improvement in the thermal efficiency means millions of dollars more income.

According to the definition of the thermal efficiency, improvement of efficiency depends on the operating conditions of the cycle such as temperature and pressure. However, changing the operational conditions is not easy always. Sometimes, it requires changing material in some parts of the power plant to tolerate the higher temperature or pressure. It may require using thicker tubes in the steam generator to tolerate higher pressure. It may require using more heating surface in the steam generator or condenser to absorb or reject more heat. As it can be seen, increasing efficiency corresponds to make some hardware changes in the power plant that is not easy at all. According to this discussion, a power plant should be optimized in the designing step to avoid unwanted changes during operation.

In the following, several techniques that are usually used for efficiency improvement are discussed in details.

4.1 Superheating

Superheating is having higher temperature at the turbine inlet at the same operational pressure, such as the 3-3' at Fig. 5.6. It is evident that achieving higher temperature requires more fuel consumption by the burner, more heat absorption in the superheater (realize that heat absorption in the economizer and boiler has not changed), and more heat loss in the condenser. In other words, the superheater and condenser heating surfaces should be increased.

Fig. 5.6 shows that the magnitudes of $\dot{Q}_{in}$, $\dot{Q}_{out}$ and also $\dot{W}_{net}$ have increased as below:

$$\Delta \dot{W}_{net} = A_{green}$$

$$\Delta \dot{Q}_{out} = A_{blue}$$

$$\Delta \dot{Q}_{in} = A_{green} + A_{blue}$$

To determine the conditions in which superheating causes efficiency improvement the differential changes of thermal efficiency would be calculated as below:

The thermal efficiency is $\eta = 1 - \frac{\dot{Q}_{out}}{\dot{Q}_{in}}$ and it can be written as Eq. (5.29):

$$\eta = \eta\left(\dot{Q}_{in}, \dot{Q}_{out}\right) \tag{5.29}$$

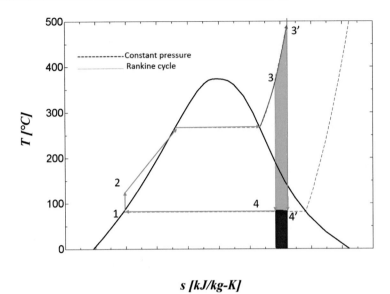

Figure 5.6 The impact of superheating on the steam power plant (SPP) cycle.

The efficiency differential due to any change in the cycle can be written as below:

$$d\eta = \frac{\partial \eta}{\partial \dot{Q}_{in}} d\dot{Q}_{in} + \frac{\partial \eta}{\partial \dot{Q}_{out}} d\dot{Q}_{out} = \frac{\dot{Q}_{out}}{\dot{Q}_{in}^2} d\dot{Q}_{in} - \frac{1}{\dot{Q}_{in}} d\dot{Q}_{out} = \frac{d\dot{Q}_{in}}{\dot{Q}_{in}} \left(\frac{\dot{Q}_{out}}{\dot{Q}_{in}} - \frac{d\dot{Q}_{out}}{d\dot{Q}_{in}} \right)$$

Since $\frac{d\dot{Q}_{in}}{\dot{Q}_{in}} > 0$, the $d\eta$ would be positive when:

$$\frac{d\dot{Q}_{in}}{\dot{Q}_{in}} > \frac{d\dot{Q}_{out}}{\dot{Q}_{out}}$$

Since both sides of the inequality are positive, integrating from both sides:

$$\int_{T_3}^{T_{3'}} \frac{d\dot{Q}_{in}}{\dot{Q}_{in}} > \int_{T_3}^{T_{3'}} \frac{d\dot{Q}_{out}}{\dot{Q}_{out}}$$

$$ln\dot{Q}_{in} > ln\dot{Q}_{out}$$

And finally:

$$\dot{Q}_{in} > \dot{Q}_{out}$$

which is always valid for an SPP, because $\dot{Q}_{in} = \dot{Q}_{out} + \dot{W}_{net}$. Hence, superheating will always increase the thermal efficiency of the SPP. Another advantage of super-heating is that steam quality at the turbine outlet becomes higher that can reduce blades erosion due to liquid water droplets.

The only problem with superheating is the metallurgical limitations in the super-heater section and also turbine blades that should tolerate the higher temperature.

Example 5.3. Consider the SPP cycle discussed the Example 5.2 for water as the work-ing fluid. If it is supposed to superheat the turbine inlet steam from 250 °C to 500 °C gradually, determine how the thermal and exergy efficiencies will change. Assume that no metallurgical problem exists for the temperature range given in the example.

Solution. The calculations are exactly the same as what was done for Example 5.2 but for different inlet steam temperature changing from 250 °C to 500 °C. The thermal and exergy efficiencies should be calculated according to the new conditions. The code which was written for Example 5.2 is updated for this example, and a parametric study in the EES is done for variable T_3. The results are plotted in Fig. 5.7. According to the results, the thermal efficiency has increases from 29% to 31.75%, while the exergy efficiency experiences bigger improvement from 69% to 75.75%. The steam quality at the turbine outlet has increased from 79% to about 94%. Care must be taken that steam quality of 79% is very low and can cause serious corrosion and erosion on the turbine internal components. Hence, the initial temperature of 250 °C is very low. Usually, the minimum allowable steam quality at the turbine exit is 88% that in this case happens at the temperature of about 400°C. It means that the steam temperature at the turbine inlet should be higher than 400°C.

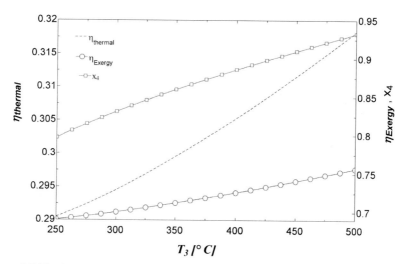

Figure 5.7 The impact of superheating on the energy/exergy efficiencies and steam quality.

4.2   Increasing maximum pressure

In this technique, the turbine inlet pressure increases while the turbine inlet temperature stays the same (Fig. 5.8). It is evident that achieving higher operating pressure requires more power consumption by the pump or even a bigger pump. In addition, according to Fig. 5.8, the new cycle loses less heat at the condenser that means a smaller condenser. In the steam generator, the economizer and boiler size should increase because the heat absorption in these two sections has increased. The superheater size may become smaller or bigger and can be determined according to the heat absorption by the superheater.

Fig. 5.8 shows that the magnitudes of $\dot{Q}_{in}$, $\dot{Q}_{out}$, and also $\dot{W}_{net}$ have changed as below:

$$\Delta \dot{W}_{net} = A_{green} - A_{yellow}$$

$$\Delta \dot{Q}_{out} = - A_{blue}$$

$$\Delta \dot{Q}_{in} = A_{green} - A_{blue} - A_{yellow}$$

Seeking the conditions in which increasing the operating pressure result in efficiency improvement, the differential of the thermal efficiency would be calculated:

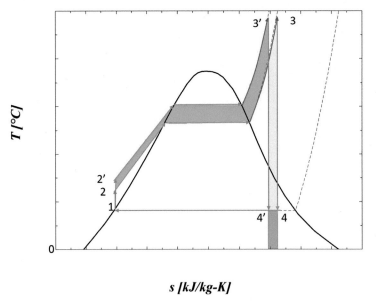

Figure 5.8 The impact of maximum operating pressure on the steam power plant (SPP) cycle.

$$dη = \frac{\partial η}{\partial \dot{Q}_{in}} d\dot{Q}_{in} + \frac{\partial η}{\partial \dot{Q}_{out}} d\dot{Q}_{out} = \frac{\dot{Q}_{out}}{\dot{Q}_{in}^2} d\dot{Q}_{in} - \frac{1}{\dot{Q}_{in}} d\dot{Q}_{out} = \frac{d\dot{Q}_{out}}{\dot{Q}_{in}} \left(\frac{\dot{Q}_{out}}{\dot{Q}_{in}} \frac{d\dot{Q}_{in}}{d\dot{Q}_{out}} - 1 \right)$$

Since $\frac{d\dot{Q}_{out}}{\dot{Q}_{out}} < 0$, the $dη$ would be positive when:

$$\frac{d\dot{Q}_{in}}{\dot{Q}_{in}} > \frac{d\dot{Q}_{out}}{\dot{Q}_{out}}$$

It means that $\frac{d\dot{Q}_{in}}{\dot{Q}_{in}}$ may be positive or negative. For the positive values $dη$ is always positive, but for the negative values it can be written as below:

$$\int_{P_2}^{P_{2'}} \frac{d\dot{Q}_{in}}{\dot{Q}_{in}} = - \int_{P_{2'}}^{P_2} \frac{d\dot{Q}_{in}}{\dot{Q}_{in}}$$

$$\int_{P_2}^{P_{2'}} \frac{d\dot{Q}_{out}}{\dot{Q}_{out}} = - \int_{P_{2'}}^{P_2} \frac{d\dot{Q}_{out}}{\dot{Q}_{out}}$$

In which $\int_{P_{2'}}^{P_2} \frac{d\dot{Q}_{in}}{\dot{Q}_{in}}$ and $\int_{P_{2'}}^{P_2} \frac{d\dot{Q}_{out}}{\dot{Q}_{out}}$ are positive on the right hand sides of the above equations, therefore:

$$- \int_{P_{2'}}^{P_2} \frac{d\dot{Q}_{in}}{\dot{Q}_{in}} < - \int_{P_{2'}}^{P_2} \frac{d\dot{Q}_{out}}{\dot{Q}_{out}}$$

$$- ln\dot{Q}_{in} < -ln\dot{Q}_{out} \rightarrow ln\left(\frac{1}{\dot{Q}_{in}} \right) < ln\left(\frac{1}{\dot{Q}_{out}} \right)$$

And finally:

$$\dot{Q}_{in} > \dot{Q}_{out}$$

which is always valid for an SPP, because $\dot{Q}_{in} = \dot{Q}_{out} + \dot{W}_{net}$. Hence, increasing the maximum operating pressure of SPP will always increase the thermal efficiency of the SPP.

A disadvantage of this technique is the steam quality reduction at the turbine outlet that increases the possibility of erosion at the last stages of the turbine. To avoid this problem, it is recommended that increasing maximum pressure and superheating to be done simultaneously.

The only problem with maximum operating pressure is the mechanical strength limitation in the steam generator components and piping that must tolerate the higher pressure.

Attention

Radical increasing the maximum operating pressure of SPP while keeping the turbine inlet temperature the same may result in having saturated steam at the turbine inlet. In this case, the thermal efficiency would reduce. Hence, to take benefit from the maximum operating pressure, we must have:

$T_{sat@P_{3'}} < T_{3'}$ in which, $T_{3'} = T_3$

This point is shown in the Example 5.4 as well.

Example 5.4. Consider the SPP cycle discussed the Example 5.2 for water as the working fluid. If it is supposed to change the pressure at the pump outlet from 30 bar to 50 bar determine how the steam quality at the turbine exit, and energy/exergy efficiencies will change.

Solution. The calculations are exactly the same as what was done for Example 5.2 but for different pump output pressures changing from 30 bar to 50 bar. The steam quality, thermal, and exergy efficiencies should be calculated according to the new pump outlet pressure and steam generator new inlet pressure. The code which was written for Example 5.2 is updated for this example, and $T_{sat@P_{3'}}$ is also calculated to check the **Attention** mentioned before this example. The results are plotted in Fig. 5.9. According to the results at pump outlet pressure from 30 to 44.1 bar (which due to pressure losses corresponds to 27 bar to 39.7 bar at the turbine inlet), the efficiency is increasing but at the turbine inlet pressure of 39.7 bar, the $T_{3'} = T_{sat@P_{3'}} = 250\,°C$, and for higher pressures the energy and exergy efficiencies of the cycle fall suddenly. The steam quality at the turbine exit decreases from about 80% to 76% at the pump exit pressure of 44.1 bar, and after that due to inappropriate steam pressure at the turbine inlet, steam starts condensation just after entering the turbine. It means that under such conditions, the whole turbine internal components would be operating with two-phase flow, and turbine failure is inescapable due to vibration, erosion, and corrosion created by the liquid droplets.

However, in such cases, superheating can be used to increase the steam quality at the turbine exit and make use of the privileges of both improvement methods. For example, if while the pump exit pressure is increased to 50 bar, steam could be superheated to 500 °C, steam quality becomes 90% at the turbine exit, which is completely safe. In addition, the thermal and exergy efficiencies increase to 33.54% and 79.74%.

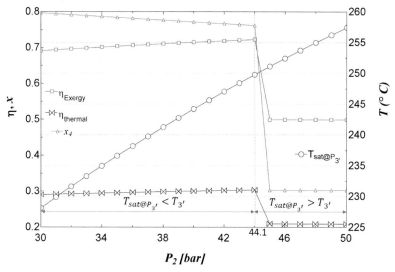

Figure 5.9 Impact of maximum pressure on the energy/exergy efficiencies and steam quality.

4.3 Decreasing minimum pressure

Decreasing the operating pressure of condenser (or turbine outlet pressure) changes the $T - s$ diagram of the cycle as shown in Fig. 5.10. It is clear that achieving lower operating pressure requires more expansion of steam by the turbine that can be done by using more turbine stages that means a bigger turbine. In addition, according to Fig. 5.10, the new cycle loses less heat at the condenser that means a smaller condenser. In the steam generator, the economizer load has increased (hence its size will be bigger as well) to some extent, but the boiler and superheater loads stay the same, and as a result, their size will remain unchanged as well.

Fig. 5.10 shows that the magnitudes of $\dot{Q}_{in}$, $\dot{Q}_{out}$, and also $\dot{W}_{net}$ have changed as below:

$$\Delta \dot{W}_{net} = A_{green} + A_{yellow}$$

$$\Delta \dot{Q}_{out} = A_{blue} - A_{green}$$

$$\Delta \dot{Q}_{in} = A_{blue} + A_{yellow}$$

To determine the conditions in which decreasing the minimum operating pressure results in efficiency improvement, the differential changes of thermal efficiency would be calculated as below:

$$d\eta = \frac{\partial \eta}{\partial \dot{Q}_{in}} d\dot{Q}_{in} + \frac{\partial \eta}{\partial \dot{Q}_{out}} d\dot{Q}_{out} = \frac{\dot{Q}_{out}}{\dot{Q}_{in}^2} d\dot{Q}_{in} - \frac{1}{\dot{Q}_{in}} d\dot{Q}_{out} = \frac{d\dot{Q}_{in}}{\dot{Q}_{in}} \left(\frac{\dot{Q}_{out}}{\dot{Q}_{in}} - \frac{d\dot{Q}_{out}}{d\dot{Q}_{in}} \right)$$

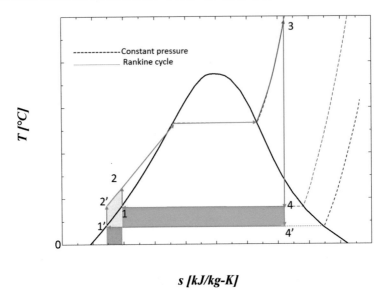

s [kJ/kg-K]

Figure 5.10 The impact of minimum operating pressure on the SPP cycle.

Since $\frac{d\dot{Q}_{in}}{\dot{Q}_{in}} > 0$, the $d\eta$ would be positive when:

$$\frac{d\dot{Q}_{in}}{\dot{Q}_{in}} > \frac{d\dot{Q}_{out}}{\dot{Q}_{out}}$$

$$\int_{P_4}^{P_{4'}} \frac{d\dot{Q}_{in}}{\dot{Q}_{in}} > \int_{P_4}^{P_{4'}} \frac{d\dot{Q}_{out}}{\dot{Q}_{out}}$$

$$ln\dot{Q}_{in} > ln\dot{Q}_{out}$$

And finally:

$$\dot{Q}_{in} > \dot{Q}_{out}$$

which is always valid for an SPP, because $\dot{Q}_{in} = \dot{Q}_{out} + \dot{W}_{net}$. Hence, reducing the minimum pressure will always increase the thermal efficiency of the SPP. A disadvantage of this technique is that steam quality at the turbine outlet decreases that can increase blades erosion due to liquid water droplets at the turbine last stages.

Attention

The only limit for this technique is the possibility of heat transfer from the steam to the cooling water in the condenser. The cooling water source temperature is out of our control and depends on the ambient temperature, season, geography,

---

Attention—cont'd

etc.; therefore, the minimum pressure should be adjusted to make the heat transfer possible from steam to the cooling water. When the condenser pressure drops below the saturation pressure of the cooling water, the steam temperature becomes smaller than the cooling water temperature and starts to receive heat from the cooling water and becoming even superheated instead of condensing. To avoid this problem:

$$P_{min} > P_{sat,water@T_{cooling\ water}}$$

For example, if the cooling water temperature is 20 °C, the $P_{sat,water@20°C} = 2.34$ kPa, it means, theoretically $P_{min} > 2.34$ kPa. This minimum limit for $T_{cooling\ water} = 35°C$ is $P_{min} > 5.63$ kPa.

In practice, a minimum temperature difference between steam and cooling water is required for effective heat transfer. This temperature difference is called approach temperature (Fig. 5.11). The smaller the approach temperature, the bigger the condenser. The approach temperature is usually about 8 °C to 12°C.

4.4 Reheating

Due to steam expansion through the turbine, its temperature and pressure drop, and steam potential for power production decreases. It is not practical to boost the steam pressure again, but the temperature can be increased yet again by reheating. Steam can be returned to the steam generator, somewhere close to the superheaters to be heated to its maximum temperature as shown in Fig. 5.12A. Reheater is an additional equipment and contains tubes very similar to the superheaters. Hence, the steam

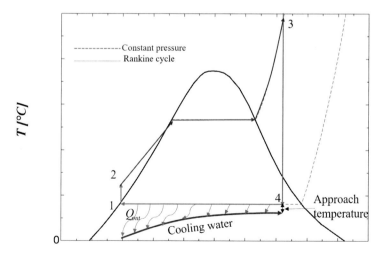

Figure 5.11 The approach temperature and heat transfer in the condenser.

(A)

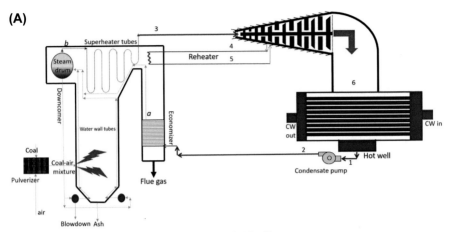

Figure 5.12A Simple cycle plus a reheater (cycle No. 2).

generator will be more expensive and bigger by using the reheater. By reheating, more power will be produced, and condenser load increases as well. This means that the condenser size should increase according to the additional heat loss load.

Fig. 5.12B shows that the magnitudes of $\dot{Q}_{in}$, $\dot{Q}_{out}$, and also $\dot{W}_{net}$ have increased as below:

$$\Delta \dot{W}_{net} = A_{yellow}$$

$$\Delta \dot{Q}_{out} = A_{blue}$$

$$\Delta \dot{Q}_{in} = A_{blue} + A_{yellow}$$

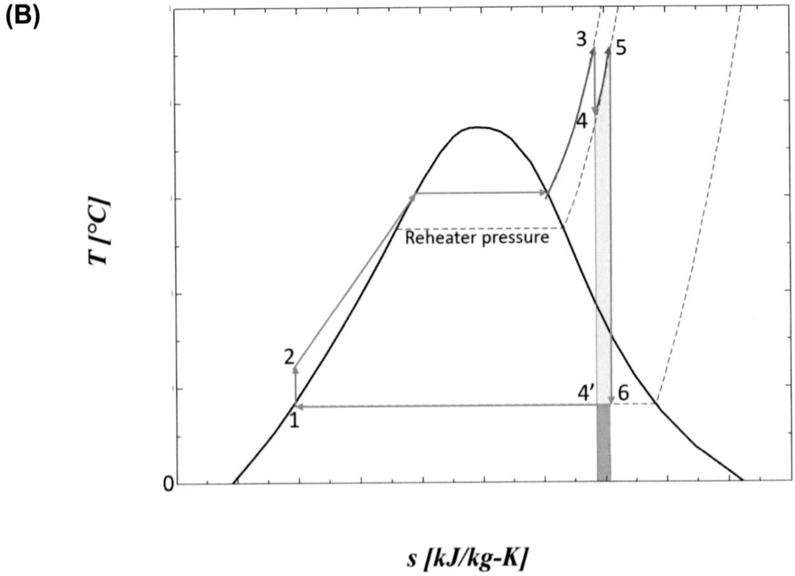

(B)

Figure 5.12B Reheating of steam in the steam generator.

To determine the conditions in which decreasing the minimum operating pressure causes efficiency improvement, the differential changes of thermal efficiency would be calculated as below:

$$d\eta = \frac{\partial \eta}{\partial \dot{Q}_{in}} d\dot{Q}_{in} + \frac{\partial \eta}{\partial \dot{Q}_{out}} d\dot{Q}_{out} = \frac{\dot{Q}_{out}}{\dot{Q}_{in}^2} d\dot{Q}_{in} - \frac{1}{\dot{Q}_{in}} d\dot{Q}_{out}$$

$$= \frac{d\dot{Q}_{in}}{\dot{Q}_{in}} \left(\frac{\dot{Q}_{out}}{\dot{Q}_{in}} - \frac{d\dot{Q}_{out}}{d\dot{Q}_{in}} \right)$$

Since $\frac{d\dot{Q}_{out}}{\dot{Q}_{out}}$ and $\frac{d\dot{Q}_{in}}{\dot{Q}_{in}}$ are positive the $d\eta$ would be positive when:

$$\frac{d\dot{Q}_{in}}{\dot{Q}_{in}} > \frac{d\dot{Q}_{out}}{\dot{Q}_{out}}$$

$$\int_{T_4}^{T_5} \frac{d\dot{Q}_{in}}{\dot{Q}_{in}} > \int_{T_4}^{T_5} \frac{d\dot{Q}_{out}}{\dot{Q}_{out}}$$

$$ln\dot{Q}_{in} > ln\dot{Q}_{out}$$

And finally:

$$\dot{Q}_{in} > \dot{Q}_{out}$$

which is always valid for an SPP, because $\dot{Q}_{in} = \dot{Q}_{out} + \dot{W}_{net}$. Hence, reheating will always increase the thermal efficiency of the SPP. An advantage of this technique is that steam quality at the turbine outlet increases that can reduce blades erosion due to liquid water droplets at the turbine last stages.

Theoretically, the maximum allowed temperature at point 5 is $T_5 = T_3$. However, since the reheater pressure is much lower than the inlet maximum pressure, the tubes and components can tolerate higher temperatures. Therefore, temperature at 5 can be higher than 3. In some designs, this difference is about $T_5 - T_3 = 25\,°F\,(13.9\,°C)$. In addition, the optimum reheater pressure may be determined according to the calculations and parametric study on the reheater pressure (P_5).

Example 5.5. Consider a basic SPP. The steam generator inlet pressure is 120 bar. The steam temperature at the turbine inlet is 540 °C, and the saturated liquid temperature at the pump inlet is 36 °C. The pressure loss due to friction in the steam generator is 10%, while in the condenser, it is 3%. The adiabatic efficiency of the turbine and pump are 0.87 and 0.85, respectively. If you are supposed to design a reheater for this cycle (Fig. 5.12):

(a) find the optimum reheat pressure ratio ($\alpha = \frac{P_4}{P_3}$) of the cycle.
(b) change the turbine inlet temperature from 540 °C to 440 °C, plot the curve $\Delta \eta$, η and x_6 vs. α for every change of 20 °C and discuss the plots.

(c) change the turbine inlet pressure from 80 bar, a subcritical pressure to 240 bar, a supercritical pressure and plot the $\Delta\eta$, η and x_6 vs. α for every change of 40 bar and discuss the plots.

Solution. (a) The working fluid properties at the pump inlet are:

$$T_1 = 36°C, x_1 = 0\% \rightarrow P_1 = 0.05945 \text{ bar}, h_1 = 150.8\frac{\text{kJ}}{\text{kg}}, s_1 = 0.5185\frac{\text{kJ}}{\text{kg.K}}$$

And at the steam generator inlet:

$$P_2 = 120 \text{ bar}, s_{s-2} = s_1 \rightarrow h_{s-2} = 162.8\frac{\text{kJ}}{\text{kg}}$$

$$h_2 = h_1 + \frac{h_{s-2} - h_1}{\eta_{ad,pump}} = 150.8 + \frac{162.8 - 150.8}{0.85} = 164.9\frac{\text{kJ}}{\text{kg}}$$

The working fluid properties at the turbine inlet are:

$$P_3 = 0.9\ P_2 = 0.9 \times 120 = 108 \text{ bar}, T_3 = 540°C \rightarrow s_3 = 6.682\frac{\text{kJ}}{\text{kg.K}}, h_3$$

$$= 3467\frac{\text{kJ}}{\text{kg}}$$

The reheat pressure ratio of $\alpha = \frac{P_{RH}}{P_3}$ can be changed from 0 to 1 numerically and calculate the cycle thermal efficiency for each α. According to the results, the optimum reheat pressure ratio (α) would be found.

For example at $\alpha = 0.2$ we have:

$$P_4 = \alpha P_3 = 0.2 \times 108 = 21.6 \text{ bar}, s_{s-4} = s_3, \rightarrow h_{s-4} = 2995\frac{\text{kJ}}{\text{kg}}$$

$$h_4 = h_3 - \eta_{ad,turbine} \times (h_3 - h_{s-4}) = 3467 - 0.87(3467 - 2995) = 3056\frac{\text{kJ}}{\text{kg}}$$

$$s_4 = 6.789\frac{\text{kJ}}{\text{kg.K}}$$

If no reheat was done, point 4′ would be the exit condition of turbine:

$$P_{4'} = P_1 + 0.03P_1 = 0.06124 \text{ bar}, s_{s-4'} = s_3, \rightarrow h_{s-4'} = 2060\frac{\text{kJ}}{\text{kg}}$$

$$h_{4'} = h_3 - \eta_{ad,turbine} \times (h_3 - h_{s-4'}) = 3467 - 0.87(3467 - 2060) = 2243\frac{\text{kJ}}{\text{kg}}$$

After reheating the steam properties would be:

$$T_5 = 540°C, P_5 = P_4 \rightarrow h_5 = 3555 \frac{kJ}{kg}, s_5 = 7.507 \frac{kJ}{kg.K}$$

And finally at the turbine outlet:

$$P_6 = P_1 + 0.03P_1 = 0.06124 \text{ bar}, s_{s-6} = s_5, \rightarrow h_{s-6} = 2315 \frac{kJ}{kg}$$

$$h_6 = h_5 - \eta_{ad,turbine} \times (h_5 - h_{s-6}) = 3555 - 0.87(3555 - 2315) = 2476 \frac{kJ}{kg}$$

$$s_6 = 6.789 \frac{kJ}{kg.K}$$

The amount of heat addition/rejection to/from the working fluid for two cases of with and without reheat would be:

$$\dot{Q}_{in,with\ RH} = \dot{m}(h_3 - h_2) + \dot{m}(h_5 - h_4) = 1 \times (3467 - 164.9) + 1$$
$$\times (3555 - 3056) = 3801 \text{ kW}$$

$$\dot{Q}_{out,with\ RH} = \dot{m}(h_6 - h_1) = 1 \times (2476 - 150.8) = 2325 \text{ kW}$$

$$\dot{Q}_{in,without\ RH} = \dot{m}(h_3 - h_2) = 1 \times (3467 - 164.9) = 3302 \text{ kW}$$

$$\dot{Q}_{out,without\ RH} = \dot{m}(h_{4'} - h_1) = 1 \times (2243 - 150.8) = 2092 \text{ kW}$$

The thermal efficiency of the cycle with and without a reheater is as below:

$$\eta_{thermal,RH} = 1 - \frac{\dot{Q}_{out}}{\dot{Q}_{in}} = 1 - \frac{2325}{3801} = 0.3883$$

$$\eta_{thermal} = 1 - \frac{\dot{Q}_{out}}{\dot{Q}_{in}} = 1 - \frac{2092}{3302} = 0.3666$$

$$\Delta\eta = (0.3883 - 0.3666) \times 100 = 2.167\%$$

It means that for the reheater pressure ratio of 0.2, the efficiency improvement with respect to the case with no reheat is 2.167%. A code is generated for the problem and α is changed from 0 to 1, and the efficiency change is calculated, and the results are plotted in Fig. 5.13. The results show that for the present cycle, the optimum reheat pressure ratio is 0.1458 and $\Delta\eta = 2.209\%$ for this pressure ratio. An important parameter that care should be taken about is the steam quality at the turbine outlet. It should not be less than 0.88 to avoid severe damage on the turbine blades due to erosion by the water droplets. The plot of steam quality variation with the reheat pressure ratio is also plotted in Fig. 5.13. Very small reheat pressure ($\alpha < 0.02$) results in efficiency reduction.

Code 5.3. The code developed for the SPP simple cycle (cycle No. 1), and simple cycle plus reheater, cycle No. 2 (refer to Table 5.6)

```
eta_adiabatic_turbine=0.87
eta_adiabatic_pump=0.85
P[1]=P_SAT(WATER, T=T[1])
T[1]=36
s[1]=entropy(water, P=P[1], X=0)
h[1]=enthalpy(water, P=P[1], X=0)
P[2]=120
h_2_s=enthalpy(water, P=P[2], s=s_2_s)
s_2_s=s[1]
s[2]=entropy(water, h=h[2], P=P[2])
eta_adiabatic_pump=(h_2_s-h[1])/(h[2]-h[1])
T[2]=TEMPERATURE(water, p=P[2], s=s[2])
T[3]=540
P[3]=P[2]-0.1*P[2]
h[3]=enthalpy(water, T=T[3], P=P[3])
s[3]=entropy(water, T=T[3], P=P[3])
alfa=0.2
P_reheat=alfa*P[3]
P[4]=P_reheat
s_4_s=s[3]
h_4_s=enthalpy(water, P=P[4], s=s_4_s)
eta_adiabatic_turbine=(h[3]-h[4])/(h[3]-h_4_s)
T[4]=TEMPERATURE(water, p=P[4], s=s[4])
s[4]=entropy(water, P=P[4], H=h[4])
T[5]=T[3]
P[5]=P[4]
h[5]=enthalpy(water, T=T[5], P=P[5])
s[5]=entropy(water, T=T[5], P=P[5])
P[6]=P[1]+0.03*P[1]
s_6_s=s[5]
h_6_s=enthalpy(water, P=P[6], s=s_6_s)
s[6]=entropy(water, h=h[6], P=P[6])
eta_adiabatic_turbine=(h[5]-h[6])/(h[5]-h_6_s)
T[6]=TEMPERATURE(water, p=P[6], s=s[6])
s_4_prime_s=s[3]
P_4_prime=P[6]
h_4_prime_s=enthalpy(water, P=P_4_prime, s=s_4_prime_s)
eta_adiabatic_turbine=(h[3]-h_4_prime)/(h[3]-h_4_prime_s)
Q_in=h[3]-h[2]+h[5]-h[4]
Q_out=h[6]-h[1]
Q_in_no_RH=h[3]-h[2]
Q_out_no_RH=h_4_prime-h[1]
eta=1-Q_out/Q_in
eta_no_RH=1-Q_out_no_RH/Q_in_no_RH
DELTA_eta=(eta-eta_no_RH)*100
```

(b) the code that was developed for finding the optimum pressure of reheater in part a) is now used for different turbine inlet temperatures, and results are plotted in Figs. 5.14 and 5.15. According to the graphs, three important results can be concluded:

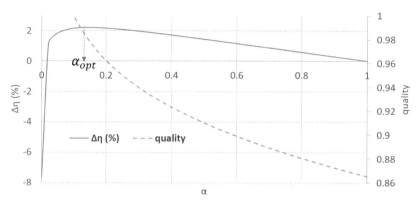

Figure 5.13 The impact of reheat pressure ratio on the efficiency change and steam quality at the turbine outlet.

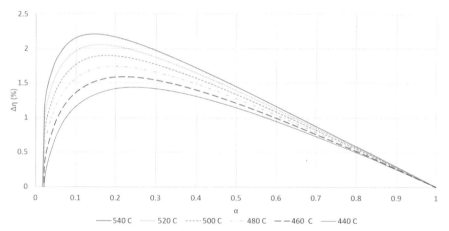

Figure 5.14 The impact of turbine inlet temperature on the efficiency change due to reheating at different reheat pressure ratio.

1. the higher the turbine inlet temperature, the more the efficiency improvement due to reheating
2. the optimum reheat pressure ratio for the lower turbine inlet temperature is bigger. According to the results, the optimum reheat pressure locates between 0.15 and 0.25.
3. for reheat pressure ratio less than 0.02, the cycle efficiency would decrease.

(c) the code that was developed for optimum pressure of a reheater in part a) is now used for different turbine inlet pressures and results are plotted in Fig.5.16 and 5.17. According to the graphs, three important results can be concluded:
1. for $0.02 < \alpha < 0.3$, the lower the turbine inlet pressure, the less the efficiency improvement due to reheating would be achieved. However, it should be reminded that higher working pressures results in higher efficiency even if no reheating is done.
2. the optimum reheat pressure ratio for the lower turbine inlet pressure is smaller. According to the results, the optimum reheat pressure locates between 0.12 and 0.18.
3. for reheat pressure ratio less than 0.02, the cycle efficiency would decrease

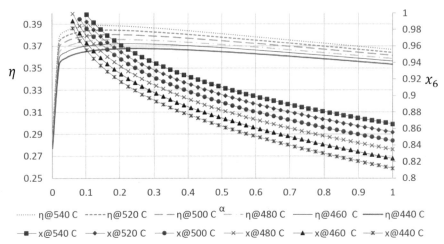

Figure 5.15 The impact of turbine inlet temperature on the efficiency due to reheating at different reheat pressure ratio.

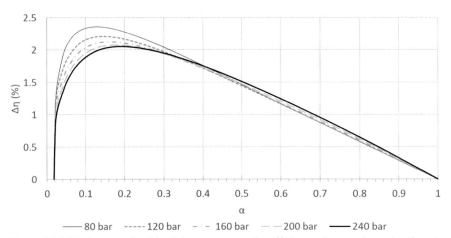

Figure 5.16 The impact of turbine inlet pressure on the efficiency change when reheating at different reheat pressure.

Care must be taken about the steam quality; at the condenser inlet, it should not be less than 0.88.

4.5 Supercritical SPP

When the operating temperature and pressure of an SPP cycle are higher than the critical point, the cycle is supercritical. A super critical SPP cycle has no evaporator or two-phase working fluid because at the critical point, subcooled liquid becomes superheated steam at once. Such cycle instead has a once-through steam generator that after

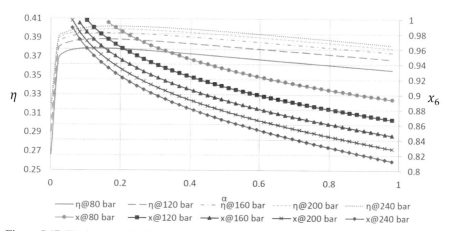

Figure 5.17 The impact of turbine inlet pressure on the efficiency and turbine exit steam quality when reheating at different reheat pressure.

a particular point steam is superheated, while before that point, working fluid is only subcooled liquid. Since the operating pressure is significantly higher, the cycle efficiency is usually more than sub-critical cycles. This is because most of heat is transferred to the working fluid at higher temperature. The supercritical cycles can have a single or multiple reheater. The critical temperature and pressure of water are 374°C and 220.6 bar, respectively. Care must be taken that as the number of reheating increases, its impact on the efficiency improvement would decrease due to diminishing return low. In addition, using more than two reheats complicates the cycle and building such cycle would increase the capital costs as well. Furthermore, due to smaller impact on the efficiency improvement, it may not be economical to use more than two reheat in the supercritical SPP.

Example 5.6. Consider a supercritical SPP cycle with the four main components. The steam generator inlet pressure is 240 bar. The steam temperature at the turbine inlet is 540 °C, and the saturated liquid temperature at the pump inlet is 36 °C. The pressure loss due to friction in the steam generator is 10%, while in the condenser it is 3%. The adiabatic efficiency of the turbine and pump are 0.87 and 0.85, respectively.

(a) find the optimum reheating pressure ratio of $\alpha = \frac{P_4}{P_3}$ for the cycle if single reheating is used.

(b) find the optimum reheating pressure ratios of $\alpha = \frac{P_4}{P_3}$ and $\beta = \frac{P_6}{P_5}$ for the cycle if double reheating is used.

(c) find the optimum reheating pressure ratios of $\alpha = \frac{P_4}{P_3}$, $\beta = \frac{P_6}{P_5}$ and $\gamma = \frac{P_8}{P_7}$ for the cycle if triple reheating is used.

Solution.

(a) the code written for Example 5.5 can be used here, and only the operating pressure should be changed to 240 bar. The optimum reheat pressure ratio can be found by two methods. One just like the Example 5.5, that a parametric table was used and α was changed from 0 to 1, and the optimum α was found for the highest value of $\eta_{thermal,RH}$. The second method is to use the max/min toolbox in the EES; it can be found under the **calculate>min/max** menu. Under this menu, a window would be opened that you should choose the cycle thermal efficiency as the function that should be maximized, and also choose α as the independent variable that its **bounds**, $0 < \alpha < 1$, also should be defined. Leave the algorithm method unchanged.

According to the results, the $\alpha_{optimum} = 0.1912$ and cycle efficiency with and without reheater is 0.402 and 0.3815, respectively, that shows 2.053% efficiency improvement.

The T-s diagram of the cycle with a single reheat is drawn in Fig. 5.18 for the optimum state.

(b) for the case of double reheat, the code used for part a is developed and according to guidance given for using **min/max** toolbox in the EES program; the cycle efficiency is maximized according to two independent variables of $0 < \alpha < 1$ and $0 < \beta < 1$.

The results show that $\alpha_{optimum} = 0.2083$ and $\beta_{optimum} = 0.1613$ and cycle efficiency with double reheat is 0.4176 that shows 1.56% efficiency improvement with respect to the cycle with single reheat. The cycle with double reheat is shown in Fig. 5.19 for the optimum state.

(c) for the case of triple reheat, the code used for part b is developed and according to guidance given for using **min/max** toolbox in the EES program; the cycle efficiency is maximized according to three independent variables of $0 < \alpha < 1$, $0 < \beta < 1$, and $0 < \gamma < 1$.

The results show that $\alpha_{optimum} = 0.2807$, $\beta_{optimum} = 0.2289$, and $\gamma_{optimum} = 0.2164$ and cycle efficiency with triple reheat is 0.4293 that shows 1.17% efficiency improvement with respect to the cycle with double reheat. The cycle with triple reheat is shown in Fig. 5.20 for the optimum state.

The code developed for the supercritical cycle with three reheaters is given in **Code 5.4**.

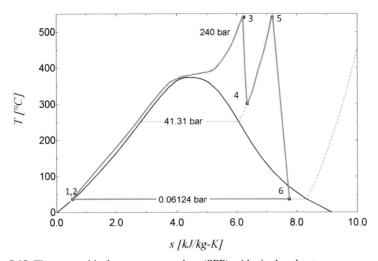

Figure 5.18 The supercritical steam power plant (SPP) with single reheat.

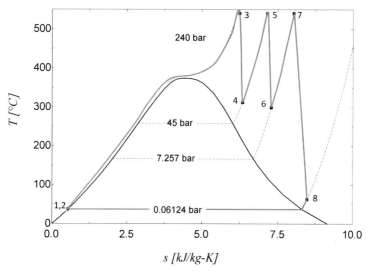

Figure 5.19 The supercritical steam power plant (SPP) with double reheat.

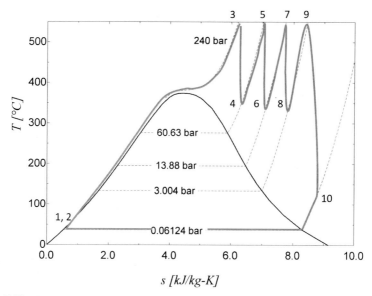

Figure 5.20 The supercritical steam power plant (SPP) with triple reheat.

Code 5.4. The code developed for triple reheater supercritical cycle

```
eta_adiabatic_turbine=0.87
eta_adiabatic_pump=0.85
P[1]=P_SAT(WATER, T=T[1])
T[1]=36
s[1]=entropy(water, P=P[1], X=0)
h[1]=enthalpy(water, P=P[1], X=0)
P[2]=240
h_2_s[1]=enthalpy(water, P=P[2], s=s_2_s[1])
s_2_s[1]=s[1]
s[2]=entropy(water, h=h[2], P=P[2])
eta_adiabatic_pump=(h_2_s[1]-h[1])/(h[2]-h[1])
T[2]=TEMPERATURE(water, p=P[2], s=s[2])
T[3]=540
P[3]=P[2]-0.1*P[2]
h[3]=enthalpy(water, T=T[3], P=P[3])
s[3]=entropy(water, T=T[3], P=P[3])
"alpha=0.2807"
P_reheat1=alpha*P[3]
P[4]=P_reheat1
s_4_s[1]=s[3]
h_4_s[1]=enthalpy(water, P=P[4], s=s_4_s[1])
eta_adiabatic_turbine=(h[3]-h[4])/(h[3]-h_4_s[1])
T[4]=TEMPERATURE(water, p=P[4], s=s[4])
s[4]=entropy(water, P=P[4], H=h[4])
T[5]=T[3]
P[5]=P[4]
h[5]=enthalpy(water, T=T[5], P=P[5])
s[5]=entropy(water, T=T[5], P=P[5])
"beta=0.2289"
P_reheat2=beta*P_reheat1
P[6]=P_reheat2
s_6_s[1]=s[5]
h_6_s[1]=enthalpy(water, P=P[6], s=s_6_s[1])
s[6]=entropy(water, h=h[6], P=P[6])
eta_adiabatic_turbine=(h[5]-h[6])/(h[5]-h_6_s[1])
T[6]=TEMPERATURE(water, p=P[6], s=s[6])
T[7]=T[6]
P[7]=P[6]
h[7]=enthalpy(water, T=T[7], P=P[7])
s[7]=entropy(water, T=T[5], P=P[7])
"gamma=0.2164"
P_reheat3=gamma*P_reheat2
P[8]=P_reheat3
s_8_s[1]=s[7]
```

```
h_8_s[1]=enthalpy(water, P=P[8], s=s_8_s[1])
s[8]=entropy(water, h=h[8], P=P[8])
eta_adiabatic_turbine=(h[7]-h[8])/(h[7]-h_8_s[1])
T[8]=TEMPERATURE(water, p=P[8], s=s[8])
T[9]=T[7]
P[9]=P[8]
h[9]=enthalpy(water, T=T[9], P=P[9])
s[9]=entropy(water, T=T[9], P=P[9])
P[10]=P[1]+0.03*P[1]
s_10_s[1]=s[9]
h_10_s[1]=enthalpy(water, P=P[10], s=s_10_s[1])
s[10]=entropy(water, h=h[10], P=P[10])
eta_adiabatic_turbine=(h[9]-h[10])/(h[9]-h_10_s[1])
T[10]=TEMPERATURE(water, p=P[10], s=s[10])
s_4_prime_s=s[3]
P_4_prime=P[8]
h_4_prime_s=enthalpy(water, P=P_4_prime, s=s_4_prime_s)
eta_adiabatic_turbine=(h[3]-h_4_prime)/(h[3]-h_4_prime_s)
s_6_prime_s=s[5]
P_6_prime=P[8]
h_6_prime_s=enthalpy(water, P=P_6_prime, s=s_6_prime_s)
eta_adiabatic_turbine=(h[5]-h_6_prime)/(h[5]-h_6_prime_s)
s_8_prime_s=s[7]
P_8_prime=P[10]
h_8_prime_s=enthalpy(water, P=P_8_prime, s=s_8_prime_s)
eta_adiabatic_turbine=(h[7]-h_8_prime)/(h[7]-h_8_prime_s)
Q_in_3RH=h[3]-h[2]+h[5]-h[4]+h[7]-h[6]+h[9]-h[8]
Q_out_3RH=h[10]-h[1]
Q_in_2RH=h[3]-h[2]+h[5]-h[4]+h[7]-h[6]
Q_out_2RH=h_8_prime-h[1]
Q_in_1RH=h[3]-h[2]+h[5]-h[4]
Q_out_1RH=h_6_prime-h[1]
Q_in_no_RH=h[3]-h[2]
Q_out_no_RH=h_4_prime-h[1]
eta_3RH=1-Q_out_3RH/Q_in_3RH
eta_2RH=1-Q_out_2RH/Q_in_2RH
eta_1RH=1-Q_out_1RH/Q_in_1RH
eta_no_RH=1-Q_out_no_RH/Q_in_no_RH
```

Table 5.4 The optimum reheat pressure ratio and efficiency for supercritical SPP.

	No reheat $n = 0$	Single reheat $n = 1$	Double reheat $n = 2$	Triple reheat $n = 3$
η_{nRH}	38.15%	40.20%	41.76%	42.93%
Optimum reheater pressure ratio	N.A.	$\alpha = 0.1912$	$\alpha = 0.2083$ $\beta = 0.1613$	$\alpha = 0.2807$ $\beta = 0.2289$ $\gamma = 0.2164$
$\Delta\eta = \eta_{nRH} - \eta_{(n-1)RH}$	N.A.	2.05%	1.56%	1.17%

Discussions. The results are briefly reported in Table 5.4; it shows that as the number of reheating steps increases its impact on efficiency improvement ($\Delta\eta$) decreases. The reason is that although the temperature is risen to its initial value of 540°C at each step, but the pressure is lower. It is clear that since pressure cannot be increases during reheating process, declining of efficiency improvement is inevitable. Using single or multi-reheating system depends on the economical evaluations. If the extra cost due to using the additional reheater is comparable with the economic benefits, more reheating is not recommended. Usually, if the payback period due to additional reheater is less than 5 years, it is welcome, otherwise it is not.

4.6   Feedwater heating

Feedwater heating refers to heating of working fluid in the SPP before entering the steam generator and after condenser by using steam extraction from turbine in different stages. More clearly, FW heating should be done between the condenser and the economizer. FW heaters are heat exchangers that use superheated or saturated steam bled from turbine for FW heating. FW heaters can be open or close. An open-FW heater is a direct contact heat exchanger in which the hot and cold streams are in direct contact and get mixed together. The condensate leaving the open-FW heater is saturated liquid. An open-FW is a multi-task heat exchanger that is used for deaerating of dissolved gases from FW (open-FW heater is sometimes called deaerator as well), providing available net positive suction head (*NPSHa*) for the FW pump, water treating by injecting chemicals to adjust potential hydrogen (PH), and sometimes water control level by injecting make-up water into it.

Fig. 5.21 shows an open-FW heater. FW is initially heated by the hot dissolved gases on its way out of deaerator and then is sprayed from top while steam is diffused from below to increase the heating surface and releasing of dissolved gases such as oxygen, carbon dioxide, etc. as well.

An open-FW is usually placed on a high altitude to provide the *NPSHa* for the pump to avoid cavitation[9] at the pump inlet. In the pump selection, the *NPSHa* must be bigger than the required net positive suction head (*NPSHr*) of the pump. *NPSHa* is the system

---

[9] If the suction pressure of pump drops below the saturation pressure of water at the pump inlet, water starts to boil and produces steam bubbles that can cause damage on the pump casing and impeller due to collapsing of bubbles, producing very high velocity micro jets of water and imping the pump inlet casing and impeller. This phenomenon is called cavitation that produces noisy sound on the pump, pressure and flow rate fluctuation at the pump discharge, and vibration of the rotor. The noise of cavitation is produced due to micro jet impingement on the pump casing and impeller. The noise sounds like marbles going through the pump casing. Cavitation can erode the pump and as a result the cavitation noise gets intensified more as the surfaces become more rough. $NPSHa = h_{static} - h_{loss} - \frac{P_{sat}}{\rho g}$. The *NPSHa* depends on the static head (h_{static}), head loss (h_{loss}), and saturation pressure (P_{sat}) at the pump suction line. To increase *NPSHa* the static head can be increased by placing the open-FW heater as much as higher than the pump inlet. In addition, by using bigger pipe diameter and less fittings at the suction line the head loss can be decrease. The only way to decrease the saturation pressure is cooling of the suction line. Cooling of the suction line needs energy consumption for the entire life of plant. Hence this method is rarely used to avoid cavitation while increasing static head and decreasing head loss is more economical and practical.

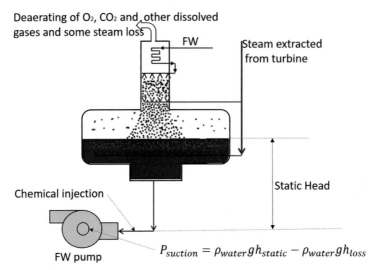

Figure 5.21 An example of open-FW heater (deaerator).

characteristics on the suction line, while *NPSHr* is the pump characteristics and is provided by the manufacturer and usually is reported in the product catalogs. Since a pump is required after each open-FW heater; therefore, to decrease the number of pumps, maintenance jobs, and electricity consumption, only one open-FW heater is usually installed in power plants and the rest of FW heaters are close type.

The main purpose of water treatment is neutralizing acid-forming slats in the feedwater, removing scales automatically, preventing scale forming, precipitating water contaminants into a sludge for removing through the blowdown, and avoiding corrosion due to oxidation.

If the PH decreases, the water becomes acidic and corrosion possibility increases as well. As a result of corrosion, the boiler tubes may be perforated. To avoid this problem, the PH should be measured periodically and be kept between 7.5 and 8.5. The mostly used chemicals to adjust PH and removing oxygen are sodium sulfite (Na_2SO_3) and hydrazine (N_2H_4). In addition, it should be well understood that for PH more than 8.5, the chance for scaling will increase.

Fig. 5.22A shows the placement of the open-FW heater, and the red dashed line in Fig. 5.22B shows the share of the open-FW heater in heating the FW. In this cycle, the pumping process is exaggerated to show the pumping process on the diagram; it should be well understood that the temperature difference due to pumping is very small, and the power consumption of pump is usually negligible in comparison with the power production of turbine or the inlet heat rate to the steam generator. For instance, pumping from 0.06 bar to 120 bar in the Example 5.5 made only 0.86°C temperature rise. In addition, the power consumption by the pump for this example is only about 1% of the turbine power production and less than 0.5% of the inlet heat rate to the cycle.

According to Fig. 5.22A the total steam mass flow rate is $1 \frac{kg}{s}$ and the extraction mass flow rate is m, the rest of steam $(1-m)$ keep passing through the turbine until

(A)

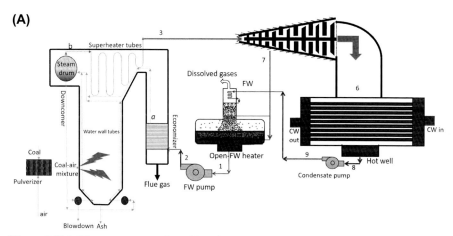

Figure 5.22A The steam power plant (SPP) simple cycle with an open-FW heater (cycle No. 3).

(B)

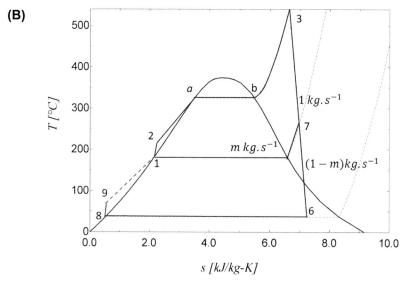

$s\ [kJ/kg\text{-}K]$

Figure 5.22B The T-s diagram of steam power plant (SPP) simple cycle with one open-FW heater.

it enters the condenser. If the open-FW heater is well insulated, its heat balance can be written as below:

$$1 \times h_1 = m \times h_7 + (1-m) \times h_9 \rightarrow m = \frac{h_1 - h_9}{h_7 - h_9}$$

According to the discussions given about the power consumption of pump and its impact on the temperature and enthalpy, the bled steam mass flow rate would be as Eq. (5.30):

$$m = \frac{h_1 - h_9}{h_7 - h_9} \cong \frac{h_1 - h_9 + w_{pump}}{h_7 - h_9 + w_{pump}} = \frac{h_1 - h_8}{h_7 - h_8} \tag{5.30}$$

A close-FW heater is a shell and tube heat exchanger that its sole duty is heat transfer from the extracted steam to the FW. Close-FW heaters may be fed with steam in the shell side from the high pressure, intermediate pressure, and low pressure stages of turbine. The FW flows inside the tube bundles. In some power plants, the number of FW heaters may exceed 10. The close-FW heater is categorized as *drain forward* and *drain backward*. In the backward type, the condensate of the bled steam is returned to the previous FW heater and flashes into it as two-phase flow (Fig. 5.23A and 5.23B). In this type, the fluid flow does not need pump and the pressure difference between the FW heaters is enough to run the condensate. Depending on the extraction pressure, temperature, and quality, the condensate leaving the heater (point 9, or 9' in Fig. 5.23B) may be saturated or subcooled liquid and the FW temperature leaving the heater at point 3 may be higher than the saturation temperature of the extraction line. The process from 9 to 10 is a flashing or throttling process, and enthalpy remains the same in this process. However, the part of liquid that is flashed into steam at point 10 should be converted to the saturated liquid at point 1 that means a bigger and more expensive condenser. In addition, due to higher load of condenser the heat loss of cycle would increase. In the backward close-FW heater a steam trap is required to be installed to avoid steam flow in the backward line.

Steam traps are equipment that let liquid flow to pass through only, and steam is trapped behind until it gets condensed. Steam traps are categorized as mechanical,

(A)

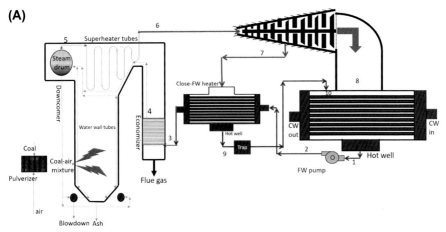

Figure 5.23A The steam power plant (SPP) simple cycle with a drain backward close-FW heater.

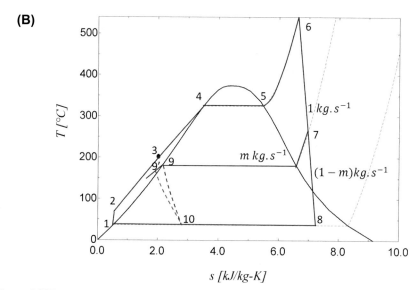

Figure 5.23B The T-s diagram of steam power plant (SPP) simple cycle with a drain backward close-FW heater.

thermostatic, and thermodynamic. The mechanical traps exist in different types of ball float, float and lever, inverted bucket, open bucket, and float-thermostatic. The thermostatic traps also exist in bellows and bi-metal types. The thermodynamic traps exist in disc and piston types. To show the operation of steam traps, a simple float and lever type is shown in Fig. 5.24. As the figure shows when the condensate level increases and Archimedes force (buoyancy force) overcomes the weight of ball, the ball floats and the plug, which is pinned to the lever, separates from its seat and lets the condensate leaves the trap.

In the drain forward type however (Fig. 5.25A), the condensate of the bled steam is pumped to the FW main line leaving the corresponding close-FW heater. In this case, a small pump is required, this pump consumes power, adds some initial cost and also yearly operation and maintenance costs. The pump is small because it is only responsible for pumping the condensate of the corresponding extraction that is a fraction of the total mass flow rate. The condensate and FW leaving the heater (points 9 and 3) are mixed together later, and the final temperature at point 11 falls between the temperature at points 10 and 3. The pressure at point 10 is the same as point 3. In addition, according to the extraction pressure, temperature, and steam quality, the condensate leaving the heater (point 9 or 9′) may be saturated or subcooled liquid (Fig. 5.25B). The FW temperature leaving the heater at point 3 may be above or below the saturation temperature of the extraction line depending on the state of point 7, if it is superheated or saturated, respectively. However, the temperature at point 3 is always below the extraction temperature at point 7.

According to the previous discussions utilizing a backward close-FW heater corresponds to:

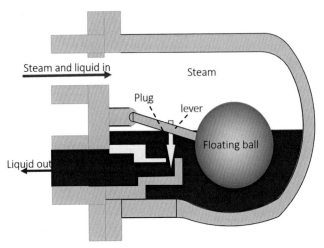

Figure 5.24 Mechanical trap, simple float and lever type.

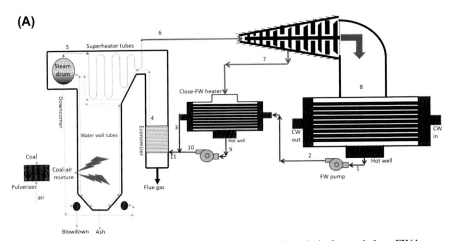

Figure 5.25A The steam power plant (SPP) simple cycle with a drain forward close-FW heater.

- Using a steam trap, steam traps are a source of steam waste in industry, they increase maintenance jobs due to having moving parts in most of trap types, in addition an initial cost should be paid for buying the steam trap.
- A bigger condenser is required to condense the additional steam generated due to flashing the backward stream.
- More heat is lost by the condenser.

And using a forward close-FW heater corresponds to:

- Using a small pump to balance the pressure between the FW line and condensate line.
- This pump consumes power that can have insignificant impact on the net power output.

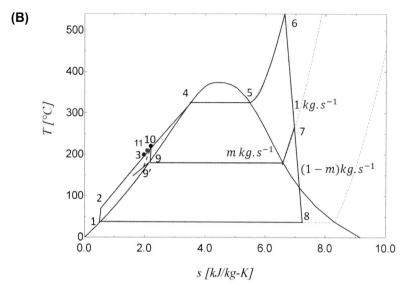

Figure 5.25B The T-s diagram of steam power plant (SPP) simple cycle with a drain forward close-FW heater.

- The pump increases the maintenance jobs due to having moving parts. The operation and maintenance costs and initial cost of a pump is more than a steam trap.

Piping costs exist in both forward and backward close heaters, but its magnitude depends on the installations and placement of components. The comparison should be done according to the existing conditions in the site.

Choosing among forward and backward close-FW heaters depends on their impact on the thermal efficiency of the cycle, operational costs, initial costs, limitations for installing components in the plant site, etc.

Too much mean temperature difference between the hot and cold streams in the close-FW heater means cheaper heat exchanger, more exergy loss, and less cycle thermal efficiency, while too small mean temperature difference means more heating surface, a bigger and more expensive heater.

The difference between the bled steam saturation temperature ($T_{extraction,sat}$) and FW temperature leaving the heater ($T_{FW,out}$) is called *terminal temperature difference, TTD*. Hence:

$$TTD = T_{extraction,sat} - T_{FW,out} \tag{5.31}$$

If steam is saturated at the extraction point, the *TTD* which is used for designing the FW heater is 5°F or 2.7 °C. But if it is superheated at the extraction point, the *TTD* which is used for designing the FW heater may vary from 0°C to −2.7 °C.

An FW heater may include three sections: condenser, desuperheater, and drain cooler as shown in Fig. 5.26. The desuperheater is the part of heater where superheated steam loses its heat and becomes saturated steam. The desuperheater exists only if the bled

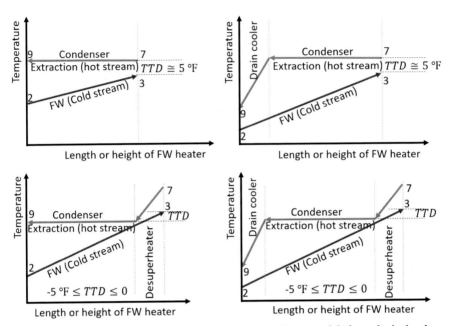

Figure 5.26 Different configurations of condenser, desuperheater, and drain cooler in the close-FW heaters, temperature variation of bled steam, and FW with respect to distance.

steam is superheated. Condenser is the section of the heater that converts saturated steam to saturated liquid. This section exists in all of the FW heaters. The drain cooler section exists if the condensate temperature is still high, and its heat can be used for heating the FW. This happens at high or intermediate pressure extractions. Four possible configurations for these three sections in the close-FW heaters are shown in Fig. 5.26.

Now that the FW-heaters, their types, and operation are discussed, it is time to explain why using an FW-heater would increase the thermal efficiency of the SPP cycle. According to the cycles presented in Figs. 5.22–5.25, using an FW-heater may decrease the net output power of the cycle (due to extraction of steam), but it will decrease the required inlet heat and heat loss in the condenser as well significantly.

According to the Carnot principal, the higher the heating temperature (T_H), the higher the efficiency of the cycle would be. Comparing the $T - s$ diagram presented in Fig. 5.25B for example with the case without FW-heater, it reveals that the mean temperature of heat addition from point 11 to 6 is higher than the average temperature of heat addition from 2 to 6. Simply, it means achieving higher efficiency due to using a single FW heater.

Now that efficiency improvement due to using FW heater is evident, an important question arises about the optimum placement and optimum number of FW heaters in an SPP cycle. In the following, the answer of these two questions is investigated in details. To start answering the two questions, solving some examples will help to understand the topic more deeply.

Example 5.7. Consider a simple SPP cycle in which the steam generator inlet pressure is 120 bar. Steam temperature at the turbine inlet is 540 °C, and the saturated liquid temperature at the condenser output is 36°C. The pressure loss due to friction and other losses in the steam generator is 10%, while in the condenser, it is 3%. The adiabatic efficiency of the turbine and pump are 0.87 and 0.85, respectively. If you are supposed to equip this cycle with an open-FW heater at the extraction pressure of P_7, determine:

(a) If $P_7 = 10$ bar how much the cycle efficiency would increase in comparison with the cycle with no FW heater?
(b) What would be the optimum extraction pressure of P_7 for the open-FW heater to maximize the cycle efficiency (use EES program toolboxes)
(c) According to the Example 5.5, if simple cycle is equipped with a reheater, the optimum reheater pressure would be $\alpha = 0.1458$. Calculate the thermal efficiency if the simple cycle with single optimized reheater could be equipped an open-FW heater at 10 bar (Fig. 5.27)?
(d) What is the optimum extraction pressure of P_7 if the reheater pressure ratio stay the same at $\alpha = 0.1458$? (single variable optimization)
(e) Optimize the cycle efficiency with respect to reheater pressure ratio and extraction pressure (two variable optimization)?
(f) Determine the TTD for the cycle with one open-FW heater.

Solution. (a) According to Fig. 5.22A and B, the working fluid properties at the condensate pump inlet are:

$$T_8 = 36°C, x_8 = 0\% \rightarrow P_8 = 0.05945 \text{ bar}, h_8 = 150.8 \frac{kJ}{kg}, s_8 = 0.5185 \frac{kJ}{kg.K}$$

And at the condensate pump outlet:

$$P_9 = P_7 = 10 \text{ bar}, s_{s-9} = s_8 \rightarrow h_{s-9} = 151.8 \frac{kJ}{kg}$$

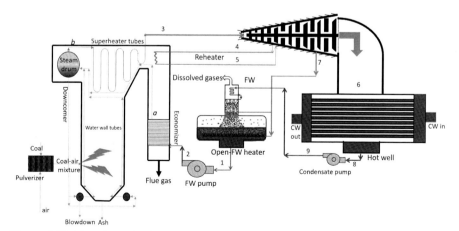

Figure 5.27 The schematic of the power plant with single reheater and an open-FW heater (cycle No. 4).

$$h_9 = h_8 + \frac{h_{s-9} - h_8}{\eta_{ad,pump}} = 150.8 + \frac{151.8 - 150.8}{0.85} = 152 \frac{kJ}{kg}$$

At the FW-pump inlet:

$$P_1 = P_7 = 10 \text{ bar}, x_1 = 0\% \rightarrow h_1 = 762.8 \frac{kJ}{kg}, s_1 = 2.139 \frac{kJ}{kg.K}$$

At the FW-pump outlet:

$$P_2 = 120 \text{ bar}, s_{s-2} = s_1 \rightarrow h_{s-2} = 775.2 \frac{kJ}{kg}$$

$$h_2 = h_1 + \frac{h_{s-2} - h_1}{\eta_{ad,pump}} = 762.8 + \frac{775.2 - 762.8}{0.85} = 777.3 \frac{kJ}{kg}$$

At the steam generator outlet:

$$P_3 = 0.9 \, P_2 = 0.9 \times 120 = 108 \text{ bar}, T_3 = 540°C \rightarrow s_3 = 6.682 \frac{kJ}{kg.K}, h_3$$

$$= 3467 \frac{kJ}{kg}$$

And finally at the turbine outlet:

$$P_6 = P_8 + 0.03P_8 = 0.06124 \text{ bar}, s_{s-6} = s_3, \rightarrow h_{s-6} = 2060 \frac{kJ}{kg}$$

$$h_6 = h_5 - \eta_{ad,turbine} \times (h_5 - h_{s-6}) = 3551 - 0.87(3551 - 2060) = 2243 \frac{kJ}{kg}$$

$$s_6 = 7.273 \frac{kJ}{kg.K}, T_6 = 36.54°C$$

$$P_7 = 10 \text{ bar}, s_{s-7} = s_3, \rightarrow h_{s-7} = 2822 \frac{kJ}{kg}$$

$$h_7 = h_3 - \eta_{ad,turbine} \times (h_3 - h_{s-7}) = 3467 - 0.87(3467 - 2822) = 2906 \frac{kJ}{kg}$$

$$s_7 = 6.854 \frac{kJ}{kg.K}$$

The heat balance in the open-FW heater is as below:

$$mh_7 + (1-m)h_9 = 1 \times h_1 \rightarrow m = \frac{h_1 - h_9}{h_7 - h_9} = \frac{762.8 - 152}{2906 - 152} = 0.2218 \frac{kg}{s}$$

The amount of heat addition, and heat rejection to/from the working fluid would be:

$$\dot{Q}_{in} = 1 \times (h_3 - h_2) = 1 \times (3467 - 777.3) = 2690 \text{ kW}$$

$$\dot{Q}_{out} = (1-m)(h_6 - h_8) = (1 - 0.2218)(2243 - 150.8) = 1628 \text{ kW}$$

The thermal efficiency of the cycle with single open-FW heater is as below:

$$\eta_{thermal} = 1 - \frac{\dot{Q}_{out}}{\dot{Q}_{in}} = 1 - \frac{1628}{2690} = 39.49\%$$

Comparing to Example 5.5 part a, the thermal efficiency has increased from 36.66% to 39.49% that shows 2.83% efficiency improvement due to using an open-FW heater with the steam turbine simple cycle.

Code 5.5. The code written for the simple cycle SPP plus an open-FW heater (cycle No. 3)

```
eta_adiabatic_turbine=0.87                    T[3]=540
eta_adiabatic_pump=0.85                       P[3]=P[2]-0.1*P[2]
T[8]=36                                       h[3]=enthalpy(water, T=T[3], P=P[3])
P[8]=P_SAT(WATER, T=T[8])                     s[3]=entropy(water, T=T[3], P=P[3])
s[8]=entropy(water, P=P[8], X=0)              P[6]=P[8]+0.03*P[8]
h[8]=enthalpy(water, P=P[8], X=0)             s_6_s[1]=s[3]
P[9]=P[7]                                     h_6_s[1]=enthalpy(water, P=P[6], s=s_6_s[1])
h_9_s[1]=enthalpy(water, P=P[9], s=s_9_s[1])  s[6]=entropy(water, h=h[6], P=P[6])
s_9_s[1]=s[8]                                 eta_adiabatic_turbine=(h[3]-h[6])/(h[3]-h_6_s[1])
s[9]=entropy(water, h=h[9], P=P[9])           T[6]=TEMPERATURE(water, p=P[6], s=s[6])
eta_adiabatic_pump=(h_9_s[1]-h[8])/(h[9]-h[8]) h_7_s[1]=enthalpy(water, P=P[7], s=s_7_s[1])
T[9]=TEMPERATURE(water, p=P[9], s=s[9])       S_7_s[1]=S[3]
P[7]=10                                       eta_adiabatic_turbine=(h[3]-h[7])/(h[3]-h_7_s[1])
P[1]=P[7]                                      T[7]=TEMPERATURE(water, p=P[7], H=H[7])
h[1]=enthalpy(water, P=P[1], X=0)              s[7]=entropy(water, T=T[7], P=P[7])
s[1]=entropy(water, P=P[1], X=0)               m*h[7]+(1-m)*h[9]=h[1]
T[1]=TEMPERATURE(water, p=P[1], s=s[1])        Q_in=h[3]-h[2]
P[2]=120                                       Q_in_without_fw[1]=h[3]-h[2]
s_2_s[1]=s[1]                                  Q_out=(1-m)*(h[6]-h[8])
h_2_s[1]=enthalpy(water, P=P[2], s=s_2_s[1])   Q_out_without_fw[1]=(h[6]-h[8]))
eta_adiabatic_pump=(h_2_s[1]-h[1])/(h[2]-h[1])          eta=1-Q_out/Q_in
T[2]=TEMPERATURE(water, p=P[2], h=h[2])
```

(b) In this section, single variable optimization is requested. Hence, the thermal efficiency should be maximized with respect to P_7. It can be done by using a **parametric table**, in the EES program. For this purpose, the **Code 5.5** is used, and P_7 is changed from 1 bar to 119 bar. According to the results of the parametric table, Fig. 5.28 is depicted. It shows that the optimum extraction pressure for the open-FW heater is $P_7 = 9.5$ bar, and the optimum efficiency for this pressure is $\eta_{thermal} = 39.5\%$.

The $T - s$ diagram of the cycle with one optimum FW-heater is shown in Fig. 5.29. Taking a close look at this figure, it reveals that the optimum FW-heater is placed at the middle temperature between the boiler and condenser saturation temperatures. The calculations also confirm this conclusion as well:

$$T_{sat,boiler} = 324.7°C$$

$$T_{sat,condenser} = 36.54°C$$

$$T_{middle} = \frac{T_{sat,boiler} + T_{sat,condenser}}{2} = \frac{324.7 + 36.54}{2} = 180.62°C$$

And:

$$T_{sat,FW} = 179.9°C$$

Hence:

$$T_{middle} \cong T_{sat,FW}$$

This means that half of the temperature elevation of FW to reach point a (economizer outlet) at the $T - s$ diagram is provided by the open-FW heater and the other half is provided by the economizer. In other words, the heat input per kg of FW, before entering the steam drum, is roughly divided equally between the economizer and the FW heater.

$$h_1 - h_9 = 601.2 \frac{kJ}{kg}$$

$$h_a - h_2 = 673.7 \frac{kJ}{kg}$$

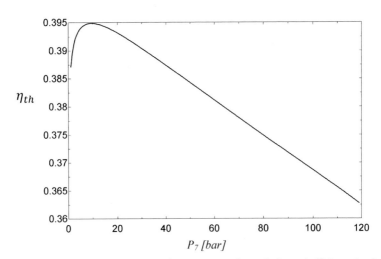

Figure 5.28 The impact of FW-heater extraction pressure on the cycle thermal efficiency (cycle No. 3).

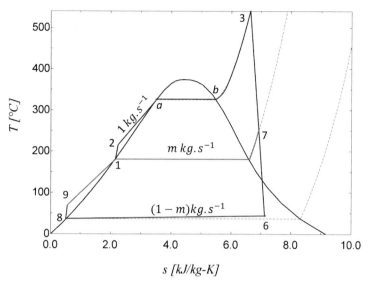

Figure 5.29 The T-s diagram of steam power plant (SPP) with an optimum open FW-heater.

(c) in this part, a simple cycle with an optimum reheater (from Example 5.5 part a) is supposed to be equipped with an open-FW heater receiving steam from turbine at $P_7 = 10$ bar. The reheater pressure ratio is kept unchanged.

The working fluid properties at the condensate pump inlet are:

$$T_8 = 36°C, \ x_8 = 0\% \rightarrow P_8 = 0.05945 \text{ bar}, h_8 = 150.8 \frac{kJ}{kg}, s_8 = 0.5185 \frac{kJ}{kg.K}$$

The working fluid properties at condensate pump exit are:

$$P_9 = P_7 = 10 \text{ bar}, s_{s-9} = s_8 \rightarrow h_{s-9} = 151.8 \frac{kJ}{kg}$$

$$h_9 = h_8 + \frac{h_{s-9} - h_8}{\eta_{ad,pump}} = 150.8 + \frac{151.8 - 150.8}{0.85} = 152 \frac{kJ}{kg}$$

At the FW-pump inlet:

$$P_1 = P_7 = 10 \text{ bar}, x_1 = 0\% \rightarrow h_1 = 762.8 \frac{kJ}{kg}, s_1 = 2.139 \frac{kJ}{kg.K}$$

At the FW-pump exit:

$$P_2 = 120 \text{ bar}, s_{s-2} = s_1 \rightarrow h_{s-2} = 775.2 \frac{kJ}{kg}$$

$$h_2 = h_1 + \frac{h_{s-2} - h_1}{\eta_{ad,pump}} = 762.8 + \frac{775.2 - 762.8}{0.85} = 777.3 \frac{kJ}{kg}$$

At the steam generator outlet:

$$P_3 = 0.9, P_2 = 0.9 \times 120 = 108 \text{ bar}, T_3 = 540°C \rightarrow s_3 = 6.682 \frac{kJ}{kg.K}, h_3 = 3467 \frac{kJ}{kg}$$

$$\alpha = \frac{P_{RH}}{P_3} = 0.1458$$

At the reheater inlet:

$$P_4 = \alpha P_3 = 0.1458 \times 108 = 24.98 \text{ bar}, s_{s-4} = s_3, \rightarrow h_{s-4} = 3031 \frac{kJ}{kg}$$

$$h_4 = h_3 - \eta_{ad,turbine} \times (h_3 - h_{s-4}) = 3467 - 0.87(3467 - 3031) = 3087 \frac{kJ}{kg}$$

$$s_4 = 6.777 \frac{kJ}{kg.K}$$

After reheating, the steam properties would be:

$$T_5 = 540°C, P_5 = P_4 \rightarrow h_5 = 3551 \frac{kJ}{kg}, s_5 = 7.436 \frac{kJ}{kg.K}$$

And finally at the turbine outlet:

$$P_6 = P_8 + 0.03P_8 = 0.06124 \text{ bar}, s_{s-6} = s_5, \rightarrow h_{s-6} = 2293 \frac{kJ}{kg}$$

$$h_6 = h_5 - \eta_{ad,turbine} \times (h_5 - h_{s-6}) = 3551 - 0.87(3551 - 2293) = 2457 \frac{kJ}{kg}$$

$$s_6 = 7.965 \frac{kJ}{kg.K}, T_6 = 36.54°C$$

At the extraction point:

$$P_7 = 10 \text{ bar}, s_{s-7} = s_5, \rightarrow h_{s-7} = 3076 \frac{kJ}{kg}$$

$$h_7 = h_5 - \eta_{ad,turbine} \times (h_5 - h_{s-7}) = 3551 - 0.87(3551 - 3076) = 3138 \frac{kJ}{kg}$$

$$s_7 = 7.54 \frac{kJ}{kg.K}$$

The heat balance for the open-FW heater is as below:

$$mh_7 + (1-m)h_9 = 1 \times h_1 \rightarrow m = \frac{h_1 - h_9}{h_7 - h_9} = \frac{762.8 - 152}{3138 - 152} = 0.1869 \frac{kg}{s}$$

The amount of heat addition and heat rejection to/from the working fluid would be:

$$\dot{Q}_{in,with\ FW} = 1 \times (h_3 - h_2) + 1 \times (h_5 - h_4)$$
$$= 1 \times (3467 - 777.3) + 1 \times (3551 - 3087) = 3259\ kW$$

$$\dot{Q}_{out,with\ FW} = (1-m)(h_6 - h_8) = (1 - 0.1869)(2457 - 150.8) = 1924\ kW$$

The thermal efficiency of the cycle with reheater and open-FW heater is as below:

$$\eta_{thermal} = 1 - \frac{\dot{Q}_{out}}{\dot{Q}_{in}} = 1 - \frac{1924}{3259} = 40.95\%$$

The cycle efficiency with one optimum reheater and no open-FW heater was 38.87%, by using an FW-heater, the efficiency is improved to 40.95% that shows 2.081% improvement. It should be mentioned that in this case, the heater placement is not optimized. The code written for the simple cycle with an open-FW heater and a reheater is given in **Code 5.6**, and the T-s diagram of the cycle is shown in Fig. 5.30.

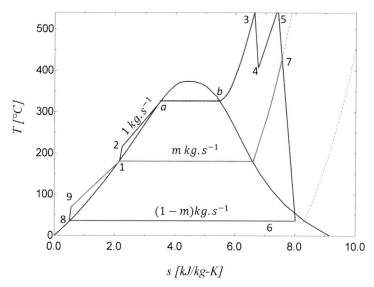

Figure 5.30 The T-S diagram of the cycle No. 4.

Code 5.6. The code written for the simple cycle SPP plus an open-FW heater and reheater (cycle No. 4).

```
eta_adiabatic_turbine=0.87                          P_4[1]=P_reheat
eta_adiabatic_pump=0.85                             s_4_s[1]=s_3[1]
T_8[1]=36                                           h_4_s[1]=enthalpy(water, P=P_4[1], s=s_4_s[1])
P_8[1]=P_SAT(WATER, T=T_8[1])                       eta_adiabatic_turbine=(h_3[1]-h_4[1])/(h_3[1]-h_4_s[1])
s_8[1]=entropy(water, P=P_8[1], X=0)                T_4[1]=TEMPERATURE(water, p=P_4[1], s=s_4[1])
h_8[1]=enthalpy(water, P=P_8[1], X=0)               s_4[1]=entropy(water, P=P_4[1], H=h_4[1])
P_9[1]=P_7[1]                                        T_5[1]=T_3[1]
h_9_s[1]=enthalpy(water, P=P_9[1], s=s_9_s[1])      P_5[1]=P_4[1]
s_9_s[1]=s_8[1]                                      h_5[1]=enthalpy(water, T=T_5[1], P=P_5[1])
s_9[1]=entropy(water, h=h_9[1], P=P_9[1])           s_5[1]=entropy(water, T=T_5[1], P=P_5[1])
eta_adiabatic_pump=(h_9_s[1]-h_8[1])/(h_9[1]-h_8[1]) P_6[1]=P_8[1]+0.03*P_8[1]
T_9[1]=TEMPERATURE(water, p=P_9[1], s=s_9[1])       s_6[1]=s_5[1]
"P_7[1]=10"                                          h_6_s[1]=enthalpy(water, P=P_6[1], s=s_6_s[1])
P_1[1]=P_7[1]                                        s_6[1]=entropy(water, h=h_6[1], P=P_6[1])
h_1[1]=enthalpy(water, P=P_1[1], X=0)               eta_adiabatic_turbine=(h_5[1]-h_6[1])/(h_5[1]-h_6_s[1])
s_1[1]=entropy(water, P=P_1[1], X=0)                T_6[1]=TEMPERATURE(water, p=P_6[1], s=s_6[1])
P_2[1]=120                                           h_7_s[1]=enthalpy(water, P=P_7[1], s=s_7_s[1])
s_2_s[1]=s_1[1]                                      S_7_s[1]=S_5[1]
h_2_s[1]=enthalpy(water, P=P_2[1], s=s_2_s[1])      eta_adiabatic_turbine=(h_5[1]-h_7[1])/(h_5[1]-h_7_s[1])
eta_adiabatic_pump=(h_2_s[1]-h_1[1])/(h_2[1]-h_1[1]) T_7[1]=TEMPERATURE(water, p=P_7[1], H=H_7[1])
T_3[1]=540                                           s_7[1]=entropy(water, T=T_7[1], P=P_7[1])
P_3[1]=P_2[1]-0.1*P_2[1]                             m*h_7[1]+(1-m)*h_9[1]=h_1[1]
h_3[1]=enthalpy(water, T=T_3[1], P=P_3[1])          Q_in[1]=h_3[1]-h_2[1]+h_5[1]-h_4[1]
s_3[1]=entropy(water, T=T_3[1], P=P_3[1])           Q_out[1]=(1-m)*(h_6[1]-h_8[1])
alfa=0.1458                                          eta=1-Q_out[1]/Q_in[1]
P_reheat=alfa*P_3[1]
```

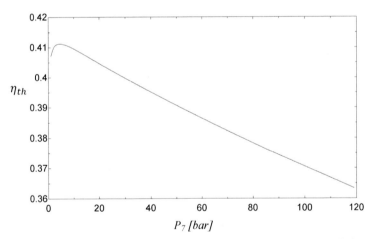

Figure 5.31 The impact of FW-heater extraction pressure on the cycle thermal efficiency (cycle No. 4).

(d) in this section, single variable optimization is requested. It can be done by using a **parametric table** or **min/max** toolbox in the EES program for the **Code 5.6**. For this purpose, according to the guidelines given for Example 5.6, the P_7 is changed from 1 bar to 119 bar. According to the parametric table, Fig. 5.31 is depicted. It shows that the optimum pressure for open-FW heater is $P_7 = 4.476$ bar, and optimum efficiency for this case is $\eta_{thermal} = 41.12\%$.

Comparing the optimum extraction pressure for the cycle with and without reheater shows that the presence of a reheater will change the optimum extraction point. It should be mentioned that although the reheater was previously optimized, by adding the open-FW heater, it is not optimum anymore and the cycle should be optimized again with respect to both reheater pressure ratio and the extraction pressure. This is done in part *e* of the present example.

(e) in this section, double-variable optimization is requested. It can be done by using **min/max** toolbox in the EES program for the **Code 5.6**. For this purpose, the P_7 is changed from 1 bar to 119 bar and $0.05 \leq \alpha \leq 1$. The optimization shows that for $\alpha = 0.2311$ and $P_7 = 5.52$ bar, the cycle efficiency reaches its maximum, which is $\eta_{thermal} = 41.23\%$. Comparing the results with part a, b, and c, it shows that the combined presence of the reheater and FW-heater will change the optimum pressure for both of the reheater and FW-heater.

(f) The *TTD*, in this case, is the temperature difference between the saturation temperature of extraction $T_{sat} = 179.9°C$ and $T_1 = 179.9°C$. Hence *TTD* $= 0$.

Example 5.8. Consider a simple SPP cycle in which the steam generator inlet pressure is 120 bar. The steam temperature at the turbine inlet is 540 °C, and the saturated liquid temperature at the condenser output is 36 °C. The pressure loss in the steam generator is 10%, while in the condenser, it is 3%. The adiabatic efficiency of the turbine and pump is 0.87 and 0.85, respectively. Consider the *TTD* $= -3$ °C for the close-FW heater.

(a) Equip this cycle with a close-FW heater and an open-FW heater (Figs. 5.32 and 5.33) at extraction pressures of 30 bar, and 20 bar, respectively, and determine the cycle efficiency.
(b) Optimize the heaters placement in the cycle to maximize the cycle efficiency.
(c) To avoid erosion and corrosion on the turbine blades steam quality at the turbine outlet is obliged to stay above 0.88. Check this criterion and propose a solution if the steam quality is low.

Solution.

(a) The working fluid properties at the FW pump inlet are:

$$P_1 = P_{10} = 20 \text{ bar}, x_1 = 0\% \rightarrow T_1 = 212.4°C, h_1 = 908.6\frac{kJ}{kg}, s_1 = 2.447\frac{kJ}{kg.K}$$

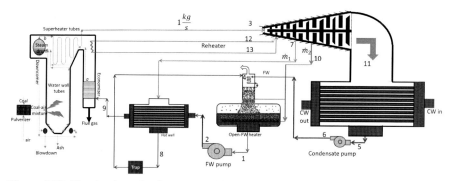

Figure 5.32 The simple cycle plus an open-FW heater and a close-FW heater (cycle No. 5).

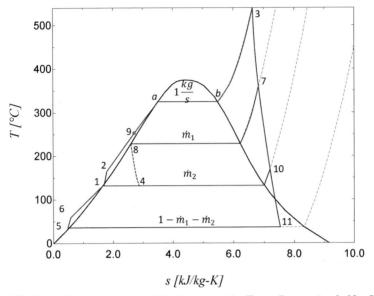

Figure 5.33 The optimum placement of the heaters on the $T - s$ diagram (cycle No. 5).

The working fluid properties at FW pump exit are:

$$P_2 = 120 \text{ bar}, s_{s-2} = s_1 \rightarrow h_{s-2} = 920.3 \frac{\text{kJ}}{\text{kg}}$$

$$h_2 = h_1 + \frac{h_{s-2} - h_1}{\eta_{ad,pump}} = 908.6 + \frac{920.3 - 908.6}{0.85} = 922.4 \frac{\text{kJ}}{\text{kg}}$$

$$T_2 = 214.7°\text{C}$$

At the turbine inlet:

$$P_3 = 108 \text{ bar}, T_3 = 540°\text{C} \rightarrow h_3 = 3467 \frac{\text{kJ}}{\text{kg}}, s_3 = 6.682 \frac{\text{kJ}}{\text{kg.K}}$$

In the open-FW heater:

$$P_4 = P_{10} = 20 \text{ bar}, h_4 = h_8 = 1008 \frac{\text{kJ}}{\text{kg}}, T_4 = 212.4°\text{C}$$

At the condenser outlet:

$$T_5 = 36°\text{C}, x_5 = 0 \rightarrow P_5 = 0.05945 \text{ bar}, s_5 = 0.5185 \frac{\text{kJ}}{\text{kg.K}}, h_5 = 150.8 \frac{\text{kJ}}{\text{kg}}$$

The working fluid properties at condenser pump exit are:

$$P_6 = P_{10} = 20 \text{ bar}, s_{s-6} = s_5 \rightarrow h_{s-6} = 152.8 \frac{kJ}{kg}$$

$$h_6 = h_5 + \frac{h_{s-6} - h_5}{\eta_{ad,pump}} = 150.8 + \frac{152.8 - 150.8}{0.85} = 153.1 \frac{kJ}{kg}$$

$$T_6 = 36.14°C$$

At the inlet of the close-FW heater:

$$P_7 = 30 \text{ bar}, s_{s-7} = s_3, \rightarrow h_{s-7} = 3078 \frac{kJ}{kg}$$

$$h_7 = h_3 - \eta_{ad,turbine} \times (h_3 - h_{s-7}) = 3467 - 0.87(3467 - 3078) = 3128 \frac{kJ}{kg}$$

$$s_7 = 6.764 \frac{kJ}{kg.K}$$

The bled steam at the outlet of the close-FW heater:

$$P_8 = P_7 = 30 \text{ bar}, x_8 = 0 \rightarrow s_8 = 2.645 \frac{kJ}{kg.K}, h_8 = 1008 \frac{kJ}{kg}, T_8 = 233.9°C$$

The FW at the outlet of the close-FW heater:

$$P_9 = P_2 = 120 \text{ bar}, T_9 = T_8 - TTD = 233.9 + 3 = 236.9°C \rightarrow h_9 = 1024 \frac{kJ}{kg}$$

The energy balance for the close-FW heater:

$$\dot{m}_1(h_7 - h_8) = h_9 - h_2 \rightarrow \dot{m}_1 = \frac{h_9 - h_2}{h_7 - h_8} = \frac{1024 - 922.4}{3128 - 1008} = 0.04783 \frac{kg}{s}$$

At the inlet of the open-FW heater:

$$P_{10} = 20 \text{ bar}, s_{s-10} = s_3, \rightarrow h_{s-10} = 2976 \frac{kJ}{kg}$$

$$h_{10} = h_3 - \eta_{ad,turbine} \times (h_3 - h_{s-10}) = 3467 - 0.87(3467 - 2976) = 3040 \frac{kJ}{kg}$$

$$s_{10} = 6.795 \frac{kJ}{kg.K}$$

The energy balance for the open-FW heater:

$$\dot{m}_1 h_4 + \dot{m}_2 h_{10} + \left(1 - \dot{m}_1 - \dot{m}_2\right) h_6 = 1 \times h_1$$

$$\rightarrow \dot{m}_2 = \frac{h_1 - h_6 + \dot{m}_1(h_6 - h_4)}{h_{10} - h_6} = \frac{908.6 - 153.1 + 0.04783(153.1 - 1008)}{3040 - 153.1}$$

$$= 0.2476 \frac{kg}{s}$$

At the turbine exit:

$$P_{11} = P_5 + 0.03 P_5 = 0.06124 \text{ bar}, s_{s-11} = s_3, \rightarrow h_{s-11} = 2060 \frac{kJ}{kg},$$

$$h_{11} = h_3 - \eta_{ad,turbine} \times (h_3 - h_{s-11}) = 3467 - 0.87(3467 - 2060) = 2243 \frac{kJ}{kg}$$

$$T_{11} = 36.54°C$$

The amount of heat addition, and heat rejection to/from the working fluid would be:

$$\dot{Q}_{in} = 1 \times (h_3 - h_9) = 1 \times (3467 - 1024) = 2444 \text{ kW}$$

$$\dot{Q}_{out} = \left(1 - \dot{m}_1 - \dot{m}_2\right)(h_{11} - h_5) = (1 - 0.04783 - 0.2476)(2243 - 150.8)$$

$$= 1474 \text{ kW}$$

The thermal efficiency of the cycle with a close-FW heater, and an open-FW heater is as below:

$$\eta_{thermal} = 1 - \frac{\dot{Q}_{out}}{\dot{Q}_{in}} = 1 - \frac{1474}{2444} = 39.68\%$$

Code 5.7. The code written for the simple cycle plus an open-FW heater and close-FW heater (cycle No. 5).

```
eta_adiabatic_turbine=0.87                          s[7]=entropy(water, h=h[7], P=P[7])
eta_adiabatic_pump=0.85                             eta_adiabatic_turbine=(h[3]-h[7])/(h[3]-h_7_s)
P[1]=P[10]                                          T[7]=TEMPERATURE(water, p=P[7], s=s[7])
T[1]=T_SAT(WATER, P=P[1])                           P[8]=P[7]
s[1]=entropy(water, P=P[1], X=0)                    T[8]=T_SAT(WATER, P=P[8])
h[1]=enthalpy(water, P=P[1], X=0)                   s[8]=entropy(water, P=P[8], X=0)
P[2]=120                                            h[8]=enthalpy(water, P=P[8], X=0)
s_2_s=s[1]                                          P[9]=P[2]
h_2_s=enthalpy(water, P=P[2], s=s_2_s)              T[9]=T[8]+3
eta_adiabatic_pump=(h_2_s-h[1])/(h[2]-h[1])         h[9]=enthalpy(water, T=T[9], P=P[9])
T[2]=TEMPERATURE(water, p=P[2], h=h[2])             m_1*(h[7]-h[8])=h[9]-h[2]
T[3]=540                                            "P[10]=20"
P[3]=P[9]-0.1*P[9]                                  s_10_s=s[3]
h[3]=enthalpy(water, T=T[3], P=P[3])                h_10_s=enthalpy(water, P=P[10], s=s_10_s)
s[3]=entropy(water, T=T[3], P=P[3])                 s[10]=entropy(water, h=h[10], P=P[10])
P[4]=P[10]                                          eta_adiabatic_turbine=(h[3]-h[10])/(h[3]-h_10_s)
T[4]=T_SAT(WATER, P=P[4])                           T[10]=TEMPERATURE(water, p=P[10], s=s[10])
h[4]=h[8]                                           x_10=quality(water, h=h[10], P=P[10])
s[4]=entropy(water, h=h[4], P=P[4])                 m_1*h[4]+m_2*h[10]+(1-m_1-m_2)*h[6]=h[1]
x_4=quality(water, h=h[4], P=P[4])                  P[11]=P[5]+0.03*P[5]
T[5]=36                                             s_11_s=s[3]
P[5]=P_SAT(WATER, T=T[5])                           h_11_s=enthalpy(water, P=P[11], s=s_11_s)
s[5]=entropy(water, P=P[5], X=0)                    s[11]=entropy(water, h=h[11], P=P[11])
h[5]=enthalpy(water, P=P[5], X=0)                   eta_adiabatic_turbine=(h[3]-h[11])/(h[3]-h_11_s)
P[6]=P[10]                                          T[11]=TEMPERATURE(water, p=P[11], s=s[11])
s_6_s=s[5]                                          x_11=quality(water, h=h[11], P=P[11])
h_6_s=enthalpy(water, P=P[6], s=s_6_s)              Q_in=h[3]-h[9]
s[6]=entropy(water, h=h[6], P=P[6])                 Q_out=(1-m_1-m_2)*(h[11]-h[5])
eta_adiabatic_pump=(h_6_s-h[5])/(h[6]-h[5])         eta=1-Q_out/Q_in
T[6]=TEMPERATURE(water, p=P[6], s=s[6])
"P[7]=30"
s_7_s=s[3]
h_7_s=enthalpy(water, P=P[7], s=s_7_s)
```

(b) optimization of this cycle requires to find the possible extraction pressures for both of the FW heaters. For this purpose, the EES toolbox is utilized by defining the **bounds** for P_7 and P_{10} from 0.06 bar to 120 bar (**Code 5.7**). The optimization shows that in the optimum conditions:

$$P_7 = 17.93 \text{ bar}, P_{10} = 3.543 \text{ bar}$$

$$\eta_{thermal,optimum} = 40.42\%$$

In comparison with the case with only one open-FW heater, the optimum efficiency has increases by $\Delta\eta = 40.42 - 39.49 = 0.93\%$.

(c) To check this problem, the steam quality at point 11 should be found. According to the data in part a:

$$T_{11} = 36.54°C, h_{11} = 2243 \frac{kJ}{kg} \rightarrow x_{11} = 0.8655$$

Since the steam quality is lower than 0.88, two possible solutions can be proposed.

1. More superheating at the turbine inlet. This solution should be used cautiously because of the metallurgical limitation of the turbine blades and superheater tubes. It can be checked thermodynamically if how much superheating should be done to increase x_{11} to above 0.88 and then consult the turbine manufacturer about the new inlet turbine temperature if it is allowed to increase the superheating temperature or not. Here, the **code 5.7** which was used in part b is

Table 5.5 The impact of superheating on the steam quality at the turbine outlet.

Cycle No.	Superheating temperature at turbine inlet	Optimum efficiency	Steam quality at turbine outlet	Independent variables in the optimum state
5	$T_3 = 540°C$	$\eta_{thermal} = 40.42\%$	$x_{11} = 0.8655$	$P_7 = 17.93$ bar $P_{10} = 3.543$ bar
	$T_3 = 545°C$	$\eta_{thermal} = 40.48\%$	$x_{11} = 0.868$	$P_7 = 17.91$ bar $P_{10} = 3.54$ bar
	$T_3 = 550°C$	$\eta_{thermal} = 40.55\%$	$x_{11} = 0.8704$	$P_7 = 17.91$ bar $P_{10} = 3.541$ bar
	$T_3 = 555$ °C	$\eta_{thermal} = 40.61\%$	$x_{11} = 0.8728$	$P_7 = 18.3$ bar $P_{10} = 3.628$ bar
	$T_3 = 560$ °C	$\eta_{thermal} = 40.68\%$	$x_{11} = 0.8751$	$P_7 = 17.99$ bar $P_{10} = 3.57$ bar
	$T_3 = 565°C$	$\eta_{thermal} = 40.68\%$	$x_{11} = 0.8775$	$P_7 = 28.47$ bar $P_{10} = 6.03$ bar
	$T_3 = 570°C$	$\eta_{thermal} = 40.81\%$	$x_{11} = 0.8798$	$P_7 = 18.32$ bar $P_{10} = 3.642$ bar
	$T_3 = 575°C$	$\eta_{thermal} = 40.87\%$	$x_{11} = 0.8822$	$P_7 = 18.36$ bar $P_{10} = 3.655$ bar
	$T_3 = 580$ °C	$\eta_{thermal} = 40.94\%$	$x_{11} = 0.8845$	$P_7 = 18.4$ bar $P_{10} = 3.668$ bar

used again for new maximum temperatures ranging from $T_3 = 540$ °C to $580°C$. The results are reported in Table 5.5. It shows that to reach $x_{11} > 0.88$ the inlet temperature should be $T_3 > 575°C$, which means 35°C higher than the temperature, which was initially used. It can be seen that efficiency improvement due to 35°C extra superheating is 0.45%.

2. Using a reheater can be the second solution. To try his suggestion, a reheater is used in the cycle, and optimization is done according to three independent variables of P_7, P_{10} and reheater pressure ratio. The new cycle and its $T - s$ diagram are shown in Figs. 5.34 and 5.35.

In the new cycle with the reheater, only the data for some state points will change, and points of 12 and 13 are new. Hence, the calculation of thermodynamic properties for these state points should be updated according to the following equations:

At the reheater extraction line, for an initial assumption of $\alpha = 0.2$.

$$\alpha = \frac{P_{12}}{P_3} \rightarrow P_{12} = 0.2 \times 108 = 21.6 \text{ bar}, s_{s-12} = s_3, \rightarrow h_{s-12} = 2995 \frac{kJ}{kg}$$

$$h_{12} = h_3 - \eta_{ad,turbine} \times (h_3 - h_{s-12}) = 3467 - 0.87(3467 - 2995) = 3056 \frac{kJ}{kg}$$

$$s_{12} = 6.789 \frac{kJ}{kg.K}$$

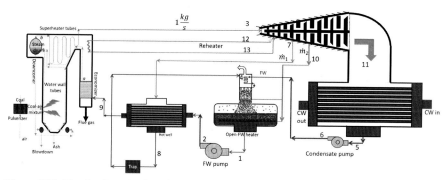

Figure 5.34 The simple cycle plus reheater, an open-FW heater and a close-FW heater (cycle No. 6).

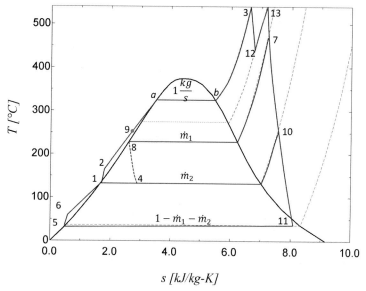

$$s \ [kJ/kg\text{-}K]$$

Figure 5.35 The $T - s$ diagram of the cycle No. 6.

After reheating of steam, its properties will be:

$$P_{13} = P_{12}, T_{13} = T_3 \rightarrow h_{13} = 3555 \frac{kJ}{kg}, s_{13} = 7.507 \frac{kJ}{kg.K}$$

$$P_7 = 30 \text{ bar}, s_{s-7} = s_{13}, \rightarrow h_{s-7} = 3681 \frac{kJ}{kg}$$

$$h_7 = h_{13} - \eta_{ad,turbine} \times (h_{13} - h_{s-7}) = 3555 - 0.87 \times (3555 - 3681) = 3664 \frac{kJ}{kg}$$

$$P_{10} = 20 \text{ bar}, s_{s-10} = s_{13}, \rightarrow h_{s-10} = 3526 \frac{\text{kJ}}{\text{kg}}$$

$$h_{10} = h_{13} - \eta_{ad,turbine} \times (h_{13} - h_{s-10}) = 3555 - 0.87 \times (3555 - 3526)$$
$$= 3530 \frac{\text{kJ}}{\text{kg}}$$

$$P_{11} = P_5 + 0.03 P_5 = 0.06124 \text{ bar}, s_{s-11} = s_{13}, \rightarrow h_{s-11} = 2315 \frac{\text{kJ}}{\text{kg}}$$

$$h_{11} = h_{13} - \eta_{ad,turbine} \times (h_{13} - h_{s-11}) = 3555 - 0.87 \times (3555 - 2315)$$
$$= 2476 \frac{\text{kJ}}{\text{kg}}$$

The energy balance for the close-FW heater:

$$\dot{m}_1 (h_7 - h_8) = h_9 - h_2 \rightarrow \dot{m}_1 = \frac{h_9 - h_2}{h_7 - h_8} = \frac{1024 - 922.4}{3664 - 1008} = 0.03818 \frac{\text{kg}}{\text{s}}$$

The energy balance for the open-FW heater:

$$\dot{m}_1 h_4 + \dot{m}_2 h_{10} + \left(1 - \dot{m}_1 - \dot{m}_2\right) h_6 = 1 \times h_1$$

$$\rightarrow \dot{m}_2 = \frac{h_1 - h_6 + \dot{m}_1 (h_6 - h_4)}{h_{10} - h_6} = \frac{908.6 - 153.1 + 0.03818(153.1 - 1008)}{3530 - 153.1}$$
$$= 0.2141 \frac{\text{kg}}{\text{s}}$$

The input heat, lost heat, and the cycle efficiency would become:

$$\dot{Q}_{in} = 1 \times (h_3 - h_2) + 1 \times (h_{13} - h_{12}) = 1 \times (3467 - 922.4) + 1$$
$$\times (3555 - 3056)$$
$$= 2942 \text{ kW}$$

$$\dot{Q}_{out} = \left(1 - \dot{m}_1 - \dot{m}_2\right)(h_{11} - h_5) = (1 - 0.03818 - 0.2141)(2476 - 150.8)$$
$$= 1739 \text{ kW}$$

$$\eta_{thermal} = 1 - \frac{\dot{Q}_{out}}{\dot{Q}_{in}} = 1 - \frac{1739}{2942} = 40.90\%$$

In comparison with part a, adding the reheater has increased the cycle efficiency by $\Delta\eta = 40.90 - 39.68 = 1.22\%$, while it has solved the problem of low-quality steam at the turbine outlet. The steam quality at point 11 is now $x_{11} = 0.9623$, which is very safe. It worth mentioning that optimizing the cycle will result in higher efficiency. For this purpose by using, the EES code (**Code 5.8**), the cycle is optimized with respect to P_7, P_{10}, and α. It was found that for $P_7 = 14.86$ bar, $P_{10} = 2.869$ bar, and $\alpha = 0.2642$, the cycle efficiency would be:

$$\eta_{thermal} = 42.05\%$$

Code 5.8. The code written for the cycle No. 6.

```
eta_adiabatic_turbine=0.87
eta_adiabatic_pump=0.85
P[1]=P[10]
T[1]=T_SAT(WATER, P=P[1])
s[1]=entropy(water, P=P[1], X=0)
h[1]=enthalpy(water, P=P[1], X=0)
P[2]=120
s_2_s=s[1]
h_2_s=enthalpy(water, P=P[2], s=s_2_s)
eta_adiabatic_pump=(h_2_s-h[1])/(h[2]-h[1])
T[2]=TEMPERATURE(water, p=P[2], h=h[2])
T[3]=540
P[3]=P[9]-0.1*P[9]
h[3]=enthalpy(water, T=T[3], P=P[3])
s[3]=entropy(water, T=T[3], P=P[3])
P[4]=P[10]
T[4]=T_SAT(WATER, P=P[4])
h[4]=h[8]
s[4]=entropy(water, h=h[4], P=P[4])
x_4=quality(water, h=h[4], P=P[4])
T[5]=36
P[5]=P_SAT(WATER, T=T[5])
s[5]=entropy(water, P=P[5], X=0)
h[5]=enthalpy(water, P=P[5], X=0)
P[6]=P[10]
s_6_s=s[5]
h_6_s=enthalpy(water, P=P[6], s=s_6_s)
s[6]=entropy(water, h=h[6], P=P[6])
eta_adiabatic_pump=(h_6_s-h[5])/(h[6]-h[5])
T[6]=TEMPERATURE(water, p=P[6], s=s[6])
"P[7]=30"
s_7_s=s[13]
h_7_s=enthalpy(water, P=P[7], s=s_7_s)
s[7]=entropy(water, h=h[7], P=P[7])
eta_adiabatic_turbine=(h[13]-h[7])/(h[13]-h_7_s)
T[7]=TEMPERATURE(water, p=P[7], s=s[7])

P[8]=P[7]
T[8]=T_SAT(WATER, P=P[8])
s[8]=entropy(water, P=P[8], X=0)
h[8]=enthalpy(water, P=P[8], X=0)
P[9]=P[2]
T[9]=T[8]+3
h[9]=enthalpy(water, T=T[9], P=P[9])
m_1*(h[7]-h[8])=h[9]-h[2]
"P[10]=20"
s_10_s=s[13]
h_10_s=enthalpy(water, P=P[10], s=s_10_s)
s[10]=entropy(water, h=h[10], P=P[10])
eta_adiabatic_turbine=(h[13]-h[10])/(h[13]-h_10_s)
T[10]=TEMPERATURE(water, p=P[10], s=s[10])
x_10=quality(water, h=h[10], P=P[10])
m_1*h[4]+m_2*h[10]+(1-m_1-m_2)*h[6]=h[1]
P[11]=P[5]+0.03*P[5]
s_11_s=s[13]
h_11_s=enthalpy(water, P=P[11], s=s_11_s)
s[11]=entropy(water, h=h[11], P=P[11])
eta_adiabatic_turbine=(h[13]-h[11])/(h[13]-h_11_s)
T[11]=TEMPERATURE(water, p=P[11], s=s[11])
x_11=quality(water, h=h[11], P=P[11])
"alfa=0.2"
alfa=P[12]/P[3]
s_12_s=s[3]
h_12_s=enthalpy(water, P=P[12], s=s_12_s)
s[12]=entropy(water, h=h[12], P=P[12])
eta_adiabatic_turbine=(h[3]-h[12])/(h[3]-h_12_s)
T[12]=TEMPERATURE(water, p=P[12], s=s[12])
P[13]=P[12]
T[13]=T[3]
h[13]=enthalpy(water, T=T[13], P=P[13])
s[13]=entropy(water, T=T[13], P=P[13])
Q_in=h[3]-h[9]+h[13]-h[12]
Q_out=(1-m_1-m_2)*(h[11]-h[5])
eta=1-Q_out/Q_in
```

Example 5.9. Consider a simple SPP cycle in which the steam generator inlet pressure is 120 bar. The steam temperature at the turbine inlet is 540 °C, and the saturated liquid temperature at the condenser output is 36°C. The total pressure loss in the steam generator is 10%, while in the condenser, it is 3%. The adiabatic efficiency of the turbine and pump is 0.87 and 0.85, respectively. Consider the $TTD = -3$ °C for the high pressure close-FW heater, and $TTD = 2.7$°C for the low pressure close-FW heater.

(a) Equip this cycle with two close-FW heaters and an open-FW heater at extraction pressures of 40 bar, 30 bar, and 20 bar, respectively, and determine the cycle efficiency (Fig. 5.36).
(b) Optimize the heaters placement in the cycle to maximize the cycle efficiency. In addition, correct the magnitude of TDD given for the part a if needed.

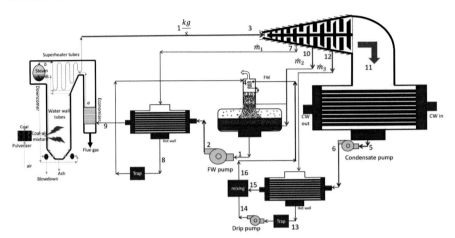

Figure 5.36 The simple cycle plus an open-FW heater and two close-FW heaters (cycle No. 7).

(c) To avoid erosion and corrosion on the turbine internal components, steam quality at the turbine outlet is obliged to stay above 0.88. Check this criterion and propose a solution if the steam quality is lower than 0.88.

Solution. The working fluid properties at the FW pump inlet are:

$$P_1 = P_{10} = 30 \text{ bar}, x_1 = 0\% \rightarrow h_1 = 1008\frac{kJ}{kg}, s_1 = 2.645\frac{kJ}{kg.K}, T_1 = 233.9°C$$

And at the FW pump outlet:

$$P_2 = 120 \text{ bar}, s_{s-2} = s_1 \rightarrow h_{s-2} = 1019\frac{kJ}{kg}$$

$$h_2 = h_1 + \frac{h_{s-2} - h_1}{\eta_{ad,pump}} = 1008 + \frac{1019 - 1008}{0.85} = 1021\frac{kJ}{kg}$$

$$T_2 = 236.3°C$$

At the turbine inlet:

$$P_3 = 0.9P_9 = 108 \text{ bar}, T_3 = 540°C \rightarrow h_3 = 3467\frac{kJ}{kg}, s_3 = 6.682\frac{kJ}{kg.K}$$

In the open FW heater:

$$P_4 = P_{10} = 30 \text{ bar}, h_4 = h_8, T_4 = 233.9°C$$

At the condenser outlet:

$$T_5 = 36°C, x_5 = 0 \rightarrow P_5 = 0.05945 \text{ bar}, \ s_5 = 0.5185 \frac{kJ}{kg.K}, h_5 = 150.8 \frac{kJ}{kg}$$

The working fluid properties at the condensate pump outlet:

$$P_6 = P_{10} = 30 \text{ bar}, s_{s-6} = s_5 \rightarrow h_{s-6} = 153.8 \frac{kJ}{kg}$$

$$h_6 = h_5 + \frac{h_{s-6} - h_5}{\eta_{ad,pump}} = 150.8 + \frac{153.8 - 150.8}{0.85} = 154.3 \frac{kJ}{kg}$$

$$T_6 = 36.21°C$$

At the inlet of the first close-FW heater:

$$P_7 = 40 \text{ bar}, s_{s-7} = s_3, \rightarrow h_{s-7} = 3156 \frac{kJ}{kg}$$

$$h_7 = h_3 - \eta_{ad,turbine} \times (h_3 - h_{s-7}) = 3467 - 0.87(3467 - 3156) = 3196 \frac{kJ}{kg}$$

$$s_7 = 6.744 \frac{kJ}{kg.K}, T_7 = 392.9°C$$

The bled steam at the outlet of the high pressure close-FW heater:

$$P_8 = P_7 = 40 \text{ bar}, x_8 = 0 \rightarrow s_8 = 2.796 \frac{kJ}{kg.K}, h_8 = 1087 \frac{kJ}{kg}$$

$$T_8 = 250.4°C$$

The FW at the outlet of the high pressure close-FW heater:

$$P_9 = P_2 = 120 \text{ bar}, T_9 = T_8 - TTD = 250.4 + 3 = 253.4°C \rightarrow h_9 = 1102 \frac{kJ}{kg}$$

The energy balance for the high pressure close-FW heater:

$$\dot{m}_1 (h_7 - h_8) = h_9 - h_2 \rightarrow \dot{m}_1 = \frac{h_9 - h_2}{h_7 - h_8} = \frac{1102 - 1021}{3196 - 1087} = 0.03819 \frac{kg}{s}$$

At the inlet of the open-FW heater:

$$P_{10} = 30 \text{ bar}, s_{s-10} = s_3, \rightarrow h_{s-10} = 3078 \frac{\text{kJ}}{\text{kg}}$$

$$h_{10} = h_3 - \eta_{ad,turbine} \times (h_3 - h_{s-10}) = 3467 - 0.87(3467 - 3078) = 3128 \frac{\text{kJ}}{\text{kg}}$$

$$s_{10} = 6.764 \frac{\text{kJ}}{\text{kg.K}}, T_{10} = 355.8°C$$

The energy balance for the open-FW heater:

$$\dot{m}_1 h_4 + \dot{m}_2 h_{10} + \left(1 - \dot{m}_1 - \dot{m}_2\right) h_{16} = 1 \times h_1$$

$$\rightarrow \dot{m}_2 = \frac{h_1 - h_{16} + \dot{m}_1(h_{16} - h_4)}{h_{10} - h_{16}} = \frac{1008 - 900.1 + 0.03819(900.1 - 1087)}{3128 - 900.1}$$

$$= 0.04531 \frac{\text{kg}}{\text{s}}$$

At the turbine exit:

$$P_{11} = P_5 + 0.03P_5 = 0.06124 \text{ bar}, s_{s-11} = s_3, \rightarrow h_{s-11} = 2060 \frac{\text{kJ}}{\text{kg}}$$

$$h_{11} = h_3 - \eta_{ad,turbine} \times (h_3 - h_{s-11}) = 3467 - 0.87(3467 - 2060) = 2243 \frac{\text{kJ}}{\text{kg}}$$

$$s_{11} = 7.273 \frac{\text{kJ}}{\text{kg.K}}, T_{11} = 36.54°C$$

At the inlet of the low pressure close-FW heater:

$$P_{12} = 20 \text{ bar}, s_{s-12} = s_3, \rightarrow h_{s-12} = 2976 \frac{\text{kJ}}{\text{kg}}$$

$$h_{12} = h_3 - \eta_{ad,turbine} \times (h_3 - h_{s-12}) = 3467 - 0.87(3467 - 2976) = 3040 \frac{\text{kJ}}{\text{kg}}$$

$$s_{12} = 6.795 \frac{\text{kJ}}{\text{kg.K}}, T_{12} = 307.4°C,$$

The bled steam at the outlet of the low pressure close-FW heater:

$$P_{13} = P_{12} = 20 \text{ bar}, x_{13} = 0 \rightarrow s_{13} = 2.447 \frac{kJ}{kg.K}, h_{13} = 908.6 \frac{kJ}{kg}$$

$$T_{13} = 212.4°C$$

The working fluid properties at drip pump outlet:

$$P_{14} = P_{10} = 30 \text{ bar}, s_{s-14} = s_{13} \rightarrow h_{s-14} = 909.8 \frac{kJ}{kg}$$

$$h_{14} = h_{13} + \frac{h_{s-14} - h_{13}}{\eta_{ad,pump}} = 908.6 + \frac{909.8 - 908.6}{0.85} = 910 \frac{kJ}{kg}$$

$$T_{14} = 212.6°C, s_{14} = 2.447 \frac{kJ}{kg.K}$$

The energy balance for the low pressure close-FW hater:

$$\left(1 - \dot{m}_1 - \dot{m}_2 - \dot{m}_3\right)(h_{15} - h_6) = \dot{m}_3(h_{12} - h_{13})$$

$$\rightarrow \dot{m}_3 = \frac{\left(1 - \dot{m}_1 - \dot{m}_2\right)(h_{15} - h_6)}{h_{12} - h_{13} + h_{15} - h_6} = \frac{(1 - 0.03819 - 0.04531)(896.7 - 154.3)}{3040 - 908.6 + 896.7 - 154.3}$$

$$\rightarrow \dot{m}_3 = 0.2368 \frac{kg}{s}$$

At point 15, the condensate properties would be:

$$P_{15} = P_{10} = 30 \text{ bar}, T_{15} = T_{13} - TTD = 212.4 - 2.7 = 209.7 \rightarrow s_{15} = 2.42 \frac{kJ}{kg.K},$$

$$h_{15} = 896.7 \frac{kJ}{kg}$$

FW pressure leaving the mixing tank would be $P_{16} = P_{10}$, and the energy balance can be written as:

$$\left(1 - \dot{m}_1 - \dot{m}_2 - \dot{m}_3\right)h_{15} + \dot{m}_3 h_{14} = \left(1 - \dot{m}_1 - \dot{m}_2\right)h_{16}$$

$$\rightarrow h_{16} = \frac{(1 - 0.03819 - 0.04531 - 0.2368)896.7 + 0.2368 \times 910}{(1 - 0.03819 - 0.04531)} = 900.1 \frac{kJ}{kg}$$

The amount of heat addition, and heat rejection to/from the working fluid would be:

$$\dot{Q}_{in} = 1 \times (h_3 - h_9) = 1 \times (3467 - 1102) = 2366 \text{ kW}$$

$$\dot{Q}_{out} = \left(1 - \dot{m}_1 - \dot{m}_2 - \dot{m}_3\right)(h_{11} - h_5) = (1 - 0.03819 - 0.04531 - 0.2368)$$

$$(2243 - 150.8) = 1422 \text{ kW}$$

The thermal efficiency of the cycle with two close-FW heaters, and an open-FW heater is as below:

$$\eta_{thermal} = 1 - \frac{\dot{Q}_{out}}{\dot{Q}_{in}} = 1 - \frac{1422}{2366} = 39.9\%$$

Code 5.9. The code written for cycle No. 7.

```
eta_adiabatic_turbine=0.87
eta_adiabatic_pump=0.85
P[1]=P[10]
T[1]=T_SAT(WATER, P=P[1])
s[1]=entropy(water, P=P[1], X=0)
h[1]=enthalpy(water, P=P[1], X=0)
P[2]=120
s_2_s=s[1]
h_2_s=enthalpy(water, P=P[2], s=s_2_s)
eta_adiabatic_pump=(h_2_s-h[1])/(h[2]-h[1])
T[2]=TEMPERATURE(water, p=P[2], h=h[2])
T[3]=540
P[3]=P[9]-0.1*P[9]
h[3]=enthalpy(water, T=T[3], P=P[3])
s[3]=entropy(water, T=T[3], P=P[3])
P[4]=P[10]
T[4]=T_SAT(WATER, P=P[4])
h[4]=h[8]
s[4]=entropy(water, h=h[4], P=P[4])
x_4=quality(water, h=h[4], P=P[4])
T[5]=36
P[5]=P_SAT(WATER, T=T[5])
s[5]=entropy(water, P=P[5], X=0)
h[5]=enthalpy(water, P=P[5], X=0)
P[6]=P[10]
s_6_s=s[5]
h_6_s=enthalpy(water, P=P[6], s=s_6_s)
s[6]=entropy(water, h=h[6], P=P[6])
eta_adiabatic_pump=(h_6_s-h[5])/(h[6]-h[5])
T[6]=TEMPERATURE(water, p=P[6], s=s[6])
"P[7]=40"
s_7_s=s[3]
h_7_s=enthalpy(water, P=P[7], s=s_7_s)
s[7]=entropy(water, h=h[7], P=P[7])
eta_adiabatic_turbine=(h[3]-h[7])/(h[3]-h_7_s)
T[7]=TEMPERATURE(water, p=P[7], s=s[7])
P[8]=P[7]
T[8]=T_SAT(WATER, P=P[8])
s[8]=entropy(water, P=P[8], X=0)
h[8]=enthalpy(water, P=P[8], X=0)
P[9]=P[2]
T[9]=T[8]+3
h[9]=enthalpy(water, T=T[9], P=P[9])
m_1*(h[7]-h[8])=h[9]-h[2]
"P[10]=30"
s_10_s=s[3]
h_10_s=enthalpy(water, P=P[10], s=s_10_s)
s[10]=entropy(water, h=h[10], P=P[10])
eta_adiabatic_turbine=(h[3]-h[10])/(h[3]-h_10_s)
T[10]=TEMPERATURE(water, p=P[10], s=s[10])
x_10=quality(water, h=h[10], P=P[10])
m_1*h[4]+m_2*h[10]+(1-m_1-m_2)*h[16]=h[1]
P[11]=P[5]+0.03*P[5]
s_11_s=s[3]
h_11_s=enthalpy(water, P=P[11], s=s_11_s)
s[11]=entropy(water, h=h[11], P=P[11])
eta_adiabatic_turbine=(h[3]-h[11])/(h[3]-h_11_s)
T[11]=TEMPERATURE(water, p=P[11], s=s[11])
x_11=quality(water, h=h[11], P=P[11])
"P[12]=20"
s_12_s=s[3]
h_12_s=enthalpy(water, P=P[12], s=s_12_s)
s[12]=entropy(water, h=h[12], P=P[12])
eta_adiabatic_turbine=(h[3]-h[12])/(h[3]-h_12_s)
T[12]=TEMPERATURE(water, p=P[12], s=s[12])
x_12=quality(water, h=h[12], P=P[12])
P[13]=P[12]
s[13]=entropy(water, x=0, P=P[13])
h[13]=enthalpy(water, x=0, P=P[13])
T[13]=TEMPERATURE(water, p=P[13], s=s[13])
P[14]=P[10]
s_14_s=s[13]
h_14_s=enthalpy(water, P=P[14], s=s_14_s)
s[14]=entropy(water, h=h[14], P=P[14])
eta_adiabatic_pump=(h_14_s-h[13])/(h[14]-h[13])
T[14]=TEMPERATURE(water, p=P[14], s=s[14])
(1-m_1-m_2-m_3)*(h[15]-h[6])=m_3*(h[12]-h[13])
P[15]=P[10]
T[15]=T[13]-2.7
s[15]=entropy(water, T=T[15], P=P[15])
h[15]=enthalpy(water, T=T[15], P=P[15])
P[16]=P[10]
(1-m_1-m_2-m_3)*h[15]+m_3*h[14]=(1-m_1-m_2)*h[16]
Q_in=h[3]-h[9]
Q_out=(1-m_1-m_2-m_3)*(h[11]-h[5])
eta=1-Q_out/Q_in
```

(b) optimization of this cycle requires maximizing the thermal efficiency for three independent variables of P_7, P_{10}, and P_{12} with the bound of 1 bar to 120 bar. The optimization can be done by the **Code 5.9** in **EES min/max toolbox.** The results show that for $P_7 = 29.04$ bar, $P_{10} = 9.61$ bar and $P_{12} = 1.237$ bar the optimum thermal efficiency would be:

$$\eta_{thermal} = 41.13\%$$

The T-s diagram of the optimized cycle is shown in Fig. 5.37. According to the optimization, the low pressure close-FW heater is completely working in the saturated condition; hence, the *TTD* for this FW heater is assumed 2.7 °C. The high-pressure close-FW heater is initially operating in superheated condition and then saturation condition. Therefore, the *TTD* for this FW heater is assumed −3 °C.

(c) in the optimum conditions, the steam quality at the turbine exit is:

$$x_{11} = 0.8655$$

That is below 0.88 and the risk for blade erosion and corrosion is high. Reviewing the Example 5.8, part c, it proposes two solutions for the low quality steam at the turbine exit, superheating, and reheating. Superheating to the temperature $T_3 = 575°C$ will result is $x_{11} = 0.8822$ that is above 0.88. The efficiency for this temperature is 41.56% that means 0.33% efficiency improvement for 35°C extra superheating. As emphasized previously, using higher superheating temperature at the turbine inlet should be consulted with the turbine and steam generator designer and manufacturers. About the second solution, some changes should be done in the cycle. The changes are

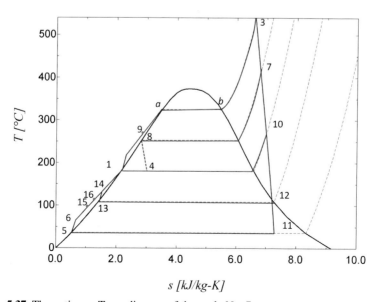

Figure 5.37 The optimum T − s diagram of the cycle No. 7.

depicted in Fig. 5.38. It should be noted that due to reheating, the low pressure FW heater inlet is superheated and the *TTD* should be changed to $-3°C$. The T-s diagram for the cycle with reheater and three FW heater in the optimum conditions is shown in Fig. 5.39. According to the optimization by using the **Code 5.10**, the placement of the reheater and FW heaters are as $\alpha = 0.2922$, $P_7 = 28.15$ bar, $P_{10} = 9.268$ bar, and $P_{12} = 1.25$ bar, which result in the optimum efficiency of:

$$\eta_{thermal} = 42.67\%$$

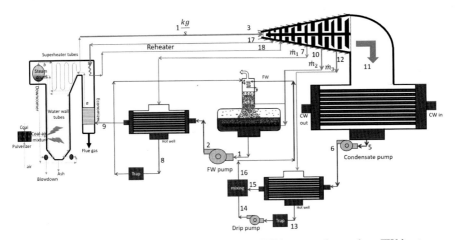

Figure 5.38 The simple cycle plus reheater, an open-FW heater and two close-FW heaters (cycle No. 8).

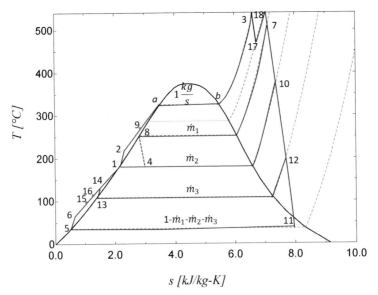

Figure 5.39 The optimum $T - s$ diagram of the cycle No. 8.

In addition, the $x_{11} = 0.9412$ that is very safe and above 0.88.

Code 5.10.   The code written for cycle No. 8.

```
eta_adiabatic_turbine=0.87
eta_adiabatic_pump=0.85
P[1]=P[10]
T[1]=T_SAT(WATER, P=P[1])
s[1]=entropy(water, P=P[1], X=0)
h[1]=enthalpy(water, P=P[1], X=0)
P[2]=120
s_2_s=s[1]
h_2_s=enthalpy(water, P=P[2], s=s_2_s)
eta_adiabatic_pump=(h_2_s-h[1])/(h[2]-h[1])
T[2]=TEMPERATURE(water, p=P[2], h=h[2])
T[3]=540
P[3]=P[9]-0.1*P[9]
h[3]=enthalpy(water, T=T[3], P=P[3])
s[3]=entropy(water, T=T[3], P=P[3])
P[4]=P[10]
T[4]=T_SAT(WATER, P=P[4])
h[4]=h[8]
s[4]=entropy(water, h=h[4], P=P[4])
x_4=quality(water, h=h[4], P=P[4])
T[5]=36
P[5]=P_SAT(WATER, T=T[5])
s[5]=entropy(water, P=P[5], X=0)
h[5]=enthalpy(water, P=P[5], X=0)
P[6]=P[10]
s_6_s=s[5]
h_6_s=enthalpy(water, P=P[6], s=s_6_s)
s[6]=entropy(water, h=h[6], P=P[6])
eta_adiabatic_pump=(h_6_s-h[5])/(h[6]-h[5])
T[6]=TEMPERATURE(water, p=P[6], s=s[6])
"P[7]=40"
s_7_s=s[18]
h_7_s=enthalpy(water, P=P[7], s=s_7_s)
s[7]=entropy(water, h=h[7], P=P[7])
eta_adiabatic_turbine=(h[18]-h[7])/(h[18]-h_7_s)
T[7]=TEMPERATURE(water, p=P[7], s=s[7])
P[8]=P[7]
T[8]=T_SAT(WATER, P=P[8])
s[8]=entropy(water, P=P[8], X=0)
h[8]=enthalpy(water, P=P[8], X=0)
P[9]=P[2]
T[9]=T[8]+3
h[9]=enthalpy(water, T=T[9], P=P[9])
m_1*(h[7]-h[8])=h[9]-h[2]
h_10_s=enthalpy(water, P=P[10], s=s_10_s)
s[10]=entropy(water, h=h[10], P=P[10])
eta_adiabatic_turbine=(h[18]-h[10])/(h[18]-h_10_s)
T[10]=TEMPERATURE(water, p=P[10], s=s[10])
x_10=quality(water, h=h[10], P=P[10])
m_1*h[4]+m_2*h[10]+(1-m_1-m_2)*h[16]=h[1]
```

```
"P[10]=30"
s_10_s=s[18]

P[11]=P[5]+0.03*P[5]
s_11_s=s[18]
h_11_s=enthalpy(water, P=P[11], s=s_11_s)
s[11]=entropy(water, h=h[11], P=P[11])
eta_adiabatic_turbine=(h[18]-h[11])/(h[18]-h_11_s)
T[11]=TEMPERATURE(water, p=P[11], s=s[11])
x_11=quality(water, h=h[11], P=P[11])
"P[12]=20"
s_12_s=s[18]
h_12_s=enthalpy(water, P=P[12], s=s_12_s)
s[12]=entropy(water, h=h[12], P=P[12])
eta_adiabatic_turbine=(h[18]-h[12])/(h[18]-h_12_s)
T[12]=TEMPERATURE(water, p=P[12], s=s[12])
x_12=quality(water, h=h[12], P=P[12])
P[13]=P[12]
s[13]=entropy(water, x=0, P=P[13])
h[13]=enthalpy(water, x=0, P=P[13])
T[13]=TEMPERATURE(water, p=P[13], s=s[13])
P[14]=P[10]
s_14_s=s[13]
h_14_s=enthalpy(water, P=P[14], s=s_14_s)
s[14]=entropy(water, h=h[14], P=P[14])
eta_adiabatic_pump=(h_14_s-h[13])/(h[14]-h[13])
T[14]=TEMPERATURE(water, p=P[14], s=s[14])
(1-m_1-m_2-m_3)*(h[15]-h[6])=m_3*(h[12]-h[13])
P[15]=P[10]
T[15]=T[13]+3
s[15]=entropy(water, T=T[15], P=P[15])
h[15]=enthalpy(water, T=T[15], P=P[15])
P[16]=P[10]
(1-m_1-m_2-m_3)*h[15]+m_3*h[14]=(1-m_1-m_2)*h[16]
"alfa=0.2"
p[17]=alfa*p[3]
s_17_s=s[3]
h_17_s=enthalpy(water, P=P[17], s=s_17_s)
s[17]=entropy(water, h=h[17], P=P[17])
eta_adiabatic_turbine=(h[3]-h[17])/(h[3]-h_17_s)
T[17]=TEMPERATURE(water, p=P[17], s=s[17])
p[18]=p[17]
T[18]=T[3]
s[18]=entropy(water, T=T[18], P=P[18])
h[18]=enthalpy(water, T=T[18], P=P[18])
Q_in=h[3]-h[9]+h[18]-h[17]
Q_out=(1-m_1-m_2-m_3)*(h[11]-h[5])
eta=1-Q_out/Q_in
```

By continuing the simulation for the SPP cycle with different numbers of FW heaters, Table 5.6 can be provided for different configurations of SPP cycle. The table shows the optimum efficiency of the SPP cycle with the FW numbers from $n = 0$ to 7 for two cases of with and without reheater. In each case, the optimum pressure of extractions and reheater are calculated.

According to the examples and discussions given, the optimum placement of the FW-heaters can be found easily by using the optimization tools. In addition, another method, which is usually used to determine the optimum placement of the FW heaters, is to use the following equation to find the FW heaters share for increasing the FW temperature:

Table 5.6 The optimization results for a simple cycle, which is equipped with different number of FW heaters and reheater. The specifications of the simple cycle are; steam generator inlet pressure of 120 bar, steam temperature at the turbine inlet of $540°C$, and the saturated liquid temperature at the condenser output is $36°C$. The total pressure loss in the steam generator is 10%, while in the condenser, it is 3%. The adiabatic efficiency of the turbine and pump is 0.87 and 0.85, respectively.

Cycle No.	Cycle configuration	$\eta_{optimum}$	Optimum variables
1	Simple cycle	36.66%	N.A
2	Simple cycle plus single reheater	38.87%	$\alpha = 0.1458$
3	Simple cycle plus an open-FW heater	39.49%	$P_7 = 4.476$ bar
4	Simple cycle plus an open-FW heater and single reheater.	41.23%	$\alpha = 0.2311$, $P_7 = 5.52$ bar
5	Simple cycle plus a close-FW heater, and an open-FW heater	40.42%	$P_7 = 17.93$ bar, $P_{10} = 3.543$ bar
6	Simple cycle plus a close-FW heater, an open-FW heater, and a reheater	42.05%	$\alpha = 0.2642$, $P_7 = 14.86$ bar, $P_{10} = 2.869$ bar
7	Simple cycle plus two close-FW heaters, and an open-FW heater	41.13%	$P_7 = 29.04$ bar, $P_{10} = 9.61$ bar, $P_{12} = 1.237$ bar
8	Simple cycle plus two close-FW heaters, an open-FW heater, and a reheater.	42.67%	$\alpha = 0.2922$, $P_7 = 28.15$ bar, $P_{10} = 9.268$ bar, $P_{12} = 1.25$ bar
9	Simple cycle plus three close-FW heaters, and an open-FW heater.	41.47%	$P_{18} = 42.09$ bar, $P_7 = 21.23$ bar, $P_{10} = 8.103$ bar, $P_{12} = 1.101$ bar
10	Simple cycle plus three close-FW heaters, an open-FW heater and one reheater.	42.90%	$\alpha = 0.3022$, $P_{18} = 32.68$ bar, $P_7 = 14.82$ bar, $P_{10} = 6.01$ bar, $P_{12} = 0.9793$ bar
11	Simple cycle plus four close-FW heaters, and an open-FW heater The last low pressure FW heater is fed with saturated mixture, the second low pressure FW is fed with saturated steam approximately, hence $TTD = 0°C$ is assumed for it.	41.82%	$P_{18} = 49.59$ bar, $P_7 = 26.15$ bar, $P_{10} = 12.8$ bar, $P_{12} = 3.251$ bar, $P_{21} = 0.6018$ bar
12	Simple cycle plus four close-FW heaters, an open-FW heater and one reheater. All FW heaters are fed with superheated steam.	43.26	$\alpha = 0.3011$, $P_{RH} = 32.51$ bar, $P_{18} = 31.72$ bar, $P_7 = 17.31$ bar, $P_{10} = 8.567$ bar, $P_{12} = 2.49$ bar, $P_{21} = 0.4594$ bar

Table 5.6 The optimization results for a simple cycle, which is equipped with different number of FW heaters and reheater. The specifications of the simple cycle are; steam generator inlet pressure of **120bar**, steam temperature at the turbine inlet of **540°C**, and the saturated liquid temperature at the condenser output is **36°C**. The total pressure loss in the steam generator is 10%, while in the condenser, it is 3%. The adiabatic efficiency of the turbine and pump is 0.87 and 0.85, respectively.—cont'd

Cycle No.	Cycle configuration	$\eta_{optimum}$	Optimum variables
13	Simple cycle plus five close-FW heaters, and an open-FW heater. Attention: Except for the last two low pressure FW heaters, the other heaters are accepting superheated steam from the extractions.	41.96%	$P_{18} = 33.24$bar, $P_7 = 19.01$ bar $P_{10} = 10.17$ bar, $P_{12} = 2.885$ bar, $P_{21} = 0.5556$ bar, $P_{26} = 51.28$ bar
14	Simple cycle plus five close-FW heaters, an open-FW heater and one reheater. Attention: The optimum placement of reheater is between the first and second high pressure close-FW heaters. All FW heaters are fed with superheated steam.	43.5%	$\alpha = 0.2918$, $P_{RH} = 31.51$ bar $P_{18} = 26.01$ bar, $P_7 = 14.37$ bar, $P_{10} = 7.694$ bar, $P_{12} = 2.227$ bar, $P_{21} = 0.4291$bar, $P_{26} = 48.1$ bar
15	Simple cycle plus six close-FW heaters, and an open-FW heater. Attention: The last two close-FW heaters accept saturated mixture at their inlet.	42.19%	$P_{18} = 35.4$bar, $P_7 = 21.99$ bar, $P_{10} = 13.56$ bar, $P_{12} = 4.564$ bar, $P_{21} = 1.45$ bar, $P_{26} = 55.1$ bar, $P_{30} = 0.3636$ bar
16	Simple cycle plus six close-FW heaters, an open-FW heater and a reheater. Attention: The optimum placement of reheater is between the first and second high pressure close-FW heaters. All FW heaters are fed with superheated steam.	43.62	$\alpha = 0.2752$, $P_{RH} = 29.73$ bar, $P_{18} = 19.34$ bar, $P_7 = 12.19$ bar $P_{10} = 6.728$ bar, $P_{12} = 2.684$ bar, $P_{21} = 0.8929$ bar, $P_{26} = 37.07$ bar, $P_{30} = 0.2579$ bar

$$\Delta T_{FW} = \frac{n}{n+1}\left(T_{sat,boiler} - T_{sat,condenser}\right) \tag{5.32}$$

In which, n is the number of FW heaters. The saturation temperature for k^{th} FW heater can be determined as below:

$$T_{sat,k} = T_{sat,condenser} + k \times \frac{\Delta T_{FW}}{n}$$

Or

$$T_{sat,k} = T_{sat,boiler} - (n + 1 - k) \times \frac{\Delta T_{FW}}{n} \qquad (5.33)$$

In which:

$$1 (lowest\ pressure\ FWH\ ) \leq k \leq n(highest\ pressure\ FWH)$$

The cycles which were presented in the previous examples with different numbers of FW heaters are examined by using above formula and also optimization toolbox of EES to find the optimum placement of FW heaters. The results are compared and plotted in Fig. 5.40. As it shows the results of both methods are very close and the biggest difference in the efficiency is 0.21% that has occurred for the case with two FW heaters. However, when we are talking about a several thousand MW power plant, the 0.21% may mean millions of dollars. This is discussed in the example 10.

Example 5.10. A 1000 MW SPP with thermal efficiency of 40% is supposed to be improved by installing some FW heaters. When the designer uses Eq. (5.33), he finds out that the optimum efficiency will be 42% while using the optimization toolbox gives him 42.1%. Evaluate the impact of 0.1% efficiency improvement difference on the

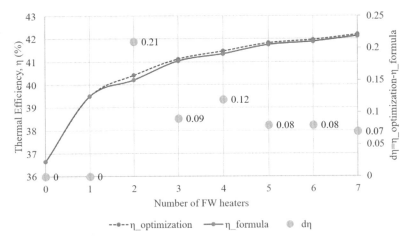

Figure 5.40 Comparison of the optimum efficiency of steam power plant (SPP) versus the number of FW heaters when using optimization techniques and Eq. (5.33) for placement of FW heaters.

cashflow of the cycle. What do you recommend to the designer? Assume that in this improvement, no change is made in the steam generator and the input heat to the power plant.

The power plant operates 7500 h of year at full load and sells electricity to the residential customers at the United States. The average electricity price for residential sector in the United States is 13.31 $\frac{\text{cents}}{\text{kWh}}$. The steam generator is fueled with natural gas with an average price of 3.67 $\frac{\$}{1000\text{ft}^3}$ in 2018. Each cubic feet of natural gas has 0.29 kWh of energy.

Solution. The efficiency differential due to any change in the cycle can be written as below:

$$dη = \frac{\partial η}{\partial \dot{Q}_{in}} d\dot{Q}_{in} + \frac{\partial η}{\partial \dot{Q}_{out}} d\dot{Q}_{out} = \frac{\dot{Q}_{out}}{\dot{Q}_{in}} \frac{d\dot{Q}_{in}}{\dot{Q}_{in}} - \frac{d\dot{Q}_{out}}{\dot{Q}_{in}} = (1 - η_1) \frac{d\dot{Q}_{in}}{\dot{Q}_{in}} - \frac{d\dot{Q}_{out}}{\dot{Q}_{in}}$$

$$→ dη = (1 - η_1) \frac{d\dot{Q}_{in}}{\dot{Q}_{in}} - \frac{d\dot{Q}_{out}}{\dot{Q}_{in}}$$

Since no change is made in the steam generator and the input heat to the power plant.

$$d\dot{Q}_{in} = 0$$

And as a result from the differential equation:

$$→ dη = -\frac{d\dot{Q}_{out}}{\dot{Q}_{in}}$$

That means the efficiency improvement is proportional to the reduction (negative sign) of the heat loss. In other words, to improve the efficiency of the cycle, the heat loss should be decreased. That means increasing the net power generated by the cycle:

$$\dot{W}_{net} = \dot{Q}_{in} - \dot{Q}_{out}$$

Hence:

$$d\dot{W}_{net} = d\dot{Q}_{in} - d\dot{Q}_{out} = -d\dot{Q}_{out} = dη\dot{Q}_{in} = \frac{dη}{η_1}\dot{W}_{net,1}$$

Therefore, the cashflow change due to more power generation can be evaluated as below:

$$dCF = \mathrm{d}\dot{W}_{net}(\mathrm{kW}) \times T\left(\frac{\mathrm{h}}{\mathrm{year}}\right) \times E_{tariff}\left(\frac{\$}{\mathrm{kWh}}\right)$$

$$= \frac{d\eta}{\eta_1}\dot{W}_{net,1}(\mathrm{kW}) \times T\left(\frac{\mathrm{h}}{\mathrm{year}}\right) \times E_{tariff}\left(\frac{\$}{\mathrm{kWh}}\right)$$

$$= \frac{0.001}{0.40} \times 1000000(\mathrm{kW}) \times 7500\left(\frac{\mathrm{h}}{\mathrm{year}}\right) \times 0.1331\left(\frac{\$}{\mathrm{kWh}}\right) = 2.5\frac{\mathrm{M\$}}{\mathrm{year}}$$

As it can be seen, 0.1% efficiency improvement means 2.5 million dollars per year that is a considerable money. According to the results, it is recommended to use the optimization methods to find the optimum placement of the FW heaters.

Now that the optimum place of the FW heaters is known, the last question, which was about the optimum number of FW heaters should be answered.

However, by plotting efficiency of the cycle for different number of feedwater heaters, it is revealed that the impact of FW heaters on efficiency improvement diminishes as the number of FW-heaters increases. This is shown clearly in Fig. 5.41. This can be justified by the *"law of diminishing return"*[10]

In order to find the optimum number of FW heaters, different criteria can be defined. Among them thermo-economic, thermo-environmental, and multicriteria

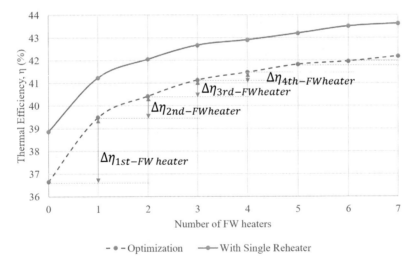

Figure 5.41 The law of diminishing return when the number of FW haters increases for two cases of with and without reheater.

---

[10] "The law of diminishing returns states that in all productive processes, adding more of one factor of production, while holding all others constant, will at some point yield lower incremental per-unit returns" [4].

evaluation method (MCEM) can be used. In addition to the thermo-economic and thermo-environmental evaluations, installing extra FW heater may be impossible due to technical problems, and placement limitations.

In the thermo-economic evaluation, the changes in the $\dot{W}_{net}$, $\dot{Q}_{in}$, and $\dot{Q}_{out}$ should be determined. Then, the change in the cash flow should be calculated by considering the fuel and electricity tariffs as Eq. (5.34):

$$\Delta CF = \Delta \dot{W}_{net}(\text{kW}) \times T\left(\frac{\text{h}}{\text{year}}\right) \times E_{tariff}\left(\frac{\$}{\text{kWh}}\right) - \Delta \dot{Q}_{in}(\text{kW}) \times T\left(\frac{\text{h}}{\text{year}}\right)$$

$$\times Fuel_{tariff}\left(\frac{\$}{\text{kWh}}\right) - I_{O\&M}$$

(5.34)

In which $I_{O\&M}$ is the annual operation and maintenance cost due to installing extra equipment, $\Delta \dot{Q}_{in} = \dot{Q}_{in,2} - \dot{Q}_{in,1}$ and $\Delta \dot{W}_{net} = \dot{W}_{net,2} - \dot{W}_{net,1}$.

On the other hand, the cost of construction and installation of the FW heater should be determined. If the cost is assumed to be $I_{FW\ heater}$, then, the payback period can be determined as Eq. (5.35):

$$PB = \frac{I_{FW\ heater}}{\Delta CF}$$

(5.35)

To estimate the price of a heat exchanger one should know the heating surface of the FW heater. The available heat from the hot stream that can be transferred to the FW stream is calculated by Eq. (5.36):

$$\dot{Q}_{hot} = \dot{m}_{hot}\left(h_{in,hot} - h_{out,hot}\right)$$

(5.36)

The FW enthalpy entering and leaving the FW heater are $h_{in,cold}$ and $h_{out,cold}$, respectively, and the heat loss from the FW heater to the surrounding is counted by its effectiveness, ε. The heat balance between the hot and cold streams can be written as below:

$$\dot{Q}_{cold} = \dot{m}_{cold}\left(h_{out,cold} - h_{in,cold}\right)$$

(5.37)

$$\dot{Q}_{cold} = \varepsilon \dot{Q}_{hot} = UA_{out}\Delta T_{LMTD}$$

$$A_{out} = \frac{\varepsilon \dot{Q}_{hot}}{U\Delta T_{LMTD}}$$

$$\Delta T_{LMTD} = \frac{\Delta T_1 - \Delta T_2}{\ln(\Delta T_1/\Delta T_2)}$$

(5.38)

In which ΔT_{LMTD} is the log-mean temperature difference of hot and cold streams, $\Delta T_1 = T_{in,hot} - T_{in,cold}$ is the temperature difference between the two fluids at one end, and $\Delta T_2 = T_{out,hot} - T_{out,cold}$ is the temperature difference between the two fluids at the other end of the FW heater. U and A_{out} are the overall mean heat transfer coefficient and effective external heating area of the tubes based on their external diameter in the FW heater. The U value can be calculated as Eq. (5.39).

$$U = \cfrac{1}{\cfrac{1}{h_{hot}} + R_{f,hot} + \cfrac{d_o ln\left(\cfrac{d_o}{d_i}\right)}{k} + \cfrac{d_o}{d_i}\cfrac{1}{h_{cold}} + \cfrac{d_o}{d_i}R_{f,cold}} \qquad (5.39)$$

In which d_o and d_i are the external and internal diameters of the tubes, and h is the convection heat transfer coefficient of the hot or cold streams according to its subscript, R_f is the fouling heat transfer resistance, and k is the tube conduction heat transfer coefficient.

According to the lifespan of the project, the PB must be less than a maximum allowed payback period of PB_{max} to let the FW heater to be used (Eq. 5.40).

$$\text{if}: PB < PB_{max} \rightarrow \text{install the FW heater} \qquad (5.40)$$

Usually $PB_{max} = 5$ is reasonable for such projects.

The thermo-environmental evaluation should be based on life cycle assessment (LCA) methods. Since installing an FW heater improves the cycle efficiency by $\Delta\eta$, it will reduce CO_2 production per kW of power generated by the cycle. According to this criterion, the life cycle CO_2 reduction due to installing the FW would be estimated. On the other hand, the CO_2 production due to manufacturing an FW heater should be determined from *cradle to grave*.[11] If the life time CO_2 reduction was greater than CO_2 production, the FW heater is recommended to be installed. Hence:

$$\text{if}: R_{CO_2} > PR_{CO_2} \rightarrow \text{install the FW heater} \qquad (5.41)$$

In which:

$R_{CO_2} = $ life time CO_2 reduction due to FW heater installation

$PR_{CO_2} = $ cradle to grave CO_2 production of FW heater

To make it more clear about how to calculate the emission production, consider an FW heater made of some tons of steel [5]. Carbone dioxide can be produced due to

---

[11] The *cradle to grave* refers to different analyses for different purposes that must be done from the birth time to the death time.

energy consumption for the following processes, which are required to be completed before having an FW heater.

1. Mining row material or scrap metal and preparation
2. Ironmaking
3. Steelmaking
4. Continuous casting and producing semi-finished products
5. Reheating
6. Hot rolling
7. Flat or long products such as plates and roll.
8. Requesting, designing, and manufacturing the FW heater.
9. Transporting
10. Installing
11. Operation and maintenance
12. Miscellaneous processes and works.
13. Death of the FW heater and recycling it as scrap metal for steel making.

For example, according to the reports by the *Worldsteel Association* on average for 2017, for every ton of steel, 1.83 tons of CO_2 were emitted to the environment [6]. This covers the CO_2 production due to steps of 1−7 described above.

To evaluate the emission reduction due to efficiency improvement of $\Delta\eta$, it require to determine the $\dot{W}_{net}$ and $\dot{Q}_{in}$.

The power plant emission index can be determined as Eq. (5.42):

$$e\left(\frac{\text{kg of } CO_2}{\text{kWh}_{power}}\right) = \frac{\dot{Q}_{in}(\text{kW}_{Fuel}) \times i_{CO_2}\left(\frac{\text{kg of } CO_2}{\text{kWh}_{Fuel}}\right)}{\dot{W}_{net}(\text{kW}_{power})} = \frac{i_{CO_2}}{\eta} \qquad (5.42)$$

In which, i_{CO_2} is the emission index for fuel burning that can be found in the standards and verified reports [7]. Power plants with lower emission index are more environment friendly.

Hence, the CO_2 reduction would be as Eq. (5.43):

$$R_{CO_2}(\text{kg of } CO_2) = \sum_{y=1}^{Y}\sum_{h=1}^{T}\left(e_1\dot{W}_{net,1} - e_2\dot{W}_{net,2}\right)_{h,y}$$
$$= \sum_{y=1}^{Y}\sum_{h=1}^{T}\left(-\Delta\dot{Q}_{in} \times i_{CO_2}\right)_{h,y} \qquad (5.43)$$

In which T is the number hours that the cycle is operating in a year and Y is the lifetime of the project. It is clear that the above formulations can be used for other emissions such as CO, NO_x, etc.

In the multicriteria decision-making methods, different approaches may be used. One may check the environmental, and economic criteria at the same time. It means installing an FW heater is allowed when both the economic and environmental criteria are verified. One may consider all of the FW heaters and their overall impact on the economical or environmental criteria of power plant and environment. However, the most conservative method is to check the thermo-economic and thermo-environmental criteria for each FW heater that is supposed to be added to the SPP.

Example 5.11. Consider you are supposed to decide about installing the fourth FW heater (Fig. 5.42) for the power plant described in Example 5.9. The thermal efficiency in the cases with 3 and 4 FW heaters is 41.13% and 41.47%, respectively. The q_{in} for the cycle with 3 and 4 FW heaters is 2452 $\frac{kJ}{kg}$ and 2351 $\frac{kJ}{kg}$. Let the power plant capacity be 200 MW, and you are supposed to use 304 stainless steel as the main material for the FW heater. The U value is 1800 $\frac{W}{m^2K}$ and CO_2 production due to burning natural gas in the steam generator is 0.2 $\frac{kg\ of\ CO_2}{kWh\ of\ natural\ gas}$. The power plant operates 7500 h of year at full load and sells electricity to the residential customers at the United States. The average electricity price for residential sector in the United States is 13.31 $\frac{cents}{kWh}$. The steam generator is fueled with natural gas with an average price of 3.67 $\frac{\$}{1000\ ft^3}$ in 2018. Each cubic feet of natural gas has 0.29 kWh of energy. The plant lifetime is 35 years. The FW heater price is estimated to be 300,000 USD, and its weight is about 50 tons. Assume that for production of each ton of FW heater, 2 tons of CO_2 is emitted to the environment. Determine:

a. The heating surface of the fourth FW heater
b. The CO_2 reduction due to installing the fourth FW heater and compare it with LCA CO_2 generated due to manufacturing of the FW heater
c. The payback period of the FW heater

Solution.

a. To determine the heating surface of the FW heater, the amount of heat transfer capacity of the FW should be determined.

Since the thermal efficiency and power plant capacity are known, then:

$$\eta_{thermal} = \frac{\dot{W}_{net}}{\dot{Q}_{in}} \rightarrow 0.4147 = \frac{200}{\dot{Q}_{in}} \rightarrow \dot{Q}_{in} = 482.27\ MW$$

So that the total mass flow rate of steam can be determined:

$$\dot{Q}_{in} = \dot{m}_{total}q_{in} \rightarrow \dot{m}_{total} = \frac{482.27}{2.351} = 205.13\ \frac{kg}{s}$$

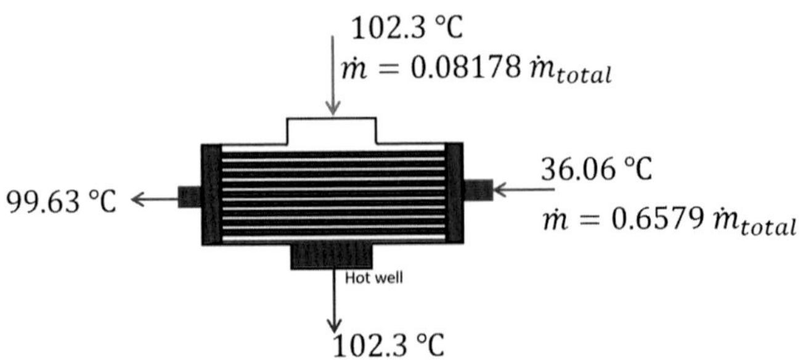

Figure 5.42 The fourth FW heater inlet and outlets.

according to the energy balance and Fig. 5.42:

$$\dot{Q}_{cold} = \dot{m}_{cold}\left(h_{in,cold} - h_{out,cold}\right) = 0.6579 \times 205.13(418 - 151.7) = 35938.52 \text{ kW}$$

$$\Delta T_1 = 102.3 - 36.06 = 66.24°C$$

$$\Delta T_2 = 102.3 - 99.63 = 2.67°C$$

$$\Delta T_{LMTD} = \frac{\Delta T_1 - \Delta T_2}{\ln(\Delta T_1/\Delta T_2)} = \frac{66.24 - 2.67}{\ln(66.24/2.67)} = 19.79 \text{ K}$$

$$\dot{Q}_{cold} = UA\Delta T_{LMTD} \rightarrow A = \frac{35938.52}{1.8 \times 19.79} = 1008.88 \text{ m}^2$$

$$\rightarrow A = 1008.88 \text{ m}^2$$

b. the reduction of carbon dioxide would be as below:

$$R_{CO_2}(kg \text{ of } CO_2) = \sum_{y=1}^{Y} \sum_{h=1}^{T}\left(-\Delta\dot{Q}_{in} \times i_{CO_2}\right)_{h,y}$$

$$= -35 \text{ years} \times 7500\frac{h}{years} \times 205.13\frac{kg}{s} \times (2351 - 2452)\frac{kJ}{kg} \times 0.2\frac{kg \text{ of } CO_2}{kWh \text{ of natural gas}}$$

$$= 1087701825 \text{ kg of } CO_2 = 1.087M \text{ tons of } CO_2$$

hence:

$$R_{CO_2}(kg \text{ of } CO_2) = 1.087M \text{ tons of } CO_2$$

The CO_2 emitted to the environment due to manufacturing of the FW heater is:

$$PR_{CO_2} = 50 \ (tons \text{ of steel}) \times 2\frac{tons \text{ of } CO_2}{tons \text{ of steal}} = 100 \text{ tons of } CO_2$$

$$\rightarrow R_{CO_2} \gg PR_{CO_2}$$

Hence the fourth FW heater is very environmental friendly.

c. To determine the payback period, we need to determine the change in the cash flow due to installing the fourth heater. For this purpose, the changes in the net power and inlet heat should be determined, hence:

$$\dot{W}_{net,2} = \eta_2\dot{Q}_{in,2} = 0.4147 \times 205.13\frac{kg}{s} \times 2351\frac{kJ}{kg} = 199.99MW$$

$$\Delta \dot{W}_{net} = 199.99 - 200 = -0.01 \text{ MW} = -10 \text{ kW}$$

$$\Delta \dot{Q}_{in} = 205.13 \frac{\text{kg}}{\text{s}} \times (2351 - 2452) \frac{\text{kJ}}{\text{kg}} = -20718.13 \text{ kW}$$

As it can be seen the change in the $\dot{W}_{net}$ is very small, because the efficiency increase is created by $\dot{Q}_{in}$ reduction, and as a result, their multiplication remained approximately the same.

$$\Delta CF = \Delta \dot{W}_{net}(\text{kW}) \times T_{operating} \left(\frac{\text{h}}{\text{year}} \right) \times E_{tariff} \left(\frac{\$}{\text{kWh}} \right)$$

$$- \frac{\Delta \dot{Q}_{in}(\text{kW}) \times T_{operating} \left(\frac{\text{h}}{\text{year}} \right) \times NG_{tariff} \left(\frac{\$}{\text{ft}^3} \right)}{0.29 \frac{\text{kWh}}{\text{ft}^3}}$$

$$= -10(\text{kW}) \times 7500 \left(\frac{\text{h}}{\text{year}} \right) \times 0.1331 \left(\frac{\$}{\text{kWh}} \right)$$

$$+ \frac{20718.13(\text{kW}) \times 7500 \left(\frac{\text{h}}{\text{year}} \right) \times 0.00367 \left(\frac{\$}{\text{ft}^3} \right)}{0.29 \frac{\text{kWh}}{\text{ft}^3}}$$

$$= -9982.5 + 1966423.96 = 1965425.46 \frac{\$}{\text{year}} = 1.95 \frac{\text{M\$}}{\text{year}}$$

$$\Delta CF = 1.95 \frac{\text{M\$}}{\text{year}}$$

$$PB = \frac{I_{FW \ heater}}{\Delta CF} = \frac{300000 \ \$}{1965425.46 \frac{\$}{\text{year}}} = 0.15 \text{ year}$$

As it can be seen, the PB is very smaller than 5 years. It means that this investment is profitable.

As a final conclusion, installing the fourth FW heater is beneficial from the thermo-economic and thermo-environmental point of views.

4.7 Air preheating

The flue gas flowing through the steam generator loses its heat as is passes over the economizer, risers, water wall tubes, superheaters, and reheaters and its temperature drops. When it approaches economizer, its temperature should be high enough to convert the subcooled liquid to the saturated liquid.

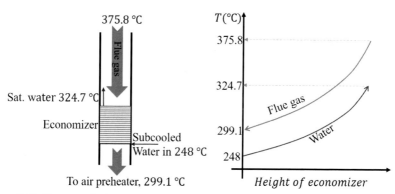

Figure 5.43 Heat transfer in the economizer.

The subcooled liquid temperature is previously elevated by using the FW heaters to some extent. During the heat transfer process, the combustion gas temperature passing over the economizer must stay higher than the subcooled liquid temperature inside the economizer tubes. The heat transfer process between the combustion gas and subcooled water is shown in Fig. 5.43. After the economizer, the flue gas can be used for preheating of combustion air by using recuperative or regenerative heat exchangers. This heat recovery saves fuel and also improves the combustion process. Fig. 5.44 shows the heat transfer in an air preheater of an ultra-supercritical double reheat coal-fired power generation system. More details about the preheaters are presented in Chapter 6.

Example 5.12. The subcooled water temperature at the inlet and outlet of the economizer for an SPP operating at 540 °C and 120 bar with seven FW heaters and a reheater is about 248 °C and 324.7 °C. If the economizer effectiveness $(\varepsilon = Q/Q_{max})$ could be 0.6, determine the flue gas inlet and outlet temperature at the economizer section. Assume that $\dot{m}_{FG}c_{p,FG} = \dot{m}_{water}c_{p,water}$.

Assumptions. Since the inlet flue gas temperature to the economizer is unknown, the minimum allowable temperature would be as the saturated water temperature leaving the economizer, i.e., 324.7°C.

Solution. According to the definition of the heat exchanger effectiveness we have:

$$\varepsilon = \frac{C_h\left(T_{h,in} - T_{h,out}\right)}{C_{min}\left(T_{h,in} - T_{c,in}\right)} = \frac{C_c\left(T_{c,out} - T_{c,in}\right)}{C_{min}\left(T_{h,in} - T_{c,in}\right)}$$

In which h, c, C, stand for hot stream, cold stream, and thermal capacity. In addition C_{min} is the smaller value among C_h and C_c. In this problem since $\dot{m}_{FG}c_{p,FG} = \dot{m}_{water}c_{p,water}$, hence $C_h = C_c$ and therefore:

$$\varepsilon = \frac{T_{h,in} - T_{h,out}}{T_{h,in} - T_{c,in}} = \frac{T_{c,out} - T_{c,in}}{T_{h,in} - T_{c,in}}$$

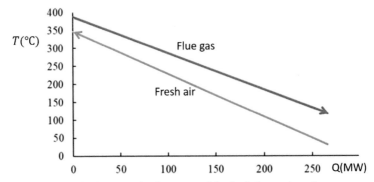

Figure 5.44 Heat transfer between the flue gas and fresh air in an air preheater.

$$\varepsilon = \frac{T_{h,in} - T_{h,out}}{T_{h,in} - 248} = \frac{324.7 - 248}{T_{h,in} - 248}$$

$$\rightarrow T_{h,out} = 299.1°C, T_{h,in} = 375.8°C$$

This increases the subcooled water temperature from 248°C to 324.7°C. Now, the flue gas with temperature of 299.1°C is available for preheating air, that is a great opportunity for heat recovery.

Example 5.13. An air preheater is installed for a steam generator of an SPP to increase the combustion air temperature from 25 °C to 180 °C. If the flue gas temperature at the inlet of the regenerator is 270 °C, determine (a) the regenerator effectiveness, (b) yearly fuel saving per kilogram of combustion air, (c) annual fuel cost saving per kilogram of combustion air, and (d) if the allowable payback period for installing the regenerator is 2.5 years determine the allowable investment for the regenerator.

Assume the same mass flow rate for the hot and cold streams. Estimate the exhaust gases as pure air in your calculations. The steam generator is fueled with natural gas with an average price of 3.67 $\frac{\$}{1000ft^3}$ in 2018. Each cubic feet of natural gas has 0.29 kWh of energy. The SPP operates 7500 h per year at full load.

Solution.

(a) Since the mass flow rate of hot and cold streams are the same and:

$$C_h = \dot{m}_h c_{p,h}, C_c = \dot{m}_c c_{p,c}, \dot{m}_c = \dot{m}_h$$

Since at higher temperature the $c_{p,c}$ is smaller than $c_{p,h}$, therefore $C_{min} = C_c$, and the effectiveness can be determined as below [8]:

$$\varepsilon = \frac{C_c(T_{c,out} - T_{c,in})}{C_{min}(T_{h,in} - T_{c,in})} = \frac{T_{c,out} - T_{c,in}}{T_{h,in} - T_{c,in}} = \frac{180 - 25}{270 - 25} = 0.63$$

(b) To determine the yearly fuel saving by using the air preheater, the yearly heat transfer to the combustion air should be calculated, hence:

$$\dot{Q}_{saving} = C_c\left(T_{c,out} - T_{c,in}\right) = \dot{m}_c c_{p@\frac{T_{c,out}+T_{c,in}}{2}}\left(T_{c,out} - T_{c,in}\right)$$

$$= 1\left(\frac{kg}{s}\right) \times 1.009\left(\frac{kJ}{kgK}\right)(180-25)(K) = 156.4\ kW$$

$$\dot{q}_{saving} = 156.4\frac{kW}{kg\ of\ air}$$

The fuel saving standard volume per year, per kg of combustion air is determined as below:

$$V_{fuel}\left(\frac{ft^3}{year.kg\ of\ air}\right) = \frac{\dot{q}_{saving}\left(\frac{kW}{kg\ of\ air}\right) \times T_{operating}\left(\frac{h}{year}\right)}{0.29\frac{kWh}{ft^3}}$$

$$= \frac{156.4\left(\frac{kW}{kg\ of\ air}\right) \times 7500\left(\frac{h}{year}\right)}{0.29\frac{kWh}{ft^3}} = 4.046 \times 10^6 \frac{ft^3}{year.\ kg\ of\ air}$$

(c) The annual cost saving due to fuel saving is also determined as below:

$$In_{fuel} = V_{fuel}\left(\frac{ft^3}{year.kg\ of\ air}\right) \times Fuel_{tariff}\left(\frac{\$}{ft^3}\right)$$

$$= 4.046 \times 10^6\frac{ft^3}{year.kg\ of\ air} \times 3.67 \times 10^{-3}\left(\frac{\$}{ft^3}\right) = 14849\frac{\$}{year.kg\ of\ air}$$

It means that due to air preheating, 14,849 USD will be saved annually per kg of combustion air.

(d) The payback period can be estimated as below:

$$PB = \frac{I}{CF}$$

In which the cash flow for the air preheater is:

$$CF = 14849\frac{\$}{year.kg\ of\ air}$$

In which the operation and maintenance costs are ignored. Hence:

$$2.5\ year = \frac{I}{14849\frac{\$}{year.kg\ of\ air}} \rightarrow I = 37122\frac{\$}{kg\ of\ air}$$

It means for each kg of combustion air the allowable investment cost for the preheater is 37,122 USD.

5. Problems

1. What are the main processes of each heat engine? What kind of equipment are required to do the processes in a SPP, a gas turbine power plant, and a refrigerator?
2. Why the start-up time of an SPP is very long?
3. What are the definitions of base load, and peak load?
4. Why SPP is the most suitable power plant for base load operation rather than peak load?
5. What is repowering? What is the purpose of repowering?
6. What is the difference between two parameters of reliability and availability of power plant?
7. What are the differences between the wet and dry cooling towers?
8. Consider two SPPs, one is fueled with natural gas and the other is fueled with coal. The average temperature of the combustion products in the steam generators are 750 °C and 800°C respectively. They both reject heat to the same environment with average temperature of 20°C. Determine the internally reversible efficiency for the cycles. Which one has higher efficiency? Why?
9. In an SPP which is equipped with seven FW heaters the combustion products maximum temperature in the steam generator is 1000°C, and they leave the stack at 350°C. Water inlet temperature to the economizer is 290 °C and maximum pressure and temperature of the cycle are 120 bar, and 540 °C. The pressure inside the combustion chamber of the steam generator is 1.05 bar. Steam mass flow rate is $1\frac{kg}{s}$. Approximate the combustion gases with pure air. Determine, a) the average temperature of the heat source b) mass flow rate of combustion gases.
10. What is the purpose of using air preheater in the steam generator of an SPP?
11. The combustion gases of problem 9 leaves the economizer section with the temperature of 350°C and is used for preheating of the fresh air before combustion. If the air preheater effectiveness is 70%, and exhaust temperature leaving the preheater would be 150°C, what would be the inlet combustion air temperature after leaving the air preheater? Assume that $T_{amb} = 20°C$, inlet air and exhaust have the same mass flow rate. And preheater effectiveness is defined as the ratio of the recovered heat to the maximum recoverable heat.
12. How using FW heaters increase the thermal efficiency of the SPP cycle?
13. Consider a basic internally reversible Rankine cycle. The steam generator operating pressure is 150 bar. The steam temperature at the turbine inlet is 535 °C and the saturated liquid temperature at the pump inlet is 36 °C. The air temperature and pressure at the inlet and outlet of the steam generator are 850 °C, 1.1 bar and 200°C, 1 bar respectively. The steam mass flow rate in the cycle is $1\frac{kg}{s}$ and ambient temperature is 20 °C. In addition the cooling water with temperature of 20°C is available for the condenser. If the combustion gases inside the steam generator could be estimated as pure air, determine, the cycle thermal efficiency, exergy efficiency, steam rate of the cycle and air mass flow rate through the steam generator.
14. For the cycle in problem 13 determine the exergy destruction by the pump, steam generator, condenser, steam turbine and the whole cycle. Which component is more exergy destructive?

15. Consider a basic SPP. The steam generator inlet pressure is 170 bar. The steam temperature at the turbine inlet is 540°C and the saturated liquid temperature at the pump inlet is 36°C. The pressure loss in the steam generator is 10% while in the condenser it is 3%. The adiabatic efficiency of the turbine and pump are 0.9 and 0.87 respectively.

The air temperature and pressure at the inlet and outlet of the steam generator are 850°C, 1.1 bar and 200 °C, 1 bar respectively. The mass flow rate of steam in the cycle is $1\frac{\text{kg}}{\text{s}}$ and ambient temperature is 20 °C. In addition the cooling water with temperature of 20 °C is available for the condenser. The combustion gases inside the steam generator can be estimated as pure air. If you are supposed to design a reheater for this cycle (Fig. 5.45):

(a) find the optimum reheat pressure ratio ($\alpha = \frac{P_4}{P_3}$) to maximize the thermal efficiency.

(b) determine if steam quality of point 6 in the optimum case is acceptable?

(c) find the optimum reheat pressure ratio ($\alpha = \frac{P_4}{P_3}$) to maximize the exergy efficiency.

16. In problem 15, consider the air mass flow rate through the steam generator to be $4\frac{\text{kg}}{\text{s}}$ while its inlet temperature is unknown, other thermodynamic parameters of the cycle remain unchanged a) What would be the inlet air temperature for $0.02 < \alpha < 1$, plot the results on a graph. b) Find the α for the maximum values of thermal and exergy efficiencies. c) what is the acceptable range for α according to x_6 limitation?

17. Why using isentropic and dry organic fluids is more appropriate for small scale power production?

18. The maximum operating pressure of an SPP is supposed to be increased in order to improve the thermal efficiency. Due to this change the steam quality at the turbine exit drops to below 0.88. a) What solutions do you recommend to solve the problem? b) What are the limitations of the solution? c) What costs increase or decrease due to using the solutions?

19. Why supercritical SPP achieves higher efficiency?

20. Consider a supercritical SPP cycle with the four main components. The steam generator inlet pressure is 240 bar. The steam temperature at the turbine inlet is 540 °C and the saturated liquid temperature at the pump inlet is 36°C. The pressure loss due to friction in the steam generator is 10% while in the condenser it is 3%. The adiabatic efficiency of the turbine and pump are 0.87 and 0.85 respectively.

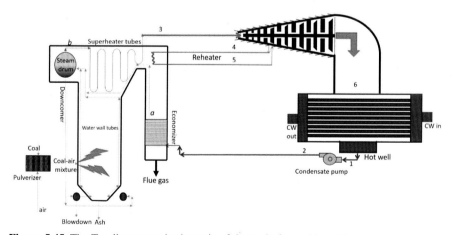

Figure 5.45 The T-s diagram and schematic of the cycle for problem 15.

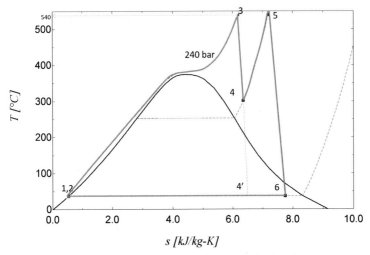

Figure 5.46 The supercritical steam power plant (SPP) with single reheat.

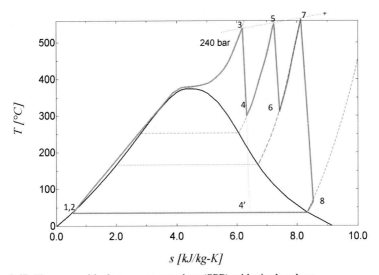

Figure 5.47 The supercritical steam power plant (SPP) with single reheat.

(a) find the optimum reheating pressure ratio of $\alpha = \frac{P_4}{P_3}$ for the cycle if single reheating is used and the temperature after reheat raised to 550°C (Fig. 5.46).

(b) find the optimum reheating pressure ratios of $\alpha = \frac{P_4}{P_3}$ and $\beta = \frac{P_6}{P_5}$ for the cycle if double reheating is used, and the temperature after first and second reheat raised to 550°C, and 560 °C respectively (Fig. 5.47).

21. Consider an SPP with thermal efficiency of 39% and electrical capacity of 1000 MW. The owner has decided to improve the thermal efficiency to 40%. There are two scenarios of: (a)

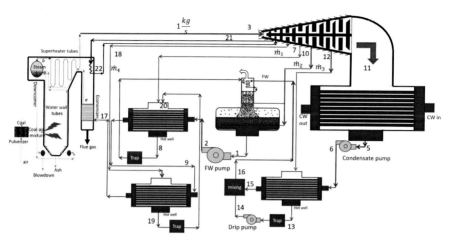

Figure 5.48 The schematic of the cycle No. 9.

Efficiency improvement by increasing the $\dot{W}_{net}$, using turbine with higher adiabatic efficiency. (b)Efficiency improvement by decreasing the $\dot{Q}_{in}$, it can be done by heat recovery from exhaust, combustion improvement, better insulation in different components, etc.

The power plant operates 7500 h of year at full load and sells electricity to the residential customers at the United States. The average electricity price for residential sector in the United States is 13.31 $\frac{cents}{kWh}$. The steam generator is fueled with natural gas with an average price of 3.67 $\frac{\$}{1000ft^3}$ in 2018. Each cubic feet of natural gas has 0.29 kWh of energy. Calculate the change in the annual net cash flow for each scenario. Assume all of other annual cash flows such as operating and maintenance costs, salaries, etc. remain the same. Is there another possible scenario to increase the power plant efficiency?

22. Consider SPP cycle in Fig. 5.48 in which the steam generator inlet pressure is 120 bar. The steam temperature at the turbine inlet is 540 °C and the saturated liquid temperature at the condenser output is 36 °C. The total pressure loss in the steam generator is 10% while in the condenser it is 3%. The adiabatic efficiency of the turbine and pump is 0.87 and 0.85 respectively. Consider the $TTD = -3°C$ for the high pressure close-FW heater, and $TTD = 2.7\,°C$ for the low pressure close-FW heater. (a) Optimize the heaters placement in the cycle to maximize the cycle efficiency. In addition, pay attention to the magnitude of TDD according to the bled steam quality. (b) By using Eqs. 5.32 and 5.33 calculate the maximum efficiency, and pressure of FW heaters. (c) To avoid erosion and corrosion on the turbine internal components, steam quality at the turbine outlet is obliged to stay above 0.88. Check this criterion and propose a solution if the steam quality is lower than 0.88.

23. Consider the SPP cycle No. 11 with 5 FW heaters in Fig. 5.49 in which the steam generator inlet pressure is 120 bar. The steam temperature at the turbine inlet is 540°C and the saturated liquid temperature at the condenser output is 36°C. The total pressure loss in the steam generator is 10% while in the condenser it is 3%. The adiabatic efficiency of the turbine and pump is 0.87 and 0.85 respectively. According to the guidelines given in the Chapter 5 determine the TTD for each FW heater. (A) Optimize the heaters placement in the cycle to maximize the cycle efficiency. In addition, pay attention to the magnitude of TDD according to the bled steam quality. (B) By using Eqs. 5.32 and 5.33 calculate the maximum efficiency, and pressure of FW heaters. (C) To avoid erosion and corrosion on the turbine

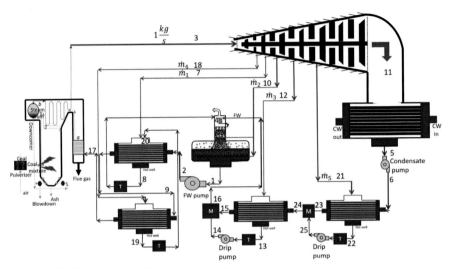

Figure 5.49 The schematic of the cycle No. 11.

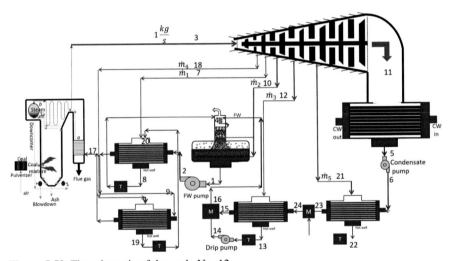

Figure 5.50 The schematic of the cycle No. 12.

internal components, steam quality at the turbine outlet is obliged to stay above 0.88. Check this criterion and use a reheater as Fig. 5.50 to solve the problem, if the steam quality was lower than 0.88.

24. Consider the SPP cycle No. 13 with 6 FW heaters in Fig. 5.51 in which the steam generator inlet pressure is 120 bar. The steam temperature at the turbine inlet is 540 °C and the saturated liquid temperature at the condenser output is 36°C. The total pressure loss in the steam generator is 10% while in the condenser it is 3%. The adiabatic efficiency of the turbine and pump is 0.87 and 0.85 respectively. According to the guidelines given in the Chapter 5 determine the TTD for each FW heater. (a) Optimize the heaters placement in the cycle to maximize the cycle efficiency. In addition, pay attention to the magnitude of TDD

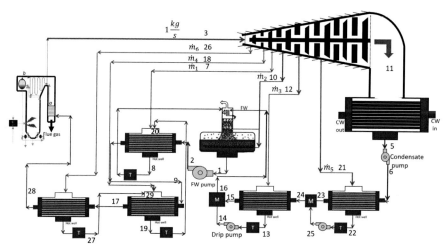

Figure 5.51 The schematic of the cycle No. 13.

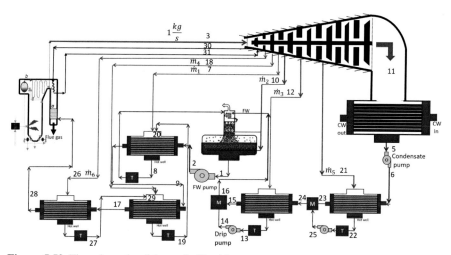

Figure 5.52 The schematic of the cycle No. 14.

according to the bled steam quality. (b) By using Eqs. 5.32 and 5.33 calculate the maximum efficiency, and pressure of FW heaters. (c) To avoid erosion and corrosion on the turbine internal components, steam quality at the turbine outlet is obliged to stay above 0.88. Check this criterion and use a reheater as Fig. 5.54 to solve the problem, if the steam quality was lower than 0.88. Find the optimum place of reheater and show it on the T-s diagram.

25. Consider the SPP cycle No. 15 with 7 FW heaters in Fig. 5.53 in which the steam generator inlet pressure is 120 bar. The steam temperature at the turbine inlet is 540 °C and the saturated liquid temperature at the condenser output is 36°C. The total pressure loss in the steam generator is 10% while in the condenser it is 3%. The adiabatic efficiency of the turbine and pump is 0.87 and 0.85 respectively. According to the guidelines given in the Chapter 5

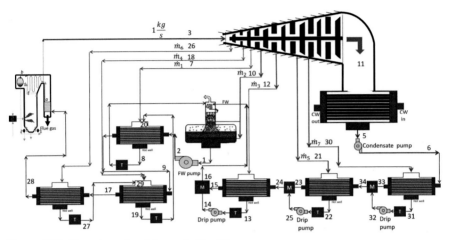

Figure 5.53 The schematic of the cycle No. 15.

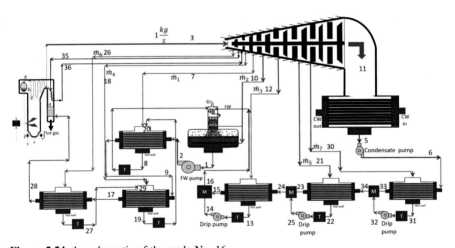

Figure 5.54 the schematic of the cycle No. 16.

determine the TTD for each FW heater. (A) Optimize the heaters placement in the cycle to maximize the cycle efficiency. In addition, pay attention to the magnitude of TDD according to the bled steam quality. (B) By using Eqs. 5.32 and 5.33 calculate the maximum efficiency, and pressure of FW heaters. (C) To avoid erosion and corrosion on the turbine internal components, steam quality at the turbine outlet is obliged to stay above 0.88. Check this criterion and use a reheater as Fig. 5.54 to solve the problem, if the steam quality was lower than 0.88. Find the optimum place of reheater and show it on the T-s diagram.

26. An air preheater is installed for a steam generator of an SPP, to increase the combustion air temperature from 20 °C to 175 °C. If the flue gas temperature at the inlet of the regenerator is 280°C, determine a) the regenerator effectiveness, (b) yearly fuel saving per kilogram of

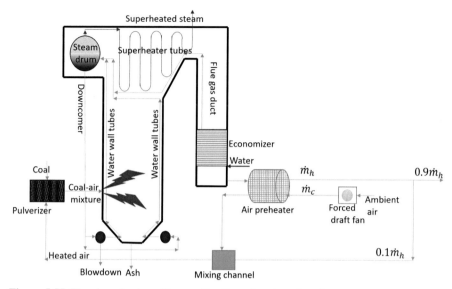

Figure 5.55 The air preheater with recycling the exhaust gas into the steam generator.

combustion air ($\dot{m}_c$), (c) annual fuel cost saving per kilogram of combustion air, and (d) if the allowable payback period for installing the regenerator is 2.5 years determine the allowable investment for the regenerator.

Assume 10% of the hot stream mass flow rate is recycled to the steam generator after passing through the preheater (Fig. 5.55). Estimate the exhaust gases as pure air in your calculations. The steam generator is fueled with natural gas with an average price of $3.67 \frac{\$}{1000\text{ft}^3}$ in 2018. Each cubic feet of natural gas has 0.29 kWh of energy. The SPP operates 7500 h of year at full load.

References

[1] EIA, Annual Energy Outlook, U.S. Energy Information Administration, United States, 2018.

[2] R.J. Campbell, Increasing the Efficiency of Existing Coal-Fired Power Plants, Congressional Research Service, 2013.

[3] USGS, The USGS Water Science School, 2016. https://water.usgs.gov.

[4] P.A. Samuelson, W.D. Nordhaus, Microeconomics, McGraw-Hill, 2001.

[5] NSC, New steel construction, An Introduction to Steelmaking, 2020. https://www.newsteelconstruction.com.

[6] Association, World Steel, Belgium, https://www.worldsteel.org/, 2020.

[7] EPA, Ctaloge of CHP Technologies, Section 4. Technology Characterization, Steam Turbines, U.S. Environmental Protection Agency, 2015.

[8] V. Quaschning, Regenerative Energiesysteme, Technologie - Berechnung - Klimaschutz, Carl Hanser Verlag München, Tabellen, 2019.

Steam power plants, components

Chapter outline

For better understanding of the concepts and principals of operation of steam power plant (SPP), its main components and also the auxiliary systems should be recognized. The main components include the steam turbine, steam generator, condenser, cooling tower, feedwater pump, electric generator, and FW heaters. In this chapter, the main components will be discussed in details; in addition, some of the critical auxiliary systems includ but not limited to the sealing, water treatment system, bearings, lubrication, flushing, combustion air systems, etc. are presented.

Power Generation Technologies. https://doi.org/10.1016/B978-0-323-95370-2.00008-9
Copyright © 2023 Elsevier Inc. All rights reserved.

1. Steam turbine

Steam turbine is a rotary energy conversion equipment that converts high-temperature and high-pressure steam to the low-pressure, low-temperature one and instead produces power due to shaft rotation. To understand, how the shaft is rotated, the internal components of a steam turbine must be introduced and explain how they convert pressure and temperature to rotation.

The main components of a steam turbine include:

1. Rotor (that includes shaft, discs, rotary buckets, or blades)
2. Casing (it is a pressure surface and holds the diaphragm and stationary nozzles in place)
3. Stationary nozzles and diaphragm (they convert pressure to velocity and guide the steam flow to the right direction)
4. Sealing system (avoids steam leakage in different places such as between casing and shroud, casing and blade tips, and from turbine to the bearing and balancing cylinder sides)
5. Bearings (they support axial and radial forces)
6. Lubrication system (that is responsible for lubricating of bearings)
7. Balancing cylinder and balancing line (not included in all of the turbines, they are used to reduce the axial forces)
8. Crossover pipes
9. Control and stop valves

Fig. 6.1A shows the rotor of a low pressure steam turbine that receives steam from the crossover pipe at the middle of the rotor, the steam flow splits into two halves and passes over seven stages before exiting the turbine. Each stage of turbine comprises of a stationary nozzle row installed on the casing and a rotary row of blades on the rotor. The stationary nozzles are shown in Fig. 6.2. Steam passes through the stationary nozzles at the entrance of turbine, between every two rows of rotary blades a stationary nozzle row should be installed on the casing side. The casing, nozzles, and diaphragms of a steam turbine are shown in Fig. 6.2.

The buckets can be classified as impulse or reaction. The reaction type operates based on pressure difference between two sides of the bucket, while the impulse one works based on the velocity difference in the inlet and outlet of the bucket. These two types of buckets and their operating principals are shown in Fig. 6.3. According to these diagrams, conversion of pressure to velocity occurs in the stationary row. Steam with high-velocity enters the rotary buckets, they absorb the kinetic energy of steam and consume it for rotating of rotor and overcoming different irreversibilities such as friction of internal components and bearings. The rotor rotation later can be used for turning an electricity generator, a pump, or a compressor.

It is worth mentioning that most of the buckets are partially impulse and partially reaction. Fig. 6.4A shows the sectional profile of rotating bucket at different heights; near the root, it is mostly impulsive, while near the tip, it is close to reactive type. In addition, different types of shroud are used at the bucket tips to avoid steam leakage at the tip (tip loss) and also stiffening the buckets and disc package and hence reducing vibration. In some cases, a damping (lashing or lacing) wire or a zig-zag pin may be used to connect the buckets of a row together for stiffening and vibration reduction (Fig. 6.4B).

Figure 6.1 The SST-4000 single shaft, with capacity of 90–380 MW (courtesy of Siemens company) [1].

The damping wire may accelerate fouling especially on the last stages of steam turbine due to low-quality steam. Failure of the blades due to the cracks in the hole of the blade can happen; it can also increase the threat of corrosion. These problems finally can increase imbalance of the rotor and vibration as well. Due to these problems damping wires and pins should be used carefully. The lashing wire at each row can be single- or multipiece. In the case of multipiece, each piece of wire fastens several buckets to each other. But in the case of single-piece, all of the buckets of the row are fastened together by a single wire. In the zig-zag pins, each pin connects two adjacent buckets together; it should be noted that in this case, due to using pins, two holes would be

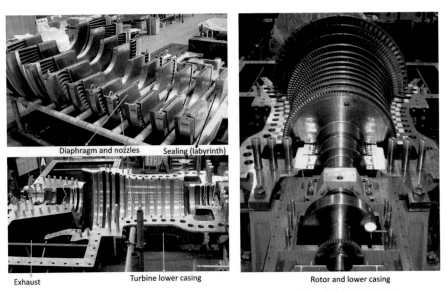

Figure 6.2 The casing, nozzles, and diaphragms of a steam turbine (MITSUBISHI heavy industries, Ltd.) [2].

created on each bucket, and as a result, the threat of cracking, corrosion, and fouling increases in comparison with the damping wire (Fig. 6.4B).

Steam takes the path of the least resistance; it means that steam would prefer to flow in the path where there is no obstacle such as buckets and nozzles. Such gaps can be found between the stationary nozzles and rotor or rotating buckets and casing. Flowing steam in such places is usually referred as the internal leakage and is unfavorable because it reduces power production and thermal efficiency of the cycle. Hence, to make the steam to flow in the right path, internal sealing is required between the rotor shroud and casing and also between stationary nozzle diaphragm and rotor. Such sealing is usually done by labyrinth that is shown in Figs. 6.5 and 6.6. Labyrinth reduces the steam leakage to a minimum allowable value by successive contraction and expansion of steam at the teeth peak and valleys and making pressure drop for restricting the steam leakage (Fig. 6.6). Labyrinth sealing exists in different patterns and configurations that will be discussed in the followings.

Labyrinth may fail sealing due to scale forming and impurity deposition in the valleys or eroding the teeth (Fig. 6.7A and B). The teeth may be eroded due to collision of external objects and steam impurities or direct contact with sealing surface that can occur in case of severe vibrations. When eroding or scale forming occurs, steam jets out because less expansion and contraction occurs, and steam energy remains high as it passes through the passage (less pressure drop).

Sealing between the shaft and casing on both ends of a steam turbine may be carried out by labyrinth, carbone rings, mechanical seals, or a combination of sealing methods. For example, carbone ring may be used with a sealing flow of steam to manage the magnitude and direction of leakage. Figs. 6.8 and 6.9 show examples of shaft sealing.

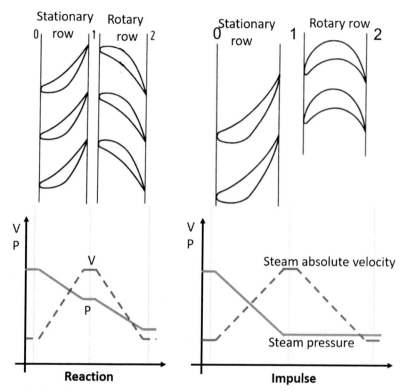

Figure 6.3 The absolute velocity and pressure of steam passing over reaction and impulse buckets.

Fig. 6.8 shows different arrangements of labyrinth for sealing of inlets and outlets of a steam turbine. It also shows that the threads can be generated on the shaft as well and be combined with labyrinth for better sealing purposes.

Fig. 6.9 shows four different combinations of labyrinth, carbone ring, and sealing steam or leak-off connections. For example, Fig. 6.9A shows the sealing of high-pressure steam by a three-step labyrinth and two leak-off connections, which are guided to the low-pressure steam section at the turbine exit and drain or gland ejector. The gland ejector (Fig. 6.9E), which is a kind of a compressor, sucks in the leak-off due to the low-pressure region generated by the high-velocity steam in the ejector. The high-velocity and leak-off steam get mixed together and leave the ejector at a moderate pressure between the two inlet pressures. Fig. 6.9B shows the sealing in the low-pressure side of a steam turbine. The purpose of sealing here is to avoid air leakage from outside to the turbine inside. For this purpose, a sealing steam with a pressure higher than both the turbine exit pressure and ambient pressure, is injected into the sealing packing. Some of the sealing steam finds its way to the turbine and the rest leaks to the drain or the ambient after passing through two labyrinth steps, and a leak-off connection, which is connected to the drain or the gland ejector. As it

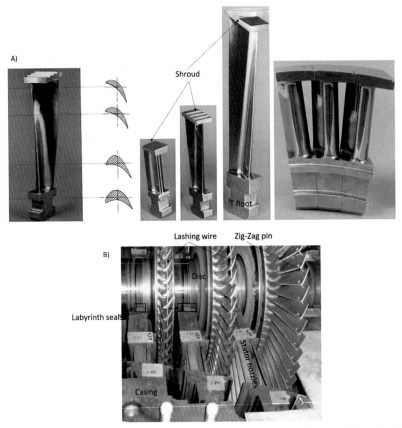

Figure 6.4 (A)The cross-section of a rotating shrouded bucket at different height of the blade, with T root (Siemens courtesy) [1] (B) The damping (lashing) wire, and zig-zag pin in the steam turbines (courtesy of Elliott Ebara turbomachinery corporation) [3].

can be seen, the operation of main labyrinth (the bigger ones) in Fig. 6.9A and B, are improved by the threaded shaft. Fig. 6.9C shows a carbon ring packing, the carbon ring packing consists of some carbon ring and a packing, which is horizontally split, a carbon ring, which is usually made of several circular pieces is fastened by a garter spring. The carbon ring packings are usually equipped with a sealing connection on top and a leak-off connection on bottom that may be connected to drain or gland ejector. One should know that the packing is stationary and connected to the casing, while shaft is rotating inside it without touching each other. Fig. 6.9D has combined a carbon ring packing with a labyrinth to seal a high-pressure steam. A leak-off connection to the low-pressure side of the turbine is placed between the labyrinth and carbon ring packing, and a leak-off connection to the drain or gland ejector is placed under the carbon ring packing. Close pictures of carbon ring packing and labyrinth are shown in Fig. 6.10A and D. The labyrinth can be equipped with a brush seal as well (Fig. 6.10D) to reduce the leakage as less as possible.

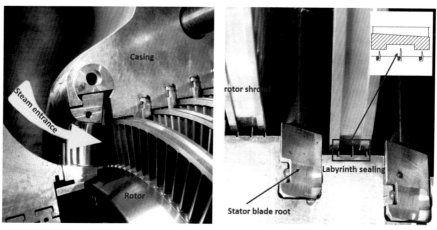

Figure 6.5 Labyrinth sealing between shroud and casing (Siemens courtesy) [1].

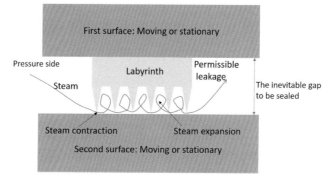

Figure 6.6 The sealing of inevitable gap between two surfaces by labyrinth.

In Fig. 6.10A, a labyrinth seal is shown, which is equipped with brush seal. The brush seal can reduce the leakage greatly. In Fig. 6.10B, a single stage steam turbine with internal components is illustrated. It shows the sealing system, bearings, lubrication system, governor, etc. The sealing system is the carbon ring packing (Figs. 6.10C and D) on the both sides of the turbine. In addition, stationary nonsparking labyrinths are used to seal the space between the carbon ring packing and bearing case. A thrust bearing is installed only on the low-pressure side to tolerate the axial movements of the shaft in both directions, and Babbitt-lined journal bearings are used on both sides of the turbine to support the radial motions of the shaft. The bearings are lubricated by using oil rings. The oil ring (or oiler ring) that acts like an oil pump (with only one element) is placed around the shaft and rotates due to the friction with the rotating shaft. Each time the oil ring spins, some oil sticks to it and it pulls the oil up from the oil reservoir to lubricate the bearings. In case of wide bearings, two oil rings may be used. Oil rings are appropriate for medium-speed applications with moderate loads. Its diameter is about 1.5–2 times the shaft diameter, and less than half of the ring diameter should

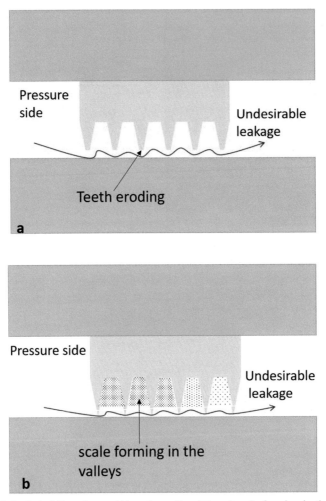

Figure 6.7 Labyrinth failure due to (A) teeth eroding (B) and scale forming in the valleys.

be submerged in the oil. Too-low or too high oil level in the oil reservoir makes the oil ring operation inefficient. Too-low oil level means that smaller piece of oil ring is in touch with oil, and as a result, smaller amount of oil will be pulled up by the oil ring. Too-high oil level means a bigger part of the ring will be submerged in the oil that produces bigger frictional forces applied to the ring from the oil. This will finally result in the reduction of the oil ring rotational speed and poor lubrication.

As the high-pressure steam flows through the turbine, different modes of motion including rotation, axial, and radial movement may occur to shaft (Fig. 6.11).

Bearings support the axial and radial movements of the rotor. The radial force is due to the rotor weight and unbalanced radial forces due to shaft rotation and steam motion and its distribution in the turbine. The axial force is mostly due to pressure difference

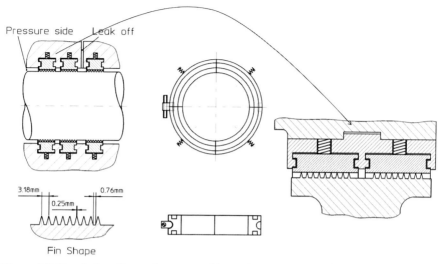

Pressure side   Leak off

3.18mm   0.76mm
0.25mm

Fin Shape

Figure 6.8 The shaft sealing at the steam turbine ends (Siemens courtesy) [1].

between the inlet and outlet of the turbine. Journal[1] or sleeve bearings and thrust bearings should be used to support the radial and axial forces of the rotor. The thrust bearing can be ball bearing (mostly for small turbine) or thrust bearing with tilting pads (mostly for large turbines). Fig. 6.12 shows some schematic drawing of bearings capable of supporting the axial and radial forces of the rotor. Fig. 6.12A shows combined radial and thrust ball bearings. In this arrangement, the axial movement of the shaft from the right to the left side is limited by installing a thrust collar (which is also called as thrust disc or thrust runner) on the shaft and the thrust ball bearings. Since the axial movement of the rotor may occur in both directions, therefore another thrust and radial bearing is required on the other side of the rotor.

Fig. 6.12B shows another combined radial and thrust bearing. In this arrangement, two stationary thrust surfaces are installed on both sides of the radial sleeve bearing, while two thrust runners are installed on the shaft. In this configuration, the axial movement of the shaft in both directions is supported by single thrust bearing on one side.

Fig. 6.12C, is a combined radial, thrust bearing that the radial bearing is a sleeve type. The thrust bearing uses thrust shoes or pads instead of stationary thrust surfaces. These thrust shoes tilt in their position; and therefore, there will be an *oil wedge* on every shoe that makes the lubrication much more efficient. This thrust bearing can only support the shaft movement from the right side to the left side. Nowadays,

---

[1] The portion of shaft which is covered by the bearing is called journal, this portion is usually hardened to resist wear against the bearing internal parts. In addition, the sleeve or pads of a journal bearing are made from softer material to avoid damaging and wearing the journal.

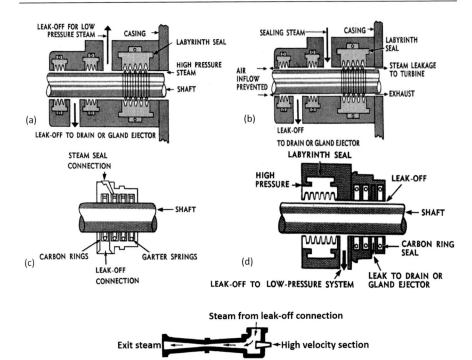

Figure 6.9 Typical gland using carbon rings with labyrinth, and gland ejector (courtesy of American petroleum institute, API, howell training company) [5].

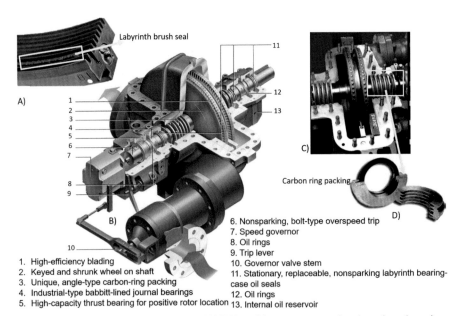

1. High-efficiency blading
2. Keyed and shrunk wheel on shaft
3. Unique, angle-type carbon-ring packing
4. Industrial-type babbitt-lined journal bearings
5. High-capacity thrust bearing for positive rotor location
6. Nonsparking, bolt-type overspeed trip
7. Speed governor
8. Oil rings
9. Trip lever
10. Governor valve stem
11. Stationary, replaceable, nonsparking labyrinth bearing-case oil seals
12. Oil rings
13. Internal oil reservoir

Figure 6.10 Single-stage turbine (SST) 500/700 and its components, brush seal, carbon ring packing (courtesy of dresser-rand) [4].

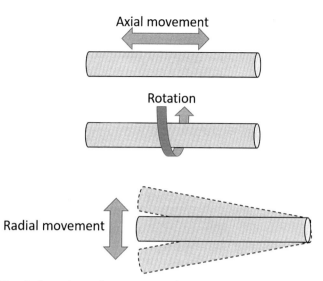

Figure 6.11 The shaft rotation and movement modes.

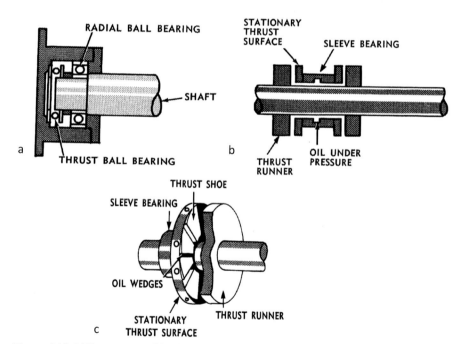

Figure 6.12 Different types of bearings, (A) combined radial and thrust ball bearings, (B) stationary thrust surface with sleeve bearing, (C) thrust shoe with sleeve bearing (courtesy of American petroleum institute, API, howell training company) [5].

advanced thrust bearings with tilting pads are designed that can support the shaft movement in both directions.

Oil wedge is the oil film with variable thickness between the two surfaces that are supposed to be lubricated. The variable oil thickness provides a pressure difference that circulates the oil and makes the cooling and lubricating more efficient. The oil wedge can be generated due to shaft rotation and its tendency to go up the internal surface of bearing, by using tilting pads, noncircular bearings such as slightly elliptical sleeve bearing, or circular sleeve with wedge shaped grooves on the sleeve surface (Fig. 6.13). In addition, Fig. 6.13 shows the oil feed hole in the journal bearing to let the pressurized oil in to lubricate the bearing and journal. The main advantage of the journal bearings with tilting pads is that they provide several oil wedges at the same time, while the plain sleeve bearing only produces two oil wedges on the journal perimeter.

Fig. 6.14 shows some details of the thrust bearings usually used for large-scale turbines. These thrust bearings are usually combined with a journal bearing and are used only at one end of the shaft. As it can be seen, a disc that is called thrust collar (or thrust runner) is installed on the shaft. The thrust collar is place in the thrust bearing, while both sides of the collar are supported by the thrust pads. The thrust pads on one side are called active, while the other side is called passive. The active pads support the shaft movements during the steady operation of the turbine, and they are usually thinner, the passive pads support the shaft movements during the emergency shut down that shaft may move in the opposite direction. During the steady operation, the dominant direction of the axial forces is from the inlet (with high pressure) to the outlet of turbine with lower pressure. Since the passive pads must tolerate the sudden loads due to emergency shut down, they are usually thicker in comparison with the active pads. The

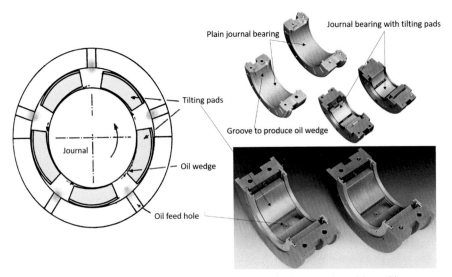

Figure 6.13 Journal or sleeve bearings, commonly used for large-scale turbines (Siemens courtesy) [1].

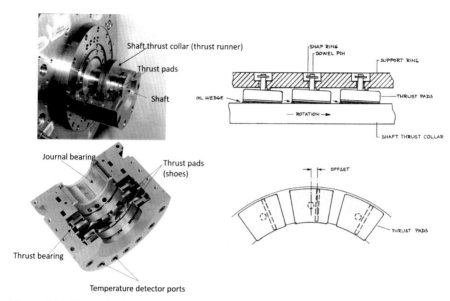

Figure 6.14 Thrust bearing, commonly used for large scale turbines (Siemens courtesy) [1].

temperature of the thrust pads must be monitored continuously to detect any possible damage on the pads. For this purpose, temperature detectors are connected to the pads through the ports on the bearing end. Since some area of the thrust pads may encounter higher pressure with thinner oil film, the temperature in this area will be higher. Therefore, placement of the temperature detector must be done with cautious. This optimum place is termed as "75/75 location" (Fig. 6.15) and can be determined by computer simulation and experiment.

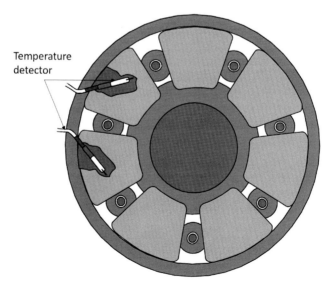

Figure 6.15 Placement of temperature detectors of thrust shoes of thrust bearings, "75/75 location" (Siemens courtesy) [1].

2. Condenser

Heat exchangers purposes are to cool, heat, condensate, or evaporate fluids. A condenser of an SPP is a shell and tube heat exchanger that rejects the heat content of the steam, leaving the turbine to the cooling medium, and produces saturated liquid. According to the TEMA,[2] the shell and tube heat exchanger type can be specified according to the front end type, shell type, and rear end type Fig. 6.16.

According to the fluid type passing through the heat exchanger, the operating temperature and pressure at the inlets and outlets, different arrangements of the front and rear heads can be combined with different shells. When the working fluid temperature is too high and the tube bundle expansion would be significant, the rear head or the tube bundle must be designed to tolerate or compensate the thermal stress due to the thermal expansion, under this condition floating heads are required. The front head is always stationary. The shell type for a condenser of power plant can be chosen among the shells presented in Fig. 6.16 except for the K type shell. The notation of the heat exchangers is designated by *Tubular Exchanger Manufacturer Association* (TEMA) (Fig. 6.17). The notation includes three letters, the first letter on the left side refers to the front head type, the middle letter shows the shell type, and the third letter shows the rear head type. For example, the AEL exchanger refers to a heat exchanger with stationary head type of A, shell type of E, and rear head of L. Fig. 6.18 shows some of the heat exchanger configurations with part numbers, which are specified in the following.

1. Stationary head-channel	21. Floating head cover-external
2. Stationary head-bonnet	22. Floating tubesheet skirt
3. Stationary head flange-channel or bonnet	23. Packing box
4. Channel cover	24. Packing
5. Stationary head nozzle	25. Packing gland
6. Stationary tubesheet	26. Lantern ring
7. Tubes	27. Tie rods and spacers
8. Shell	28. Transverse baffles or support plates
9. Shell cover	29. Impingement plate
10. Shell flange-stationary head end	30. Longitudinal baffle
11. Shell flange-rear head end	31. Pass partition
12. Shell node	32. Vent connection
13. Shell cover flange	33. Drain connection
14. Expansion joint	34. Instrument connection
15. Floating tubesheet	35. Support saddle
16. Floating head cover	36. Lifting lug
17. Floating head cover flange	37. Support bracket
18. Floating head backing device	38. Weir
19. Split shear ring	39. Liquid level connection
20. Slip-on backing flange	40. Floating head support

---

[2] Tubular Exchanger Manufacturer Association.

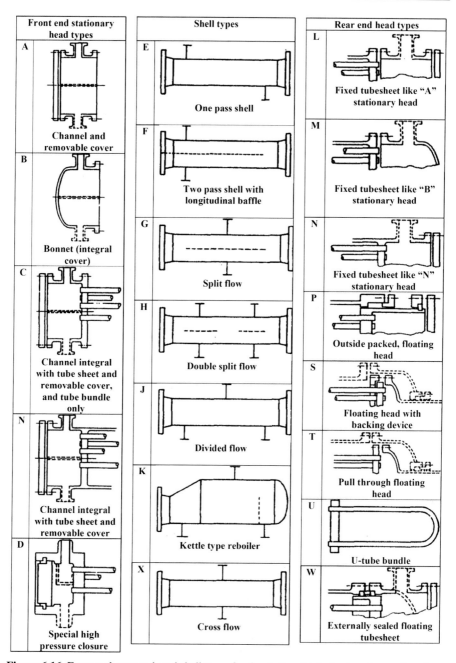

Figure 6.16 Front end, rear end, and shell types for the shell and tube heat exchanger based on TEMA [6].

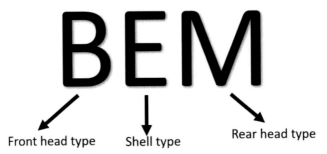

Figure 6.17 The notation designated by TEMA for heat exchangers.

The functions of the most important components of heat exchangers are explained below:

2.1 Tubes

Tubes are the main internal parts of a shell and tube heat exchanger that one of the cold or hot streams flow inside it. The total tubes inside a heat exchanger are called core or tube bundle. Tubes in a shell and tube heat exchanger may be straight or U type. The tube ends are usually supported by a tube sheet.

2.2 Shell

The outer casing containing the tubes, baffles, tie rods, spacers, transverse baffles, or support plates, impingement plate, longitudinal baffle, etc. is called shell. Its main duty is to let the shell-side fluid flow around the tubes and make the heat transfer as much as possible.

2.3 Stationary head-channel

It is a kind of front head with a horizontal pass partition that let the tube-side fluid pass the shell length twice (two-pass heat exchanger). Usually, instrument connections are installed on the head nozzles to monitor temperature and pressure of the fluid by using the thermometer and pressure gauge. The channel cover of this head is removable that lets the operator to explore the inside of front head in case of need for troubleshooting. It has two head nozzles for the input and output of tube-side fluid.

2.4 Stationary head bonnet

It is also a stationary front head type, with only on head nozzle; it is also flanged to the stationary tube sheet.

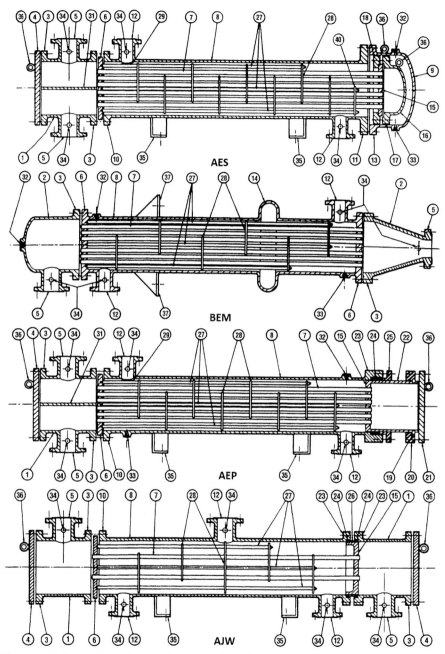

Figure 6.18 Four different heat exchanger configurations (AES, BEM, AEP, and AJW) with their defined part numbers [6].

2.5 Tube sheet

A tube sheet (Fig. 6.19) is a perforated circular plate that supports the tubes in a shell and tube heat exchanger. To avoid leakage and undesirable mixing of the hot and cold streams, the tube's ends must be sealed at the tubesheet holes. The tubesheet can be stationary or floating. The floating type (AJW, and AEP in Fig. 6.18) is used to avoid thermal stress due to thermal expansion of the tube bundle. The stationary tube sheet can be welded to the shell (BEM, in Fig. 6.18) or the head (CFU, in Fig. 6.16) and or as a separate part that is placed between the head and shell (AJW, in Fig. 6.18).

During the operation of a heat exchanger, some tubes may fail for different reasons such as erosion, corrosion, thermal stress, fouling, etc. In this case, the tube ends are plugged (Fig. 6.19) to avoid leakage from the high-pressure side to the low-pressure side and mixing of the two streams. Less than 10% of the tubes can be plugged; if you must plug more than 10% of the tubes, you are recommended to retube the failed tube of the heat exchanger because plugging more than 10% may disrupt the heat transfer between the cold and hot streams. For example, in a condenser, plugging more than 10% of the tubes may result in steam/liquid mixture at the condenser outlet that results in cavitation in the condensate pump. The reason is that the available heating surface would be less than the minimum required heating surface for complete condensation of the steam. In the condenser of an SPP, the cooling water (cold stream) is in the tube side, while steam (hot stream) is in the shell side.

To avoid leakage at the tubesheet holes, the clearance between the tube and tubesheet must be sealed by welding, rolling, tube expansion, etc.

Tubes can be attached to the tubesheet in four patterns of triangular with angle of 60°, triangular with angle of 30°, rectangular with angle of 90°, and rectangular with angle of 45° (Fig. 6.19).

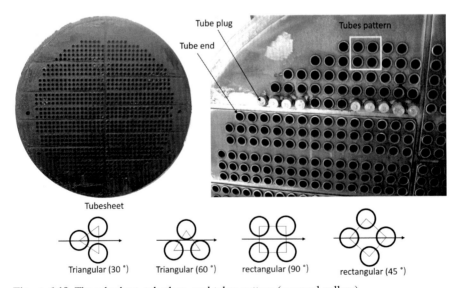

Figure 6.19 The tubesheet, tub plugs, and tubes pattern (personal gallery).

2.6 Baffle

Baffles exist in different types and shapes. They can be used as transverse baffles, support plates, impingement plate, and longitudinal baffle. The transverse baffles types include, single segmental, double segmental, triple segmental, (Fig. 6.20) disk and doughnut, with no-tube-in-window segmental, rod baffles, helical baffles, twisted tubes, etc.

The main purposes of using transverse baffles (Fig. 6.21) are to:

- Make the shell-side fluid flow turbulent to increase convective heat transfer coefficient and improve heat transfer.
- Eliminate dead spots (zones) in the heat exchanger, dead spots such as corners are places where fluid velocity is small and can lead to fouling and deposition in the heat exchanger.
- Support the tubes, keep then in place, avoid sagging, and reduce tube vibration.
- Direct the shell side flow along the tubes.

Fig. 6.21 shows different components of a Kettle-type reboiler heat exchanger used for heating bitumen by steam.

The longitudinal baffle is installed inside the shell to generate two passes for the shell-side fluid on the baffle sides. In cases when multipass shell is required, the number of longitudinal baffles increases accordingly. Fig. 6.22 shows two simulated condensers, in which, Fig. 6.22A is a single-pass condenser, while Fig. 6.22B is a two-pass condenser.

The impingement baffle is a small plate, which is located inside the shell, on the tubes at the inlet channels to protect the tubes from direct strike of the inlet flow. This plate is sacrified to protect the tubes from erosion and induced vibration by the inlet shell fluid.

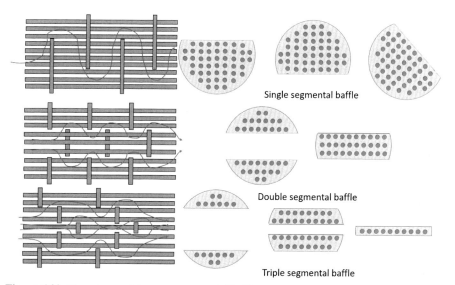

Figure 6.20 Three types of transvers segmental baffles.

Figure 6.21 The components of a kettle-type reboiler heat exchanger used for heating bitumen by steam (personal gallery).

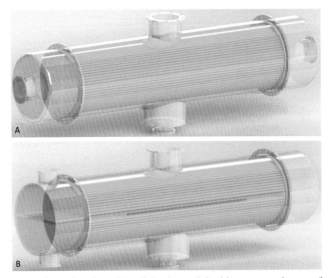

Figure 6.22 Three-dimensional models of single- and double-pass condensers (by Alireza Heidari, mechanical engineering student at University of Kurdistan).

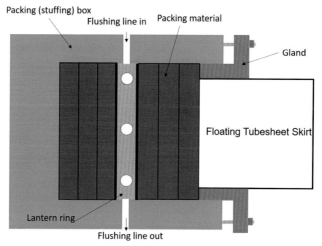

Figure 6.23 The sealing system of rear floating head (type P) heat exchanger.

2.7   Tie rods and spacers

Tie rod is a long screw and nut, which is used to fasten the tube bundle and the baffles together. The tie rods usually go through the first and the last baffles and the other baffles that locate between them. Between every two baffles, the tie rod is covered by a piece of tube that is called a spacer. The spacer maintains the distance between the baffles, and as a result, the flow pattern inside the shell does not change. In addition, movement of the baffles may scratch the tube surface that can result is corrosion; therefore, baffles must stay still by using the tie rods and spacers.

2.8   Vent connection

It is a valve that should be opened during the priming[3] of the heat exchanger to exhaust the noncondensable gases to the atmosphere.

2.9   Drain connection

It is a valve usually placed at the lowest place of the heat exchanger to empty the fluid inside the heat exchanger before the maintenance jobs.

2.10   Packing box

Packing box is a box that contains packing, lantern ring, and packing gland to provide sealing of the floating head (Fig. 6.23).

---

[3] Priming is the process of removing air from the equipment and suction line, and fill it with the process liquid.

2.11 Packing

It is the sealing material usually covered with graphite or polytetrafluoroethylene (PTFE), placed between floating tubesheet skirt and packing box. The packing must be adjusted by the packing gland. The gland must be neither very loose nor very tight because it can cause packing damage and extra leakage, respectively.

2.12 Packing gland

The gland is an L-shaped ring, which is used for adjusting the packing to avoid leakage. Some screws are used to fasten gland to the packing box.

2.13 Lantern ring

It is a perforated hollow ring placed between packings to provide lubrication and cooling for the packings on both sides. The cooling/lubricating liquid should be injected to the packing box (stuffing box) through a flushing line exactly at the place where the lantern ring is located. Fig. 6.24 shows a metal lantern ring. Today, nonmetallic lantern rings made of polytetrafluoroethylene (FTFE) are also available for rotating applications such as pumps that avoid shaft damage due to wearing between the lantern ring and the shaft or its sleeve.

2.14 Expansion joint

The expansion joint is a U-shaped part of shell that is used to provide expansion of the shell. The shell expansion may be required due to thermal expansion of the tube bundle. It protects the tubes from bending during the thermal expansion.

3. Feedwater pump

Feedwater pump is a rotary energy conversion equipment that converts electrical or mechanical power to hydraulic power. The purpose of using a feedwater pump is to

Figure 6.24 Metallic lantern ring, (by Alireza Heidari, mechanical engineering student at the university of Kurdistan).

provide a mass flow rate of liquid with the required pressure to enter the steam generator unit. To understand how the pump works, it is essential to introduce the internal components of the pump. Usually, feedwater pumps are multistage radial flow (centrifugal) pumps that at least include the following components.

1. Shaft
2. Impellers
3. Diffusers
4. Casing
5. Volute
6. Balancing drum
7. Bearings
8. Sealing system
9. Sleeve
10. Wearing rings
11. Flushing system
12. Suction and discharge line
13. Coupling
14. Driver
15. Filter and strainer

Fig. 6.25 is a single-stage centrifugal pump driven by an electric motor. Several components of the pumping system including the coupling, driver, bearing housing, discharge, and suction flanges, and volute are demonstrated.

To understand how a pump increases the liquid pressure, we need to find out more about the impeller, diffuser, and volute (Fig. 6.26). Liquid enters the pump impeller through the suction eye, and then it splits between the diffuser-shaped spaces between the impeller vanes. Since the impeller is rotating, both pressure and velocity of liquid

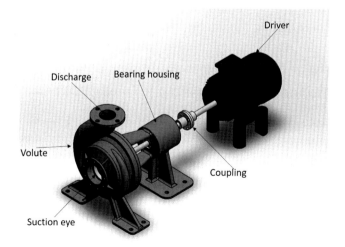

Figure 6.25 Single-stage centrifugal pump (by Amir Hossien Pajooh, mechanical engineering student at University of Kurdistan).

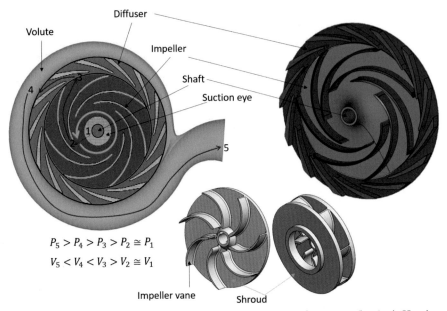

$$P_5 > P_4 > P_3 > P_2 \cong P_1$$
$$V_5 < V_4 < V_3 > V_2 \cong V_1$$

Figure 6.26 Impeller, diffuser, volute, and pressurizing process in a pump (by Amir Hossien Pajooh, mechanical engineering student at University of Kurdistan).

would increase from point 2 to point 3. The pressure increase is due to increasing cross-section area from point 2 to point 3, while the velocity increase is due to impeller rotation. From the diffuser inlet to its outlet, the cross-section area increases, and according to the Bernoulli equation, velocity would be converted to pressure. Again at point 4, where the liquid enters the volute, the cross-section passage through the volute would increase from point 4 to point 5, and as a result, the liquid velocity would be converted to pressure.

3.1 Impeller, diffuser, and volute

Impellers, diffusers, and volute are responsible to convert the mechanical power of the driver to the hydraulic power and transfer it to the liquid.

The pressurizing process starts when the driver rotates, and its rotation transfers to the pump shaft through the coupling. Before starting the pump, the lubrication and flushing systems must be checked; it should be primed and vented to avoid damaging the pump. When the pump starts, it creates vacuum pressure at the pump inlet (suction eye). Liquid enters the impeller and splits between the diffuser type channels of the impeller. Due to the impeller rotation, liquid velocity increases, and also because of the diffuser-shaped channels, its pressure rises as well (Fig. 6.27). Since the main purpose of a pump is to elevate pressure, the high velocity of liquid is first converted to pressure in the stationary diffusers located just after the impeller. In the multistage pumps, between every two neighbor impellers a row of nozzles exists to convert

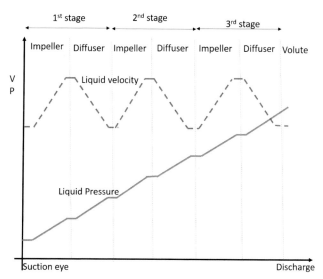

Figure 6.27 Velocity and pressure diagram in the multistage pumps.

velocity to pressure, and guide the liquid to the suction eye of the next impeller. Fig. 6.27 shows the velocity and pressure trends in a multistage pump.

The volute which exists in the single-stage centrifugal pumps or at the last stage of the multistage centrifugal pumps is a passage with increasing cross-sectional area toward the pump discharge. Its duty is to provide enough space to discharge exit liquid from the stationary diffusers, to balance the radial forces, and to convert fluid velocity to pressure.

3.2 Sealing system

Sealing systems can be categorized as static or dynamic. The static sealing refers to the sealing of clearance between two stationary surfaces and can be done by glues, washers, O-, C-, U-, and V-shaped rings. The dynamic sealing refers to the sealing of clearance between two surfaces in which at least one of them is moving. The moving mode can be either reciprocating or rotational. For example, sealing between the cylinder and its piston in a reciprocating pump or sealing the gap between the rotational shaft of a centrifugal pump and its casing are two examples of dynamic sealing for reciprocating and rotational motions. Flow chart in Fig. 6.28 shows the classification of the sealing types.

The sealing system of the centrifugal pumps is designed to minimize the leakage between shaft and casing. For small water pumps, usually packing box is used for sealing but for big pumps such as those used as feedwater pump of SPP, or the pumps containing expensive, or dangerous liquids, mechanical seals must be used.

The packing box of a pumps is illustrated in Fig. 6.29; it usually comprises of several packing rings, a lantern ring, sealing gland, and stuffing box. Adjusting the gland is very important for proper sealing. In addition, the lantern ring must be located

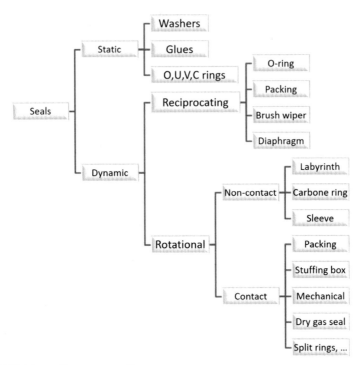

Figure 6.28 The sealing category diagram.

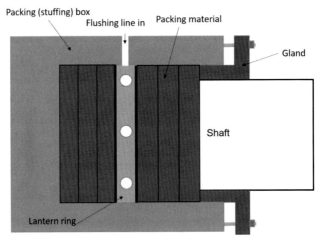

Figure 6.29 The packing box of a pump and its components.

exactly below the flushing line for proper distribution of the cooling liquid. The lantern ring may be metallic or nonmetallic such as PTFE (Fig. 6.30). When using packing box, one droplet per second of leakage is natural and essential for cooling. Less leakage because of over tightening the gland, can cause drying, overheating and damaging the packing, and finally more leakage. Over loosening the gland results in undesired leakage that is a waste of working fluid and energy resources.

The mechanical seals, which are used for larger pumps such as the SPP feedwater pumps, or the applications with toxic, flammable, expensive, or dangerous liquids, minimize the leakage to about 10 droplets or even less per hour per seal. This leakage is usually invisible because it evaporates, or in case of dangerous liquids, it would be drained to a safe basin. The mechanical seal comprises of two stationary and rotating faces, collar (thrust unit), some O-rings as secondary seals, and adjusting spring (Fig. 6.31). The stationary and rotating faces are flat, annular sliding surfaces that are pushed together. The rotating face is fixed to the shaft, while the stationary face is fixed to the pump casing. The cooling liquid for the mechanical seal, which is in contact with faces, can be either the process liquid or liquid from the external sources. When the process liquid is extremely hot or cold, contaminant, corrosive, etc., an external source should be used for the cooling purpose. The cooling plan can be chosen

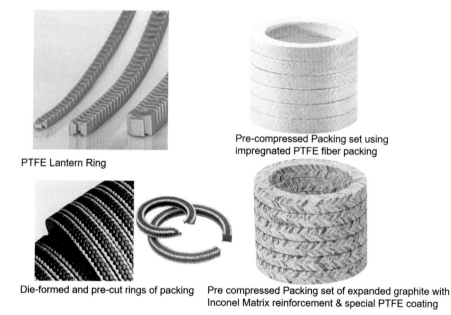

PTFE Lantern Ring

Pre-compressed Packing set using impregnated PTFE fiber packing

Die-formed and pre-cut rings of packing

Pre compressed Packing set of expanded graphite with Inconel Matrix reinforcement & special PTFE coating

Figure 6.30 Packings and lantern ring made of PTFE (AESSEAL global technology center) [7].

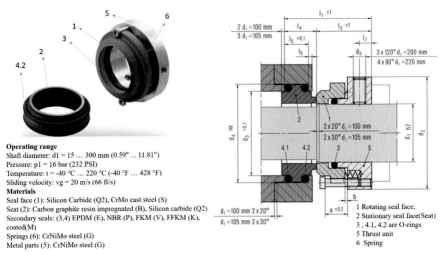

Operating range

Shaft diameter: d1 = 15 … 300 mm (0.59" … 11.81")
Pressure: p1 = 16 bar (232 PSI)
Temperature: t = -40 °C … 220 °C (-40 °F … 428 °F)
Sliding velocity: vg = 20 m/s (66 ft/s)

Materials

Seal face (1): Silicon Carbide (Q2), CrMo cast steel (S)
Seat (2): Carbon graphite resin impregnated (B), Silicon carbide (Q2)
Secondary seals: (3,4) EPDM (E), NBR (P), FKM (V), FFKM (K), coated(M)
Springs (6): CrNiMo steel (G)
Metal parts (5): CrNiMo steel (G)

1 Rotating seal face,
2 Stationary seal face(Seat)
3 , 4.1, 4.2 are O-rings
5 Thrust unit
6 Spring

Figure 6.31 A balanced mechanical seal, its components, operating range, and material (EagleBurgmann courtesy) [8].

according to the API flushing plans.[4] The cooling liquid penetrates the gap (which is about one micro meter) between the faces and tends to leak while lubricating the sliding faces to avoid wear. Dry operation of seal can result in immediate damage of the seal (Fig. 6.32).

Figure 6.32 Mechanical seal failure due to dry operation (personal gallery).

---

[4] American Petroleum Institute has recommended several cooling plans for cooling the mechanical seal systems.

Fig. 6.31 shows a mechanical seal by EagleBurgmann company that can be used for different applications including thermal power plants. Part number one is the main seal surface made of silicon carbide and CrMo cast steel; it is integrated with the thrust unit with some springs between them. The thrust unit position is fixed by pinning it to the shaft. In addition, the position of the seal surface can be adjusted according to the balance between the forces exerted by the cooling liquid pressure in the sealing chamber and the spring. The part number three is a static O-ring, which seals the space between the shaft and seal face; furthermore the static O-rings numbered with 4.1 and 4.2 are used to seal the gap between the stationary seal face and the stuffing box. Fig. 6.33 shows an O-ring. O-ring is a static seal when both surfaces in which their gap is sealed by the O-ring are stationary. An O-ring can also be dynamic seal, such as rings used for sealing the gap between the cylinder and piston. O-rings are mostly made of nitrile, neoprene, or fluoro-elastomers. They are light and flexible, and under compression will deform to follow the contours of the component parts to be sealed. "O" rings can be utilized in rotary, reciprocating, and oscillating applications. Commonly troubles that may occur for O-rings include extrusion (nibbling), over compression, heat hardening, thermal degradation, spiral failure, chemical degradation, explosive decompression, abrasion, plasticizer extraction, installation damage, and weather or ozone cracking. It should be noted that the "O" in the O-ring name refers to the "O" shape of the cross-section. In some applications, the C-ring, D-ring, U-ring, etc. may be used.

According to the application and operational conditions, mechanical seals can be used as single seal, tandem seal, and dual seal (Fig. 6.34). The tandem and dual configurations are mostly suitable for the applications with toxic, hazardous, flammable, or expensive process fluid. They can minimize the leakage and also separate the

Figure 6.33 O-ring for sealing purposes (by Sadegh Daryaei, mechanical engineering student at university of kurdistan).

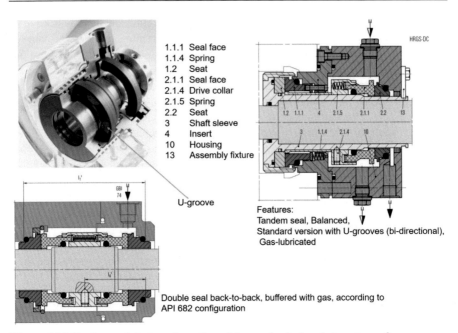

1.1.1 Seal face
1.1.4 Spring
1.2   Seat
2.1.1 Seal face
2.1.4 Drive collar
2.1.5 Spring
2.2   Seat
3   Shaft sleeve
4   Insert
10   Housing
13   Assembly fixture

U-groove

Features:
Tandem seal, Balanced,
Standard version with U-grooves (bi-directional),
Gas-lubricated

Double seal back-to-back, buffered with gas, according to
API 682 configuration

Figure 6.34 Tandem and dual configuration of the mechanical seals (courtesy of EagleBurgmann) [8].

process fluid from outside. In addition, when cooling medium leakage to the process fluid is inadmissible, using the tandem and dual configurations is highly recommended because by using a buffer line, the process fluid can be protected against mixing with the cooling fluid. The U-grooves on the seal face help to provide better film cooling between the sliding faces. These grooves can also be in different patterns, and according to their shape (to be symmetric or not to be), they can be classified as bi- or unidirectional.

The cooling liquid that is used for flushing the sealing system can be taken from the pump inside, pump discharge, or an external source. API 682 has recommended several plans for the piping of flushing line for the seal cooling. Some selective plans are graphically presented in Fig. 6.35. In addition, the description, advantages, and general points of these API plans are presented in Table 6.1.

Fig. 6.36 shows a five stage centrifugal pump specially designed for feedwater pumping in thermal power plants. The operating conditions and material of a typical feedwater pump are given in Table 6.2 [10].

The performance range of the pump presented in Table 6.2 is shown in Fig. 6.37.

The balance drum shown in Fig. 6.36 is used to reduce the axial force applied to the thrust bearings. The main axial force is generated due to pressure difference between suction and discharge; hence, the axial force direction would be from the discharge to the suction side (From left to right side). The balance drum should generate an axial force in the opposite direction. For this purpose, its right side is pressurized by the

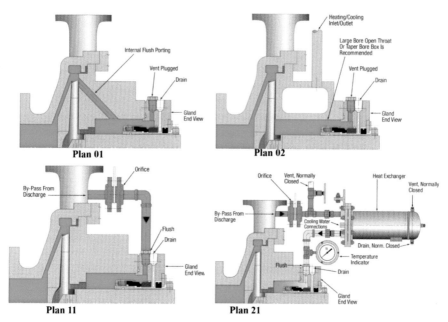

Figure 6.35 Four API plans for flushing of the mechanical seals (John Crane courtesy) [9].

discharge pressure, while the left side is vacuumed by the balance line that is connected to the suction line. The pressure difference between both sides of the balance drum generates an axial force from right to the left side that reduces the overall axial force applied to the thrust bearings.

4. Steam generator

A steam generator provides high-quality superheated steam required to be fed into the steam turbine for power generation. Depending on the cycle maximum pressure, it may be subcritical or supercritical. The supercritical SPP has a once-through steam generator without the boiling section, while a subcritical steam generator includes the boiling section as well.

SPP with supercritical steam generator reaches higher thermal efficiency than the SPP with subcritical steam generator. Supercritical steam generators need pipes with special alloys in the superheaters and reheaters because they have to withstand much higher pressures at high temperatures, which makes these steam generators more expensive than the subcritical type. A complete package of coal-fired subcritical boiler proposed by Babcock and Wilcox Company is illustrated in Fig. 6.38.

The main components of this steam generator include furnace, tube sections (superheaters, reheaters, risers, downcomers, water wall tubes, economizer), steam drum, low NO_x burners, coal silo, roll wheel coal pulverizer, bottom ash handling system,

Table 6.1 The description, advantages and general points of some of the API plans (John Crane courtesy).

Plan 01

Description: Plan 01 is an internal recirculation from the pump discharge area of the pump into the seal chamber, similar to plan 11 but with no exposed piping.

Advantages: No product contamination and no external piping. Advantageous on highly viscous fluids at lower temperatures to minimize the risk of freezing that can occur with exposed piping arrangements.

General: This flush plan should only be used for clean products as dirty products can clog the internal line. Not recommended on vertical pumps.

Plan 02

Description: Plan 02 is a noncirculating flush plan recommended only where adequate vapor suppression can be assured.

Advantages: Solids are not continually introduced into the seal chamber, no external hardware is required, and natural venting occurs when used with a tapered bore seal chamber.

General: Ideal with large bore/tapered bore ANSI/ASME B73.1 or ISO 3069 type "C" seal chambers or with hot process pumps utilizing a cooling jacket. On the latter services, a plan 62 with steam quench can also provide some additional cooling.

Plan 11

Description: Plan 11 is the most common flush plan in use today. This plan takes fluid from the pump discharge (or from an intermediate stage) through an orifice(s) and directs it to the seal chamber to provide cooling and lubrication to the seal faces.

Advantages: No product contamination and piping is simple.

General: If the seal is setup with a distributed or extended flush, the effectiveness of the system will be improved.

Plan 21

Description: Plan 21 is a cooled version of plan 11. The product from pump discharge is directed through an orifice, then to a heat exchanger to lower the temperature before being introduced into the seal chamber.

Advantages: Process fluid cools and lubricates the seal, therefore, no dilution of process stream. Cooling improves lubricity and reduces the possibility of vaporization in the seal chamber.

General: Plan 21 is not a preferred plan, either by API or many users, due to the high heat load put on the heat exchanger. A plan 23 is preferred.

---

forced draft fan, induced draft fan, air heater, selective catalytic reduction system (SCR), flue work, fly ash handling system, electrostatic precipitator, wet flue gas desulfurization system (FGD), and stack. A brief description with close picture for each component is provided below:

Furnace: is the combustion zone, where fuel burns and heat is then transferred to the steam generation sections.

Ash handling systems: A system to collect and dispose of discharged ash, when it cools down to a convenient temperature and then utilizing it in different industries such as cement plant, construction, and brick manufacturing. Ash should be collected at

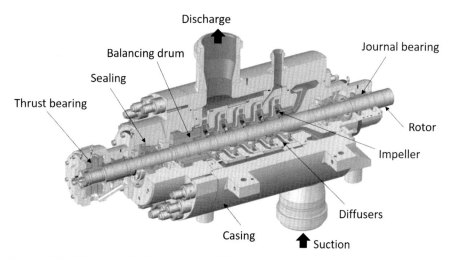

Figure 6.36 A five stage feedwater pump and its main components [10].
Courtesy of Elsevier.

Table 6.2 The specifications of a seven stage feedwater pump [11].

Operating conditions		Components material	
Operating temperature	Up to 220°C	Casing	Chrome steel
Operating pressure	Up to 430 bar	Impellers	Chrome steel
Head	Up to 4000 m	Shaft	Chrome steel
Capacity	Up to 1750 $\frac{m^3}{h}$	Balancing system	Chrome steel
Size	Up to 300 mm		

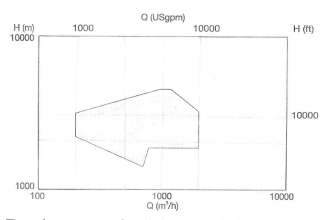

Figure 6.37 The performance range of a typical seven stage feedwater pump, regenerated graph by author [11].
Courtesy of Sulzer.

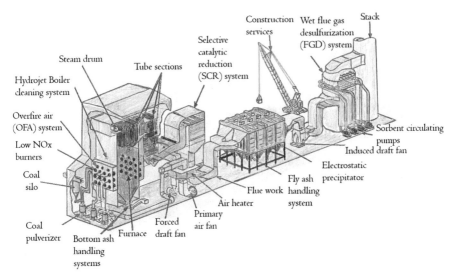

Figure 6.38 A complete package of subcritical boiler and its auxiliary systems, regenerated figure by author [12].
Courtesy of Babcock & Wilcox.

specified points along the flue gas stream. Fig. 6.39 shows the ash collecting points and approximate percentage of ash at each point. In a coal-fired steam generator, ash can be categorized according to its characteristics. The *bottom ash* is mostly heavy slag and is collected at the bottom of the furnace with a temperature of about 2400°F (1316°C). The ash drops into a circulating water and conveyed to a dewatering system or a pond. Fig. 6.40 shows a hydraulic bottom ash handling system. The *mill rejects* are the heavy pieces of stone and iron pyrite that are collected from the pulverizer discharge. *Economizer ash or popcorn ash*, are heavy particles that deviate from the path in a sudden change of path. This happens in places like air heaters and back pass. *Fly ash* is the fine and soft particles that are collected in the particulate control systems such as fabric filter and electrostatic precipitator system. About 50%−70% of the total ash is collected

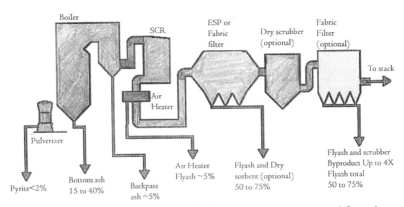

Figure 6.39 The ash collecting points along the flue gas stream, regenerated figure by author.
Courtesy of Babcock & Wilcox [12].

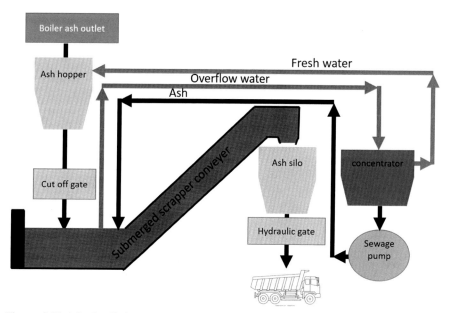

Figure 6.40 A hydraulic bottom ash handling system.

by the fly ash handling system. Since fly ash are very small particles, the hydraulic system used for the bottom ash cannot be used here. *Scrubber byproduct* is the reacted lime discharged from desulfurization devices.

Roll wheel coal pulverizer: Coal pulverizer is used to pulverize pieces of coal into fine particles (about $100 \mu m$) of coal before entering the furnace. Pulverizer also dries the coal by forcing hot air through the bottom of it to remove any water content from the coal, and conveying the coal dust to the burners. The third function of pulverizer is coal classification by separating the large pieces using a rotating piece and centrifugal force. Fig. 6.41 shows a drawing of a roll wheel coal pulverizer manufactured by Babcock & Wilcox company and a coal pulverizer manufactured by Mitsubishi Heavy Industry (MHI).

Coal silo: It is a storage that keeps raw coal before entering the pulverizer and is placed just before the pulverizer.

Low NOx burner: Low NOx burners create larger and more branched flames and reduce the NOx generation during the combustion process by controlling air to fuel ratio in different stages and reducing the flame temperature. Low NOx burners can be combined with other primary methods such as over fire air, and flue gas recirculation, to achieve higher NOx reduction. Fig. 6.42 shows two low NOx burners for firing coal. PM burner is used for swirl combustion method and NR burner is used for opposed firing method.

Over fire air system (OFA): It is an air nozzle used to maximize NOx reduction and minimization of combustible losses during the combustion process. It diverts a portion of air from the combustion zone to make the region fuel rich/oxygen lean, and as a result, less air will be available for combustion that results in lower flame

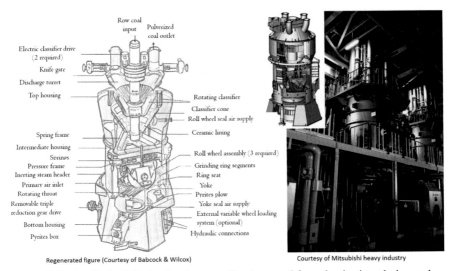

Regenerated figure (Courtesy of Babcock & Wilcox)   Courtesy of Mitsubishi heavy industry

Figure 6.41 A roll wheel Coal pulverizer specifications used for pulverization, drying and classification of coal [12,2].

temperature. The remaining fuel will be combusted with the diverted air later, and as a result, a stepped-like combustion occurs (Fig. 6.43).

Hydrojet boiler cleaning system: It is a tool for cleaning any surface inside the furnace that is exposed to the water jet line up to a distance of 30 m. The inside of the furnace is divided into cleaning zones where the hydrojet system can clean a desired zone or the whole system automatically, according to a predetermined pattern. Fig. 6.44 shows a cleaning pattern by the hydrojet and also the hydrojet components including valves, plumbing, pump, pressure gages, etc.

4.1 Steam drum

Steam drum is a cylindrical reservoir that provides water for the boiling section of the steam generator. Water flows to the boiling section by the downcomer tubes connected

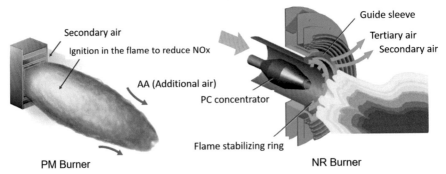

Figure 6.42 Low NOx burners for coal firing [2].
Courtesy of Mitsubishi Heavy Industry.

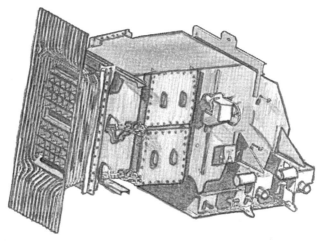

Figure 6.43 Over fire system used for NOx reduction, regenerated figure by author. Courtesy of Babcock & Wilcox, [12].

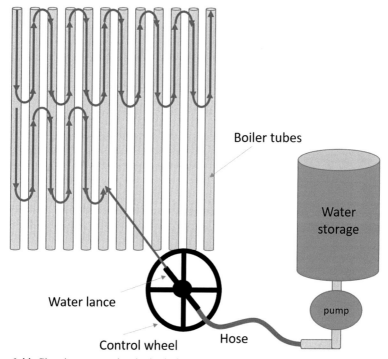

Figure 6.44 Cleaning pattern by the hydrojet and also the installations, regenerated figure by author [12].
Courtesy of Babcock & Wilcox.

to the bottom of steam drum. The boiling water and steam then return to the steam drum by the tubes called riser, which are also connected to the drum. Inside the drum, steam is separated from the liquid by using steam separator systems such as cyclones and scrubbers and dry saturated steam leaves the drum from its top to the superheaters.

Fig. 6.45 shows a steam drum with its internal components and also a low-weight vertical steam separator. Cyclone separator is a device with no moving parts that guide the liquid (heavy material) down and steam (light material) up. Scrubber is also a kind of a mist eliminator that stops small liquid particles to leave the drum. Steam drum also has tube connections for the feedwater coming from the economizer. The vertical steam separator also acts the same as the horizontal steam drum.

4.2 Air preheater

The hot flue gas before being exhausted through the stack can be used for preheating of the combustion air. For this purpose, a heat exchanger is used. This heat exchanger can be of two types:

* *Recuperative* type
* *Regenerative* type

The recuperator is a direct heat transfer exchanger in which both fluids remain un-mixed, and they are separated by a conduction wall. The wall can be tubular, plate, with or without fins. Since cleaning internal surface of passages is easier, the flue

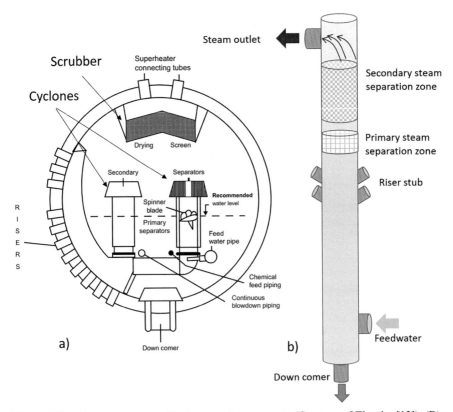

Figure 6.45 (A) A steam drum with its internal components (Courtesy of Elsevier [13]), (B) a low weight vertical steam separator, drawing by author.

gas flows inside the tube. However, this type of a preheater is recommended for clean flue gases. These preheaters can be single- or multipass. The most important advantages of recuperates are ease of manufacture, ease of cleaning, being silent due to having no moving parts, uniform temperature of heated air, less thermal shock, and no sealing problems. The only limit for manufacturing recuperators is the high temperature of the flue gas. The material should tolerate this high temperature. The main threats for recuperator are fouling due to volatiles of exhaust gas and dust. Three different configurations of the recuperator are presented in Fig. 6.46. It shows that flue gas is flowing through the tubes, while air is crossing the tubes.

The regenerative type of reheaters consists of three main parts: hoods, matrix, and a motor. If the matrix stays still, but the hoods rotate continuously, it is called

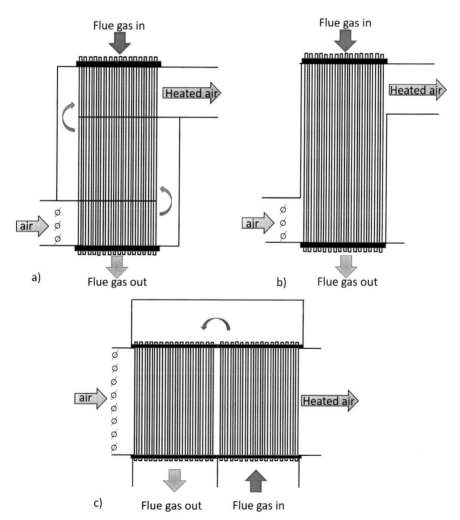

Figure 6.46 Gas through tube recuperator configurations, (A) air triple pass, flue gas single pass, (B) air single pass, flue gas single pass, and (C) flue gas double pass, air single pass.

Rothemuhle type, but if the hoods remain stationary while the matrix rotates, it is called *Ljungstrom* type (Fig. 6.47).

Fig. 6.48 shows a Ljungstrom-type regenerator. The rotational speed is about 4 rpm. Usually half of the cylindrical grid is heated by the hot flue gas, while the other half is giving up its heat to the combustion air.

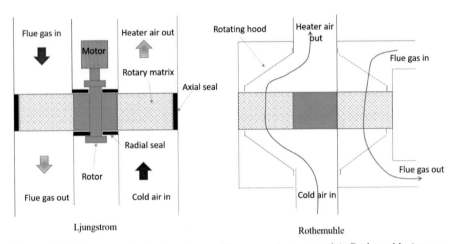

Figure 6.47 The regenerator heat exchanger Ljungsrom (rotary matrix), Rothemuhle (rotary hood).

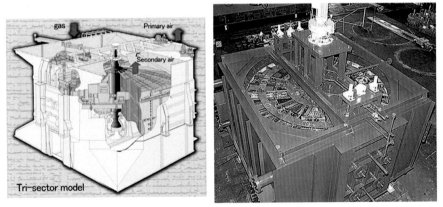

Rotating reproduction type air preheater

Figure 6.48 The Ljungstrom type regenerator.
Courtesy of Mitsubishi Heavy Industry [2].

Fans: Fresh air is usually forced to enter the furnace through the burners by using forced draft fan, and the flue gas is induced from the furnace to the stack by using a forced induced fan.

Selective catalytic reduction system (SCR): It is an effective system to reduce the nitrogen oxides in the flue gas by converting them to nitrogen and water. The catalyst is the main component of this system.

Electrostatic precipitator (ESP): An electrostatic precipitator is a kind of filter to separate soot and ash from the flue gas before they exit the stack. The ESP utilizes a high voltage electrostatic field to separate soot, dust, ash, etc.

Wet flue gas desulfurization system (FGD): It is a technology used to remove sulfur dioxide from the flue gas before entering the stack. In the wet flue gas desulfurization system, a slurry of alkaline sorbent, such as limestone, lime, or seawater, is used to scrub flue gases. The gas is forced through the liquid pool or liquid spray to bring them in contact. In addition to the wet scrubbing, there are some other desulfurization methods such as spray dry, wet sulfuric acid process, SNOX flue gas desulfurization, and dry sorbent injection systems.

4.3 Tube sections

The tube section of a steam generator includes waterwall tubes, superheaters, attemperators, reheaters, and economizers (Fig. 6.49). The feedwater first enters the economizer, such as air heater, recovers lo-quality heat from the flue gas. For each $22°C$

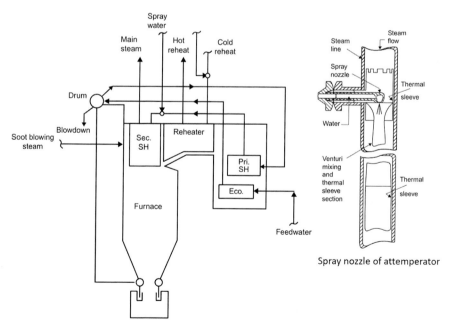

Figure 6.49 The economizer, superheaters, reheater and attemperator [13]. Courtesy of Elsevier.

cooling of the flue gas, the overall boiler efficiency increases by approximately 1%. Economizers also reduce the thermal shock and large water temperature fluctuations as the feedwater enters the steam drum or waterwall tubes. After the economizer, in a subcritical steam generator, water enters the steam drum, and from there, it flows downward by the downcomer tubes and then distributing in the waterwall tubes to become saturated mixture. The saturated mixture of liquid and steam then goes back to the steam drum through the risers. In the drum, steam is separated and enters the superheaters. Attemperators which are placed between the superheaters are used to adjust the superheated steam temperature before entering the turbine, also protecting the next superheater tubes from overheating. Adjustment is usually done by injecting water spray flow in the steam. A control valve adjusts the water flow according to the steam temperature at the turbine inlet and/or at the second superheater inlet. Attemperators may also be installed at the exit of reheaters before steam inlet to the turbine.

4.4   Stack

Stack is a vertical hollow column placed just after the flue gas desulfurization (FGD) system. Stacks which can also have an altitude of up to 400 m exhaust and disperse the combustion products at high altitude over a greater area, so the concentration of environmental pollutants in the air does not exceed the limitations of governmental and environmental policies and regulations. The high altitude of the stack also provides a pressure difference between the bottom and top of the stack, and there will be no need for a forced induced fan. The main reason for generating such pressure difference is the temperature difference and hence density difference of the flue gas inside the stack and atmospheric air outside the stack. A problem with the tall stacks is the vortex shedding due to air flowing around the stack that may cause a tall stack to collapse. Vortex shedding is an oscillating flow that forms when air is flowing past a bluff body such as stack. It produces an oscillating force that can cause a huge stack to collapse. In Figs. 6.50−6.52 the firing methods in steam generators and their characteristics are discussed and Figs. 6.53−6.56 present three different attemperators usually used in the steam generators.

5.   Cooling towers

The cooling tower duty is to reject the heat, taken from the steam in the condenser to the heat sink (which can be ambient, river, sea, or ocean). The temperature variation and definition of *approach* and *range* temperature in a cooling tower are shown in Fig. 6.56. The temperature change occurs due to evaporation cooling (latent heat transfer) and also direct heat transfer due to temperature difference of air and water (sensible heat transfer).

In the wet cooling towers, water loss mailnly occurs due to evaporation, evaporation rate at typical design conditions is approximately 1% of the water flow rate for each $6.9°C$ ($12.5°F$) of water temperature range. Water loss also occurs due to drifting, which is water droplets carryover by the air stream to the atmosphere. Drift is usually

For horizontal combustion of coal with medium and high volatility, it is appropriate to use short-flame turbulent burners. The burners may be arranged in the front wall, front and back walls or both side walls based on ease of design and arrangement. It can accommodate any type of fuel such as gas, oil and pulverized fuel. A compound burner, can be used for burning any combination of the mentioned fuels.

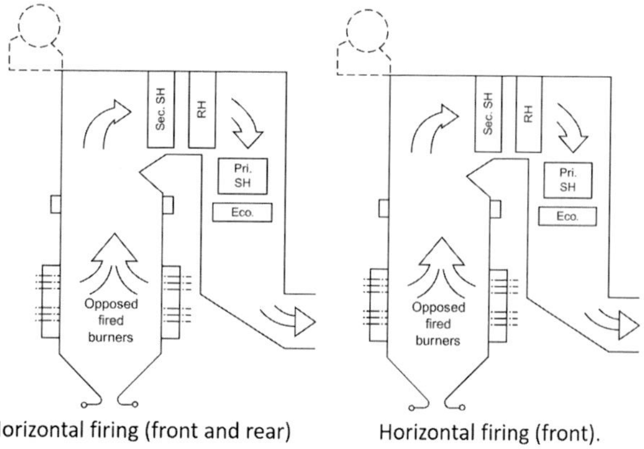

Horizontal firing (front and rear)

Horizontal firing (front).

Figure 6.50 Horizontal firing in the steam generators [13]. Courtesy of Elsevier.

The vertical firing boiler is used in places where the volatile matter of coal is less than 18 percent. Coal with less volatile matter takes longer to burn, so it doesn't need to turbulate the combustion flow as much as in horizontal fire boilers. Pulverized coal and primary air are injected downwards from the upper corner of the boiler, and secondary air is simultaneously injected in its vicinity and in the lower part. This flow causes the combustion products to be directed to the chimney in a U-shaped path, and the flame also becomes a W-shape.

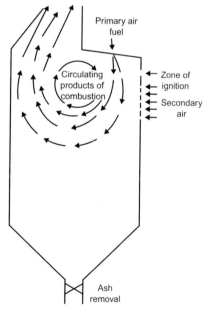

Figure 6.51 Vertical firing in the steam generators [13]. Courtesy of Elsevier.

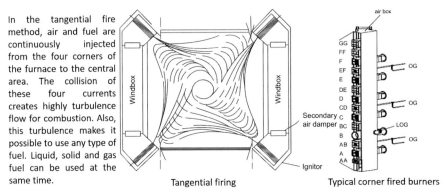

In the tangential fire method, air and fuel are continuously injected from the four corners of the furnace to the central area. The collision of these four currents creates highly turbulence flow for combustion. Also, this turbulence makes it possible to use any type of fuel. Liquid, solid and gas fuel can be used at the same time.

Tangential firing

Typical corner fired burners

Figure 6.52 Tangential firing, coal burners are located at levels "A, B, C, D, E, and F." Oil burners are located between coal burners at levels "AB, CD, and EF." Air is supplied at levels "AA, AB, BC, CD, DE, EF, and FF." Over-fire air is supplied at level "GG." [13]. Courtesy of Elsevier.

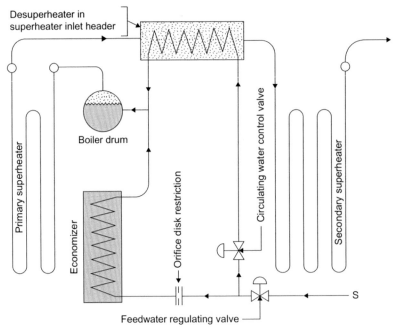

Figure 6.53 Header type attemperation [13]. Courtesy of Elsevier.

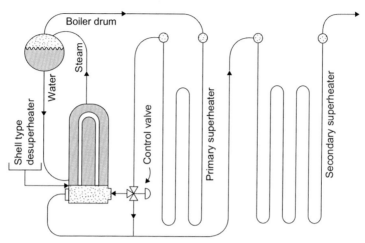

Figure 6.54 Shell and tube type attemperation [13].
Courtesy of Elsevier.

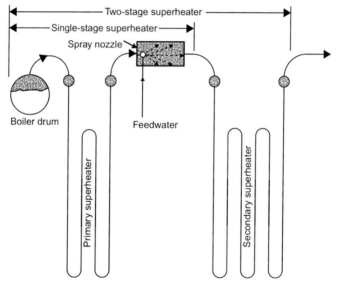

Figure 6.55 Direct contact type attemperation [13].
Courtesy of Elsevier.

about 0.2% of the water flow rate for each 6.9°C. Blowdown is another water loss reason that should be drained from the tower basin to maintain the water quality at an acceptable condition. Blowdown is mandatory because due to evaporation, the concentration of dissolved solid material increases. The amount of blowdown flow rate,

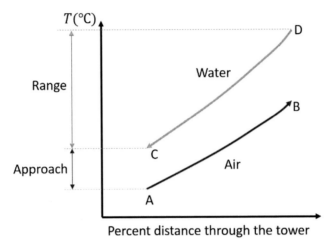

Figure 6.56 The temperature variation between water and air in counter flow cooling tower.

$(F_{blowdown}(\text{kg.s}^{-1}))$, is usually determined according to the evaporation rate and cycle of concentration[5] as Eq. (6.1):

$$F_{blowdown} = \frac{F_{evaporation}}{C - 1} \tag{6.1}$$

The water losses should be compensated by a make-up water line. The make-up water flow rate is determined by to the evaporation, drift, and blowdown flow rates as Eq. 6.2.

$$F_{make-up} = F_{blowdown} + F_{evaporation} + F_{drift} \tag{6.2}$$

Cooling towers used in the SPP can be categorized as wet, dry and hybrid, direct and indirect systems, and as well as natural and mechanical draft cooling tower. In the wet cooling towers, the cooling water returning from the condenser would be in direct contact with the cooling air stream. The air stream can be created by a natural draft due to chimney effect in a tall tower or by using fan in a short tower. In both cases, water is sprayed by nozzles over fill packings to increase the contact area between water and air streams. In the dry cooling towers, the cooling water, which is demineralized (for the indirect type) or the exit steam from the turbine (in the direct type), returns into some air cooled condenser (ACC). The cooling air can be provided by forced/induced draft fans in a short tower or by the chimney effect of a tall tower. The hybrid cooling tower combines dry and wet cooling towers technologies and

---

[5] Cycles of concentration (C) is the ratio of total dissolved solids (ppm) in the blowdown to the total dissolved solids (ppm) in the make-up water. C usually ranges from 5 to 10, the smaller the C the more the make-up water flow rate.

eliminates the thick plume above the tower, especially on cold and humid times. Hybrid cooling towers are especially suitable for situations where the power plant is close to residential areas, airports, etc.

The main components of a cooling tower include film fills, distribution nozzles, film fill protection device, fan stacks, drift eliminators, noise attenuation system, air inlet louvers, and multirow bundles. Fig. 6.57 presents most of these components.

Distribution nozzles are used to achieve optimal water distribution on the fill pack at various water flow rates. To evaluate the nozzles, different tests including discharge coefficients, reduced loss-of-pressure, and wear tests are conducted by the manufacturers. Fig. 6.58 shows different types of water spray nozzles.

The film fills provide an effective great heat surface area for the water and cooling air. They also convert the water droplets produced by the nozzles into thin film of water flowing over the fills surface. Fills exist in different patterns, and they should be selected based on operating parameters such as total suspended solids, water chemistry, cycles of concentration, and the presence of contaminants. Fig. 6.59 shows different types of packings.

Film fill protection device is used in the applications, which water contains sand and debris. It protects nozzles and the upper layer of the film fills. It is made of PVC-extruded pipes and polypropylene (PP) spacers and installed at the upper surface of the film fills.

Figure 6.57 The main components of cooling tower [14]. Mahtab Gostar, https://cooling-tower.ir/.

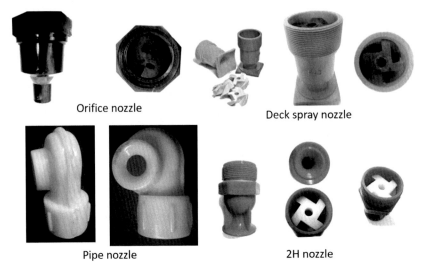

Figure 6.58 Different types of spray nozzles used in cooling towers.
Courtesy of Mahtab Gostar, https://cooling-tower.ir/.

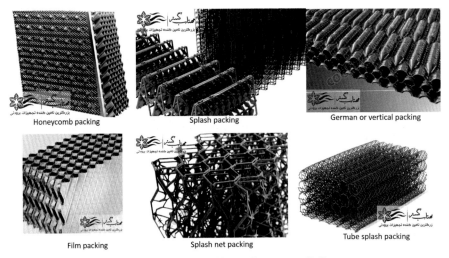

Figure 6.59 Different types of packings used in cooling towers [14].
Courtesy of Mahtab Gostar, https://cooling-tower.ir/.

Drift eliminators are installed above the distribution level (wet deck) at the exit way of air stream from the cooling tower. They are made of PVC or polypropylene and are assembled in packs. They create a maze channel patterns for the air stream. Water droplets on their leaving the tower must pass through these maze path, but they are mostly entrapped due to having higher inertia and stick to the drift eliminator surfaces and then fall on the fills. They play an important role in reduction of water waste. Generally, an efficient eliminator design reduces drift loss to between 0.001% and 0.005% of the water circulation rate. Fig. 6.60 shows different types of drift eliminators.

Inlet louvers are installed at the air inlets to the tower; they reduce the risk of debris and external objects to enter the cooling tower; they also provide shade for the water basin and reduce the possibility of algae and bacteria generation due to direct sunlight on the basin. Today, they are mostly made of PVC.

In situations where the tower components and processes may create a sound problem, several solutions can be recommended. Placing the tower as far as possible from the sound-sensitive areas is a simple solution. Two-speed fan motors can be used to reduce tower sound levels to a nominal 12 db during light-load periods. Minimizing the motor cycling to reduce noise fluctuations and using low-sound fans are other recommendations. Noise attenuators specially designed for the cooling tower are available from most manufacturers. Noise attenuation systems are some baffle form components that can be installed at the air inlet and/or on top of fan stacks. They can also be installed to reduce high-frequency noise produced by water fall from fills to water in the basin. The system is composed of high-quality PVC-extruded elements packed by polypropylene spacers and is installed over the water level in the basin, typically on FRP (Fiber reinforced plastic) beams.

Multi row bundles are the externally finned tube heat exchangers used in the hybrid (dry-wet) cooling towers to avoid plume generation on the tower exit.

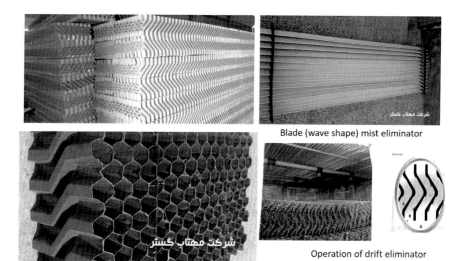

Blade (wave shape) mist eliminator

Operation of drift eliminator

Honeycomb drift eliminator, different grades

Figure 6.60 Different types of drift eliminators used in cooling towers, [14]. Courtesy of Mahtab Gostar, https://cooling-tower.ir/.

In addition to the main components, the accessories of a cooling tower include electric motors, gear, backstop devices, lubrication systems, and sensors to monitor mechanical components boxes, driving shafts, and fans. Firewalls to prevent fire propagation between cells, FRP risers to feed water in each cells, valves, vacuum systems, dampers, and mixing devices for dry section of hybrid cooling towers, firefighting system, cleaning system for heat exchangers, and monitoring system.

Figure 6.61 shows some different configuration of the mechanical draft (forced/induced) wet cooling towers presented by ASHRAE [12]. Some of these towers are very popular especially for air conditioning systems and the cooling tower of chillers as well as steam power plants.

Figure 6.62 shows two towers, on the left is a hybrid (wet-dry) cooling tower with finned tube bundle heat exchanger for the dry section, installed in parallel with the wet section. The heated air passing from the wet and dry sections is mixed together, and as a result, the plume over the tower decreases. On the right side is a natural draft hyperbolic wet cooling tower. The tall structure provides the chimney effect that results in continuous air stream through the tower and over the fills. The main advantage of this tower is to eliminate electricity consumption required by the fans, but water loss due to evaporation, drift, and blowdown is still critical.

Figure 6.63 shows mechanical draft dry cooling tower. Water flows through the finned tubes of the heat exchangers that may be installed in different patterns of A, V, and Δ frames. Air is forced/induced by fans over the tubes surfaces and water flowing inside the tubes gets cooled. The main advantage of the dry cooling tower is to decreasing water loss to a minimum value that may be due to leakage. However, the electricity consumption by the fans is still high that is the main disadvantage of this type of cooling tower.

Figure 6.64 shows a dry natural draft cooling tower used in the combined cycle power plant of Sanandaj, Iran. Photos were taken while the author was visiting this power plant with mechanical engineering students. This tower is built over some X-legs columns, that can be seen from the inside view, the finned tube heat exchangers are installed in V-frame pattern behind the X-legs on the outside, and natural draft air stream passes over them. As a result, the electricity consumption by the fans is omitted, while water consumption is also minimized. In the hot seasons, if required, water will be sprayed by some nozzles over the surface of heat exchangers to increase the heat transfer rate. Hence, water consumption is also minimized to the leakage and occasional sprays over the heat exchangers in the hot days.

6. Water treatment system

Water treatment system provides water with the required standards for the steam generator and the cooling tower as well. Water treatment system of a steam generator should be able to remove the harmful contaminants before entering the steam generator, control the water chemistry, maximize the use of steam condensate, minimize

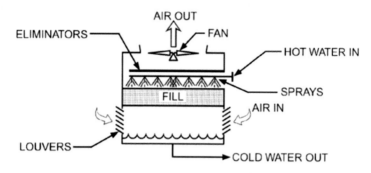

INDUCED-DRAFT COUNTERFLOW

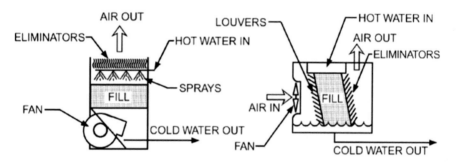

FORCED-DRAFT COUNTERFLOW   FORCED-DRAFT CROSSFLOW

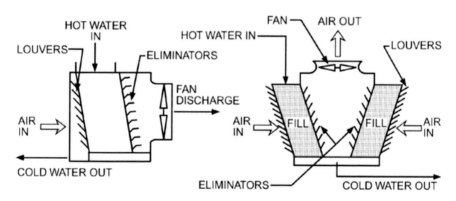

INDUCED-DRAFT CROSSFLOW   INDUCED-DRAFT CROSSFLOW
(SINGLE FLOW TOWER)   (DOUBLE AIR ENTRY)

Figure 6.61 Conventional mechanical (induced/forced) draft wet cooling towers (ASHRAE inc.) [15].

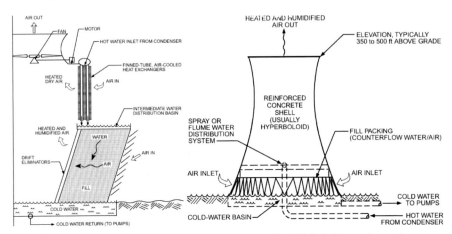

Figure 6.62 Natural draft, wet (*right*) cooling tower, (*right side*), hybrid wet-dry cooling tower (*left side*) (ASHRAE inc.) [15].

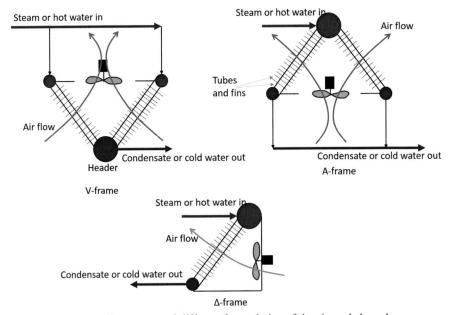

Figure 6.63 Dry cooling tower, and different frame design of the air cooled condenser.

the plant shutdown, increase the components lifetime, lengthen the maintenance periods, and control the return-line corrosion. A basic water treatment system may include filtration, ultrafiltration, ion exchange, softening, reverse osmosis, deaeration, and coagulations (Fig. 6.65).

Students taking photo inside of tower

Students, visiting dry cooling tower   X-legs, students inside of cooling tower

Figure 6.64 The author and students from University of Kurdistan visiting the combined cycle power plant of Sanandaj, the picture shows the dry, natural draft cooling tower from inside and outside.

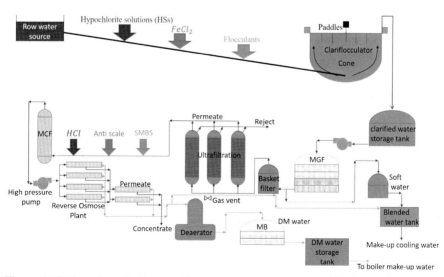

Figure 6.65 The schematic diagram of a water treatment system used for the cooling systems and boiler (KeloEco environmental limited) [16].

In Fig. 6.65, hypochlorite solutions (HSs), which is also called liquid chlorine, are mostly used for disinfecting water, liquid ferrous chloride ($FeCl_2$) is a chemical used for odor control. Removing fine particles from the water is not easy, but flocculants can be used to cluster the fine particles; as a result, the flocs may float to the surface or settle to the bottom that are easier to remove. Clariflocculator flocculates and clarifies the water stream entering it. The flocculation is improved by using some rotating paddles, while a cone shape baffle helps the heavy particles to settle, and the water flows radially upward. Multigrade filter (MGF) is a complex of coarse and fine filters for filtering both large and small suspended particles. Reverse osmosis (RO) is a process that removes impurities from water by using pressure to push the water through a membrane. Permeate is water treatment for the product water of the reverse osmosis unit. Deaerator is used to remove gas molecules from the water by spraying water in to a low pressure vessel; gas is then vented into the atmosphere. The mixed bed unit (MB) is used to further refinement the water and contains two types of resins, cation and anion, in the same column. Micron cartridge filters (MCFs) remove the contaminates bigger than the micron size of the filter. This is achieved by specially designed cartridge elements placed inside the filter housing. Ultrafiltration refers to water purification process in which water is pushed through a semipermeable membrane. During this process, heavy contaminants remain on the pressure side, and pure water and low-weight contaminates pass through the membrane to the other side. Sodium metabisulfite (SMBS) is frequently used for removing of free chlorine; theoretically, 1.34 mg of sodium metabisulfite will remove 1.0 mg of free chlorine. Hydrochloric acid (HCl) is a neutralizing substance and is well suited for water treatment applications. Water PH must decrease by using HCl to reduce the chance of calcium carbonate scaling in the RO concentrate.

Most of water sources contain some impurities that can be harmful for the steam generator, condenser, and cooling tower. These impurities may include alkalinity, calcium, carbon dioxide, chloride, magnesium, oxygen, high or low PH, silica, sodium, and sulfate. As the working pressure of the steam generator increases, the purity of water must increase as well. ASME[6] has provided a guideline for acceptable impurities of water in different working pressure of steam generators (Table 6.3).

7. Electric generator

The magnets can be used to generate electricity; for this purpose, the magnetic field should be moved, and as a result, the electrons flow (that means electricity production). Some metals such as copper and aluminum have loosely held electrons. Moving a magnet around a coil of copper wire, or moving a coil of copper wire around a magnet, pushes the electrons in the wire and generate electrical current. Electric generators convert kinetic energy into electrical energy.

---

[6] American Society of Mechanical Engineering.

Table 6.3 Guidelines for water quality in industrial water-tube steam generators for reliable continuous operation by ASME [17].

Boiler feedwater				Boiler water		
Drum pressure (psi)	Iron (ppm Fe)	Copper (ppm Cu)	Total hardness (ppm $CaCO_3$)	Silica (ppm SiO_2)	Total alkalinity (ppm $CaCO_3$)	Specific conductance (micro-ohms/ cm), (unneutralized)
0−300	0.1	0.05	0.3	150	700	7000
301−450	0.05	0.025	0.3	90	600	6000
451−600	0.03	0.02	0.2	40	500	5000
601−750	0.025	0.02	0.2	30	400	4000
751−900	0.02	0.015	0.1	20	300	3000
901−1000	0.02	0.015	0.05	8	200	2000
1001−1500	0.01	0.01	0.0	2	0.0	150
1501−2000	0.01	0.01	0.0	1	0.0	100

Electric generators can be classified as an AC or DC[7] generator. They both generate electricity by using mechanical power. Electric generator of a power plant converts the rotational motion of the turbine's shaft into electricity. In some cases, for example, in some kind of wave powered power plants, instead of rotational motion, a reciprocating motion is used to generate electricity. The main components of an electric generator are stator and rotor. The rotor of generator receives its rotation from the turbine through a coupling.

The stator is a series of insulated coils of wire that form a stationary cylinder, the stator surrounds an electromagnetic shaft, which is called rotor. Turning the rotor produce an electric current in each part of the coils, which becomes a separate electric conductor. The currents of the coils get together and form the final current that flows through the power lines for consumption.

The generators also need to be cooled to reduce thermal gradient between critical components. Most of this heat is generated due to electric current through conductors. Sometimes, it is also due to air molecule interactions due to shaft rotation. In addition, heat is generated in the bearings. Mitsubishi Power is one of main producer of electric generators in the world that produces air cooled, hydrogen cooled, and water cooled generators in different capacities for thermal power plants including gas turbines, steam turbines, and combined cycles. Fig. 6.66 shows these three types of generators manufactured by Mitsubishi Power [15].

[7] Direct current, or alternative current.

Water/hydrogen cooled turbine generator, efficiency at least 99%, Capacity: 500-1400 MVA, Voltage Up to 27kV

Air cooled turbine generator, Capacity: Up to 350 MVA, Voltage 10 to 18kV, or specified voltage, efficiency at least 98.5%

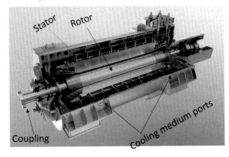

Figure 6.66 Three types of generators [18].
Courtesy of Mitsubishi Power.

8. Problems

1. Who was the first inventor of steam turbines?
2. What would be the impact of scale forming on the last stages of a steam turbine? When they should be cleaned?
3. How the steam quality is tested before entering the steam turbine?
4. Why the first stages of a steam turbine are mostly impulsive, while the last stages are mostly reactive type?
5. What is a deaerator? How it works? Draw it schematically and find a real picture of it in power plants.
6. What are the applications of steam ejector in steam power plants?
7. What kind of sealing systems are usually used for steam turbine?
8. Due to improper deaeration, the pressure inside condenser increases, what would be the impact of this problem on the cycle performance?
9. How the internal parts of a condenser should be cleaned from scales and fouling?
10. What equipment is usually used to keep the condenser pressure low? How it works?
11. The cooling water temperature leaving the condenser has increased and is out of range. As an engineer how do you troubleshoot this problem? What are the possible causes?
12. Draw the schematic diagram of a lubrication system. What components it has?
13. What are the main properties of the oil used for lubricating of bearings in steam turbine?
14. The oil temperature returning to the oil reservoir is abnormally high. As an engineer how do you troubleshoot this problem? What are the possible causes?

15. Imagine you are standing by the feedwater pump of an SPP and you are hearing a noise similar to sand impingement on the internal parts of the pump. Can you guess what is happening? What are the possible causes of this strange sound?

16. The pressure gauge at the pump discharge detects fluctuations, what are the possible causes for pressure fluctuations?

17. The pressure at the feedwater pump discharge is falling, what are the possible causes?

18. Where does the make-up water of steam generator is usually injected into the cycle?

19. Where does the blowdown is taken from the steam generator cycle?

20. The water level in the steam drum must be controlled to stay in a safe level range. How it is usually controlled in a steam generator?

21. Compare the yearly water consumption of dry cooling tower with wet cooling tower?

22. In a wet cooling tower, the water is not well cooled. What are the possible causes?

23. What is *carry over* in a steam generator, and what are the possible causes?

24. What is the *"test pipe"* in a steam generator?

References

[1] https://www.siemens.com/global/en.html, accessed on 4/11/2022.

[2] MITSUBISHI heavy industries, LTD, https://www.mhi.com/products, accessed on 4/11/2022.

[3] Elliott Ebara Turbomachinery Corporation, https://www.elliott-turbo.com/, accessed on 4/11/2022.

[4] Dresser-Rand, https://www.siemens-energy.com/global/en/offerings/compression/contact.html, accessed on 4/11/2022.

[5] American Petroleum Institute, API, https://www.api.org/products-and-services/training, accessed on 4/11/2022.

[6] TEMA, http://kbcdco.tema.org/, accessed on 4/11/2022.

[7] AESSEAL Global Technology Centre, https://www.aesseal.com/en/locations/europe/gb/syk/aesseal-plc-headquarters, accessed on 4/11/2022.

[8] EagleBurgmann, https://www.eagleburgmann.com/en, accessed on 4/11/2022.

[9] J. Crane, https://www.johncrane.com/en/products/mechanical-seals, accessed on 4/11/2022.

[10] Sulzer Pumps Ltd, Winterthur, Switzerland, Centrifugal Pump Handbook, third edition, 2010, Elsevier Ltd.

[11] Sulzer Ltd, https://www.sulzer.com/en, accessed on 4/11/2022.

[12] Babcock & Wilcox, https://www.babcock.com/, accessed on 4/11/2022.

[13] D.K. Sarkar, Thermal power plant design and operation, Elsevier, 2015.

[14] Mahtab Gostar Company, https://cooling-tower.ir/, accessed on 1/30/2023.

[15] ASHRAE Inc, https://www.ashrae.org/, accessed on 4/11/2022.

[16] KeloEco Environmental Limited, https://kelopure.en.alibaba.com/, accessed on 4/11/2022.

[17] ASME, https://www.asme.org/, accessed on 4/11/2022.

[18] Mitsubishi Power, https://power.mhi.com/products/generators, accessed on 1/30/2023.

Gas turbine power plant

7

Chapter outline

Gas turbines (GTs) are rotary, combustion engine power generators that burn liquid or gaseous fuels with the cleaned and compressed air (coming from the compressor) in the combustion chamber(s). Then, the high-temperature and high-pressure products enter the turbine to expand while passing through the stages. As the gas molecules move forward, their thermal and potential energy convert to kinetic energy that will be expended for rotating the turbine shaft. The turbine rotor is connected to an electric generator for turning the electromagnetic rotor of generator in front of its stator for electricity generation. Fig. 7.1 shows the main components of a GT and its main specifications manufactured by Mitsubishi Power, as one of the leading companies in this field.

Gas turbines have several advantages over the steam turbines such as fast start up, needing less space for installation, fast installation, smaller installation costs, approximately zero water consumption, and easier to control. On the other hand, they have lower thermal efficiency unless being used as combined cycle; they have shorter

Power Generation Technologies. https://doi.org/10.1016/B978-0-323-95370-2.00009-0
Copyright © 2023 Elsevier Inc. All rights reserved.

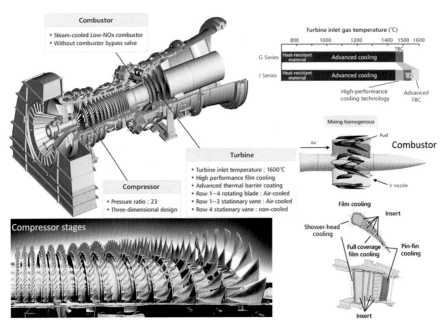

Figure 7.1 Technical features of J-series gas turbine, compressor with 15 stages and pressure ratio of 23, 4-stage turbine with inlet temperature of 1600 °C, steam cooled combustor, and air cooled turbine blades and nozzles [1].
Courtesy of Mitsubishi power.

lifetime. According to the report published by Allied Market Research [2], the global GT market was estimated at $18.5 billion in 2020 and is expected to hit $25.4 billion by 2030. Leading GT manufacturers between 2016 and 2025, based on global market share, is reported in Fig. 7.2. It shows that General Electric (U.S), Siemens (Germany), and MHPS (Japan) are the three leading companies according their share in the global market.

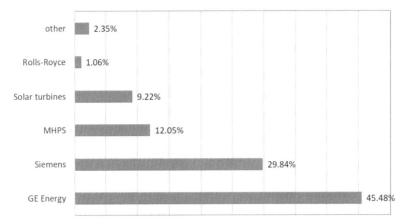

Figure 7.2 Leading gas turbine manufacturers between 2016 and 2025, based on global market share [2].

Gas turbines play an important role in the electricity generation worldwide. The market forecasts also show that this market is growing very fast in the next decade. Therefore, students will need to find out more about GT, its technologies, operation, maintenance, troubleshooting, designs, challenges, and its market as well. In the following, the thermodynamic principles and also components of GTs will be introduced.

In the present chapter, the operation and processes of GTs are discussed; it is analyzed thermodynamically, and different efficiency improvement techniques are applied. Practical examples are presented. The EES code for each example is also attached to the solution of examples for the students and readers to extend the code and do more analyses on the cycles. In addition, the main and auxiliary components of a GT are presented with very clear and high-quality pictures from the main manufacturers worldwide. At the end, several exercises are designed for better understanding of the GT cycle and its components.

1. Technology description

According to the thermodynamics principals, a GT cycle like every other heat engines comprises of four essential processes of *compression, heat addition, expansion,* and *heat rejection* (Fig. 7.3). The first three processes occur in the compressor, combustion chamber, and turbine sections of the GT, respectively. But the heat rejection can happen by exhausting the combustion gases into the atmosphere (for open cycle GT) or by using a cooler heat exchanger and cooling medium that can be air or water (for close cycle GT). Most of the industrial GTs are open cycle, internal combustion,

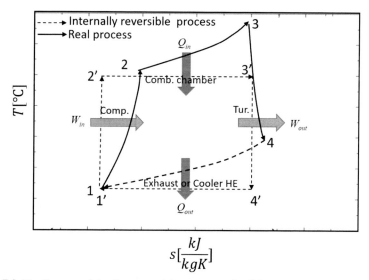

Figure 7.3 T-s diagram of the four essential processes of a GT.

and use fresh air as the working fluid. However, there are some closed cycle GTs with external combustion system that use other gases such as CO_2 as the working fluid.

1.1 GT thermodynamics

An internally reversible GT cycle comprises an isentropic compression, isobaric heating, an isentropic expansion, and an isobaric heat rejection. These for processes together are called the Brayton cycle, named after George Brayton (1830–1892), the American engineer who developed it originally for use in piston engines. The real GT cycles have internal irreversibilities due to friction, mixing, combustion, etc.; hence, the real cycle deviates from the reversible Brayton cycle. In the Brayton cycle, the *cold air standard assumptions* are also used. It is assumed that the working fluid is pure air, air is perfect gas, the combustion process is replaced by a constant pressure heating process, exhaust is replaced by a constant pressure heat rejection process, and air specific heat remains constant.

Example 7.1. Consider the Siemens GT model of SGT-600 (Fig. 7.4). It is a double shaft turbine that in ISO conditions[1] generates **net power** of 24.5 MWe. The exhaust

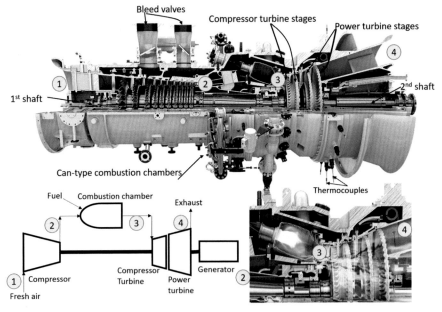

Figure 7.4 Double shaft Siemens gas turbine model of SGT-600 and the schematic diagram [3]. Courtesy of Siemens.

---

[1] The standard conditions used by the gas turbine industry are $59°F/15°C$, 14.7 psia/1.013 bar and 60% relative humidity, which are established by the International Standards Organization (ISO) and frequently referred to as ISO conditions.

temperature and mass flow rate are 543°C and 81.3 $\frac{kg}{s}$. The compression ratio of compressor is 14, its adiabatic efficiency is 0.87, and the nominal rotational speed is 7700 rpm. The ambient temperature and pressure are $P_1 = 101.325$ kPa, $T_1 = 15\,°C$.

a. Draw the T-s thermodynamic diagram of the real and reversible GT on the same graph.
b. The fuel energy rate burned in the combustion chamber ($\dot{Q}_{in}$), the power generated by turbine ($\dot{W}_{tur}$), and the power consumption of compressor($\dot{W}_{comp}$).
c. Determine the GT thermal efficiency and validate your answer with the SGT-600 net thermal efficiency reported by the Siemens ($\eta_{net,th} = 33.6\%$). What are the possible reasons for the difference?
d. Determine the heat rate of the GT.
e. Determine the exergy efficiency of the cycle.

Solution. Prior to the thermodynamic evaluation of the cycle, it is convenient to learn more about the GT used in this example. This GT is a double-shaft GT, one shaft connects the compressor to the *compressor turbine* and the other shaft connects the *power turbine* to the generator. The power turbine and compressor turbine are not connected mechanically. The power produced by the compressor turbine is completely consumed by the compressor, while the power turbine only feeds the electric generator. In general, the GTs can be single shaft, double shaft, and tipple shaft. In the double and triple shafts, the shafts can have different speeds.
The input data are as below:

$$\dot{W}_{net} = 24.5 \times 10^3 \text{ kW}, \ \dot{m}_4 = 81.3\frac{kg}{s}, \ T_4 = 543°C, \ P_1 = 101.325 \text{ kPa}, \ T_1$$

$$= 15\,°C, \ \frac{P_2}{P_1} = 14, \ N = 7700 \text{ rpm}, \ \eta_{ad,comp} = 87\%$$

Assumptions. It is assumed the working fluid through the cycle is pure dry air, and the air mass flow rate everywhere is constant, that means every extraction from the GT, or fuel mass flow rate are negligible in comparison with the air mass flow rate at point 1. The pressure loss in the combustion chamber is neglected.

a. To draw the T-s diagram, the state point properties such as temperature and entropy should be calculated ate points 1 to 4.

Starting from point 1:

$$P_1 = 101.325 \text{ kPa}, \ T_1 = 15\,°C \rightarrow h_1 = 288.5\frac{kJ}{kg}, \ s_1 = 5.661\frac{kJ}{kg.K}$$

Points 2 and 2s:

$$P_2 = 14 P_1 = 1419 \text{ kPa}, \ s_{2s} = s_1 \rightarrow h_{2s} = 613.2\frac{kJ}{kg}$$

$$\eta_{ad,comp} = \frac{h_{2s} - h_1}{h_2 - h_1} \rightarrow 0.87 = \frac{613.2 - 288.5}{h_2 - 288.5} \rightarrow h_2 = 661.7\frac{kJ}{kg} \rightarrow T_2$$

$$= 378.6\,°C \rightarrow s_2 = 5.738\frac{kJ}{kg.K}$$

Points 4 and 4s:

$$T_4 = 543\,°C, \quad P_4 = P_1 \rightarrow h_4 = 839.8\frac{kJ}{kg}, \quad s_4 = 6.739\frac{kJ}{kg.K}, \quad s_{4s} = s_3$$

Point 3:

$$\dot{W}_{net} = \dot{W}_{tur} - \dot{W}_{comp} = \dot{m}_4(h_3 - h_4) - \dot{m}_2(h_2 - h_1) \rightarrow$$

$$24.5 \times 10^3 = 81.3(h_3 - 839.8) - 81.3(661.7 - 288.5) \rightarrow h_3 = 1514\frac{kJ}{kg} \rightarrow T_3$$

$$= 1126\,°C \rightarrow s_3 = 6.603\frac{kJ}{kg.K}$$

According to the above state points the T-s diagram is provided in Fig. 7.5:

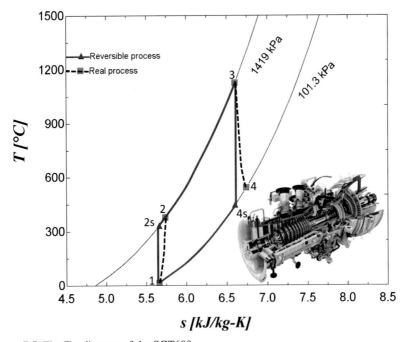

Figure 7.5 The T-s diagram of the SGT600.

b.

$$\dot{Q}_{in} = \dot{m}_4(h_3 - h_2) = 81.3(1514 - 661.7) = 69322 \, kW$$

$$\dot{W}_{comp} = \dot{m}_2(h_2 - h_1) = 81.3(661.7 - 288.5) = 30344 \, kW$$

$$\dot{W}_{tur} = \dot{m}_4(h_3 - h_4) = 81.3(1514 - 839.8) = 54844 \, kW$$

c. The thermal efficiency is as below:

$$\eta_{th} = \frac{\dot{W}_{net}}{\dot{Q}_{in}} = \frac{24.5 \times 10^3}{69322} = 35.34\%$$

The net thermal efficiency reported by the manufacturer is 33.6%. The net efficiency is smaller because it subtracts the electricity consumption by the auxiliary systems such lube oil pumps, air conditioning systems, control and monitoring, cooling water pumps, cooling tower fans, etc. from the power generated by the turbine, in other words:

$$\eta_{net,th} = \frac{\dot{W}_{net}}{\dot{Q}_{in}} = \frac{\dot{W}_{tur} - \dot{W}_{comp} - \dot{W}_{auxiliary}}{\dot{Q}_{in}}$$

Another reason for less efficiency of the real cycle is that about 5% of air mass flow rate is usually extracted from the compressor stages for cooling of the turbine blades and stationary nozzles without entering the combustion chamber. This cooling air then gets mixed with the hot combustion gases, and as a result, it lowers the average temperature of gas flow inside the turbine, and finally it reduces the thermal efficiency.

d. The heat rate can be calculated as below:

$$HR = \frac{\dot{Q}_{in}\left(\frac{kJ}{h}\right)}{\dot{W}_{net}(kW)} = \frac{69322\left(\frac{kJ}{s}\right) \times 3600\left(\frac{s}{h}\right)}{24.5 \times 10^3 (kW)} = \frac{3600}{\eta_{th}} = \frac{3600}{0.3534}$$

$$= 10186.76\left(\frac{kJ}{kWh}\right)$$

e. The exergy efficiency is also determined as below, firstly, the mean temperature of heat addition, T_H should be determined:

$$T_H = \frac{\dot{Q}_{in}}{\dot{m}_4(s_3 - s_2)} = \frac{69322}{81.3(6.603 - 5.738)} = 986 \, K, \, \eta_{II} = \frac{X_{recovered}}{X_{supplied}}$$

$$= \frac{\dot{W}_{net}}{\dot{Q}_{in}\left(1 - \frac{T_0}{T_H}\right)} = \frac{24.5 \times 10^3}{69292\left(1 - \frac{25 + 273.15}{986}\right)} = 50.66\%$$

The EES code written for the example 7.1 is presented in **Code 7.1**.

Code 7.1. The code for simple cycle of SGT600 in Example 7.1.

```
"siemense SGT600"                          h2s=enthalpy(air, P=P[2], S=s[1])
W_net=24.5*10^3                            ETA_COMP=0.87
m_dot=81.3                                 ETA_COMP=(h[1]-h2s)/(h[1]-h[2])
T[4]=543                                   T[2]=TEMPERATURE(Air,h=h[2])
P[4]=101.325                               T[3]=TEMPERATURE(Air,h=h[3])
P[1]=P[4]                                  s[2]=ENTROPY(Air,T=T[2],P=P[2])
T[1]=15                                    s[3]=ENTROPY(Air,T=T[3],P=P[3])
h[1]=ENTHALPY(Air,T=T[1])                  T_H=Q_in/(m_dot*(s[3]-s[2]))
s[1]=ENTROPY(Air,T=T[1],P=P[1])            EX_supplied=Q_in*(1-(25+273.15)/T_H)
h[4]=ENTHALPY(Air,T=T[4])                  EX_recovered=W_net
s[4]=ENTROPY(Air,T=T[4],P=P[4])            eta_II=EX_recovered/EX_supplied
ETA_th=W_net/Q_in                          Ts[1]=T[1]
Q_in=m_dot*(h[3]-h[2])                     Ts[2]=TEMPERATURE(air, h=h2s)
Q_out=m_dot*(h[4]-h[1])                    Ts[3]=T[3]
P[2]/P[1]=14                               Ts[4]=TEMPERATURE(air, s=s[3], P=P[4])
P[3]=P[2]                                  HR=1/ETA_th*3600
W_net=m_dot*(h[3]-h[4])-m_dot*(h[2]-h[1])  Ss[1]=s[1]
W_comp=m_dot*(h[2]-h[1])                   Ss[2]=s[1]
w_tur=m_dot*(h[3]-h[4])                    Ss[3]=s[3]
                                           Ss[4]=s[3]
```

1.2 Efficiency improvement techniques

Efficiency improvement opportunities in the GTs exist in the air intake system, compressor, combustion chamber, turbine, and exhaust as well.

Air intake

To avoid scale forming and also erosion on the compressor blades, air must be filtered before entering the compressor. Inside the air intake, different systems are used to provide clean air with proper temperature (Fig. 7.6A and B1). Filters, self-cleaning system of the filters, dust removal system, silencers, anti-icing for winters, and inlet air chilling system for hot summers are the main components of an air intake system. Less pressure drop in the air intake means less power consumption by the compressor; hence, the GT manufacturers have established self-cleaning system for continuous cleaning of the filters by using predefined cleaning algorithms to avoid large pressure loss in the filters. They usually use reverse high-velocity air jets against the filters to clean them from dust and debris. The removed dust and debris are then removed from the air intake by the dust removal fan. In winters, to avoid ice forming at the compressor inlet and on the filters, anti-icing system is usually used. It can be some steam or hot water coils at the air intake inlet, or a hot pressurized air extracted from the compressor stages and inserted at the air intake inlet.

During the hot seasons, the ambient temperature may reach above 50°C, compressing hot air consumes more power, and as a result, the efficiency declines. To overcome

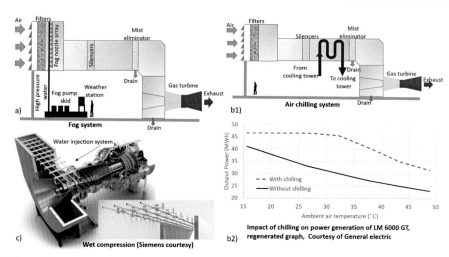

Figure 7.6 The inlet air cooling systems, (A) fogging, (B1) mechanical chilling, and (C) wet compression [3,4].

this problem, air should be cooled before or during compression. This can be done by typical fog system, wet compression, or mechanical air chillers. In the fog system, demineralized water is sprayed into the air after the filters, and due to evaporative cooling, air temperature drops. Mist eliminators are then used for trapping water droplets before entering compressor. This process reduces air temperature that results in reduction of compressor power consumption and increases the mass flow rate that results in rising the power production of turbine. The wet compression is a process in which micro droplets of water at a rate higher than fully saturation condition is injected into air just at the inlet of compressor. The additional water droplets enter the compressor at a uniform distribution around the rotor and play the role of intercooler inside the compressor. Hence, this method increases the compressor performance by three mechanisms of inlet cooling, intercooling, and adding the mass flow rate. The air chillers can reject heat by using chilled water from cooling towers or air-cooled heat exchangers. Water chillers require a continuous supply of cool water. The coils used for air chillers can be used for the anti-icing systems working with hot water or steam as well. Fig. 7.6 shows the inlet air cooling systems of GT including fogging, mechanical chilling, and wet compression. In addition, Fig. 7.6B2 shows the impact of air chilling system installed on an LM6000-PC GT. For example at ambient temperature of 49°C (120°F) it shows about 8000 kWh increase in power generation. Table 7.1 compares the three main air inlet chiller systems from thermo-economic point of views. It shows that air chillers are the most expensive method, while wet compression is the cheapest and most effective method.

Inlet air chilling requires different energy consumption rates depending on the chilling type. Hence, in real cases, the power or heat consumed by the chiller must be included in the calculations of the cycle. In addition to the inlet air chilling, some auxiliary systems such as air conditioning, lubrication system, lighting, oil cooling, etc.

Table 7.1 Comparison of the inlet air cooling systems [5].

Method	Fogging	Wet compression	Chilling
Compressor inlet air temp (°C)	26	26	10
Net power output change (%)	6	15 to 20	11
Heat rate improvement (%)	1	1 to 3	−2 to +2
Installed cost ($/kW)	80 to110	45 to 80	1000 to 1150
Plant integration	Inlet duct/silencing system treatment (coating, lining, drains), nozzle placement. Deionized water supply and drains	Inlet duct treatment, control system integration for combustion and emergency response, cooling system modification, rotor grounding (all included in above pricing) deionized water supply and drains	Coil integration with inlet air system, cooling tower and make up water required, chilling units to be installed
Maintenance	Pumps, valves, nozzles, and inlet system. Compressor blades if carryover is excessive	Pumps, valves, nozzles	Mechanical chillers, inlet coils and cooling tower. Similar to operating CT in a cold climate

require power. They also should be considered in the net power calculations. Hence, the net power would be:

$$\dot{W}_{net} = \dot{W}_{tur} - \dot{W}_{comp} - \dot{W}_{auxiliary} - \dot{W}_{inlet\ cooling} \qquad (7.1)$$

The net power and input heat of the GT should be corrected according to the power consumption by the auxiliary systems and inlet air chilling system type (Table 7.2).

Example 7.2. Consider the SGT600 (Fig. 7.4), the GT in Example 7.1. It is summer, and ambient temperature has reached 45°C. Assume that the turbine inlet/outlet temperatures stay the same as the Example 7.1 by adjusting fuel injection. In addition, the compression ratio and adiabatic efficiency of the compressor, and air mass flow rate remains unchanged. Determine the following terms in comparison with their magnitude at ISO conditions in Example 7.1.

Table 7.2 Power and heat consumption modifications according to the inlet air chilling systems.

Chilling type	Considerations
Cooling water only in coils	Pumping power and fan power in the cooling tower should be subtracted from the net power $\dot{W}_{net} = \dot{W}_{tur} - \dot{W}_{comp} - \dot{W}_{auxiliary} - \dot{W}_{pump} - \dot{W}_{fan}$
Absorption/ adsorption chillers	Fuel consumption should be added to the input heat and pumping power should be subtracted from the net power $\dot{W}_{net} = \dot{W}_{tur} - \dot{W}_{comp} - \dot{W}_{auxiliary} - \dot{W}_{pump}$ $\dot{Q}_{in,new} = \dot{Q}_{in} + \dot{Q}_{chiller}$
Compression chiller	Power consumption of the compressor and condenser fan should be subtracted from the net power $\dot{W}_{net} = \dot{W}_{tur} - \dot{W}_{comp} - \dot{W}_{auxiliary} - \dot{W}_{chiller\ comp} - \dot{W}_{cond.\ fan}$
Fog system	The pumping power of the demineralized water must be considered $\dot{W}_{net} = \dot{W}_{tur} - \dot{W}_{comp} - \dot{W}_{auxiliary} - \dot{W}_{pump}$
Wet compression	The pumping power of the demineralized water must be considered $\dot{W}_{net} = \dot{W}_{tur} - \dot{W}_{comp} - \dot{W}_{auxiliary} - \dot{W}_{pump}$

a. the thermal efficiency drop, b) discuss the impact of changing compression work and input heat on the thermal efficiency drop, c) the net power generation reduction, and d) if by using an inlet air chiller the temperature could be reduced, plot the inlet temperature reduction (ΔT_1) impact on the $\Delta \eta_{th} = \eta_{th@T_{chilled}} - \eta_{th@45°C}$, and net power generation.

Assumptions. It is assumed the working fluid through the cycle is pure dry air, and the air mass flow rate everywhere is constant, that means every extraction from the GT, or fuel mass flow rate are negligible in comparison with the air mass flow rate at point 1. The pressure loss in the combustion chamber is neglected. The power consumption by the auxiliary and chilling system are neglected.

Solution. **a.** to determine the thermal efficiency, the enthalpy of all four state points should be determined, hence:

Starting from point 1:

$$P_1 = 101.325 \text{ kPa}, \quad T_1 = 45°C \rightarrow h_1 = 318.7 \frac{\text{kJ}}{\text{kg}}, \quad s_1 = 5.761 \frac{\text{kJ}}{\text{kg.K}}$$

Points 2 and 2s:

$$P_2 = 14P_1 = 1419 \text{ kPa}, \quad s_{2s} = s_1 \rightarrow h_{2s} = 676.6 \frac{\text{kJ}}{\text{kg}}$$

$$\eta_{ad,comp} = \frac{h_{2s} - h_1}{h_2 - h_1} \rightarrow 0.87 = \frac{676.6 - 318.7}{h_2 - 318.7} \rightarrow h_2 = 730 \frac{\text{kJ}}{\text{kg}} \rightarrow T_2$$

$$= 442.4 \,°C \rightarrow s_2 = 5.838 \frac{\text{kJ}}{\text{kg.K}}$$

Points 4 and 4s:

$$T_4 = 543\ ^\circ\text{C}, \quad P_4 = P_1 \rightarrow h_4 = 839.8\,\frac{\text{kJ}}{\text{kg}}, \quad s_4 = 6.739\,\frac{\text{kJ}}{\text{kg.K}}, \quad s_{4s} = s_3$$

Point 3:

$$T_3 = 1126^\circ\text{C}, \quad P_3 = P_2 \rightarrow h_3 = 1514\,\frac{\text{kJ}}{\text{kg}} \rightarrow s_3 = 6.603\,\frac{\text{kJ}}{\text{kg.K}}$$

$$\dot{Q}_{in} = \dot{m}_4(h_3 - h_2) = 81.3(1514 - 730) = 63730\ \text{kW}$$

$$\dot{W}_{comp} = \dot{m}_2(h_2 - h_1) = 81.3(730 - 318.7) = 33441\ \text{kW}$$

$$\dot{W}_{tur} = \dot{m}_4(h_3 - h_4) = 81.3(1514 - 839.8) = 54805\ \text{kW}$$

$$\dot{W}_{net} = \dot{W}_{tur} - \dot{W}_{comp} = 54805 - 33441 = 21364\ \text{kW} \rightarrow$$

$$\eta_{th} = \frac{\dot{W}_{net}}{\dot{Q}_{in}} = \frac{21364}{63730} = 33.52\%$$

According to the results of the Example 7.1, the efficiency drop due to higher ambient temperature equals to:

$$\Delta\eta_{th} = \eta_{th@45^\circ C} - \eta_{th@ISO} = 33.52 - 35.34 = -1.82\%$$

b. the percent changes of $\dot{W}_{comp}$ and $\dot{Q}_{in}$ are calculated below:

$$\frac{\Delta\dot{W}_{comp}}{\dot{W}_{comp@ISO}} \times 100 = \frac{\dot{W}_{comp@45^\circ C} - \dot{W}_{comp@ISO}}{\dot{W}_{comp@ISO}} \times 100$$

$$= \frac{33441 - 30344}{30344} \times 100 = 10.2\%$$

$$\frac{\Delta\dot{Q}_{in}}{\dot{Q}_{in@ISO}} \times 100 = \frac{\dot{Q}_{in@45^\circ C} - \dot{Q}_{in@ISO}}{\dot{Q}_{in@ISO}} \times 100 = \frac{63730 - 69292}{69292} \times 100$$

$$= \frac{-5660}{69292} \times 100 = -8.03\%$$

As the results show, the percent increase of compression work is bigger than reduction percent of the input heat; as a result, the thermal efficiency will experience reduction of 1.82%.

c. the net power reduction:

$$\Delta\dot{W}_{net} = \dot{W}_{net@45^\circ C} - \dot{W}_{net@ISO} = 21364 - 24.5 \times 10^3 = -3136\ \text{kW}$$

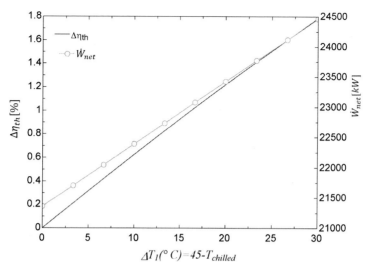

Figure 7.7 The impact of inlet air chilling by using air chillers on the efficiency improvement, and net power generation.

d. by using the parametric table analysis in the EES and **Code 7.2**, the T_1 will be commented and it will be changed from 45°C to 15°C to see its impact on the thermal efficiency change $\Delta\eta_{th} = \eta_{th@T_{chilled}} - \eta_{th@45°C}$. The results are plotted in Fig. 7.7 according to the graph for every 10°C of inlet air cooling the efficiency improvement will be about 0.6%. It means $\eta_{th@T_{chilled}=15°C} - \eta_{th@45°C} = 1.79\ \%$.

Code 7.2. The code for simple cycle of SGT600 in Example 7.2, plus inlet air chiller.

```
"Siemens SGT600"                              h[3]=ENTHALPY(Air,T=T[3])
m_dot=81.3                                    W_net=Q_in-Q_out
T[4]=543                                      ETA_ISO=0.3534
P[4]=101.325                                  W_comp_ISO=30344
P[1]=P[4]                                     Q_in_ISO=69292
"T[1]=45"                                     W_net_ISO=24.5*10^3
h[1]=ENTHALPY(Air,T=T[1])                     "a)"
s[1]=ENTROPY(Air,T=T[1],P=P[1])               ETA_th=W_net/Q_in
h[4]=ENTHALPY(Air,T=T[4])                     DELTAETA=(ETA_th-ETA_ISO)
s[4]=ENTROPY(Air,T=T[4],P=P[4])               "b)"
h2s=enthalpy(air, P=P[2], S=s[1])             DELTAQ_in_PERCENT=(Q_in-
ETA_COMP=0.87                                 Q_in_ISO)/Q_in_ISO*100
ETA_COMP=(h[1]-h2s)/(h[1]-h[2])               DELTAW_COMP_PERCENT=(W_comp-
T[2]=TEMPERATURE(Air,h=h[2])                  W_comp_ISO)/W_comp_ISO*100
s[2]=ENTROPY(Air,T=T[2],P=P[2])               "c)"
P[2]/P[1]=14                                  DELTAW_net=W_net-W_net_ISO
T[3]=1126                                     "d)"
W_comp=m_dot*(h[2]-h[1])                      DELTAT_1=45-T[1]
W_tur=m_dot*(h[3]-h[4])                       ETA_45C=0.3352
Q_in=m_dot*(h[3]-h[2])                        DELTAETA_chilling=(ETA_th-ETA_45C)*100
Q_out=m_dot*(h[4]-h[1])
```

Example 7.3. If the inlet, air cooling of the Example 7.2 is supposed to be done by a fogging system (Fig. 7.6A), such that air at the compressor inlet becomes saturated (relative humidity of $\phi_1 = 100\%$) with water vapor, and its temperature reaches the water temperature of 15°C, by neglecting the pumping power of the fog system, determine, a) the water consumption rate of the fog system, b) the efficiency improvement due to temperature reduction from 45°C to 15°C, c) the turbine power generation, compressor power consumption, and net power produced in comparison with using inlet air chiller for the same temperature reduction discussed in part c of Example 7.2. d) plot the water consumption rate of the fog system if the compressor inlet temperature is supposed to change from 15°C to 45°C.

Assumptions. Pressure drop due to filters in the air intake is neglected, ambient air is assumed dry. The water content in the inlet air to the compressor is treated as air.

Solution. **a.** to determine the water consumption rate, the specific humidity at the compressor inlet should be determined.

$$T_1 = T_{fog} = 15\,°C \rightarrow P_{sat,1} = 1.706\ kPa \rightarrow w_1 = \frac{0.622\phi_1 P_{sat,1}}{P_1 - \phi_1 P_{sat,1}} \rightarrow$$

$$w_1 = \frac{0.622 \times 1 \times 1.706}{101.325 - 1.706} = 0.01065\ \frac{kg\ water}{kg\ dry\ air}$$

$$w_1 = \frac{\dot{m}_{fog}}{\dot{m}_{air}} \rightarrow \dot{m}_{fog} = 0.01065 \times 81.3 = 0.866\ \frac{kg\ water}{s}$$

b. to determine the thermal efficiency, the enthalpy of all four state points should be determined, hence:

Starting from point 1:

$$P_1 = 101.325\ kPa,\ \ T_1 = 15\,°C \rightarrow h_1 = 288.5\ \frac{kJ}{kg},\ \ s_1 = 5.661\ \frac{kJ}{kg.K}$$

Points 2 and 2s:

$$P_2 = 14P_1 = 1419\ kPa,\ \ s_{2s} = s_1 \rightarrow h_{2s} = 613.2\ \frac{kJ}{kg}$$

$$\eta_{ad,comp} = \frac{h_{2s} - h_1}{h_2 - h_1} \rightarrow 0.87 = \frac{613.2 - 288.5}{h_2 - 288.5} \rightarrow h_2 = 661.7\ \frac{kJ}{kg} \rightarrow T_2$$

$$= 378.6\,°C \rightarrow s_2 = 5.738\ \frac{kJ}{kg.K}$$

Points 4 and 4s:

$$T_4 = 543\,°C,\ \ P_4 = P_1 \rightarrow h_4 = 839.8\ \frac{kJ}{kg},\ \ s_4 = 6.739\ \frac{kJ}{kg.K},\ \ s_{4s} = s_3$$

Point 3:

$$T_3 = 1126\,°C\,,\ P_3 = P_2 \rightarrow h_3 = 1514\frac{kJ}{kg} \rightarrow s_3 = 6.603\frac{kJ}{kg.K}$$

The mass flow rate should be modified by adding the water injection from fog system to the inlet air mass flow rate; hence:

$$\dot{m}_1 = \dot{m}_{air} + \dot{m}_{fog} = 81.3 + 0.866 = 82.17\frac{kg}{s},\ \ \dot{m}_1 = \dot{m}_4 = \dot{m}_2$$

$$\dot{Q}_{in} = \dot{m}_4(h_3 - h_2) = 82.18(1514 - 661.7) = 70021\ kW$$

$$\dot{W}_{comp} = \dot{m}_2(h_2 - h_1) = 82.18(661.7 - 288.5) = 30667\ kW$$

$$\dot{W}_{tur} = \dot{m}_4(h_3 - h_4) = 82.18(1514 - 839.8) = 55388\ kW$$

$$\dot{W}_{net} = \dot{W}_{tur} - \dot{W}_{comp} = 55388 - 30667 = 24721\ kW\ \rightarrow$$

$$\eta_{th} = \frac{\dot{W}_{net}}{\dot{Q}_{in}} = \frac{24721}{70021} = 35.31\%$$

According to the results of the Example 7.2, the efficiency improvement equals to:

$$\Delta\eta_{th} = \eta_{th@15°C\ by\ fog} - \eta_{th@45°C} = 35.31 - 33.52 = 1.79\%$$

c.

$$\Delta\dot{W}_{net} = \dot{W}_{net\ by\ fog@15\ °C} - \dot{W}_{net\ by\ chiller\ @15\ °C} = 24721 - 24461 = 260\ kW$$

$$\Delta\dot{W}_{tur} = \dot{W}_{tur\ by\ fog@15°C} - \dot{W}_{tur\ by\ chiller\ @15°C} = 55388 - 54805 = 590\ kW$$

$$\Delta\dot{W}_{comp} = \dot{W}_{tur\ by\ fog@15\ °C} - \dot{W}_{tur\ by\ chiller\ @15°C} = 30667 - 30344 = 326\ kW$$

It means that by adding fog to the air stream, the overall compression work increases, but the power generation by turbine increases more, and as a result, the final net power will increase about 0.260 MW, while the thermal efficiency stayed the same.

a. The water mass flow rate required by the fog system for $15 \le T_1(°C) \le 45$, and $\phi_1 = 100\%$ at the compressor inlet is determined by using the EES **Code 7.3**, and plotted in Fig. 7.8.

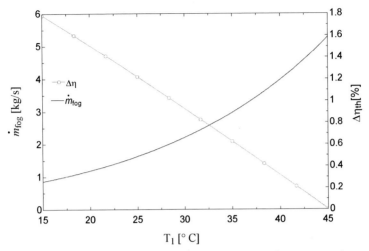

Figure 7.8 Water mass flow rate of the fog system and efficiency improvement in comparison with efficiency at 45°C.

Code 7.3. The code for simple cycle of SGT600 in Example 7.2, plus fog system.

```
"Siemens SGT600"                              W_comp=m_dot*(h[2]-h[1])
m_dot=m_dot_air+m_dot_fog                     W_tur=m_dot*(h[3]-h[4])
m_dot_air=81.3                                Q_in=m_dot*(h[3]-h[2])
T_fog=15                                      Q_out=m_dot*(h[4]-h[1])
P[1]=P[4]                                     h[3]=ENTHALPY(Air,T=T[3])
T[1]=T_fog                                    W_net=Q_in-Q_out
h[1]=ENTHALPY(Air,T=T[1])                     "a)"
s[1]=ENTROPY(Air,T=T[1],P=P[1])               omega_1=HUMRAT(AirH2O,T=T_fog,P=P[1],R=1)
T[4]=543                                      Psat1=P_SAT(Water,T=T[1])
P[4]=101.325                                  m_dot_fog=m_dot_air*omega_1
h[4]=ENTHALPY(Air,T=T[4])                     "b)"
s[4]=ENTROPY(Air,T=T[4],P=P[4])               ETA_th=W_net/Q_in
h2s=enthalpy(air, P=P[2], S=s[1])             ETA_45C=0.3352
ETA_COMP=0.87                                 DELTAETA=(ETA_th-ETA_45C)
ETA_COMP=(h[1]-h2s)/(h[1]-h[2])               "c)"
T[2]=TEMPERATURE(Air,h=h[2])                  W_net_45C=21364
s[2]=ENTROPY(Air,T=T[2],P=P[2])               DELTAW_net=W_net-W_net_45C
P[2]/P[1]=14
T[3]=1126
```

Regenerative/recuperative gas turbine

The exhaust gas temperature of GTs is considerably high and can be reused for different heat requiring processes such as heating, cooling, steam generation, power generation, etc. The exhaust gas can also be reused for efficiency improvement of the GT by preheating of the compressed air before entering the combustion chamber. A turbine which uses exhaust gas for preheating of the compressed gas is called a *regenerative/recuperative GT* (R-GT). The schematic of such turbine and its $T- s$

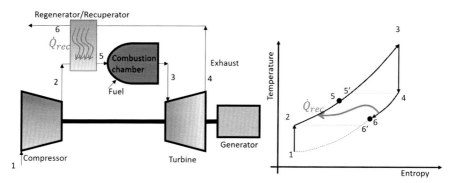

Figure 7.9 A regenerative/recuperative gas turbine.

diagram is shown in Fig. 7.9. In this figure, points 5′ and 6′ are for ideal heat recovery, and 5 and 6 for real heat recovery. The thermodynamical analysis of an R-GT is as follow:

$$\dot{W}_{comp} = \dot{m}(h_2 - h_1), \; \dot{W}_{tur} = \dot{m}(h_3 - h_4), \dot{Q}_{in} = \dot{m}(h_3 - h_5),$$

$$\dot{Q}_{out} = \dot{m}(h_6 - h_1), \eta_{th} = \frac{\dot{W}_{net}}{\dot{Q}_{in}} = 1 - \frac{\dot{Q}_{out}}{\dot{Q}_{in}} \tag{7.2}$$

And the recovered heat($\dot{Q}_{rec}$) for an ideal and real heat recovery system are as below:

$$\dot{Q}_{rec-ideal} = \dot{m}(h_{5'} - h_2), \; \dot{Q}_{rec-real} = \dot{m}(h_5 - h_2) \tag{7.3}$$

The effectiveness of the regenerator/recuperator is determined as below:

$$\varepsilon = \frac{\dot{Q}_{rec-real}}{\dot{Q}_{rec-ideal}} = \frac{h_5 - h_2}{h_4 - h_2} \cong \frac{T_5 - T_2}{T_4 - T_2} \tag{7.4}$$

The effectiveness of GT regenerators is typically in the range $0.8 \leq \varepsilon \leq 0.95$. However, as discussed previously in Chapter 6, although the structures of recuperator and regenerator are different, but they both transfer heat from the hot stream to the cold stream. In a regenerator, the matrix, or the hood rotates, and each time one part of the heating surface is exposed to the hot and cold gas alternately (Fig. 7.10), but in a recuperator, heat is transferred through walls that separate hot and cold streams (Fig. 7.11).

Regenerators have some advantages over recuperators; they are more compact, their manufacturing and material costs are lower, and due to periodic, hot and cold streams they have less fouling problems. The main disadvantages of regenerators are their sealing problem between hot and cold streams and power consumption required for rotating matrix and hood.

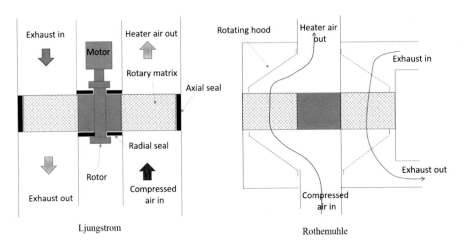

Figure 7.10 The regenerator heat exchanger Ljungsrom (rotary matrix), Rothemuhle (rotary hood).

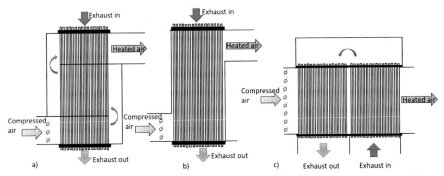

Figure 7.11 Gas through tube recuperator configurations, (A) air triple pass, flue gas single pass, (B) air single pass, flue gas single pass, and (C) flue gas double pass, air single pass.

Fig. 7.12 shows the mercury 50 recuperative single shaft GT. It has a ten-stage compressor with compression ratio of 9.9, a two-stage turbine, and 8 fuel injector. It produces $\dot{W}_{net} = 4.6$ MW with final exhaust temperature of $T_6 = 365\,°\text{C}$ and exhaust mass flow rate of $\dot{m}_6 = 17.7\,\frac{7\text{kg}}{\text{s}}$. It has a heat rate of $HR = 9350\,\frac{\text{kJ}}{\text{kWh}}$ and thermal efficiency of 38.5%.

Intercooler/reheater

During the compression process, the air temperature increases as well. As the air gets hotter, the compression work increases, and as a result, the overall efficiency decreases. To decrease the temperature rise during compression, it is recommended to use an intercooler between high-pressure and low-pressure compressors (HPC and

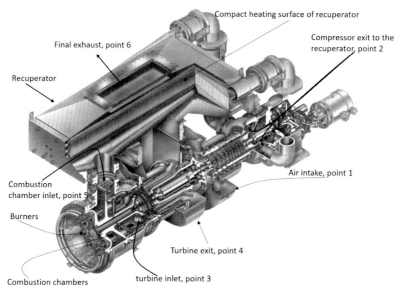

Figure 7.12 Mercury 50, recuperated gas turbine [6,7].
Courtesy of Solar Turbines Incorporated.

LPC). Intercooling may be done by cold water misting into the hot compressed air (Fig. 7.13) or by using an external heat exchanger (Fig. 7.14). Intercooling by an external heat exchanger needs bigger installations for the heat exchanger inlets and outlets; it needs continues cooling medium (usually water), and as a result, it needs pumping station and cooling tower, if the heat exchanger is an air cooler, it will need some fans and more power consumption for intercooling (Fig. 7.15). It decreases

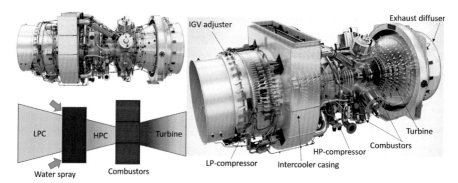

Figure 7.13 SGT-A65 three-shaft, axial flow, aeroderivative industrial and marine 50/60 Hz gas turbine, with spry intercooling, gross output of 67.4 MWe (50 Hz), heat rate of 8724 kJ/kWh, gross simple cycle efficiency of 41.3%, shaft speed of 3000 rpm, exhaust mass flow of 178 kg/s, exhaust temperature of 431 °C, NOx emission of ≤ 25 ppmvd, [3].
Courtesy of Siemens.

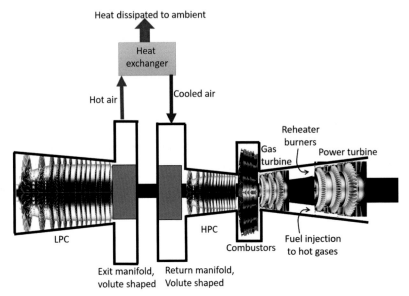

Figure 7.14 The schematic of a gas turbine with intercooler and reheater.

the compression power. However, using spry or mist needs smaller installation, smaller pump, and no cooling tower at all. The water used for spraying must be demineralized to avoid scale forming on the compressor blades. Due to evaporative cooling, the cooling process happens faster, and compression power reduces on the compressor side. In addition, due to adding water, the potential for power generation increases on the turbine side as well. Although using intercooler may increase $\dot{W}_{net}$, and improve the back work ratio of the GT ($r_{bw} = \frac{\dot{W}_{comp}}{\dot{W}_{tur}}$), but it decreases the mean temperature of heat addition (T_H) because of heat dissipation in the intercooler. As a result, it will decrease the thermal efficiency of the cycle.

During the expansion process, the gas pressure and temperature drop. Practically, nothing can be done to pressurize a hot gas again while flowing inside the turbine section, but its temperature can be elevated by placing some burners as a reheater. It should be noted that due to metallurgical limitations, and also NO_x reduction, extremely excess air is always used in the combustion chamber; hence, the gases leaving the combustion chamber always have enough oxygen for a new combustion process in the reheater. Reheating increases the potential for generating more power in the turbine section and hence more $\dot{W}_{net}$. Care must be taken, although the back work ratio improves, the thermal efficiency will not necessarily increase. The reason is that reheating will increase the mean temperature of heat rejection that will result in lower efficiency. However, using a regenerator with intercooler and reheater in a GT can improve the thermal efficiency as well because the mean temperature of heat addition will increase due to reheater, while at the same time, the mean temperature of heat rejection will decrease due to using regenerator (Fig. 7.16).

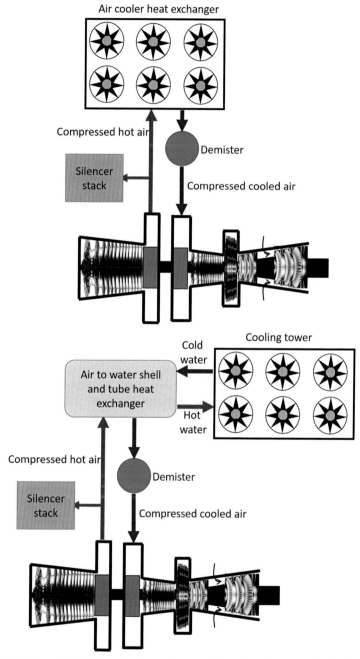

Figure 7.15 Schematic of intercooler types of gas turbine, air to air intercooler (up) and air to water intercooler (down).

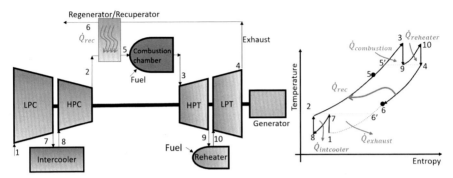

Figure 7.16 A gas turbine with an intercooler, reheater, and exhaust heat recovery system.

For a GT equipped with an intercooler, reheater, and exhasut heat recovery the energy balance equations will be as below:

$$\dot{W}_{comp} = \dot{m}(h_7 - h_1) + \dot{m}(h_2 - h_8), \ \dot{W}_{tur} = \dot{m}(h_3 - h_9) + \dot{m}(h_{10} - h_4), \ \dot{Q}_{in}$$

$$= \dot{m}(h_3 - h_5) + \dot{m}(h_{10} - h_9), \ \dot{Q}_{out} = \dot{m}(h_6 - h_1) + \dot{m}(h_7 - h_8), \ \eta_{th}$$

$$= \frac{\dot{W}_{net}}{\dot{Q}_{in}} = 1 - \frac{\dot{Q}_{out}}{\dot{Q}_{in}}$$

$$(7.5)$$

Example 7.4. If the intercooler in Fig. 7.16 is supposed to cool the compressed air from temperature T_7 to $T_8 = T_1$, what would be the optimum intercooling pressure for minimum compression power? what is the optimum reheating pressure for achieving maximum turbine power if it is supposed to reheat the gas from T_9 to $T_{10} = T_3$? Assume the cycle is internally reversible and working fluid in the cycle is pure air as perfect gas.

Solution. By using cold air standard assumptions, for the optimum intercooling pressure:

$$\dot{W}_{comp} = \dot{m}(h_7 - h_1) + \dot{m}(h_2 - h_8) = \dot{m}C_p(T_7 - T_1 + T_2 - T_8)$$

$$= \dot{m}C_pT_1\left(\frac{T_7}{T_1} + \frac{T_2}{T_8} - 2\right)$$

$$= \dot{m}C_pT_1\left(\left(\frac{P_7}{P_1}\right)^{\frac{k-1}{k}} + \left(\frac{P_2}{P_8}\right)^{\frac{k-1}{k}} - 2\right)$$

$$= \dot{m}C_pT_1\left(r_{LPC}^{\frac{k-1}{k}} + r_{HPC}^{\frac{k-1}{k}} - 2\right)$$

$$R = r_{LPC} \times r_{HPC} \rightarrow \dot{W}_{comp} = \dot{m}C_pT_1\left(r_{LPC}^{\frac{k-1}{k}} + \left(\frac{R}{r_{LPC}}\right)^{\frac{k-1}{k}} - 2\right) \rightarrow \frac{d\dot{W}_{comp}}{dr_{LPC}}$$

$$= 0 \rightarrow R = r_{LPC}^2 \rightarrow r_{LPC} = r_{HPC} = \sqrt{R}$$

in which r is the compression ratio of LPC and HPC, and R is the total compression ration of the compressor.to find the optimum reheat pressure, the power generation by the turbine section is optimized as below:

$$\dot{W}_{tur} = \dot{m}(h_3 - h_9) + \dot{m}(h_{10} - h_4) = \dot{m}C_p(T_3 - T_9 + T_{10} - T_4)$$

$$= \dot{m}C_pT_3\left(2 - \frac{T_9}{T_3} - \frac{T_4}{T_{10}}\right)$$

$$= \dot{m}C_pT_3\left(2 - \left(\frac{P_9}{P_3}\right)^{\frac{k-1}{k}} - \left(\frac{P_4}{P_{10}}\right)^{\frac{k-1}{k}}\right)$$

$$= \dot{m}C_pT_3\left(2 - \left(\frac{1}{r_{HPT}}\right)^{\frac{k-1}{k}} - \left(\frac{1}{r_{LPT}}\right)^{\frac{k-1}{k}}\right)$$

$$R = r_{LPT} \times r_{HPT} \to \dot{W}_{tur} = \dot{m}C_pT_3\left(2 - \left(\frac{1}{r_{HPT}}\right)^{\frac{k-1}{k}} - \left(\frac{r_{HPT}}{R}\right)^{\frac{k-1}{k}}\right)$$

$$\frac{d\dot{W}_{tur}}{dr_{HPT}} = 0 \to R = r_{HPT}^2 \to r_{LPT} = r_{HPT} = \sqrt{R}$$

In other words, the optimum intercooling and reheating pressures of a two-stage compression/expansion GT to achieve minimum compression power and maximum turbine power are as below:

$$P_{opt, intercooler} = P_7 = P_8 = \sqrt{P_1 P_2}$$

$$P_{opt, reheater} = P_9 = P_{10} = \sqrt{P_3 P_4}$$

In which $T_8 = T_1$ and $T_{10} = T_3$.

Example 7.5. Consider the Siemens GT model of SGT600 (Fig. 7.4) The turbine inlet temperature and mass flow rate are 1126°C and 81.3 $\frac{kg}{s}$. The compression ratio of compressor is 14; the adiabatic efficiency of turbine, and compressor sections is 0.87 and the nominal rotational speed is 7700 rpm. The ambient temperature and pressure are $P_1 = 101.325$ kPa, $T_1 = 15°C$. You are supposed to determine a) the optimum intercooling and reheating pressures of the cycle; b) the $\dot{W}_{comp}$, $\dot{W}_{tur}$, $\dot{Q}_{in}$, $\dot{Q}_{out}$, and thermal efficiency of the new GT. c) Compare the results with SGT600 without reheater and intercooler. d) What would happen for the thermal efficiency if a regenerator with effectiveness of 0.85 was installed at the turbine exhaust?

Assumptions. The working fluid through the cycle is pure dry air, the air mass flow rate everywhere is constant, that means every extraction from the GT, or fuel mass flow rate are negligible in comparison with the main air mass flow rate at point 1. The pressure loss in the combustion chamber, intercooler, and regenerator is neglected.

a. The optimum reheater and intercooler pressures can be determined according to the Example 7.4:

$$P_{opt, intercooler} = P_7 = P_8 = \sqrt{P_1 P_2} = \sqrt{P_1 P_1 R} = P_1 \sqrt{R} = 101.325\sqrt{14}$$
$$= 379.1 \text{ kPa}$$

$$P_{opt, reheater} = P_9 = P_{10} = \sqrt{P_3 P_4} = P_4 \sqrt{R} = 101.325\sqrt{14} = 379.1 \text{ kPa}$$

b. The state point properties should be determined as below:

$$P_1 = 101.325 \text{ kPa}, \quad T_1 = 15\,^{\circ}\text{C} \rightarrow h_1 = 288.5\frac{\text{kJ}}{\text{kg}}, \quad s_1 = 5.661\frac{\text{kJ}}{\text{kg.K}}$$

$$P_7 = 379.1 \text{ kPa}, \quad s_{7s} = s_1 \rightarrow h_{7s} = 421\frac{\text{kJ}}{\text{kg}}, \quad \eta_{ad,comp} = \frac{h_{7s} - h_1}{h_7 - h_1} \rightarrow 0.87$$

$$= \frac{421 - 288.5}{h_7 - 288.5}$$

$$\rightarrow h_7 = 440.8\frac{\text{kJ}}{\text{kg}} \rightarrow T_7 = 165.8\,^{\circ}\text{C}$$

$$P_8 = 379.1 \text{ kPa}, \quad T_8 = T_1 \rightarrow h_8 = 288.5\frac{\text{kJ}}{\text{kg}}, \quad s_8 = 5.282\frac{\text{kJ}}{\text{kg.K}}$$

$$P_2 = R \times P_1 = 1419 \text{ kPa}, \quad s_{2s} = s_8 \rightarrow h_{2s} = 421\frac{\text{kJ}}{\text{kg}}$$

$$\eta_{ad,comp} = \frac{h_{2s} - h_8}{h_2 - h_8} \rightarrow 0.87 = \frac{421 - 288.5}{h_2 - 288.5} \rightarrow h_2 = 440.8\frac{\text{kJ}}{\text{kg}} \rightarrow T_2$$

$$= 165.8\,^{\circ}\text{C} \rightarrow s_2 = 5.328\frac{\text{kJ}}{\text{kg.K}}$$

$$P_3 = 1419 \text{ kPa}, \quad T_3 = 1126\,^{\circ}\text{C} \rightarrow h_3 = 1514\frac{\text{kJ}}{\text{kg}}, \quad s_3 = 6.603\frac{\text{kJ}}{\text{kg.K}}$$

$$P_9 = 379.1 \text{ kPa}, \quad s_{9s} = s_3 \rightarrow h_{9s} = 1061\frac{\text{kJ}}{\text{kg}}, \quad \eta_{ad,tur} = \frac{h_3 - h_9}{h_3 - h_{9s}} \rightarrow 0.87$$

$$= \frac{1514 - h_9}{1514 - 1061}$$

$$\rightarrow h_9 = 1119 \, \frac{kJ}{kg} \rightarrow T_9 = 790.9 \, °C$$

$$P_{10} = 379.1 \, kPa, \quad T_{10} = T_3 \rightarrow h_{10} = 1514 \frac{kJ}{kg}, \quad s_{10} = 6.981 \frac{kJ}{kg.K}$$

$$P_4 = P_1 = 101.325 \, kPa, \quad s_{4s} = s_{10} \rightarrow h_{4s} = 1061 \frac{kJ}{kg}$$

$$\eta_{ad,tur} = \frac{h_{10} - h_4}{h_{10} - h_{4s}} \rightarrow 0.87 = \frac{1514 - h_4}{1514 - 1061} \rightarrow h_4 = 1119 \frac{kJ}{kg} \rightarrow T_4$$

$$= 790.9°C \rightarrow s_4 = 7.038 \frac{kJ}{kg.K}$$

$$\dot{W}_{comp} = \dot{m}(h_7 - h_1) + \dot{m}(h_2 - h_8) = 81.3\{440.8 - 288.5 + 440.8 - 288.5\}$$
$$= 24760 \, kW$$

$$\dot{W}_{tur} = \dot{m}(h_3 - h_9) + \dot{m}(h_{10} - h_4) = 81.3\{1514 - 1119 + 1514 - 1119\}$$
$$= 64136 \, kW$$

$$\dot{W}_{net} = \dot{W}_{tur} - \dot{W}_{comp} = 64136 - 24760 = 39376 \, kW$$

$$\dot{Q}_{in} = \dot{m}(h_3 - h_2) + \dot{m}(h_{10} - h_9) = 81.3\{1514 - 440.8 + 1514 - 1119\}$$
$$= 119314 \, kW$$

$$\dot{Q}_{out} = \dot{m}(h_4 - h_1) + \dot{m}(h_7 - h_8) = 81.3\{1119 - 288.5 + 440.8 - 288.5\}$$
$$= 79938 \, kW$$

$$\eta_{th} = \frac{\dot{W}_{net}}{\dot{Q}_{in}} = \frac{39376}{119314} = 33\%$$

In comparison with the cycle with no reheater and no intercooler, the thermal efficiency has decreased from 35.34% to 33%. This is due to heat loss at higher temperature of $T_4 = 790.9°C$ while heat addition at lower temperature of $T_2 = 165.8 \, °C$. Attention must be paid that although compression work has decreased and turbine generates more power, but input heat and heat loss have increased as well, that result in efficiency reduction. Since the exhaust temperature is considerably high, ($T_4 = 790.9 \, °C$) using a regenerator would be beneficial.

c. The regenerator effectiveness is 0.85, hence:

$$\varepsilon \cong \frac{T_5 - T_2}{T_4 - T_2} \to 0.85 = \frac{T_5 - 165.8}{790.9 - 165.8} \to T_5 = 697.1\,^\circ C \to h_5 = 1012\,\frac{kJ}{kg},\ s_5$$

$$= 6.175\,\frac{kJ}{kg.K}$$

To calculate the gas properties at point 6 a heat balance would be written for the regenerator as below:

$$\dot{m}(h_4 - h_6) = \dot{m}(h_5 - h_2) \to 1119 - h_6 = 1012 - 440.8 \to h_6 = 548\,\frac{kJ}{kg},\ s_6$$

$$= 6.305\,\frac{kJ}{kg.k},\ T_6 = 270.3^\circ C$$

The compression work and turbine power generation will not change by installing a regenerator, but the input heat and heat loss will decrease as below:

$$\dot{Q}_{in} = \dot{m}(h_3 - h_5) + \dot{m}(h_{10} - h_9) = 81.3\{1514 - 1012 + 1514 - 1119\}$$
$$= 72852\ kW$$

$$\dot{Q}_{out} = \dot{m}(h_6 - h_1) + \dot{m}(h_7 - h_8) = 81.3\{548 - 288.5 + 440.8 - 288.5\}$$
$$= 33475\ kW$$

And as a result the thermal efficiency will jump from 33% to 54.05%.

$$\eta_{th} = \frac{\dot{W}_{net}}{\dot{Q}_{in}} = \frac{39376}{72852} = 54.05\%$$

the EES code developed for the example 7.5 is presented in **Code 7.4** for further practices and evaluations by the readers.

Important conclusion

Installing intercooler, and reheater for GT are recommended if and only if a regenerator is installed at the exhaust of the GT. Otherwise, the efficiency will drop.

Code 7.4. The EES code written for the Example 7.5.

```
"siemens SGT600 with intercooler and          T[9]=TEMPERATURE(Air,h=h[9])
reheater"                                      s[9]=ENTROPY(Air,T=T[9],P=P[9])
m_dot=81.3                                     T[10]=T[3]
P[1]=101.325                                   P[10]=P[9]
T[1]=15                                        h[10]=ENTHALPY(Air,T=T[10])
h[1]=ENTHALPY(Air,T=T[1])                      s[10]=ENTROPY(Air,T=T[10],P=P[10])
s[1]=ENTROPY(Air,T=T[1],P=P[1])                P[4]=P[1]
R_p=14                                         s_4s=s[10]
P[7]=sqrt(R_p)*P[1]                            h_4s=enthalpy(air, P=P[4], S=s_4s)
s_7s=s[1]                                       ETA_TUR=(h[10]-h[4])/(h[10]-h_4s)
ETA_COMP=0.87                                  T[4]=TEMPERATURE(Air,h=h[4])
h_7s=enthalpy(air, P=P[7], S=s_7s)             s[4]=ENTROPY(Air,T=T[4],P=P[4])
ETA_COMP=(h[1]-h_7s)/(h[1]-h[7])               epsilon=0.85
T[7]=TEMPERATURE(Air,h=h[7])                   epsilon=(T[5]-T[2])/(T[4]-T[2])
s[7]=ENTROPY(Air,T=T[7],P=P[7])                P[5]=P[2]
P[8]=P[7]                                      h[5]=ENTHALPY(Air,T=T[5])
T[8]=T[1]                                       s[5]=ENTROPY(Air,T=T[5],P=P[5])
h[8]=ENTHALPY(Air,T=T[8])                      W_comp=m_dot*(h[7]-h[1])+m_dot*(h[2]-h[8])
s[8]=ENTROPY(Air,T=T[8],P=P[8])                W_tur=m_dot*(h[3]-h[9])+m_dot*(h[10]-h[4])
P[2]=R_p*P[1]                                  W_net=W_tur-W_comp
s_2s=s[8]                                       Q_in_no_regenerator=m_dot*(h[3]-h[2])+m_dot*(h[10]-
h_2s=enthalpy(air, P=P[2], S=s_2s)             h[9])
ETA_COMP=(h[8]-h_2s)/(h[8]-h[2])               Q_out_no_regenerator=m_dot*(h[7]-h[8])+m_dot*(h[4]-
T[2]=TEMPERATURE(Air,h=h[2])                   h[1])
s[2]=ENTROPY(Air,T=T[2],P=P[2])                ETA_th_no_regenerator=W_net/Q_in_no_regenerator
T[3]=1126                                       Q_in_with_regenerator=m_dot*(h[3]-h[5])+m_dot*(h[10]-
P[3]=P[2]                                      h[9])
h[3]=ENTHALPY(Air,T=T[3])                      Q_out_with_regenerator=m_dot*(h[7]-h[8])+m_dot*(h[6]-
s[3]=ENTROPY(Air,T=T[3],P=P[3])                h[1])
P[9]=P[3]/sqrt(R_p)                            ETA_th_with_regenerator=W_net/Q_in_with_regenerator
s_9s=s[3]                                       h[4]-h[6]=h[5]-h[2]
h_9s=enthalpy(air, P=P[9], S=s_9s)             T[6]=TEMPERATURE(Air,h=h[6])
ETA_TUR=0.87                                    P[6]=P[4]
ETA_TUR=(h[3]-h[9])/(h[3]-h_9s)                s[6]=ENTROPY(Air,T=T[6],P=P[6])
```

Example 7.6. If the intercooling of the final cycle of Example 7.5 is supposed to be done with water spray with temperature of 15°C under adiabatic cooling ($\phi_8 = 100\%$), what would be the new $\dot{W}_{comp}$, $\dot{W}_{tur}$, $\dot{Q}_{in}$, $\dot{Q}_{out}$ and thermal efficiency of the new GT. Determine the water mass flow consumption in the intercooler.

Assumptions. The added water to the air stream will be treated as pure air, and only its mass flow rate will be added to the stream after injection. The pressure loss in the combustion chamber, intercooler, and regenerator is neglected. Fuel mass flow rate and extractions from the compressor are ignored. The pressure change due to injecting mist at the intercooler is ignored.

Solution. The only difference of the present example with the previous example is that intercooling is done with mist spry similar to that shown in Fig. 7.13 for the LM6000 gas turbine.

$$P_1 = 101.325 \text{ kPa}, \quad T_1 = 15\,°C \rightarrow h_1 = 288.5 \frac{\text{kJ}}{\text{kg}}, \quad s_1 = 5.661 \frac{\text{kJ}}{\text{kg.K}}$$

$$P_7 = 379.1 \text{ kPa}, \quad s_{7s} = s_1 \rightarrow h_{7s} = 421 \frac{\text{kJ}}{\text{kg}}, \quad \eta_{ad,comp} = \frac{h_{7s} - h_1}{h_7 - h_1} \rightarrow 0.87$$

$$= \frac{421 - 288.5}{h_7 - 288.5}$$

$$\rightarrow h_7 = 440.8 \, \frac{kJ}{kg} \rightarrow T_7 = 165.8 \,^{\circ}C$$

$$P_8 = 379.1 \text{ kPa}, \quad T_8 = T_1 \rightarrow h_8 = 288.5 \frac{kJ}{kg}, \quad s_8 = 5.282 \frac{kJ}{kg.K}$$

$$\phi_8 = 100\%,$$

$$P_{sat,15^{\circ}C} = 1.706 \text{ kPa} \rightarrow w_8 = \frac{0.622\phi_8 P_{sat,8}}{P_8 - \phi_8 P_{sat,8}} \rightarrow$$

$$w_8 = \frac{0.622 \times 1 \times 1.706}{379.1 - 1.706} = 0.002811 \, \frac{kg \text{ water}}{kg \text{ dry air}}$$

$$w_8 = \frac{\dot{m}_{mist}}{\dot{m}_{air}} \rightarrow \dot{m}_{mist} = 0.002811 \times 81.3 = 0.2285 \frac{kg \text{ water}}{s}$$

$$\dot{m}_8 = \dot{m}_{fog} + \dot{m}_{air} = 0.2285 + 81.3 = 81.53 \, \frac{kg}{s}$$

$$P_2 = R \times P_1 = 14 \times 101.325 = 1419 \text{ kPa}, \quad s_{2s} = s_8 \rightarrow h_{2s} = 421 \frac{kJ}{kg}$$

$$\eta_{ad,comp} = \frac{h_{2s} - h_8}{h_2 - h_8} \rightarrow 0.87 = \frac{421 - 288.5}{h_2 - 288.5} \rightarrow h_2 = 440.8 \frac{kJ}{kg} \rightarrow T_2$$

$$= 165.8^{\circ}C \rightarrow s_2 = 5.328 \frac{kJ}{kg.K}$$

$$P_3 = 1419 \text{ kPa}, \quad T_3 = 1126^{\circ}C \rightarrow h_3 = 1514 \frac{kJ}{kg}, \quad s_3 = 6.603 \frac{kJ}{kg.K}$$

$$P_9 = 379.1 \text{ kPa}, \quad s_{9s} = s_3 \rightarrow h_{9s} = 1061 \frac{kJ}{kg}, \quad \eta_{ad,tur} = \frac{h_3 - h_9}{h_3 - h_{9s}} \rightarrow 0.87$$

$$= \frac{1514 - h_9}{1514 - 1061}$$

$$\rightarrow h_9 = 1119 \, \frac{kJ}{kg} \rightarrow T_9 = 790.9 \,^{\circ}C$$

$$P_{10} = 379.1 \text{ kPa}, \quad T_{10} = T_3 \rightarrow h_{10} = 1514 \frac{kJ}{kg}, \quad s_{10} = 6.981 \frac{kJ}{kg.K}$$

$$P_4 = P_1 = 101.325 \text{ kPa}, \quad s_{4s} = s_{10} \rightarrow h_{4s} = 1061 \frac{kJ}{kg}$$

$$\eta_{ad,tur} = \frac{h_{10} - h_4}{h_{10} - h_{4s}} \rightarrow 0.87 = \frac{1514 - h_4}{1514 - 1061} \rightarrow h_4 = 1119 \frac{kJ}{kg} \rightarrow T_4$$

$$= 790.9\,°C \rightarrow s_4 = 7.038 \frac{kJ}{kg.K}$$

The regenerator effectiveness is 0.85, hence:

$$\varepsilon \cong \frac{T_5 - T_2}{T_4 - T_2} \rightarrow 0.85 = \frac{T_5 - 165.8}{790.9 - 165.8} \rightarrow T_5 = 697.1°C \rightarrow h_5 = 1012 \frac{kJ}{kg}, \ s_5$$

$$= 6.175 \frac{kJ}{kg.K}$$

To calculate the gas properties at point 6 a heat balance would be written for the regenerator as below:

$$\dot{m}_4(h_4 - h_6) = \dot{m}_5(h_5 - h_2) \rightarrow 1119 - h_6 = 1012 - 440.8 \rightarrow h_6 = 548 \frac{kJ}{kg}, \ s_6$$

$$= 6.305 \frac{kJ}{kg.K}, \quad T_6 = 270.3°C$$

$$\dot{W}_{comp} = \dot{m}_1(h_7 - h_1) + \dot{m}_2(h_2 - h_8) = 81.3(440.8 - 288.5)$$
$$+ 81.53(440.8 - 288.5)$$
$$= 24795 \ kW$$

$$\dot{W}_{tur} = \dot{m}_3(h_3 - h_9) + \dot{m}_{10}(h_{10} - h_4) = 81.53\{1514 - 1119 + 1514 - 1119\}$$
$$= 64317 \ kW$$

$$\dot{W}_{net} = \dot{W}_{tur} - \dot{W}_{comp} = 64317 - 24795 = 39522 \ kW$$

$$\dot{Q}_{in} = \dot{m}_3(h_3 - h_5) + \dot{m}_{10}(h_{10} - h_9) = 81.53\{1514 - 1012 + 1514 - 1119\}$$
$$= 73056 \ kW$$

$$\dot{Q}_{out} = \dot{Q}_{in} - \dot{W}_{net} = 72852 - 64317 = 8535 kW$$

$$\eta_{th} = \frac{\dot{W}_{net}}{\dot{Q}_{in}} = \frac{39522}{73056} = 54.1\%$$

Code 7.5. The EES code for a GT with water spray intercooling and reheater.

```
"siemens SGT600 with water spry intercooler        s_9s=s[3]
and reheater"                                       h_9s=enthalpy(air, P=P[9], S=s_9s)
m_dot=81.3                                          ETA_TUR=0.87
P[1]=101.325                                        ETA_TUR=(h[3]-h[9])/(h[3]-h_9s)
T[1]=15                                             T[9]=TEMPERATURE(Air,h=h[9])
h[1]=ENTHALPY(Air,T=T[1])                           s[9]=ENTROPY(Air,T=T[9],P=P[9])
s[1]=ENTROPY(Air,T=T[1],P=P[1])                     T[10]=T[3]
R_p=14                                              P[10]=P[9]
P[7]=sqrt(R_p)*P[1]                                 h[10]=ENTHALPY(Air,T=T[10])
s_7s=s[1]                                           s[10]=ENTROPY(Air,T=T[10],P=P[10])
ETA_COMP=0.87                                       P[4]=P[1]
h_7s=enthalpy(air, P=P[7], S=s_7s)                  s_4s=s[10]
ETA_COMP=(h[1]-h_7s)/(h[1]-h[7])                    h_4s=enthalpy(air, P=P[4], S=s_4s)
T[7]=TEMPERATURE(Air,h=h[7])                        ETA_TUR=(h[10]-h[4])/(h[10]-h_4s)
s[7]=ENTROPY(Air,T=T[7],P=P[7])                     T[4]=TEMPERATURE(Air,h=h[4])
w[8]=m_dot_mist/m_dot                               s[4]=ENTROPY(Air,T=T[4],P=P[4])
T_water=15                                          epsilon=0.85
RH[8]=1                                             epsilon=(T[5]-T[2])/(T[4]-T[2])
w[8]=HUMRAT(AirH2O,T=T_water,P=P[7],R=R             P[5]=P[2]
H[8])                                               h[5]=ENTHALPY(Air,T=T[5])
T[8]=T_water                                        s[5]=ENTROPY(Air,T=T[5],P=P[5])
PG=P_SAT(Water,T=T_water)                           W_comp=m_dot*(h[7]-h[1])+m_dot8*(h[2]-h[8])
P[8]=P[7]                                           W_tur=m_dot8*(h[3]-h[9])+m_dot8*(h[10]-h[4])
m_dot8=m_dot+m_dot_mist                             W_net=W_tur-W_comp
h[8]=ENTHALPY(AIR,T=T[8])                           Q_in_no_regenerator=m_dot8*(h[3]-
s[8]=ENTROPY(AIR,T=T[8], P=P[8])                    h[2])+m_dot8*(h[10]-h[9])
P[2]=sqrt(R_p)*P[8]                                 ETA_th_no_regenerator=W_net/Q_in_no_regenera
s_2s=s[8]                                           tor
h_2s=enthalpy(air, P=P[2], S=s_2s)                  Q_in_with_regenerator=m_dot8*(h[3]-
ETA_COMP=(h[8]-h_2s)/(h[8]-h[2])                    h[5])+m_dot8*(h[10]-h[9])
T[2]=TEMPERATURE(Air,h=h[2])                        ETA_th_with_regenerator=W_net/Q_in_with_regen
s[2]=ENTROPY(Air,T=T[2],P=P[2])                     erator
T[3]=1126                                           h[4]-h[6]=h[5]-h[2]
P[3]=P[2]                                           T[6]=TEMPERATURE(Air,h=h[6])
h[3]=ENTHALPY(Air,T=T[3])                           P[6]=P[4]
s[3]=ENTROPY(Air,T=T[3],P=P[3])                     s[6]=ENTROPY(Air,T=T[6],P=P[6])
P[9]=P[3]/sqrt(R_p)
```

Important conclusion

Due to mist injection and increasing the mass flow rate, the net power and the thermal efficiency have increased slightly. For example, the thermal efficiency has increased from 54.05% to 54.1%. This rise of thermal efficiency is coincided with less costs in comparison with other intercooling systems such as using heat exchangers with cooling tower. In addition, in mist intercooling system, very small pumping power is consumed, while in the other systems, the power consumption in the heat exchanger intercooler, and cooling tower fans and pumps is much more than the mist system. Furthermore, some pressure loss in the heat exchanger intercooler exists. In other words, if the auxiallary power consumption by the intercooler's subsystems could be considered, the difference in the efficiency could be even more. Hence mist intercooling system is highly recommended.

2. Gas turbine components

The main components of a GT include the air intake, compressor, combustion chamber, turbine, exhaust, and electric generator. There are some auxiliary systems such as starter, lubrication system, compressor washing system, fuel system, and monitoring and control system. In the following, the main components are described and also some brief data is given for the auxiliary systems.

2.1 Air intake system

Airborne contaminants such as foreign objects, dust, smoke, oil mists, carbon, and sea salts if they enter the compressor without proper filtration will probably cause foreign object damage (FOD) fouling, erosion, or corrosion in different parts of GT. FOD can happen due to poor maintenance programs that let big objects to enter the compressor and can cause severe damage to the internal parts. Particles sizes from 2 to 10 μm can adhere to the blades or other internal surfaces. They change the airfoil profiles and result in efficiency reduction. The generated fouling produce mass unbalance on the rotor that will cause vibrations and will also reduce the amount of air flowing through the compressor. Vibration can cause dangerous damage to bearings and sealing systems if they exceed the allowable range. Vibration can also cause flow separation from the blade surfaces, which in turn causes stall. Stall, if not controlled, in turn will intensify the vibration, increase the gas temperature, and even break the internal parts of the turbine. Fouling can be removed by compressor washing according to the manufacturer guidelines.

Liquid or solid particles bigger than 10 μm can cause erosion on the internal parts of the GT. Contrary to fouling, erosion is a nonreversible phenomenon. Erosion can also cause roughness on the blades, changing airfoil profile, flow separation, and vibration.

Corrosion can happen due to adherence of chemically reactive particles to the internal surfaces of the GT. The corrosion in the compressor is called *cold corrosion*, and in the combustion chamber, it is called *hot corrosion*. Corrosion is a nonreversible phenomenon, and the corroded components must be replaced.

Filtration of particles with sizes bigger than 2 μm has a huge impact on reducing fouling, erosion, and corrosion in the GT. It can greatly improve the performance of the GT as well as its lifespan. In addition, programmed *online and offline compressor washing* with demineralized water and recommended detergents according to the manufacturer instructions and manual can clean the GT especially the compressor section. Since the air contaminants have different sizes, using only high-efficiency filters with very small mesh sizes result in quick choking of the filters. However, using prefiltration to remove the particles with sizes bigger than 10 μm can postpone choking of the high-efficiency filters. Bag filters are generally used for prefiltration. An important advantage of prefilters is that they can be mostly replaced without shutting down the plant. Cartridge or plate filters are commonly used for filtering particles smaller than 10 μm. The filters of an air intake system can be static and/or self-cleaning pulse filter. The static filters when clogged with contaminants and their pressure drop exceeded the allowable magnitude must be removed and cleaned. If the static filters were undamaged after cleaning and their pressure drop according to the test procedures was within the allowable range, they can be used again; otherwise, they must be replaced. Self-cleaning filters are generally cartridge cylindrical filters that are cleaned by a high-pressure pulse air jet that blows in the opposite direction of the main stream into the filter, removing dust and other contaminants from the filter surface. The self-cleaning system can be activated and clean the filters based on the amount of pressure drop of the filters or a time period of for example 5 min. A small compressor supplies this high-pressure air. And a dust removal fan collects the dust and evacuated to outside of the air intake system. The pressure drop of clean filters must not exceed 5 mbar (500 Pa), and they must be replaced or cleaned if pressure drop exceeded 13 mbar (1300 Pa).

Fig. 7.17 shows photos of the air-intake system, prefilters, high-efficiency filters, mist separator, and evaporative cooling media.

Ice can be generated on the filters due to the freezing temperature of ambient. Ice can also be generated at the bell mouth (compressor inlet) even when the ambient temperature is not freezing. This happens due to increasing velocity of air in the air intake, dropping temperature, and freezing water vapor content in the inlet air. Ice on the filters can cause excessive pressure drop, decreasing air flow through the compressor, and even revers flow from turbine to the compressor exit. This reverse flow that is called surge, happens due to excessive reduction of air flow through the compressor and

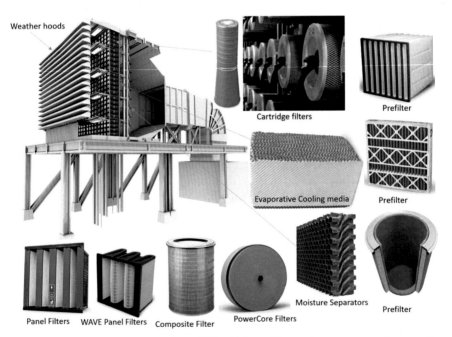

Figure 7.17 Air-intake system, prefilters, high-efficiency filters, mist separator, and evaporative cooling media [8].
Courtesy of Donaldson Company, Inc.

higher pressure at the compressor downstream (at the turbine inlet or combustion chamber). If the pressure drop across the filters exceeds the allowable range, to protect the GT from damaging due to possible stall or surge the emergency gates (or bypass doors) will automatically open, and let the nonfiltered air enter the compressor, and then the control system will start the emergency shutdown of the GT.

Ice generation due to high velocity of air in the air intake, on the bell mouth can cause serious damages to the compressor blades. Both filter freezing and bell mouth freezing can be avoided by using different anti-icing systems. The anti-icing tools include:

- Compressor bleed air system: Hot compressed air extracted from the compressor stages and distributed before the filters by using a distribution tubes and nozzles (Fig. 7.18A).
- Steam coils installed before the filter banks and after the weather hoods.
- Hot water coils installed before the filter banks and after the weather hoods.
- Heated vanes with electrical resistance heating elements before the filter banks and after the weather hoods.

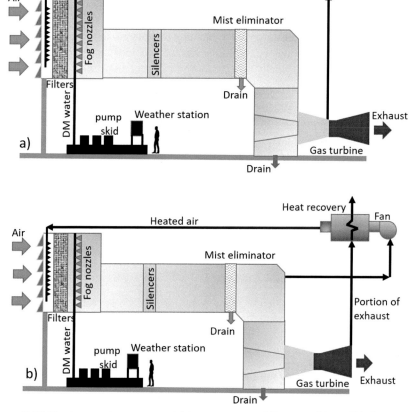

Figure 7.18 The two most common anti-icing systems used for gas turbines, (A) Compressor bleed air system and (B) heated air by exhaust heat recovery.

- Heated fresh air by using heat recovery from the exhaust and injecting it into the air intake entrance (Fig. 7.18B).

The main disadvantage of compressor bleed air system for anti-icing is the reduction of 2%−5% of the total net power and thermal efficiency.

The inlet air cooling systems including fog system, wet compression, air chillers, etc. are previously discussed in Section 1.2.1. air intake.

The silencers are used to reduce noise and making sure about providing a healthy and safe environment for the operators. The materials used in the silencers include carbon, galvanized, and stainless steel. Some manufacturers offer silencers to reduce noise to 35 dB. Noise above 85 dB is harmful for human ear, and below 70 dB is safe. Fig. 7.19 shows rectangular- and elbow-shaped silencers manufactured by dB Noise Reduction Inc [7].

Example 7.7. The ambient temperature has dropped to $-2°C$, and freezing over the filters is possible. The GT is equipped with a compressor bleed from the compressor outlet for anti-icing system. The turbine inlet temperature and the exhaust mass flow rate at normal operation are $1126°C$ and $81.3 \frac{kg}{s}$. The compression ratio of compressor is 14, the adiabatic efficiency of turbine and compressor sections is 0.87, and the nominal rotational speed is 7700 rpm. The ambient pressure is $P_{amb} = 101.325$ kPa. Determine a) the bleed mass flow rate to increase the inlet air temperature from $-2°C$ to $10°C$. b) the net power and the efficiency change due to using compressor bleed for anti-icing.

Assumptions. Pressure drop in the air intake and combustion chamber is neglected, and working fluid is pure air. The fuel mass flow rate is negligible in comparison with the air mass flow rate at point 1.

Solution. Starting from point 1 at the compressor inlet:

$$P_1 = 101.325 \text{ kPa}, \quad T_1 = 10\,°C \rightarrow h_1 = 283.5\,\frac{kJ}{kg}, \quad s_1 = 5.643\,\frac{kJ}{kg.K}$$

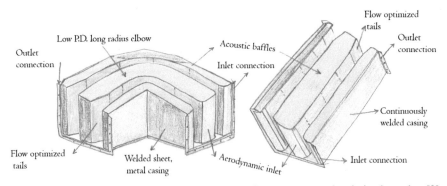

Figure 7.19 Rectangular and elbow shaped silencers, figure regenerated, painting by author [9]. Courtesy of dB Noise Reduction Inc.

Points 2 and 2s:

$$P_2 = 14P_1 = 1419 \text{ kPa}, \quad s_{2s} = s_1 \rightarrow h_{2s} = 602.6 \frac{\text{kJ}}{\text{kg}}$$

$$\eta_{ad,comp} = \frac{h_{2s} - h_1}{h_2 - h_1} \rightarrow 0.87 = \frac{602.6 - 283.5}{h_2 - 283.5} \rightarrow h_2 = 650.3 \frac{\text{kJ}}{\text{kg}} \rightarrow T_2 = 367.8 \,^{\circ}\text{C}$$

Point 3:

$$T_3 = 1126 \,^{\circ}\text{C}, \quad P_3 = P_2 \rightarrow h_3 = 1514 \frac{\text{kJ}}{\text{kg}}, \quad s_3 = 6.603 \frac{\text{kJ}}{\text{kg.K}},$$

Points 4 and 4s:

$$P_4 = P_1, \quad s_{4s} = s_3 \rightarrow h_{4s} = 735 \frac{\text{kJ}}{\text{kg}} \rightarrow \eta_{ad,tur} = \frac{h_4 - h_3}{h_{4s} - h_3} \rightarrow 0.87 = \frac{h_4 - 1514}{735 - 1514}$$

$$\rightarrow h_4 = 836.3 \frac{\text{kJ}}{\text{kg}}, \quad T_4 = 539.8 \,^{\circ}\text{C}$$

Energy balance of mixing the compressor bleed with the inlet air to the air intake:

$$T_{amb} = -2 \,^{\circ}\text{C} \rightarrow h_{amb} = 271.4 \frac{\text{kJ}}{\text{kg}}$$

$$\dot{m}_{normal} h_{amb} + \dot{m}_{bleed} h_2 = \left(\dot{m}_{normal} + \dot{m}_{bleed} \right) h_1$$

$$81.3 \times 271.4 + \dot{m}_{bleed} \times 650.3 = \left(81.3 + \dot{m}_{bleed} \right) \times 283.5$$

$$\rightarrow \dot{m}_{bleed} = 2.676 \frac{\text{kg}}{\text{s}}$$

The net power and efficiency calculation in the normal operation with inlet temperature of 10°C:

$$\dot{Q}_{in} = \dot{m}_{normal}(h_3 - h_2) = 81.3(1514 - 650.3) = 70211 \text{ kW}$$

$$\dot{W}_{tur} = \dot{m}_{normal}(h_3 - h_4) = 81.3(1514 - 836.3) = 55093 \text{ kW}$$

$$\dot{W}_{comp} = \dot{m}_{normal}(h_2 - h_1) = 81.3(650.3 - 283.5) = 29825 \text{ kW}$$

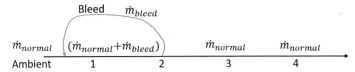

Figure 7.20 Mass flow diagram of the gas turbine due to activating anti icing system.

$$\dot{W}_{net} = \dot{W}_{tur} - \dot{W}_{comp} = 55093 - 29825 = 25268 \text{ kW}$$

$$\eta_{th,normal} = \frac{\dot{W}_{net}}{\dot{Q}_{in}} = \frac{25268}{70211} = 35.99\%$$

The net power and efficiency calculation when anti-icing is activated:

The power generation by the turbine and heat input to the combustion chamber stay the same because their mass flow rates are the same as the normal operation, but the compressor mass flow rate has increased to $\dot{m}_{normal} + \dot{m}_{bleed}$ (Fig. 7.20).

$$\dot{W}_{comp,with\ anti-icing} = \left(\dot{m}_{normal} + \dot{m}_{bleed}\right)(h_2 - h_1)$$
$$= (81.3 + 2.676)(h_2 - 650.3 - 283.5) = 30807 \text{ kW}$$

$$\dot{W}_{net,with\ anti-icing} = \dot{W}_{tur} - \dot{W}_{comp,with\ anti-icing} = 55093 - 30807 = 24286 \text{ kW}$$

$$\eta_{th,with\ anti-icing} = \frac{\dot{W}_{net,with\ anti-icing}}{\dot{Q}_{in}} = \frac{24286}{70211} = 34.59\%$$

$$\Delta\dot{W}_{net} = \frac{\dot{W}_{net,with\ anti-icing} - \dot{W}_{net}}{\dot{W}_{net}} \times 100 = \frac{24286 - 25268}{25268} \times 100 = -3.88\%$$

$$\Delta\eta_{th} = \frac{\eta_{th,with\ anti-icing} - \eta_{th,normal}}{\eta_{th,normal}} \times 100 = \frac{34.59 - 35.99}{35.99} \times 100 = -3.88\%$$

As it can be seen due to activating the anti-icing system, the net power and efficiency have decreased by 3.88%. The code used for the calculations of the GT equipped with the anti-icing system is presented as **Code 7.6.**

Code 7.6. The code of a simple cycle equipped with compressor bleed anti-icing system.

"anti-icing" m_dot_normal=81.3 P[4]=101.325 P[1]=P[4] R=14 P[2]=R*P[1] T[1]=10 T_amb=-2 h[1]=ENTHALPY(Air,T=T[1]) s[1]=ENTROPY(Air,T=T[1],P=P[1]) h2s=enthalpy(air, P=P[2], S=s[1]) ETA_COMP=0.87 ETA_COMP=(h[1]-h2s)/(h[1]-h[2]) T[2]=TEMPERATURE(Air,h=h[2]) h_amb=ENTHALPY(Air,T=T_amb) m_dot_normal*h_amb+m_dot_bleed*h[2]=(m _dot_normal+m_dot_bleed)*h[1] T[3]=1126 P[3]=P[2] s[3]=ENTROPY(Air,T=T[3],P=P[3]) h[3]=ENTHALPY(Air,T=T[3])	h4s=enthalpy(air, P=P[4], S=s[3]) ETA_TUR=0.87 ETA_TUR=(h[3]-h[4])/(h[3]-h4s) T[4]=TEMPERATURE(Air,h=h[4]) W_comp=(m_dot_normal+m_dot_bleed)*(h[2]- h[1]) W_tur=(m_dot_normal)*(h[3]-h[4]) W_net=W_tur-W_comp "UNDER NORMAL CONDITION WITH INLET TEMPERATURE OF 10 C" W_comp_normal=(m_dot_normal)*(h[2]-h[1]) W_tur_normal=(m_dot_normal)*(h[3]-h[4]) W_net_normal=W_tur_normal-W_comp_normal DW_net=(W_net- W_net_normal)/W_net_normal*100 Q_in=m_dot_normal*(h[3]-h[2]) ETA_NORMAL=W_net_normal/Q_in ETA_ANTIICING=W_net/Q_in DETA_th=(ETA_ANTIICING- ETA_NORMAL)/ETA_NORMAL*100

2.2 Compressor

The compressor main duty is to increase the gas pressure before heat addition. In the GT applications, it can be axial or centrifugal. The centrifugal compressors are mostly proper for small-scale power generation such as those used in aircrafts and turboprops, while the axial compressors are used for large-scale stationary power plants. In axial compressors, the flow direction is kept parallel to the rotor, while in the centrifugal types, it is perpendicular to the axis of rotor. The axial compressors can provide gas flow rate of several times bigger than the centrifugal type.

Both centrifugal and axial compressors increase gas pressure based on the same principals. They both have some stators at the inlet that provides diffuser passages and installed on the casing, their job is to adjust flow direction, and guide it to the rotary impeller (in the centrifugal compressor) or to the rotating blades (in the axial compressors). In addition, they convert gas velocity to pressure according to the Bernoulli concept. Most of compressors have some rows of adjustable stators at the inlet of compressor, they are called Inlet Guide Vane (IGV), or Variable Stator Vanes (VSV) (Fig. 7.13), and their main duties are controlling gas flow rate at the startup and shutdown, correcting flow direction, and converting gas velocity to pressure. The angle of attack of IGVs can be adjusted at their fixed location on the casing by some gear and pinion, hydraulic mechanisms, and electric actuators (See Figs.7.1, 7.2, 7.13, and 7.21A). The number of IGV stages, depends on the compressor size, and application. For applications with variable inlet gas density, more IGV stages is required. After each row of stator, there is an impeller (in the centrifugal compressor) or a row of rotary blades installed on the rotor. The main duty of impeller or rotary blades is to transfer kinetic energy to the gas flow. Due to energy transfer and also the diffuser shape passages created by the impeller's vanes or rotary blades, both pressure and velocity of gas stream will increase (Fig. 7.21B). During normal operation, compressor receives its rotational energy from the turbine shaft, because they are installed on the same rotor (See Figs. 7.1, and 7.21A). However, in the startup, when the combustion chamber is not ignited; yet, the compressor must be turned with a starter such as electric motor, a small steam turbine, diesel engine, or other drivers. This will continue until when the combustion chamber is ignited, and turbine is nearly operating at nominal speed and load. After that, the starter will be disconnected, and a turbine will provide the compression power required. Fig. 7.21 shows the diffuser, impeller, rotary blades of compressor, and a qualitative graph of pressure and velocity variation in a compressor. Fig. 7.21A shows a drawing of a gas turbine. Its compressor has 9 stages, air after passing through the air intake, enters the compressor. It first passes over the first row of IGVs, its velocity decreases and its pressure increases. Then air gets close to the rotary blades, and after passing through the diffuser passage created by the neighbor blades, casing, and rotor surfaces, its velocity and pressure increase at the same time. The same processes will happen for all of the 14 stages (Fig. 7.21B).

Compressor provides compressed air through the shaft for cooling of the rotating blades, and anti-icing system as well. The air required for cooling the turbine nozzles is provided by a bleed from the compressor casing to the turbine casing (Fig. 7.13).

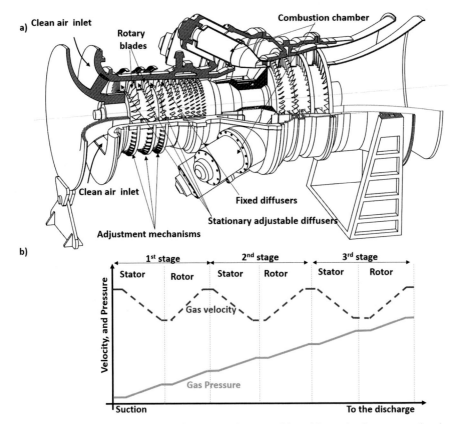

Figure 7.21 (A) Axial compressor with 9 stages, 4-stage turbine with annular Can type combustion chamber, Drawing by Amir Hossain Pajooh, Mechanical engineering student at University of Kurdistan (B) the general trend of pressure and velocity of air inside the compressor.

Figure 7.22 The bleed valves of SGT-400, and SGT-600 [3].
Courtesy of Siemens.

The cooling air of the rotor blades is provided through the compressor hub to the turbine hub. The air required by the anti-icing system is extracted from the compressor
casing as well. On the compressor casing, there are some bleeds equipped with valves
(butterfly valves usually) that are used during startup and shut down process to avoid
stall and surge (Fig. 7.4). Both bleed valves and IGVs (VSVs) are mostly used to avoid
stall, surge, and safe start up and shut down. During normal operation, bleed valves are
closed and IGVs are fully open. The bleed valves may evacuate the bled air into
exhaust or at the compressor inlet. Injecting the bled air into the compressor inlet increases the mass flow rate and can reduce the risk of surge due to low flow rate. However, this task can be done by more opening the IGVs. Injecting high-temperature bled
air into compressor inlet will increase the compression temperature and consequently
compression power consumption as well. Fig. 7.22 shows the single bleed valve location of SGT-400 taken from the seventh stage of the compressor, and also low pressure
and high pressure bleed valves of the SGT-600.

2.3 Combustion chamber

A combustor is responsible for heating the high-pressure air before the turbine section.
It exists in several configurations of tubular (side combustors), can-annular, annular,
and external. The temperature inside the combustion chamber may range from
$3400°F$ ($1871°C$) to $3500°F$ ($1927°C$). About 10% of the total air entering the combustion chamber is usually used for combustion; the rest of it is used for cooling, mixing, and dilution to reach the permissible operational temperature of the turbine. The
combustion chambers may be straight through flow or reverse flow. In the first type,
flow directly enters the chamber, while in the second type, it enters the chamber in

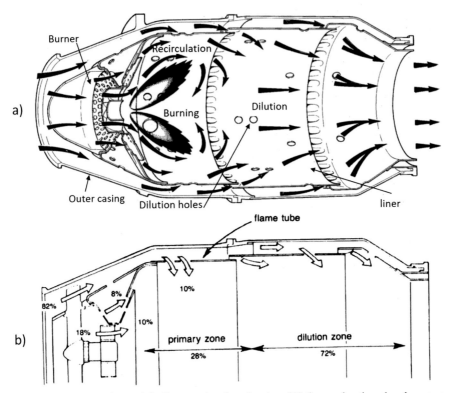

Figure 7.23 An annular straight flow combustion chamber, (A) the combustion chamber zones and (B) air distribution inside a combustion chamber [10].
Courtesy of Rolls-Royce Limited.

a reverse direction. Inside of each combustion chamber can be divided to three main regions of recirculation zone, burning zone with a recirculation zone, and dilution zone. Fig. 7.23A shows these three regions for a straight through flow combustor. In addition, the approximate air distribution inside the combustion chamber is shown in Fig. 7.23B. Combustion systems of GTs have some components in common such as igniter, flame detector, fuel nozzles, flow sleeve, swirl or mixing element to mix air and fuel, liner, casing, and transition piece to the turbine. The pressure drop in combustion chambers is about 2−10% of the compressor outlet pressure.

The air velocity at the compressor exit is about 122 m/s to 183 m/s, and the permissible velocity inside the combustion chamber must stay below 15.2 m/s to avoid carrying the flame to the turbine inlet nozzles and protecting them from burning. Theoretically, the air velocity through the maximum cross section of a combustion chamber for a reverse-flow is 3.6 m/s and for a straight-through flow is 10.7−18.3 m/s.

Fig. 7.24, shows a Can annular combustion system of the SGT-400 and the main components of the combustion system. In addition, Fig. 7.25 shows a can-annular combustion system of the SGT-50 GT of Siemens. The flame is created inside the liner; flow sleeve is between the liner and outer casing. This combustion system is a

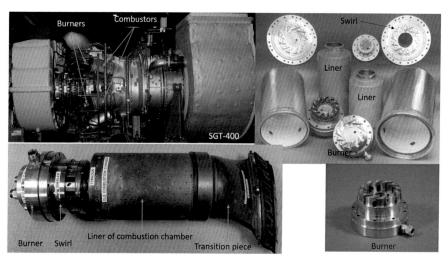

Figure 7.24 Can annular combustion chamber and its components, SGT-400 [3]. Courtesy of Siemens.

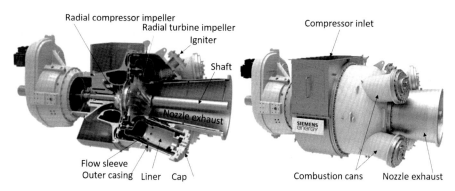

Figure 7.25 SGT-50, the radial gas turbine and compressor and the can-annular combustion chambers, 2 MW, gross efficiency of 26%, exhaust flow rate of 15 kg/s for using dry low emission (DLE) combustor, and pressure ratio of 7 [3]. Courtesy of Siemens.

reverse flow type. All the cans are connected together through some crossfire tubes placed between them. The main purpose of crossfire tubes is to make a pressure balance between all cans, transfer fire to the neighbor combustion cans and also reduce the need for installing igniter and flame detector on every can. Usually, two to three cans are equipped with igniter (spark plug), and two to three other cans are equipped with flame detectors. During the startup time, at least two out of three flame detectors must approve flame detection after a short time (usually less than 1 minute); otherwise, the safety system closes the main fuel valves and start purging the GT to avoid explosion

High temperature resistance ceramic liner

Inside the silo combustion chamber for repair

Figure 7.26 SGT 2000, with double-silo combustion chamber, capacity 187 MW, efficiency 36.5% for simple cycle, pressure ratio 12.8, exhaust mass flow 558 kg/s, and exhaust temperature 536°C [3].
Courtesy of Siemens.

or undesired combustion. If only two flame detectors are installed both of them must approve flame detection to continue the startup process.

Gas turbines with side combustors are also available; they usually occupy more space. They have single or double silo chambers with several burners. The liner of these combustion chambers is usually made of high-temperature resistance ceramic. Fig. 7.26 shows the SGT-2000 with double-side silo combustion chambers.

Gas turbines with external combustion chambers are designed to heat pure air. In these systems, different types of fuel can be used while no combustion gas enters the GT. Examples of such cycles are the closed cycle gas turbines working with CO_2 or other gases.

2.4 Turbine

Turbine is designed to convert the pressure and thermal energy of gasses leaving the combustion chamber to power. For this purpose, the nozzle stators and rotor blades (or nozzles and impellers in radial turbines) of a GT convert the pressure and thermal energy to velocity. The nozzles are stationary and placed on the casing inside, but the buckets (the rotary blades) are connected to the rotor. In the first stage of a GT, stator nozzles are placed at the turbine inlet. They convert pressure to velocity and guide the high-velocity gas on the rotational buckets. The buckets absorb the kinetic energy of gas and consume it to make rotation on the shaft. Due to this energy transfer, the velocity and pressure drop at the bucket row output. When gas enters the second stage, it passes over the nozzles again, its pressure drops and converts to velocity. Again in the rotary buckets, the kinetic energy transfers to the rotor that results in velocity and pressure drop at the bucket output (Fig. 7.27). The nozzles and buckets of the Ansaldo GT frame 6B are shown in this figure. As it can be seen, the first and second stages utilize cooling air to decrease the nozzles and buckets temperature from inside when they are facing the high-temperature and high-pressure gases from outside. The cooling air flows continuously. The cooling air for nozzles comes from casing, and for the buckets, it comes from rotor inside. The third stage does not need cooling because both temperature and pressure have dropped enough after two stages. Cooling of nozzles and buckets can be more improved by using computer simulations. Fig. 7.28 shows one of these simulations by the Siemens Digital Industries Software.

The material and properties of the nozzles and buckets are given below [11]:

For the nozzles:

First stage: Base Alloy-414, Chordal hinge design to minimize cooling air leakage with optional TBC/MCrAlY coating depending on operating conditions.

Second stage: Nozzle - Base Alloy-939-mod, Diaphragm—AISI 410, Brush seal configuration will generate an increased power output and improved heat rate. Standard configuration

Nozzles are coated with Al—Si diffusion coating to enhance oxidation resistance.

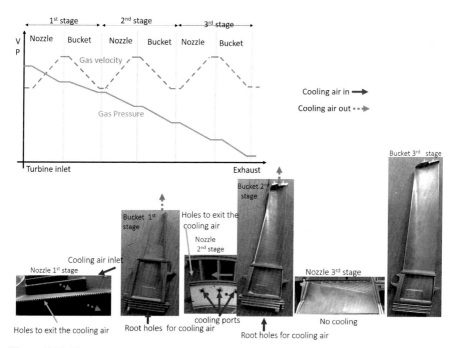

Figure 7.27 The pressure and velocity change through the turbine section, photos from author gallery.
Courtesy of Ansaldo.

Simulation of   Cooling holes, air out,   No cooling,
cooling, 1st stage   2nd stage   third stage

Figure 7.28 Simulation of cooling by Simcenter STAR-CCM+ [3].
Courtesy of Siemens Digital Industries Software.

Third stage: Base Alloy-738-LC.

And for the buckets:

First stage: Base Alloy-111-DS, MCrAlY airfoil coating plus internal aluminide coating, with16-cooling hole design to allow for increased inlet temperatures.

Second stage: Base Alloy-738-LC with 6 turbulated cooling hole design with co-balt base hardface material on the z-notches.

Third stage: Base Alloy-738-LC, no cooling system.

2.5   Exhaust

Combustion gases after leaving the last stage of the GT, enter a diffuser to partially increase the gas pressure before entering the exhaust chimney. This extra pressure is required to overcome the ambient pressure when leaving the chimney. Such diffuser can be seen in Fig. 7.4 for the double shaft Siemens gas turbine model of SGT600.

On the diffuser there are several thermocouples to measure temperature annularly. The temperature difference between two neighbor thermocouples should not exceed a certain value (usually 30°C) if this happens it means there is a problem in flow distri-bution, or fuel injection in the burners. Such thermocouples can be seen in Fig. 7.4 on the diffuser. One of these thermocouples is shown in Fig. 7.29.

2.6   Starter

At the startup process of a GT, turbine is not generating power; hence, an external source must be used to rotate the compressor. This external source is called starter. It is evident that when turbine reaches nominal load, the starter will be disconnected. A starter can be an electric motor, a steam turbine, and a static starter. Static starter technology is a lower cost, alternative technology used to "start" GTs by turning the generator into a motor. Generally, the startup process of a GT can be divided into

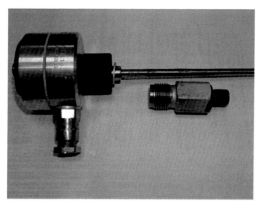

Figure 7.29 Exhaust thermocouple and adaptor [3].
Courtesy of Siemens.

two steps of cold and warm startup. During the cold startup, the GT is first accelerated using a starter and reaches cranking speed, then spends some time at this speed for purging the GT. The warm startup begins when speed is lowered and igniters start to generate steady flame in the combustion chamber(s). Fig. 7.30 shows the starter

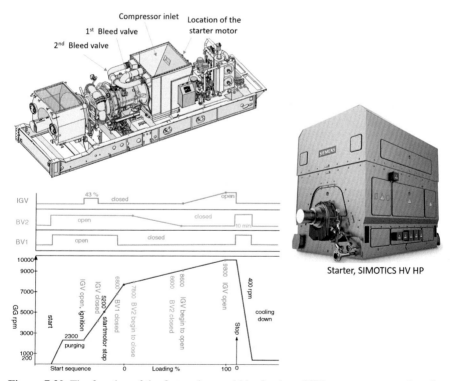

Figure 7.30 The function of the first and second bleed valves, IGV, starter motor, and cool down process for SGT-600 [3].
Courtesy of Siemens.

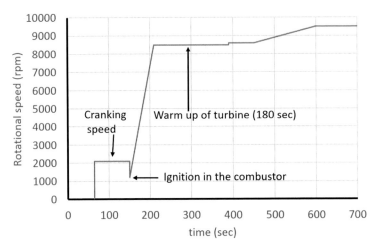

Figure 7.31 The standard startup of GE LM6000 gas turbine, regenerated graph [4]. Courtesy of General Electric.

location and also the function of the first and second bleed valves, IGV, starter motor, and shut down process for the SGT-600. During the shutdown process, when the fuel injection stops, the starter keeps the GT running at turning speed for cooling purposes. It shows that at 5200 rpm the starter is disconnected and power generator provides the power required by the gas generator (GG). Fig. 7.31 also shows the standard startup of GE LM6000 GT. Cranking is essential for rotary equipment such as GT during the startup and shutdown because it prevents hogging and sagging in the turbine rotor.

2.7 Generator

A generator is coupled with the GT shaft and converts the shaft rotation to electricity. Its main components include rotor, stator, bearings, and cooling system. The rotor is an electromagnetic shaft connected to the turbine shaft through a coupling. The stator is a series of insulated coils of wire that form a stationary cylinder; the stator surrounds an electromagnetic shaft, which is called rotor. Turning the rotor produces an electric current in each part of the coils, which becomes a separate electric conductor. The currents of the coils get together and form the final current that flows through the power lines for consumption. Due to the electric current in the conductors, heat will be generated inside the generator; air molecules between rotor and stator also get hot due to their friction with the rotating shaft and also intermolecular interactions. To remove this heat a cooling system should be utilized. Different fluids such as hydrogen, air, or water can be used for cooling. The air-cooled generators are excellent for power plant applications that demand simple, and flexible operation. The hydrogen-cooled generators are the right choice for high-efficiency applications and can work in simple and combined-cycle power plants. They are more economic than many water-cooled generators due to their high power output. The water-cooled generators are the best choice for large power plants where output requirements exceed the cooling capabilities of air-cooled or hydrogen-cooled units. Fig. 7.32 shows two air-cooled and water generators of Siemens that can be used for gas or steam turbines. The air-cooled generators produce more noise, while the water-cooled units are silent. The water-cooled

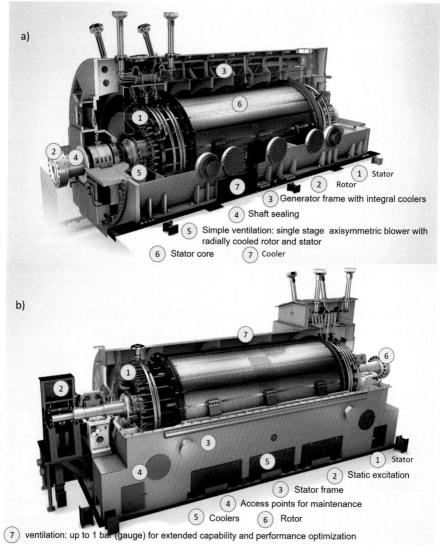

a)

1) Stator
2) Rotor
3) Generator frame with integral coolers
4) Shaft sealing
5) Simple ventilation: single stage axisymmetric blower with radially cooled rotor and stator
6) Stator core   7) Cooler

b)

1) Stator
2) Static excitation
3) Stator frame
4) Access points for maintenance
5) Coolers   6) Rotor
7) ventilation: up to 1 bar (gauge) for extended capability and performance optimization

Figure 7.32 (A) Water-cooled generator, model: SGen5-3000W, Capacity: 540—1300 MVA, Efficiency up to 99%, weight: 350 tons (B) air cooled generators, SGen5-2000P, Capacity: 370—560 MVA, Efficiency up to 99%, weight: 350 tons [3].
Courtesy of Siemens Energy.

generators are more complex for maintenance in comparison with the air cooled-units. Due to the complexity of water-cooled systems, they are more expensive.

2.8   Compressor washing

Air pollutions that scape the air intake filters can develop deposit on the compressor blades. These deposits can result in flow reduction, surging, efficiency loss, and vibration. To keep the GT efficiency high, and avoid corrosion, surging, etc. it is recommended to determine the washing periods of compressors by evaluating the compressor performance. GT manufacturers usually propose instructions for washing compressors. For example, an instruction given by Siemens includes following items:

1. The time period of washing.
2. The number of washing cycles to make the compressor fully clean.
3. The cleaning agent dilution ratio.
4. The number of rinsing after the last injection of washing liquid.

Compressor washing will be mandatory when one of the followings happen:

1. The compressor polytropic efficiency has dropped by 1.5%.
2. The compressor mass flow rate has decreased by 3%.
3. Unexplainable vibrations or compressor surging happens.

Fig. 7.33A shows the washing system of a GT recommended by Siemens. As it can be seen, it has two storage tanks, one for demineralized water, and the other for detergent. Solenoid valves adjusted according to the dilution ration, each tank has an electric heater to increase temperature to 60°C; it also has a filter after the tanks. A pump increases the liquid pressure to 50−60 bar (g). During the liquid injection for washing or rinsing, the GT speed is 2300 rpm, and during soaking, it is 215 rpm. Fig. 7.33B shows the washing cycles of the GT proposed by Siemens. Before starting the off-line washing, the GT must be shut down and cooled.

2.9   Lube oil system

The lube oil system duty is to provide oil with the designed pressure, temperature, and quality for the bearings, gears, jack oil pumps, etc. The main components of lube oil system include oil tank, pump, oil cooler, oil filter, and heater. Except for the tank, the other main components have main and auxiliary components. The main and auxiliary components must always be ready for operation and switching during the GT operation. When oil returns from the lubricated components, its temperature is high and may contain some impurities. It enters the oil tank; the oil tank has a sloped V-shaped bottom to move the impurities to the lowest point of tank for draining purposes.

While oil is settling in the tank, some oil vapor and other gases will be vented by a fan to the atmosphere. The pump sucks the oil and pushes it to the cooler; if the temperature is high, it enters the cooler; otherwise, the temperature control valve (TCV) operates and bypasses the oil and lets it go to the filter. On the filter, there is also pressure control valve (PCV), if the pressure drop across the filter is in the specified range,

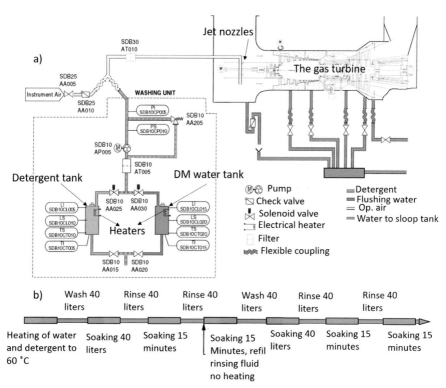

Figure 7.33 (A) The washing system components and diagram and (B) the washing cycles for SGT-600 [3].
Courtesy of Siemens.

oil enters the operating filter; otherwise, the PCV operates and send the oil to the auxiliary standby filter. If the pressure drop of filter exceeds the allowed magnitude, it must be removed, cleaned, or replaced. After the filter, oil goes to the bearings, gearboxes, etc., some components may require high-pressure oil for lubrication, in such cases, gear pumps increase the oil pressure before entering the component such as bearing number 2 in Fig. 7.34. This figure shows the lube oil diagram of the SGT-600. The jack oil pump is used before the startup time; it provides high-pressure oil to elevate the shaft from the bearing surfaces and provide initial lubrication. Heaters are mostly used before startup time; if the oil temperature is very low, the oil pump will not start until the heaters increase the oil temperature to the proper magnitude. The heaters can be electrical or steam/hot water coils. The oil type for lubrication is specified by the GT manufacturer, for example, the SGT-600 uses mineral-based turbine oil ISO VG46.

2.10 Fire extinguishing system

The purpose of the fire extinguishing system is to detect fire in the turbine room and extinguish it. It can operate with water or CO_2. Fig. 7.35 shows a CO_2-based

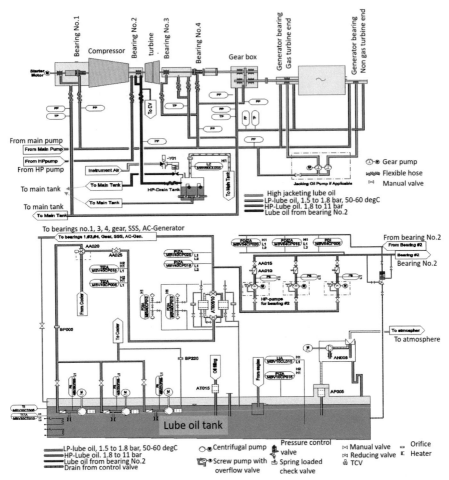

Figure 7.34 Lube oil system of SGT-600 [3].
Courtesy of Siemens.

extinguishing system for the SGT-600. Its main components include CO_2 bottles, sirens, sensors, weighing system, blocking system, warning lights, hose, and nozzles. The water-based system main components include water source, pump, sensors, alarms, water distribution system, and nozzles.

3. Hydrogen turbines

Hydrogen turbine is a GT that burns hydrogen or hydrogen blend instead of natural gas. Leading companies such as General Electric and Siemens are producing commercialized hydrogen turbines. These turbines will no longer release pollutants into the

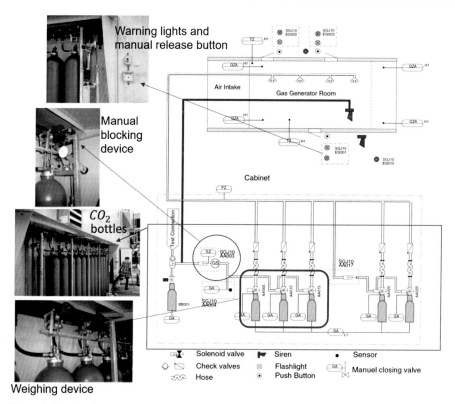

Figure 7.35 CO_2 fire Extinguishing system of SGT-600 [3].
Courtesy of Siemens.

atmosphere by burning hydrogen. Figs. 7.36 and 7.37 show the volume percent of the hydrogen used in different GT classes of General Electric and Siemens. It shows some models are developed to use 100% hydrogen, and some use hydrogen blend with lower volume fraction.

The use of hydrogen turbines in a power plant may cause changes in the overall balance of the power plant, auxiliary systems such as the combustion system, ventilation system of the GT enclosure, fuel accessories, safety, and firefighting systems. Most of these changes are due to the fact that hydrogen is a much more flammable fuel than natural gas and liquefied fuels. The sensors, gas detection systems, and safety equipment are all based on other fossil fuels, so changes must be made based on hydrogen.

Because GTs are inherently resilient to fuel, they can be configured to run on green hydrogen or similar fuels as a new unit or even upgrade after long service on traditional natural gas fuels. The range of changes required to configure a GT to run on hydrogen depends on the initial configuration of the GT and the overall balance of the plant, as well as the desired hydrogen concentration in the fuel. For the existing units, upgrading can be done with scheduled downtime to minimize plant outage, but for the new units, it can be a part of initial design and configuration.

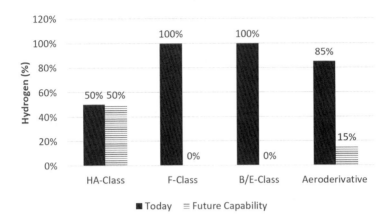

Figure 7.36 The volume percent of the hydrogen used in different GT classes, regenerated graph [4].
Courtesy of General Electric.

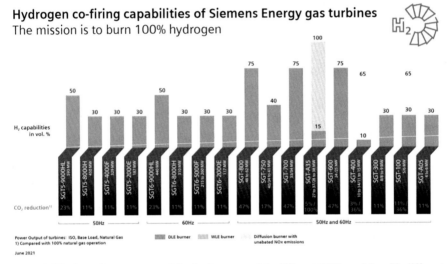

Figure 7.37 The volume percent of the hydrogen used in different SGT models with different burners and percent of CO_2 reduction [3].
Courtesy of Siemens.

Depending on the hydrogen source/method of production, international color—based classification is defined as following:

Gray (or black): Gasification of coal or reforming of natural gas without carbon capture.

Blue: Steam reforming of methane with carbon capture and storage.

Green: Electrolysis of water using renewable power.

Pink (Red): Electrolysis of water using nuclear power.

Turquoise: Pyrolysis of methane which produces hydrogen and solid carbon as a by-product.

White: Gasification or other process using 100% biomass as a feedstock.

The hydrogen production cost by gray (or black) method is the least. Hydrogen storage and transportations is very complicated because of it thermodynamic properties. It has extremely low density and flammable. For pipeline transmission, it must be compressed to $35-150$ bar, and for storage, the required pressure is about 350 bar.

According to the extremely high pressures, the transmission and storage equipment need special design and constructions. In addition, providing these pressures means high energy consumption. Hydrogen liquefying is even more difficult because it needs to reach temperatures less than $-250°C$ by using cryogenic methods. Liquid hydrogen must be kept in special double-walled cryogenic tanks.

4. Problems

1. According to the drawing presented in Fig. 7.38, identify 10 main components and number them on the figure.
2. Look through the newest GTs and find out, a) the maximum compression ratio of the compressor and b) the maximum temperature admitted by the turbine.
3. What are the emission reduction techniques used for emission control of GT?
4. What is the air to fuel mass ratio range used for GTs?
5. Find the emission production by the GTs of the leading companies such as Siemens and General Electric.

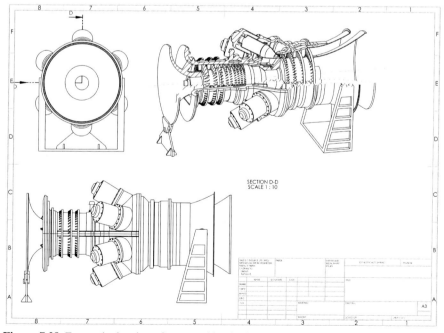

Figure 7.38 Free style drawing of a gas turbine in three views.
By Amir Hossain Pajooh, mechanical engineering student, University of Kurdistan.

6. Why the GT inlet stages at the turbine section need cooling? How they are cooled?
7. What are the cooling techniques used for turbine blade cooling?
8. Why using reheaters without using exhaust regenerator reduces thermal efficiency of GT?
9. What are the intercooling methods? Which method is more effective?
10. What are the compressor inlet air cooling methods? Compare them from economic point of view.
11. Why the bleed valves must be open during the startup process?
12. What are the possible reasons for stall in the compressor of a GT?
13. Under what condition the compressor can surge?
14. Under what conditions the compressor off line washing must be done?
15. Draw the schematic diagram of the lubrication system of a GT.
16. What is the purpose of using jack oil pump in the lubrication system of a GT?
17. In the lubrication systems, the oil coolers are placed before the oil filters, does it make any difference if oil could be first filtered and then cooled? Discuss your answer.
18. The exhaust temperature of a GT is very high; in some cases, it may be higher than 600°C. If you are asked to use this heat for more power generation, what do you recommend?
19. Draw the schematic diagram of the liquid fuel system of a GT.
20. Consider the SGT5-9000HL (50 Hz) (Fig. 7.39). It is a single shaft turbine that in ISO conditions generates net power of 593 MWe. The exhaust temperature and mass flow rate are 670°C and 1050 $\frac{kg}{s}$. The compression ratio of compressor is 24, its adiabatic efficiency is 0.87, and the nominal rotational speed is 3000 rpm. The ambient temperature and pressure are $P_1 = 101.325$ kPa, $T_1 = 15°C$.
    a. Draw the T-s thermodynamic diagram of the real and reversible GT on the same graph.
    b. The fuel energy rate burned in the combustion chamber ($\dot{Q}_{in}$), the power generated by turbine ($\dot{W}_{tur}$), the power consumption of compressor ($\dot{W}_{comp}$).
    c. Determine the GT thermal efficiency and validate your answer with the SGT600 net thermal efficiency reported by the Siemens ($\eta_{net,th} = 43\%$). What are the possible reasons for the difference?

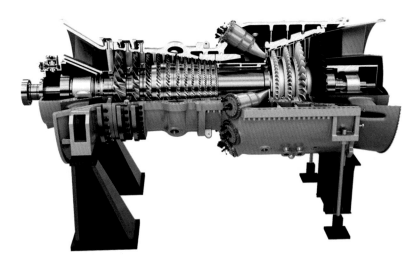

Figure 7.39 The SGT5-9000HL (50 Hz) with 4-stage air-cooled turbine, can annular combustion system [3].
Courtesy of Siemens.

    d. Determine the heat rate of the GT.

    e. Determine the exergy efficiency of the cycle.

    f. Perform a parametric analysis to understand the impact of exhaust temperature on the thermal efficiency when it changes from 640°C to 700°C.

    g. Perform a parametric analysis to understand the impact of pressure ratio on the thermal efficiency when it changes from 20 to 28.

21. Consider the SGT5-9000HL (50 Hz) (Fig. 7.36) the GT in problem 20. It is summer, and ambient temperature has reached 45°C. Assume that the turbine inlet/outlet temperatures stay the same as the example 20 by adjusting the fuel injection. In addition, the compression ratio adiabatic efficiency of the compressor, and air mass flow rate remain unchanged. Determine the following terms in comparison with their magnitude at ISO conditions in problem 20.

    a. The thermal efficiency drop

    b. Discuss the impact of changing compression work and input heat on the thermal efficiency drop

    c. The net power generation reduction

    d. If by using an inlet air chiller the temperature could be reduced to 15°C, plot the inlet temperature reduction (ΔT_1) impact on the $\Delta \eta_{th} = \eta_{th@T_{chilled}} - \eta_{th@45°C}$, and net power generation.

22. If the inlet air cooling of problem 21 is supposed to be done by a fogging system (Fig. 7.6A), such that air at the compressor inlet becomes saturated (relative humidity of $\phi_1 = 100\%$) with water vapor and its temperature reaches the water temperature of 15°C, by neglecting the pumping power of the fog system, determine,

    a. The water consumption rate of the fog system,

    b. The efficiency improvement due to temperature reduction from 45°C to 15°C,

    c. The turbine power generation, compressor power consumption and net power produced in comparison with using inlet air chiller for the same temperature reduction discussed in part c) of problem 20.

    d. Plot the water consumption rate of the fog system if the compressor inlet temperature is supposed to change from 15°C to 45°C.

23. Consider the Siemens GT model of SGT5-9000HL (50 Hz). The turbine inlet temperature and mass flow rate are 1126°C and 1050 $\frac{kg}{s}$. The compression ratio of compressor is 24, the adiabatic efficiency of turbine and compressor sections is 0.87 and the nominal rotational speed is 3000 rpm. The ambient temperature and pressure are $P_1 = 101.325$ kPa, $T_1 = 15\ °C$. You are supposed to determine

    a. the optimum intercooling and reheating pressures of the cycle

    b. the $\dot{W}_{comp}$, $\dot{W}_{tur}$, $\dot{Q}_{in}$, $\dot{Q}_{out}$ and thermal efficiency of the new GT.

    c. Compare the results with SGT600 without reheater and intercooler.

    d. What would happen for the thermal efficiency if a regenerator with effectiveness of 0.85 was installed at the turbine exhaust?

24. If the intercooling of the final cycle of problem 23 is supposed to be done with water spray with temperature of 15°C under adiabatic cooling ($\phi_8 = 100\%$), what would be the new $\dot{W}_{comp}$, $\dot{W}_{tur}$, $\dot{Q}_{in}$, $\dot{Q}_{out}$, and thermal efficiency of the new GT. Determine the water mass flow consumption in the intercooler.

25. The ambient temperature has dropped to $-2°C$, and freezing over the filters is possible. The GT is equipped with a bleed from the compressor outlet for anti-icing system. The turbine inlet temperature and the exhaust mass flow rate at normal operation are 1126°C and 1050 $\frac{kg}{s}$. The compression ratio of compressor is 24, the adiabatic efficiency of turbine and compressor sections is 0.87, and the nominal rotational speed is 3000 rpm. The ambient pressure is $P_{amb} = 101.325$ kPa. Determine

a. The anti-icing bleed mass flow rate to increase the inlet air temperature from $-2°C$ to $15°C$. Determine

b. The net power and the efficiency change due to using compressor bleed for anti-icing.

References

[1] Mitsubishi heavy industry, https://power.mhi.com/products/gasturbines/technology/steam-cooled-combustor.
[2] https://www.statista.com/.
[3] www.siemens-energy.com.
[4] https://www.ge.com/.
[5] E.D.N.CO − fogging, Wet compression system.
[6] Solar Turbines Incorporated, https://www.caterpillar.com/en/brands/solar-turbines.html.
[7] M.P. Boyce, Gas Turbine Engineering Handbook, third ed., Gulf Professional Publishing, An imprint of Elsevier, 2006.
[8] https://www.donaldson.com/en-us/.
[9] https://www.dbnoisereduction.com.
[10] https://www.rolls-roycemotorcars.com/en_GB/home.html.
[11] https://www.ansaldoenergia.com/.

Further reading

[1] M. Sammak, Anti-Icing in Gas Turbines, Thesis for the Degree of Master of Science, Lund University, 2006.

Micro gas/steam turbines power plants

Chapter outline

Micro gas turbines (MGTs) are rotary dynamic power generators that produce power in the scales of 100 *kW*. They usually have a rotor with a single-stage radial flow impeller for the turbine and the compressor as well. The thermal efficiency of MGTs is smaller than that of their counterparts in large-scale power plants. One reason for the lower thermal efficiency of MGT is that leaks from the internal clearances (for example impeller tip loss) are inevitable to ensure the safe operation of the MGT. Despite the fact that the MGTs have become much smaller, the clearances are still of the order of permissible clearances in the large gas turbines. The second reason is that the maximum operating pressure and temperature is smaller than that in the large turbines. The third reason is the higher friction loss due to smaller Reynolds number. The Reynolds number is smaller because of smaller components sizes in the MGT. Hence, as it can be seen, there are fundamental restrictions to increase the thermal efficiency of MGTs.

The rotational speed of microturbines is usually very high and greater than 40,000 rpm, while in the large turbines, it is usually less than 20,000 rpm. It is not possible to use oil-lubricated bearings due to the high speed shaft, which is why air bearings have been developed for microturbines. The efficiency of microturbines is less than 20% without the use of a recuperator, but using a recuperator to recover, the exhaust heat increases the thermal efficiency of MGT by 20%–30%. The recuperator preheats the compressed air before entering the combustion chamber.

MGT is derived from turbocharger technologies, but it has its own combustion chamber instead of using the exhaust gases of reciprocating engines. In addition, it has an integrated generator to convert the mechanical power to electricity.

MGTs are well fitted for distributed electricity generation, known as DG, and smart micro-grid purposes. They are used in hotels, campuses, universities, complexes, towers, small industries, etc. to provide combined heat and power (electricity) at the

Power Generation Technologies. https://doi.org/10.1016/B978-0-323-95370-2.00003-X
Copyright © 2023 Elsevier Inc. All rights reserved.

same time (CHP). In some projects, the heat or a part of electricity are also used to provide the cooling demand beside the heat and power, which is called trigeneration or CCHP[1]. MGTs have significant advantages that make them very popular, and they are going to capture a considerable share of distributed power generation market. The main advantages of the MGT can be listed as below [1]:

- They produce electricity at the place of demand, so the grid and network losses are omitted.
- They reduce the costs for developing the public electricity grid.
- The high-quality exhaust of MGTs can be utilized for heating, cooling, desalination, and drying purposes.
- Waste heat of MGT can be utilized as the heat source of small scale Organic Rankine Cycle (ORC) as the bottoming cycle.
- They have a very fast and easy startup and shutdown procedures.
- They are compact and need a very small footprint for installations.
- Their noise is well controlled for using in the buildings, it is about 65−70 db at the distance of 10 m, which is safe for human ear.
- Their modular design allows for easy and low cost maintenance and installation.
- The modular design provides the possibility of increasing the capacity for the future development.
- Modular design provides the possibility of turning down some units, while the other units still operate with high efficiency. In other words, in the part load, the efficiency remains high because several units can operate independently.
- They can operate with different liquid and gaseous fuels.
- They can be used in on-grid and off-grid modes.
- They can be used to run pumps, compressors, and fans in small industries or air conditioning systems.
- No lubricating oil system is required because of using air bearings.
- Synchronization system is integrated with no external switchgear.
- Minimum moving parts reduces the maintenance jobs and unscheduled downtimes.
- Low maintenance costs.

The main manufacturers of MGT are Capstone, Elliott Energy Systems, AlliedSignal, Ingersoll-Rand Energy Systems and Power Recuperators WorksTM, Turbec, Browman Power and ABB Distributed Generation and Volvo Aero Corporation.

Micro steam turbines (MST) are designed to produce electricity in the scales of 100 *kW*. They can utilize waste heat in a small steam generator for producing steam or use high-pressure process seam to produce electricity. These turbines can be installed in an existing plant to utilize the extra steam when it is available for intermittent power generation, or they can be used in a separate cycle to produce continuous electricity. In addition, they can be used instead of pressure reducing valves (PRV) in the steam generation units to make use of the extra pressure for power generation [2].

There are several manufacturers of MSTs such as Bestech India, Energent Corporation, and Mizun Consultants and Engineers.

---

[1] Combined cooling heating and power.

1. Thermodynamic analyses

The thermodynamics of power generators play an important role to understand, design, and optimize these cycles. For this purpose, in the following, the thermodynamics of basic MGT and MST cycles are presented.

1.1 Thermodynamics of MGT

Micro gas turbines like the industrial gas turbines follow the basic Brayton cycle, which was discussed in the chapter 7. Today, most of MGTs are equipped with recuperator to increase the thermal efficiency. Fig. 9.1 shows Brayton cycle for a recuperated MGT.

In the basic thermodynamic analysis, the standard air assumptions are utilized these assumptions include:

(a) The working fluid everywhere is pure air.
(b) The combustion process is replaced by a heat addition process.
(c) The exhaust process is replaced by a heat rejection process in a heat exchanger.
(d) All of the processes are internally reversible.
(e) Air is assumed as perfect gas.

The main processes shown in Fig. 8.1 include:

1−2 Isentropic and adiabatic compression by the compressor (mostly radial compressor).

2−5 Constant pressure preheating of the compressed air in the recuperator.

5−3 Constant pressure heat addition process in the combustion chamber.

3−4 Isentropic and adiabatic expansion in the turbine (mostly radial flow turbine).

4−6 Constant pressure heat recovery from the exhaust gases in the recuperator.

6−1 constant pressure heat rejection to the atmosphere through the exhaust.

The governing equations and assumptions for these six processes can be formulated as follow:

For the adiabatic and isentropic compression in the process 1−2:

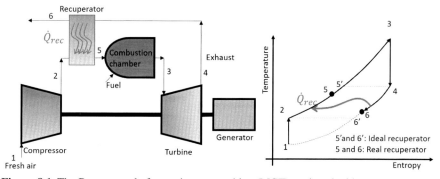

Figure 8.1 The Brayton cycle for a micro gas turbine (MGT) equipped with a recuperator

$$s_1 = s_2$$

$$\dot{W}_{comp} = \dot{m}_{air}(h_2 - h_1) \tag{8.1}$$

$$r_p = \frac{P_2}{P_1}$$

In which, r_p is the pressure ratio of the compressor. Under additional assumption of constant specific heat (cold air standard assumption), the following equations stand for the compression process:

$$r_p = \left(\frac{v_1}{v_2}\right)^k = \left(\frac{T_2}{T_1}\right)^{\frac{k}{k-1}} \tag{8.2}$$

in which k is the specific heat ratio of air at standarad conditions which is 1.4.
For the isobar preheating process of 2–5:

$$P_2 = P_5$$
$$\dot{Q}_{rec-ideal} = \dot{m}_{air}(h_{5'} - h_2) = \dot{m}(h_4 - h_2) \tag{8.3}$$
$$\dot{Q}_{rec-real} = \dot{m}_{air}(h_5 - h_2)$$

The effectiveness of the recuperator is determined as below:

$$\varepsilon = \frac{\dot{Q}_{rec-real}}{\dot{Q}_{rec-ideal}} = \frac{h_5 - h_2}{h_4 - h_2} \cong \frac{T_5 - T_2}{T_4 - T_2} \tag{8.4}$$

The constant pressure heat addition process 5–3:

$$P_5 = P_3$$
$$\dot{Q}_{in} = \left(\dot{m}_{air} + \dot{m}_{fuel}\right)(h_3 - h_5) \tag{8.5}$$

After the heat addition, the hot pressurized gases would be expanded in the process 3–4:

$$s_3 = s_4$$

$$\dot{W}_{tur} = \left(\dot{m}_{air} + \dot{m}_{fuel}\right)(h_3 - h_4) \tag{8.6}$$

$$r_p = \frac{P_3}{P_4}$$

In which, r_p is the pressure ratio of the turbine. Under additional assumption of constant specific heat (Cold Air Standard assumption), the following equation stands for the expansion process:

$$r_p = \left(\frac{v_4}{v_3}\right)^k = \left(\frac{T_3}{T_4}\right)^{\frac{k}{k-1}} \tag{8.7}$$

In the adiabatic and constant pressure conditions, the heat transfer from the exhaust gases to the compressed air would be:

$$\dot{Q}_{rec-real} = \left(\dot{m}_{air} + \dot{m}_{fuel}\right)(h_4 - h_6) \tag{8.8}$$

Which means the heat given by the exhaust is received by the compressed air.

The heat rejection in the process 6-1 also can be calculated as below:

$$\dot{Q}_{out} = \left(\dot{m}_{air} + \dot{m}_{fuel}\right)(h_6 - h_1) \tag{8.9}$$

And the thermal and electrical efficiencies would be:

$$\eta_{th} = 1 - \frac{\dot{Q}_{out}}{\dot{Q}_{in}}$$
$$\eta_e = \frac{E_{net}}{\dot{Q}_{in}} \tag{8.10}$$

Example 8.1. MGT model of C200S manufactured by the Capstone company produces 200 kW of electricity with an electrical efficiency of 33%. The pressure ratio of the compressor is 3.5, and the turbine inlet temperature is 875 °C. This MGT accepts sour gas fuels with up to 5000 ppm H_2S. Fig. 8.2 shows the MGT components. The exhaust mass flow is 1.3 kg/s. The landfill gas HHV used in the MGT is 42,000 kJ/kg [3]. Nominal full power performance is measured at ISO conditions of 15 °C, 101.325 kPa, and 60% RH. The electric generator efficiency is 0.98, also only 0.95 of the total wasteheat in the exhaust is absorbed by the compressor exit flow, and the rest is dissipated to the environment through the recuperator body ($\eta_{recuperator} = 0.95$). Assume the process to be reversible in the compressor, and the pressure losses in the recuperator and the combustion chamber are neglected. The process in the turbine is real, and it has irreversibilities. Use the standard air assumptions wherever needed and determine:

(a) The mass flow of fuel consumption.
(b) The air to fuel mass flow ratio.
(c) The state point properties and the T-s diagram of the MGT.
(d) The recuperator effectiveness.

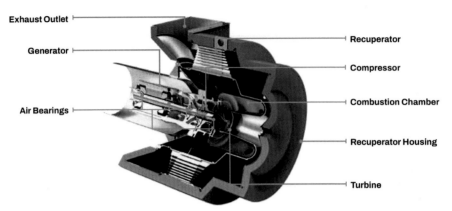

Exhaust Outlet

Generator

Air Bearings

Recuperator

Compressor

Combustion Chamber

Recuperator Housing

Turbine

Figure 8.2 The components of C200S, the micro gas turbine (MGT) manufactured by Capstone [1].
(Courtesy of Capstone).

(e) The adiabatic efficiency of the turbine section.
(f) The exergy destruction of the cycle and the exergy efficiency of the cycle.

Solution. The standard air assumptions are used in the cycle except for the turbine, which it expands the working fluid irreversibly. The standard air assumptions include:

The working fluid everywhere is pure air; the combustion process is replaced by a heat addition process. The exhaust process is replaced by a heat rejection process in a heat exchanger. All of the processes are internally reversible, and air is assumed as perfect gas.

a.

To determine the fuel mass flow, the electrical efficiency equation can be used.

$$\eta_e = \frac{E_{net}}{\dot{Q}_{in}} = \frac{E_{net}}{\dot{m}_{fuel} HV_{fuel}} \rightarrow$$

$$\dot{m}_{fuel} = \frac{E_{net}}{\eta_e HV_{fuel}} = \frac{200}{0.33 \times 42000} = 0.01443 \frac{kg}{s}$$

b.

According to the mass conservation principal, the exhaust mass flow equals to the mass flow of air plus the mass flow of fuel.

$$\dot{m}_{fuel} + \dot{m}_{air} = \dot{m}_{ex} \rightarrow 0.01443 + \dot{m}_{air} = 1.3 \rightarrow \dot{m}_{air} = 1.286 \frac{kg}{s} \rightarrow$$

The air to fuel mass flow rate is:

$$AF = \frac{\dot{m}_{air}}{\dot{m}_{fuel}} = \frac{1.286}{0.01443} = 89.09$$

More discussion:

Metallurgical limitation in the thermal power generator units has always been a challenge. Since the combustion temperature under stoichiometric conditions is much higher than the melting point of material used in the engines; therefore, the engine must be protected against melting by different methods. The very high air to fuel ratio guarantees the safe operation of MGT by decreasing the combustion gas temperature. An important difference between the gas turbine families with the reciprocating engines is that there is no jacket cooling in the gas turbines and MGT, while the reciprocating engines are continuously cooled by the jacket cooling. The lack of jacket cooling is compensated by bigger air to fuel ratio that lowers the maximum temperature, dilutes the combustion gases, and cools the MGT internal components as well. In the jacket cooling, a cooling medium like water is circulating around the engine body, cylinders, and cylinder heads. As a result, the air to fuel ratio in the reciprocating engines is much lower than that in the MGT and gas turbines.

c.

Starting from point 1:

$$T_1 = 15\,^\circ C,\ P_1 = 101.325\ kPa \rightarrow h_1 = 288.5\frac{kJ}{kg},\ s_1 = 5.661\frac{kJ}{kg.K}$$

Point 2: Since the compression is reversible:

$$s_2 = s_1 = 5.661\frac{kJ}{kg.K},\ P_2 = r_P \times P_1 = 3.5 \times 101.325 = 354.6\ kPa \rightarrow h_2$$
$$= 413\frac{kJ}{kg}$$

$$\dot{W}_{comp} = \dot{m}_{air}(h_2 - h_1) = 1.286(413 - 288.5) = 160.1\ kW$$

Point 3:

$$T_3 = 875\,^\circ C,\ P_3 = P_2 = 354.6\ kPa \rightarrow h_3 = 1217\frac{kJ}{kg},\ s_3 = 6.767\frac{kJ}{kg.K}$$

Point 5:

$$\dot{Q}_{in} = \frac{E_{net}}{\eta_e} = \frac{200}{0.33} = 606\ kW$$

$$\dot{Q}_{in} = \dot{m}_{ex}(h_3 - h_5) \rightarrow 1.3(1217 - h_5) = 606 \rightarrow h_5 = 750.7 \frac{kJ}{kg} \rightarrow T_5 = 461.5\,^\circ C$$

$$P_5 = P_2 = 354.6\ kPa,\ s_5 = 6.265 \frac{kJ}{kg.K}$$

Point 4:
the net power is used to determining the point 4 properties:

$$\dot{W}_{net} = \frac{E_{net}}{\eta_{gen}} = \frac{200}{0.98} = 204.1\ kW$$

$$\dot{W}_{net} = \dot{W}_{tur} - \dot{W}_{comp} = \dot{m}_{ex}(h_3 - h_4) - \dot{W}_{comp} \rightarrow 204.1 = 1.3(1217 - h_4) \rightarrow h_4$$

$$= 936.8 \frac{kJ}{kg} \rightarrow T_4 = 630.2\,^\circ C$$

$$P_4 = P_1 = 101.325\ kPa \rightarrow s_4 = 6.852 \frac{kJ}{kg.K}$$

Point 6:
Since the recuperator loses 5% of the exhaust energy from its body and the rest is transferred to the compressed air, hence:

$$\eta_{recuperator} = \frac{\dot{Q}_{absorbed}}{\dot{Q}_{available}} = \frac{\dot{m}_{air}(h_5 - h_2)}{\dot{m}_{ex}(h_4 - h_6)} \rightarrow 0.95 = \frac{1.286(750.7 - 413)}{1.3(936.8 - h_6)} \rightarrow$$

$$h_6 = 585.3 \frac{kJ}{kg} \rightarrow T_6 = 306.1\,^\circ C$$

$$P_6 = P_1 \rightarrow s_6 = 6.371 \frac{kJ}{kgK}$$

The high-quality exhaust is very suitable for bottoming cycles specially ORC, or as the heat source of an absorption chiller, or the heat required by a small desalination unit. It is worth mentioning that in the real data reported by the manufacturer, the exhaust temperature is 280 $^\circ C$. The reason for higher temperature in the calculations is probably because of considering reversible process in the compressor, combustor, and recuperator and also making some assumptions due to lack of input data. The T-s diagram of the MGT is shown in Fig. 8.3.

d:

The recuperator effectiveness is determined as below:

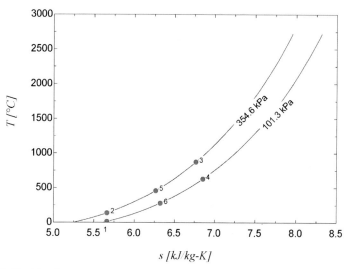

Figure 8.3 The T-s diagram of the recuperated micro gas turbine (MGT) in Example 8.1.

$$\varepsilon = \frac{h_5 - h_2}{h_4 - h_2} = \frac{750.7 - 413}{936.8 - 413} = 64.47\%$$

e:

The adiabatic efficiency of the turbine is the ratio of the actual power to the isentropic power generated by the turbine as below:

$$\eta_{ad,tur} = \frac{\dot{W}_{a,tur}}{\dot{W}_{s,tur}} = \frac{h_3 - h_4}{h_3 - h_{4s}}$$

To determine the h_{4s}:

$$s_{4s} = s_3, \ P_{4s} = P_4 = P_1 \rightarrow h_{4s} = 862.5 \frac{kJ}{kg} \rightarrow$$

$$\eta_{ad,tur} = \frac{1217 - 936.8}{1217 - 862.5} = 79.04\%$$

f:

The exergy destruction can be estimated as below:

To determine the exergy efficiency of the engine, the average heating temperature, T_{H-mean}, must be determined as below:

$$T_{H-mean} = \frac{h_3 - h_5}{s_3 - s_5} = \frac{936.8 - 558.1}{6.767 - 6.265} = 928.6 \ K$$

And the sink temperature which is ambient temperature is:

$$T_L = T_1 = 15\,^\circ C = 288.15\ K$$

Therefore, the Carnot cycle efficiency can be determined:

$$\eta_{carnot} = 1 - \frac{T_L}{T_{H-mean}} = 1 - \frac{288.15}{928.6} = 68.97\%$$

And the exergy efficiency or the second law efficiency is:

$$\eta_{II} = \frac{\eta_{th}}{\eta_{carnot}}$$

The thermal efficiency can be determined by dividing the electrical efficiency by the generator efficiency as below:

$$\eta_{th} = \frac{\eta_e}{\eta_{gen}} = \frac{0.33}{0.98} = 33.67\%$$

Hence:

$$\eta_{II} = \frac{\eta_{th}}{\eta_{carnot}} = \frac{33.67}{68.97} = 48.82\%$$

Or by using the following equation:

$$\eta_{II} = \frac{exergy\ recovered}{exergy\ supplied} = \frac{\dot{W}_{net}}{\dot{Q}_{in}\left(1 - \dfrac{T_L}{T_H}\right)} = \frac{E_{net}}{\dot{Q}_{in}}\left(\frac{1}{\eta_{carnot}\eta_{gen}}\right) = \frac{\eta_e}{\eta_{carnot}\eta_{gen}}$$

$$= \frac{\eta_{th}}{\eta_{carnot}}$$

Which is exactly the same as the previous calculation.

The exergy destruction:

$$\underbrace{\left(\dot{X}_{in} - \dot{X}_{out}\right)}_{net\ rate\ of\ entropy\ transfer} - \dot{X}_{destroyed} = \frac{dX_{system}}{dt}$$

for steady state operation $\frac{dX_{system}}{dt} = 0$, therefore:

$$\dot{X}_{destroyed} = \dot{X}_{in} - \dot{X}_{out} = \dot{Q}_{in}\left(1 - \frac{T_L}{T_{H-mean}}\right) - \dot{W}_{net}$$

$$= 606\left(1 - \frac{288.15}{928.6}\right) - \frac{200}{0.98} = 213.9\ kW$$

The exergy destruction can also be determined as below:

$$\dot{X}_{destroyed} = \dot{Q}_{in}(\eta_{carnot} - \eta_{th}) = 606(0.6897 - 0.3367) = 213.9 \, kW$$

The code written for this example is given in the Code 8.1 in the following:

Code 8.1. the code developed for the MGT with recuperator, MGT model C200S

```
"MGT C200S"                              Q_in=m_ex*(h[3]-h[5])
E_net=200                                T[5]=temperature(air,H=H[5])
eta_e=0.33                               P[5]=P[2]
HV_fuel=42000                            s[5]=entropy(air, T=T[5], P=P[5])
T[1]= 15                                 Q_available=m_ex*(h[4]-h[6])
P[1]=101.325                             Q_absorbed=m_air*(h[5]-h[2])
T[3]=875                                 eta_recuperator= Q_absorbed/Q_available
r_p=3.5                                  T[6]=temperature(air, H=H[6])
eta_recuperator=0.95                     P[6]=P[1]
eta_gen=0.98                             s[6]=entropy(air, T=T[6], P=P[6])
Q_in=E_net/eta_e                         T[4]=temperature(air, H=H[4])
m_ex=1.3                                 P[3]=P[2]
Q_in=m_fuel*HV_fuel                      s[3]=entropy(air, P=P[3], T=T[3])
m_air=m_ex-m_fuel                        S[4]=entropy(air, P=P[4], T=T[4])
AF=m_air/m_fuel                          P[4]=P[1]
s[1]=entropy(air, P=P[1], T=T[1])        EPSILON=(h[5]-h[2])/(h[4]-h[2])
h[1]=enthalpy(air, T=T[1])               eta_ad_tur=(h[3]-h[4])/(h[3]-h4s)
P[2]=r_p*P[1]                            S4s=s[3]
s[2]=s[1]                                h4s=enthalpy(air, s=s4s, P=p[4])
T[2]=temperature(air, s=s[2], p=p[2])    T_H_mean=(h[3]-h[5])/(s[3]-s[5])
h[2]=enthalpy(air, t=t[2])               T_L=T[1]+273.15
W_comp=m_air*(h[2]-h[1])                  ETA_carnot=1-T_L/T_H_mean
W_net=E_net/eta_gen                      ETA_th=eta_e/eta_gen
W_net=m_ex*(h[3]-h[4])-m_air*(h[2]-h[1]) eta_II=ETA_th/ETA_carnot
h[3]=enthalpy(air, T=T[3])               X_destroyed=Q_in*ETA_carnot-W_net
                                         X_DESTROYED2=Q_in*(ETA_carnot-ETA_th)
```

Example 8.2. In C30 MGT manufactured by Capstone, it produces net 29.433 kW of electricity; the turbine inlet temperature is 850 $°C$. The electrical efficiency is 25.34%, and the heating value of natural gas as fuel is 47,000 $\frac{kJ}{kg}$. The pressure drop of the cold and hot streams in recuperators is 18.5 and 178.6 mbar. The pressure drop in the combustor is also 0.3 bar. The compressor pressure ratio is 3.39. The air flow through the compressor is 0.30,745 $\frac{kg}{s}$. The adiabatic efficiency of the turbine and compressor is 0.85 and 0.8265, respectively. It is assumed that the MGT is well insulated and no heat loss occur from the its body. The generator efficiency is also 0.98. determine:

(a) The fuel consumption mass flow rate.
(b) The air to fuel mass flow ratio.
(c) The state point properties and the T-s diagram.
(d) The turbine adiabatic efficiency.
(e) The exergy destruction and exergy efficiency.
(f) The recuperator effectiveness.

Solution.

a.

To determine the fuel mass flow, the electrical efficiency equation can be used.

$$\eta_e = \frac{E_{net}}{\dot{Q}_{in}} = \frac{E_{net}}{\dot{m}_{fuel} HV_{fuel}} \rightarrow$$

$$\dot{m}_{fuel} = \frac{E_{net}}{\eta_e HV_{fuel}} = \frac{29.433}{0.2534 \times 47000} = 0.002471 \frac{kg}{s}$$

b.

The air to fuel mass flow rate is:

$$AF = \frac{\dot{m}_{air}}{\dot{m}_{fuel}} = \frac{0.3075}{0.002471} = 124.4$$

Care must be taken that all of the air do not take part in the combustion process. Very high AF may extinguish the flame. Most of this air is used for reducing flame temperature and is mixed with combustion gases for cooling the internal components and dilution. The temperature must reach a point, which is tolerable with the internal components of the combustion chamber and turbine.

c.

Starting from point 1:

$$T_1 = 15\,^{\circ}C, \; P_1 = 101.325 \; kPa \rightarrow h_1 = 288.5 \frac{kJ}{kg}, \; s_1 = 5.661 \frac{kJ}{kg.K}$$

Point 2:

$$\eta_{ad,comp} = \frac{\dot{W}_{s,comp}}{\dot{W}_{a,comp}} = \frac{h_1 - h_{2s}}{h_1 - h_2}$$

$$s_{2s} = s_1 = 5.661 \frac{kJ}{kg.K}, \; P_2 = r_P \times P_1 = 3.39 \times 101.325 = 343.5 \; kPa \rightarrow h_{2s}$$

$$= 409.3 \frac{kJ}{kg}$$

$$0.8265 = \frac{288.5 - 434.6}{288.5 - h_2} \rightarrow h_2 = 434.6 \frac{kJ}{kg}$$

Point 3:

$$T_3 = 850\,^{\circ}C, \; P_3 = P_2 - \Delta P_{rec,cold} - \Delta P_{combustor} = 343.5 - 1.85 - 30$$

$$= 311.6 \; kPa \rightarrow h_3 = 1188 \frac{kJ}{kg}, \; s_3 = 6.778 \frac{kJ}{kg.K}$$

Point 5:

$$\dot{Q}_{in} = \frac{E_{net}}{\eta_e} = \frac{29.433}{0.2534} = 116.2 \ kW$$

$$\dot{m}_{ex} = \dot{m}_{air} + \dot{m}_{fuel} = 0.3075 + 0.002471 = 0.3099 \ \frac{kg}{s}$$

$$\dot{Q}_{in} = \dot{m}_{ex}(h_3 - h_5) \rightarrow 0.3099(1188 - h_5) = 116.2 \rightarrow h_5 = 813.1 \ \frac{kJ}{kg} \rightarrow T_5$$
$$= 518.7 \ ^{\circ}C$$

$$P_5 = P_2 - \Delta P_{rec,cold} = 343.5 - 1.85 = 341.6 \ kPa, \ s_5 = 6.357 \frac{kJ}{kg.K}$$

Point 4:
The net power is used to determining the point 4 properties:

$$\dot{W}_{net} = \frac{E_{net}}{\eta_{gen}} = \frac{29.433}{0.98} = 30.03 \ kW$$

$$\dot{W}_{net} = \dot{W}_{tur} - \dot{W}_{comp} = \dot{m}_{ex}(h_3 - h_4) - \dot{m}_{air}(h_2 - h_1)$$
$$= 0.3099(1188 - h_4) - 0.3075(434.6 - 288.5)$$
$$= 30.03 \rightarrow h_4 = 946 \ \frac{kJ}{kg} \rightarrow T_4 = 638.4 \ ^{\circ}C$$

$$P_4 = P_1 + \Delta P_{rec,hot} = 101.325 + 17.86 = 119.2 \ kPa \rightarrow s_4 = 6.816 \frac{kJ}{kg.K}$$

Point 6:
Since the MGT is well insulated:

$$\dot{Q}_{available} = \dot{Q}_{absorbed} \rightarrow \dot{m}_{ex}(h_4 - h_6) = \dot{m}_{air}(h_5 - h_2) \rightarrow 0.3099(946 - h_6)$$
$$= 0.3075(813.1 - 434.6) \rightarrow$$

$$h_6 = 570.6 \frac{kJ}{kg} \rightarrow T_6 = 292 \ ^{\circ}C$$

$$P_6 = P_1 = 101.3 \ kPa \rightarrow s_6 = 6.346 \frac{kJ}{kgK}$$

And the T-s diagram of the MGT is drawn below in Fig. 8.4:

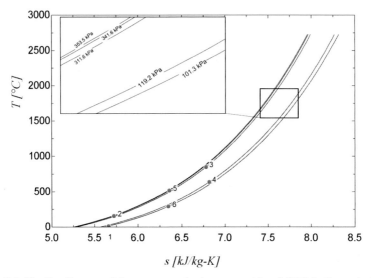

Figure 8.4 The T-s diagram of the recuperated micro gas turbine (MGT) in Example 8.2.

d.

$$\eta_{ad,tur} = \frac{\dot{W}_{a,tur}}{\dot{W}_{s,tur}} = \frac{h_3 - h_4}{h_3 - h_{4s}}$$

$$s_{4s} = s_3 = 6.778 \ \frac{kJ}{kg.K}, \ P_4 = 119.2 \ kPa \rightarrow h_{4s} = 912.4 \frac{kJ}{kg}$$

$$\eta_{ad,tur} = \frac{1188 - 946}{1188 - 912.4} = 87.81 \ \%$$

e.

The exergy destruction and exergy efficiency can be estimated as below:
To determine the exergy efficiency of the engine, the average heating temperature, T_{H-mean}, must be determined as below:

$$T_{H-mean} = \frac{h_3 - h_5}{s_3 - s_5} = \frac{1188 - 813.1}{6.778 - 6.357} = 890 \ K$$

And the sink temperature which is ambient temperature is:

$$T_L = T_1 = 15 \ ^\circ C = 288.15 \ K$$

Therefore, the Carnot cycle efficiency can be determined:

$$\eta_{carnot} = 1 - \frac{T_L}{T_{H-mean}} = 1 - \frac{288.15}{890} = 67.62\%$$

And the exergy efficiency or the second law efficiency is:

$$\eta_{II} = \frac{\eta_{th}}{\eta_{carnot}}$$

The thermal efficiency can be determined by dividing the electrical efficiency by the generator efficiency as below:

$$\eta_{th} = \frac{\eta_e}{\eta_{gen}} = \frac{0.2534}{0.98} = 25.86\%$$

Hence:

$$\eta_{II} = \frac{\eta_{th}}{\eta_{carnot}} = \frac{25.86}{67.62} = 38.24\%$$

Or by using the following equation:

$$\eta_{II} = \frac{exergy\ recovered}{exergy\ supplied} = \frac{\dot{W}_{net}}{\dot{Q}_{in}\left(1 - \frac{T_L}{T_H}\right)} = \frac{E_{net}}{\dot{Q}_{in}}\left(\frac{1}{\eta_{carnot}\eta_{gen}}\right) = \frac{\eta_e}{\eta_{carnot}\eta_{gen}}$$

$$= \frac{\eta_{th}}{\eta_{carnot}}$$

Which is exactly the same as the previous calculation.
The exergy destruction:

$$\underbrace{\left(\dot{X}_{in} - \dot{X}_{out}\right)}_{net\ rate\ of\ entropy\ transfer} - \dot{X}_{destroyed} = \frac{dX_{system}}{dt}$$

For steady-state operation $\frac{dX_{system}}{dt} = 0$, therefore:

$$\dot{X}_{destroyed} = \dot{X}_{in} - \dot{X}_{out} = \dot{Q}_{in}\left(1 - \frac{T_L}{T_{H-mean}}\right) - \dot{W}_{net}$$

$$= 116.2\left(1 - \frac{288.15}{890}\right) - 30.03 = 45.81\ kW$$

The exergy destruction can also be determined as below:

$$\dot{X}_{destroyed} = \dot{Q}_{in}(\eta_{carnot} - \eta_{th}) = 116.2(0.6762 - 0.2586) = 45.81\ kW$$

The code written for this example is given in the Code 8.2 in the following:

Code 8.2. the code developed for the MGT with recuperator, MGT model C30 in Example 8.2

```
"MGT C30"                                  h[3]=enthalpy(air, T=T[3])
E_net=29.433                               P[3]=P[2]-DELTAP_rec_cold-DELTAP_comb
eta_e=0.2534                               S[3]=entropy(air, t=t[3], p=p[3])
HV_fuel=47000                              Q_in=m_ex*(h[3]-h[5])
T[1]= 15                                   T[5]=temperature(air,H=H[5])
P[1]=101.325                               W_net=E_net/eta_gen
T[3]=850                                   W_net=m_ex*(h[3]-h[4])-m_air*(h[2]-h[1])
r_p=3.39                                   T[4]=temperature(air, H=H[4])
eta_ad_comp=0.8265                         P[4]=P[1]+DELTAP_rec_hot
m_air=0.30745                              S[4]=entropy(air, t=t[4], p=p[4])
DELTAP_rec_cold= 18.5*0.1                  eta_ad_tur=(h[3]-h[4])/(h[3]-h_4s)
DELTAP_rec_hot=178.6*0.1                   s_4s=s[3]
DELTAP_comb=0.3*100                        h_4s=enthalpy(air, P=P[4], S=s_4s)
eta_gen=0.98                               P[5]=P[2]-DELTAP_rec_cold
Q_in=E_net/eta_e                           s[5]=entropy(air, T=T[5], P=P[5])
Q_in=m_fuel*HV_fuel                        m_ex*(h[4]-h[6])=m_air*(h[5]-h[2])
AF=m_air/m_fuel                            T[6]=temperature(air,H=H[6])
m_ex=m_air+m_fuel                          P[6]=P[1]
s[1]=entropy(air, P=P[1], T=T[1])          s[6]=entropy(air, T=T[6], P=P[6])
h[1]=enthalpy(air, T=T[1])                 EPSILON=(h[5]-h[2])/(h[4]-h[2])
P[2]=r_p*P[1]                              T_H_mean=(h[3]-h[5])/(s[3]-s[5])
s_s2=s[1]                                  T_L=T[1]+273.15
h_2s=enthalpy(air, P=P[2], S=s_s2)         ETA_carnot=1-T_L/T_H_mean
eta_ad_comp=(h[1]-h_2s)/(h[1]-h[2])        ETA_th=eta_e/eta_gen
T[2]=temperature(air, h=h[2])              eta_II=ETA_th/ETA_carnot
W_comp=m_air*(h[2]-h[1])                   X_destroyed=Q_in*ETA_carnot-W_net
s[2]=entropy(air, t=t[2], p=p[2])          X_DESTROYED2=Q_in*(ETA_carnot-ETA_th)
```

Tables 8.1−8.3 present a complete data about the Capstone MGTs. These tables give the electrical, thermal, and geometrical specifications of the MGTs. All of the data are gathered from the published catalogs by the Capstone.

1.2   Thermodynamics of MST

The MST can be used in two ways to produce power and electricity.

1. It can be used as a pressure reducer in a steam utility unit. In this case, no specific pump, condenser, and steam generator are needed for the MST cycle. The MST itself installed as a pressure regulator. In this case, the MST operation depends on the steam production and steam pressure in the steam utility unit. Such steam turbine is also called backpressure turbine because the steam pressure at the turbine outlet is above the atmospheric pressure (Fig. 8.5).
2. It can be used as a complete Organic Rankine Cycle (ORC) that means having a steam generator, steam turbine, feed pump, and condenser. In such a case, the heating source can be provided from burning different fuels, using solar heat or other renewable energies, or waste heat from reciprocating engines, MGT, fuel cells, etc (Fig. 8.6).

In the case of using low-quality waste heat, water may not be superheated at operating high pressures, and using saturated steam in turbines can cause corrosion and erosion. To avoid this problem, other organic fluids with smaller critical pressure

Table 8.1 Electrical characteristics of different MGTs [4].

MGT model	$\dot{W}_{net}$ (kW) HP (LP)	$\eta_{e,LHV}$ HP (LP)	Voltage (VAC)	Frequency (Hz)		Max. Output current (A)	
				Grid connection	Stand alone	Grid connection	Stand alone
C15	15(−)	23(−)	400−480	50/60	10−60	23	54
C30	30(28)	26(25)	400−480	50/60	10−60	46	46
C65	65(−)	28(−)	400−480	50/60	10−60	100	100
C200	200(−)	33(−)	400−480	50/60	10−60	290A RMS@400V, 240A RMS @480V	310A RMS
C600	600(−)	33(−)	400−480	50/60	10−60	870A RMS@400V, 720A RMS @480V	930A RMS
C800	800(−)	33(−)	400−480	50/60	10−60	1160A RMS@400V, 960A RMS @480V	1240A RMS
C1000	1000(−)	33(−)	400−480	50/60	10−60	1450A RMS@400V, 1200A RMS @480V	1550A RMS

HP (LP) refers to the high pressure (low pressure) fuel system of the MGT.
Nominal full power performance is given at ISO conditions: 59F, 14.696 psia, 60% RH.
Some utilities may require additional equipment for grid interconnectivity.

Table 8.2 Fuel and exhaust characteristics of MGTs [4].

MGT model	Fuel/engine characteristics					Exhaust characteristics		
	Natural gas HHV (MJ/m³)	Inlet gage pressure, HP (LP) (kPa)	Fuel flow HHV, HP(LP) (MJ/hr)	Net heat rate LHV, HP(LP) (MJ/kWh)	NOₓ at 15% O₂ (mg/m³)	NOₓ (lb/MWhe)	Exhaust gas flow (kg/s)	Exhaust temperature (C)
C15	30.7–47.5	379–414 (–)	255 (–)	15.5 (–)	N/A	N/A	N/A	N/A
C30	30.7–47.5	379–414 (1.4–6.9)	457 (444)	13.8 (14.4)	18	0.64	0.31	275
C65*	30.7–47.5	517–552(–)	919(–)	12.9(–)	8	0.17	0.51	311
C200	30.7–47.5	517–552(–)	2400(–)	10.9(–)	18**	0.4***	1.3	280
C600	30.7–47.5	517–552(–)	7200(–)	10.9(–)	18**	0.4***	4	280
C800	30.7–47.5	517–552(–)	9600(–)	10.9(–)	18**	0.4***	5.3	280
C1000	30.7–47.5	517–552(–)	12,000(–)	10.9(–)	18**	0.4***	6.7	280

Exhaust emissions for standard Natural Gas at 39.4 MJ/Nm3 (1000 BTU/scf) (HHV).

The low pressure models can use natural gas as fuel while the high pressure models can use natural gas, landfill gas, digester gas, propane, diesel, and kerosene.

*This C65 has an integrated copper core heat recovery system that produces 120 kW hot water with total efficiency LHV of 80%. Heat recovery for water takes place at inlet temperature of 57°C (135°F) and flow rate of 2.5 L/s (40 GPM).

**Low NOx version produces 8 mg/m³, or 0.14 lb/MWhe.

Table 8.3 Dimensions, weight, installation clearances, and noise of MGTs [4].

| Model of MGT | Dimensions W×D×H (m)* | Weight (kg) | Min. Clearance requirements (m)** | | | | Noise in full load at 10 m (db) |
			Above	Left and right	Front	Rear	
C15	0.76 × 1.5 × 1.9	578	0.61	0.76	0.93	0.92	65
C30	0.76 × 1.5 × 1.8	578	0.61	0.76	0.93	0.92	65
C65	0.76 × 2.2 × 2.6	1450	0.61	0.76	1.7	0.76	65
C200	1.7 × 3.8 × 2.5	3413	0.6	1.1	1.1	1.8	65
C600	2.4 × 9.1 × 2.9	15,014	0.6	1.5 and 0	1.5	2	65
C800	2.4 × 9.1 × 2.9	15,558	0.6	1.5 and 0	1.5	2	65
C1000	2.4 × 9.1 × 2.9	20,956	0.6	1.5 and 0	1.5	2	65

*W×D×H stands for Width×Depth×Height.
**Height dimensions are to the roof line. Exhaust outlet extends at least 7 in above the roof line. Clearance requirements may increase due to local code considerations.

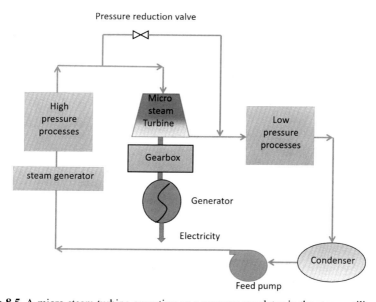

Figure 8.5 A micro steam turbine operating as a pressure regulator in the steam utility unit.

and temperature can be used. Such fluids must also be nonflammable, nontoxic, nonexplosive, and nonhazardous.

Fig. 8.7 shows a MST manufactured by the Energent corporation, capable of producing 275 *kW* of electricity.

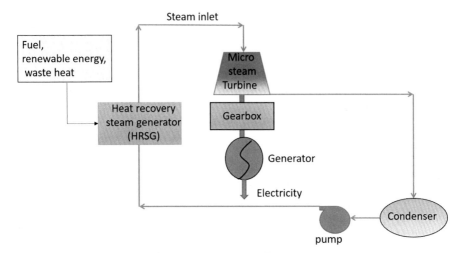

Figure 8.6 A micro steam turbine operating stand alone and fueled with different energy sources.

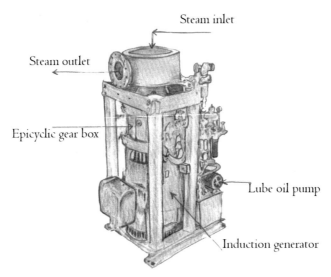

Figure 8.7 Micro steam turbine, regenerated figure, painting by Sholeh Ebrahimi [2]. Courtesy of Energent corporation.

The working fluid of the cycle presented in Fig. 8.5 is water, but the working fluid of the cycle in Fig. 8.6 can be other organic fluids depending on the available heat source quality. The working fluids can be *wet*, *dry*, or *isentropic*. For example, water and ammonia are wet fluids, R113 and cyclohexane are dry fluids, and R11 and R123 are isentropic working fluids. Fig. 8.8 shows the T-s diagram of these three types of fluids. As it can be seen, feeding a turbine with saturated steam of wet fluid results

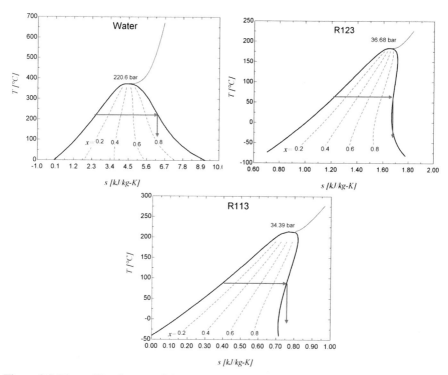

Figure 8.8 The mollier diagram of the wet, dry, and isentropic working fluids with the constant steam quality lines.

in two-phase flow through the turbine, corrosion, and erosion through the turbine. Hence, superheating is always required for the wet working fluids. For the isentropic fluids, saturated steam at the inlet of turbine remains saturated as it expands isentropically through the turbine. Since the real expansion process is irreversible and entropy increases the isentropic fluid becomes superheated; therefore, the danger of two-phase flow, erosion, and corrosion inside the turbine is very low. However, for dry working fluids, saturated steam becomes superheated as it expands through the turbine, and there is no danger of two-phase flow, corrosion, and erosion.

Example 8.3. An ORC is supposed to be designed based on the Energent MST. The cycle is to be designed for three working fluids of water, R123, and R113. The steam flow rate is $1.7 \frac{kg}{s}$ and pressure at the inlet and discharge of the MST are 1034 kPa and 206.8 kPa. The gearbox and the generator efficiencies are 97% and 96%. The inlet steam temperature to the MST is 250°C. Assume the cycle is internally reversible.

(a) Compare the net power and thermal efficiency of the cycle for different working fluids.
(b) Draw the T-s diagram of the ORC for the three working fluids.

Solution.

a.

By considering water as the working fluid, the state point properties for the cycle presented in Fig. 8.9 would be determined:

$$T_1 = 250\,^{\circ}C,\ P_1 = 1034\ kPa \rightarrow h_1 = 2941\frac{kJ}{kg},\ s_1 = 6.906\frac{kJ}{kg.K}$$

since the cycle is internally reversible:

$$s_2 = s_1 = 6.906\frac{kJ}{kg.K},\ P_2 = 206.8\ kPa \rightarrow h_2 = 2625\frac{kJ}{kg},\ x_2 = 96.23\%$$

Since the steam quality is more than 88%, it is acceptable.

$$P_3 = P_2 = 206.8\ kPa,\ x_3 = 0 \rightarrow h_3 = 509.2\frac{kJ}{kg},\ s_3 = 1.542\frac{kJ}{kg.K}$$

$$P_4 = P_1 = 1034\ kPa,\ s_4 = s_3 = 1.542\frac{kJ}{kg.K} \rightarrow h_4 = 510.1\frac{kJ}{kg}$$

And the net power and thermal efficiency would be:

$$\dot{W}_{net} = \dot{W}_{MST} - \dot{W}_{pump} = \dot{m}(h_1 - h_2) - \dot{m}(h_4 - h_3)$$
$$= 1.7(2941 - 2625) - 1.7(510.1 - 509.2) = 535.1kW$$

$$\dot{W}_{net} = 535.1kW$$

$$\dot{Q}_{in} = \dot{m}(h_1 - h_4) = 1.7(2941 - 510.1) = 4134\ kW$$

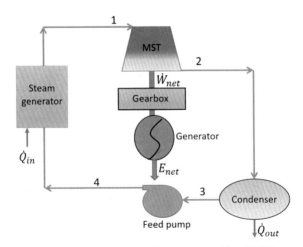

Figure 8.9 The organic rankine cycle for the micro steam turbine (MST).

$$\eta_{th} = \frac{\dot{W}_{net}}{\dot{Q}_{in}} = \frac{535.1}{4134} = 12.94\%$$

The electrical efficiency and net electricity generation by the ORC would be calculated by considering the gearbox and generator efficiencies.

$$\eta_e = \eta_{th} \times \eta_{gearbox} \times \eta_{gen} = 0.1297 \times 0.97 \times 0.96 = 12.05\%$$

$$E_{net} = \dot{W}_{net} \times \eta_{gearbox} \times \eta_{gen} = 535.1 \times 0.97 \times 0.96 = 498.3 \ kWe$$

For the R123 as the working fluid:

$$T_1 = 250\ ^\circ C, \ P_1 = 1034 \ kPa \rightarrow h_1 = 580.2 \frac{kJ}{kg}, \ s_1 = 1.986 \frac{kJ}{kg.K}$$

$$s_2 = s_1 = 1.986 \frac{kJ}{kg.K}, \ P_2 = 206.8 \ kPa \rightarrow h_2 = 537.5 \frac{kJ}{kg}$$

$$P_3 = P_2 = 206.8 \ kPa, \ x_3 = 0 \rightarrow h_3 = 251.6 \frac{kJ}{kg}, \ s_3 = 1.173 \frac{kJ}{kg.K}$$

$$P_4 = P_1 = 1034 \ kPa, \ s_4 = s_3 = 1.173 \frac{kJ}{kg.K} \rightarrow h_4 = 252.2 \frac{kJ}{kg}$$

And the net power and thermal efficiency would be:

$$\begin{aligned}
\dot{W}_{net} = \dot{W}_{MST} - \dot{W}_{pump} &= \dot{m}(h_1 - h_2) - \dot{m}(h_4 - h_3) \\
&= 1.7(580.2 - 537.5) - 1.7(252.2 - 251.6) \\
&= 71.66 kW
\end{aligned}$$

$$\dot{W}_{net} = 535.1 kW$$

$$\dot{Q}_{in} = \dot{m}(h_1 - h_4) = 1.7(580.2 - 252.2) = 558 \ kW$$

$$\eta_{th} = \frac{\dot{W}_{net}}{\dot{Q}_{in}} = \frac{71.66}{558} = 12.84\%$$

The electrical efficiency and net electricity generation by the ORC would be calculated by considering the gearbox and generator efficiencies.

$$\eta_e = \eta_{th} \times \eta_{gearbox} \times \eta_{gen} = 0.1284 \times 0.97 \times 0.96 = 11.96\%$$

$$E_{net} = \dot{W}_{net} \times \eta_{gearbox} \times \eta_{gen} = 71.66 \times 0.97 \times 0.96 = 66.73 \; kWe$$

For the R113 as the working fluid:

$$T_1 = 250 \,°C, \; P_1 = 1034 \; kPa \rightarrow h_1 = 372.9 \frac{kJ}{kg}, \; s_1 = 0.9964 \frac{kJ}{kg.K}$$

$$s_2 = s_1 = 0.9964 \frac{kJ}{kg.K}, \; P_2 = 206.8 \; kPa \rightarrow h_2 = 338.9 \frac{kJ}{kg}$$

$$P_3 = P_2 = 206.8 \; kPa, \; x_3 = 0 \rightarrow h_3 = 100.1 \frac{kJ}{kg}, \; s_3 = 0.3488 \frac{kJ}{kg.K}$$

$$P_4 = P_1 = 1034 \; kPa, \; s_4 = s_3 = 0.3488 \frac{kJ}{kg.K} \rightarrow h_4 = 100.8 \frac{kJ}{kg}$$

And the net power and thermal efficiency would be:

$$\begin{aligned}
\dot{W}_{net} = \dot{W}_{MST} - \dot{W}_{pump} &= \dot{m}(h_1 - h_2) - \dot{m}(h_4 - h_3) \\
&= 1.7(372.9 - 338.9) - 1.7(100.8 - 100.1) \\
&= 56.56 kW
\end{aligned}$$

$$\dot{W}_{net} = 56.56 kW$$

$$\dot{Q}_{in} = \dot{m}(h_1 - h_4) = 1.7(372.9 - 100.8) = 462.8 \; kW$$

$$\eta_{th} = \frac{\dot{W}_{net}}{\dot{Q}_{in}} = \frac{56.56}{462.8} = 12.22\%$$

The electrical efficiency and net electricity generation by the ORC would be calculated by considering the gearbox and generator efficiencies.

$$\eta_e = \eta_{th} \times \eta_{gearbox} \times \eta_{gen} = 0.1222 \times 0.97 \times 0.96 = 11.38\%$$

$$E_{net} = \dot{W}_{net} \times \eta_{gearbox} \times \eta_{gen} = 56.56 \times 0.97 \times 0.96 = 52.67 \; kWe$$

b.

The T-s diagram of the ORC for three different working fluids, with the same inlet temperature and pressure to the MST and the same outlet pressure from the MST are shown in Fig. 8.10. The state points of 3 and 4 are very close to each other because the temperature elevation due to pressurizing an incompressible fluid is very small.

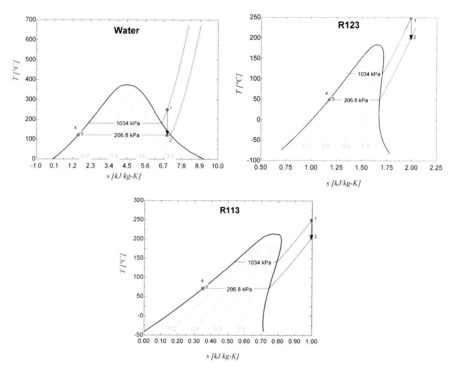

Figure 8.10 The T-s diagram of the ORC for three different working fluids, with the same steam flow rate, inlet temperature, and pressure to the micro steam turbine (MST) and the same outlet pressure from the MST.

Conclusions. According to Fig. 8.8, the critical temperature and pressure of the dry and isentropic organic fluids are smaller than the wet fluids. In addition, while the wet fluids have a bell-type Mollier diagram shape, the dry and isentropic Mollier diagrams are half a bell or even less than half a bell shape. These have caused the smaller area (net power) under the Mollier diagrams of the dry and isentropic fluids. In the present example, the steam mass flow rate, the inlet and outlet pressure of the MST, and the maximum temperature of the cycles are the same, but the net power of wet fluids according to the discussions is much bigger. The efficiency of the cycle with three working fluids are close, but as it can be seen the outlet steam from the MST for the isentropic and dry fluids is still highly superheated (SH). This means that there is a higher potential for power extraction if the exit pressure could be lower. In the catalog of the Energent company, the minimum allowable pressure is reported to be 2 psig (13.79 kPa). By changing the exit pressure from 13.79 kPa to 206.8 kPa, the changes of the steam quality at the turbine exit, the saturation temperature in the condenser (T_3), the thermal efficiency, and the net power of the cycle are presented for the three working fluids in Table 8.4. The calculations are carried out by using the Code 8.3 of the present example and by using the parametric table in the EES software.

Table 8.4 The impact of the minimum pressure on the ORC cycle characteristics.

Water					R123				
P_2 (kPa)	T_3(°C)	x_2	η_{th}	$\dot{W}_{net}$	P_2 (kPa)	T_3(°C)	x_2	η_{th}	$\dot{W}_{net}$
13.79	52.22	0.8453	0.2617	1211	13.79	-17.48	SH	0.2682	180.7
35.23	72.83	0.8801	0.223	999.5	35.23	1.641	SH	0.2259	145
56.67	84.47	0.8996	0.2009	883.8	56.67	12.71	SH	0.2024	126
78.12	92.84	0.9138	0.1849	802.1	78.12	20.81	SH	0.1855	112.9
99.56	99.48	0.925	0.1721	738.4	99.56	27.31	SH	0.1722	102.8
121	105	0.9344	0.1613	685.8	121	32.78	SH	0.1611	94.62
142.5	109.8	0.9426	0.152	640.9	142.5	37.54	SH	0.1515	87.7
163.9	114	0.9498	0.1437	601.6	163.9	41.77	SH	0.143	81.7
185.3	117.8	0.9563	0.1362	566.7	185.3	45.59	SH	0.1354	76.4
206.8	121.3	0.9623	0.1294	535.1	206.8	49.08	SH	0.1284	71.66

R113				
P_2 (kPa)	T_3(°C)	x_2	η_{th}	$\dot{W}_{net}$
13.79	-1.714	SH	0.2523	145.7
35.23	19.14	SH	0.2128	116.1
56.67	31.2	SH	0.1909	100.6
78.12	40.01	SH	0.1752	89.92
99.56	47.08	SH	0.1629	81.73
121	53.05	SH	0.1526	75.08
142.5	58.24	SH	0.1437	69.48
163.9	62.85	SH	0.1358	64.64
185.3	67.03	SH	0.1287	60.38

The saturation temperature in the condenser must be higher than the cooling water temperature for efficient heat transfer and condensation. As Table 8.4 shows for water the saturation temperature at the minimum pressure of 13.79 kPa, is 52.22 °C that is higher than the cooling water temperature, which is usually in the range of 15−30 °C. If the cooling water temperature could be 30 °C, the saturation temperature should be at least 10 °C higher for proper heat transfer and condensation inside the condenser. This means the saturation temperature of higher than 40 °C is acceptable (the acceptable range from this point of view is highlighted with green color). Another constraint is the minimum steam quality at the MST exit that must be higher than 88% to avoid corrosion and erosion in the turbine. For R113 and R123, the steam quality at the MST exit remains superheated (SH) in the considered pressure range, while for water, it falls below 0.88 at the pressure of 13.79 kPa. The allowable range of steam quality is highlighted with yellow color. The common range between these two constraints gives us the minimum allowable pressure for the MST exit. The allowable range is highlighted with blue color.

The code written for this example is presented in the Code 8.3.

Code 8.3. The code written for the ORC with three working fluids in Example 8.3

```
R$='water'
m_steam=1.7   "kg/s"
P[1]=1034
T[1]=250
s[1]=entropy(R$,P= P[1], T=T[1])
S[2]=S[1]
P[2]=206.8
H[2]=ENTHALPY(R$, S=S[2], P=P[2])
T[2]=Temperature(R$, s=s[2], P=P[2])
X[2]=QUALITY(R$, P=P[2], s=s[2])
P[3]=P[2]
eta_gen=0.96
eta_gerabox=0.97
h[1]=ENTHALPY(R$, P=P[1], T=T[1])
h[3]=enthalpy(R$,P= P[2], X=0)
s[3]=entropy(R$,P= P[2], X=0)
T[3]=Temperature(R$, s=s[3], p=p[3])
s[4]=s[3]
p[4]=p[1]
h[4]=enthalpy(R$, P=P[4], S=S[4])
T[4]=TEMPERATURE(R$, P=P[4], S=S[4])
W_pump=m_steam*(h[4]-H[3])
W_turb=m_steam*(h[1]-h[2])
W_net=W_turb-W_pump
Q_in=m_steam*(h[1]-h[4])
eta_th=W_net/Q_in
eta_e=eta_th*eta_gerabox*eta_gen
E_net=W_net*eta_gerabox*eta_gen
```

Example 8.4. The ORC described in Example 8.3 is supposed to be coupled with the MGT model C200S, which presented in Example 8.1 to build a micro scale combined cycle. The working fluid of the MST is supposed to be R113, the inlet/outlet pressure of the MST are still the same as the Example 8.3, but the mass flow rate is reduced to 1.5 kg/s, and the inlet temperature of the turbine depends on the amount and quality of the available waste heat. The exhaust gas of MGT enters the steam generator as the only heat source, and it is cooled to 105 °C at point 11 in Fig. 8.11. No supplementary fuel is used in the steam generator. Determine:

(a) The inlet temperature of the steam turbine.
(b) The total net power generated by the combined cycle.
(c) The overall efficiency of the combined cycle.

Solution. Since the MGT described in Example 8.1 is used in the combined cycle without any change, all of the results for the MGT remain the same (Table 8.5). Hence

$$\eta_{th,MGT} = 33.67\%, \; \dot{W}_{net,MGT} = 204.1 \; kW, \; E_{net,MGT} = 200 \; kWe, \; \dot{Q}_{in,MGT}$$
$$= 116.2 \; kW$$

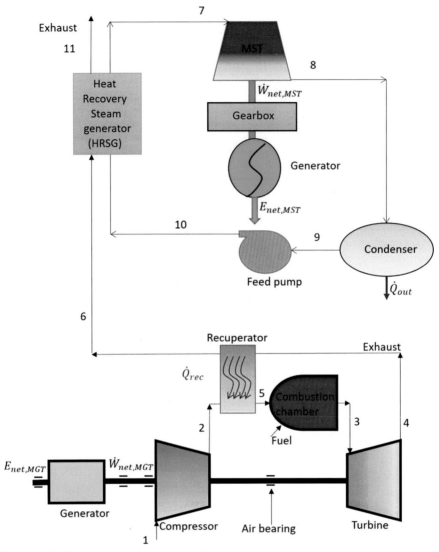

Figure 8.11 The micro combined cycle in Example 8.4.

Table 8.5 The state point properties of the MGT in the Example 8.1

Point	$h\left(\frac{kJ}{kg}\right)$	$P(kPa)$	$s\left(\frac{kJ}{kgK}\right)$	$T(°C)$
1	288.5	101.3	5.661	15
2	413	354.6	5.661	138.5
3	1217	354.6	6.767	875
4	936.8	101.3	6.852	630.2
5	750.7	354.6	6.265	461.5
6	585.3	101.3	6.371	306.1

However, for the ORC cycle, since the input temperature of the MST is designed according to the available waste heat, the results of the ORC need to be updated.

a.

Starting from point 11 at the exhaust of the HRSG unit:

$$T_{11} = 105\ ^\circ C,\ P_{11} = P_6 = 101.3 \rightarrow h_{11} = 379.2\frac{kJ}{kg},\ s_{11} = 5.935\frac{kJ}{kgK}$$

The energy balance in the HRSG is written to determine the inlet enthalpy to the MST. The HRSG is assumed well insulated and no heat loss occurs from its body.

$$\dot{m}_{ex}(h_6 - h_{11}) = \dot{m}_{steam}(h_7 - h_{10})$$

The properties at points 9 and 10:

$$P_9 = P_8 = 206.8\ kPa,\ x_3 = 0 \rightarrow h_9 = 100.1\frac{kJ}{kg},\ s_9 = 0.3488\frac{kJ}{kg.K}$$

$$P_{10} = P_7 = 1034\ kPa,\ s_{10} = s_9 = 0.3488\frac{kJ}{kg.K} \rightarrow h_{10} = 100.8\frac{kJ}{kg}$$

Hence:

$$1.3(585.3 - 379.2) = 1.5(h_7 - 100.8)$$

$$\rightarrow h_7 = 279.4\frac{kJ}{kg} \rightarrow T_7 = 142.9\ ^\circ C$$

In addition:

$$x_7 = superheated,\ s_7 = 0.7966\frac{kJ}{kgK}$$

Point 8:

$$s_8 = s_7 = 0.7966\frac{kJ}{kgK},\ P_8 = 206.8\ kPa \rightarrow h_8 = 254.8\frac{kJ}{kg},\ T_8 = 95.51\ ^\circ C$$

b.

The net power of the combined cycle is the summation of the net power produced by the MGT and MST:

$$\dot{W}_{net} = \dot{W}_{net,MGT} + \dot{W}_{net,MGT} = 204.1 + \dot{W}_{tur,MST} - \dot{W}_{pump,MST}$$
$$= 204.1 + \dot{m}_{steam}(h_7 - h_8) - \dot{m}_{steam}(h_{10} - h_9)$$
$$= 204.1 + 1.5(279.4 - 254.8)$$
$$- 1.5(100.8 - 100.1)$$
$$= 204.1 - 36.96 - 1.039 = 240 \; kW$$

$$\dot{W}_{net} = 240 \; kW$$

That shows the cycle is able to produce 35.92 kW of power just by using the heat recovery form the exhaust.

c.

The overall thermal efficiency of the cycle is determined is determined as below:

$$\eta_{th} = \frac{\left(\dot{W}_{net,MGT} + \dot{W}_{net,MGT}\right)}{\dot{Q}_{in,MGT}} = \frac{\dot{W}_{net}}{\dot{Q}_{in,MGT}} = \frac{240}{606.1} = 39.6\%$$

Which means about 6% of efficiency improvement without burring extra fuel. The ORC efficiency can be further improved by using the efficiency improvement methods, such as increasing the operating pressure, decreasing the condenser pressure, more superheating by burning extra fuel in the HRSG.

The T-s diagram of the ORC is depicted in Fig. 8.12:

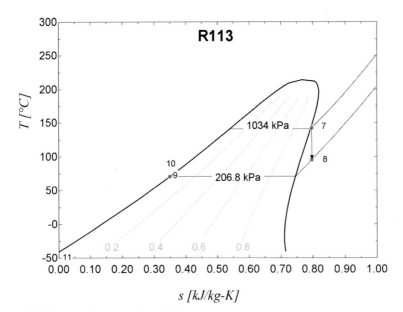

Figure 8.12 The T-s diagram of the ORC as the bottoming cycle.

The code developed for the Example 8.4 is presented in the Code 8.4.

Code 8.4. The code developed for the micro combined cycle in the Example 8.4

```
"MGT C200S"
E_net_MGT=200
eta_e_MGT=0.33
HV_fuel=42000
T[1]= 15
P[1]=101.325
T[3]=875
r_p=3.5
eta_recuperator=0.95
eta_gen_MGT=0.98
Q_in_MGT=E_net_MGT/eta_e_MGT
m_ex=1.3
Q_in_MGT=m_fuel*HV_fuel
m_air=m_ex-m_fuel
AF=m_air/m_fuel
s[1]=entropy(air, P=P[1], T=T[1])
h[1]=enthalpy(air, T=T[1])
P[2]=r_p*P[1]
s[2]=s[1]
T[2]=temperature(air, s=s[2], p=p[2])
h[2]=enthalpy(air, t=t[2])
W_comp=m_air*(h[2]-h[1])
W_net_MGT=E_net_MGT/eta_gen_MGT
W_net_MGT=m_ex*(h[3]-h[4])-m_air*(h[2]-
h[1])
h[3]=enthalpy(air, T=T[3])
Q_in_MGT=m_ex*(h[3]-h[5])
T[5]=temperature(air,H=H[5])
P[5]=P[2]
s[5]=entropy(air, T=T[5], P=P[5])
Q_available=m_ex*(h[4]-h[6])
Q_absorbed=m_air*(h[5]-h[2])
eta_recuperator=Q_absorbed/Q_available
T[6]=temperature(air, H=H[6])
P[6]=P[1]
s[6]=entropy(air, T=T[6], P=P[6])
T[4]=temperature(air, H=H[4])
P[3]=P[2]
s[3]=entropy(air, P=P[3], T=T[3])
S[4]=entropy(air, P=P[4], T=T[4])
P[4]=P[1]

T[11]=105
P[11]=P[6]
H[11]=enthalpy(air, T=T[11])
s[11]=entropy(air, t=t[11], p=p[11])

R$='R113'
m_steam=1.5   "kg/s"
P[7]=1034
m_steam*(h[7]-h[10])=m_ex*(h[6]-h[11])
T[7]=TEMPERATURE(R$, P=P[7], h=h[7])
s[7]=entropy(R$,P= P[7], H=H[7])
X[7]=QUALITY(R$, P= P[7], H=H[7])
S[8]=S[7]
P[8]=206.8
H[8]=ENTHALPY(R$, S=S[8], P=P[8])
T[8]=Temperature(R$, s=s[8], P=P[8])
X[8]=QUALITY(R$, P=P[8], s=s[8])
P[9]=P[8]
eta_gen_MST=0.96
eta_gerabox=0.97
"h[7]=ENTHALPY(R$, P=P[7], T=T[7])"
h[9]=enthalpy(R$,P= P[8], X=0)
s[9]=entropy(R$,P= P[8], X=0)
T[9]=Temperature(R$, s=s[9], p=p[9])
s[10]=s[9]
p[10]=p[7]
h[10]=enthalpy(R$, P=P[10], S=S[10])
T[10]=TEMPERATURE(R$, P=P[10], S=S[10])
W_pump_MST=m_steam*(h[10]-H[9])
W_turb_MST=m_steam*(h[7]-h[8])
W_net_MST=W_turb_MST-W_pump_MST
Q_HRSG=m_steam*(h[7]-h[10])
eta_th_MST=W_net_MST/Q_HRSG
eta_e_MST=eta_th_MST*eta_gerabox*eta_gen_MST
E_net_MST=W_net_MST*eta_gerabox*eta_gen_MST

W_net=W_net_MST+W_net_MGT
eta_th=W_net/Q_in_MGT
```

2. Components of micro gas/steam turbines

Unlike the large scale gas turbines and steam turbines, micro gas/steam turbines have some limited components. An MGT for example has no blade cooling, no air intake system, no VIGV, no bleed valves, no lube oil system, etc.

The main components of an MGT include the rotor, combustion chamber, generator, air bearings, recuperator, generator cooling fins, and control system (Fig. 8.13).

The compressor is a radial flow impeller, flow enters from the suction eye at the center of impeller, then distributes between the diffuser type channels created by the

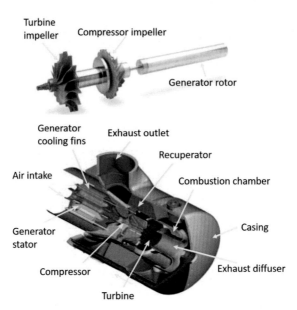

Figure 8.13 The MGT and its components (courtesy of Capstone) [1].

impeller vanes, and its shroud. Due to the shaft rotation, velocity and pressure of air increase. The air velocity then is converted to pressure as it passes through diffuser shape between the impeller vanes and compressor volute.

After the compressor, the recuperator, which is a counter flow compact heat exchanger, preheats the compressed air by the exhaust gas.

The preheated air goes to the combustor. Since the flame temperature is beyond the tolerable temperature of the combustor and turbine components, the air to fuel mass flow ratio is designed to be very high (usually bigger than 100) to avoid overheating, burning, and melting the internal parts. The combustion air burns with the fuel, and the extra air is sequentially added to the combustion gases to decrease the temperature and to cool the internal components of the combustor.

The compressed hot gasses enter the turbine peripherally and flows between the nozzle shape channels created by the impeller vanes and the shroud. As it passes through the nozzle shape channels the pressure converts to velocity and its momentum turns the turbine impeller. The expanded gas leaves the turbine impeller from its center of rotation and goes to the exit diffuser where its pressure is partially elevated to pass through the recuperator.

The high speed MGT rotor's motions must be controlled by radial and thrust air bearings. In the air bearings, instead of oil, compressed clean air is used to keep the clearance between the bearing surface and shaft surface in both radial and thrust bearings. There are two main types of air bearing technologies, the aerostatic air bearing and aerodynamic air bearing.

In the aerostatic air bearings, pressurized air from an external source is utilized to fill the gap between the two surfaces of the shaft and bearing surface. Pressurized air is continuously flowing through the restrictors into the bearing clearance and then escapes to the atmosphere. There are several types of restrictors such as simple orifice fed, pocketed orifice, slot fed, and porous surfaces. Fig. 8.14 shows an example of aerostatic air typical orifice bearing.

In the aerodynamic air bearings, the air film between the bearing and the shaft is generated due to the relative motion of the surfaces. They exist in several types including simple cylinders, tri-lobe, spirally grooved (Fig. 8.15), axially grooved, herringbone grooved, and stepped [5].

Micro steam turbines that are mostly designed for utilizing in the ORCs may have radial (centrifugal) flow turbines, Euler turbine, single or double wheel Curtis turbines. In addition to the ORC applications, small-scale turbines in industries are used as the driver of pumps and fans. Fig. 8.16 shows and Euler turbine designed for accepting saturated steam at the inlet and moisture content of 10% at the turbine exit.

An important advantage of Euler turbine is that unlike the radial turbine, it pulls away moisture content and contaminants from the rotor and protects the turbine from erosion. Due to the radial outflow of the Euler turbine, the operating speed is reduced to approximately half of the radial inflow centrifugal turbines. The smaller speed results in smaller gearbox.

The pressure ratio of higher than 4:1 in the centrifugal micro turbines is possible when multistage expander is used that result in more complexity and costs. In the Euler turbines, two rows of blades can be machined with a pressure ratio of 10:1.

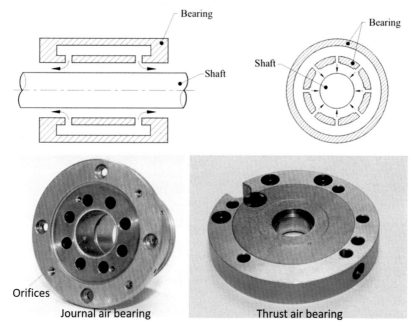

Figure 8.14 Aerostatic air bearing (courtesy of celera motion, a novanta company) [5].

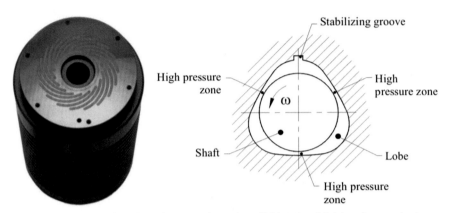

Figure 8.15 Cross section through an aerodynamic radial bearing (*right*), celera motion's westwind aerodynamic thrust bearing showing spiral grooves (*left*) (courtesy of celera motion, a novanta company) [5].

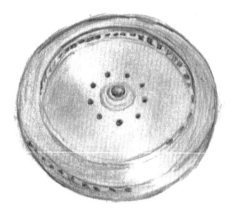

Figure 8.16 Turbine wheel of the Euler turbine, regenerated figure, painting by Sholeh Ebrahimi [2,6]. Courtesy of Energent corporation.

3. Problems

1. How do you evaluate the impact of increasing ambient temperature on the net power and thermal efficiency of an MGT?
2. How do you evaluate the impact of ambient pressure on the net power and thermal efficiency of an MGT?
3. The electricity demand of a consumer varies from 50 *kW* to 200 *kW* during 24 h. From the energy point of view, what do you recommend? Installing four 50 *kWe* units or a 200 *kWe* unit? Why?
4. What are the main reasons for low thermal efficiency of MGT in comparison with the large scale gas turbines?

Table 8.6 The specifications of some organic working fluids.

$P_{min,ORC}$	Type	P_{cr} (barg)	T_{cr}(K)	Working fluid
	Wet	220.64	674.15	Water
	Wet	40.59	374.2	R134a
	Wet	113.3	405.4	Ammonia
	Wet	42.47	369.7	Propane
	Dry	40.75	535.5	Cyclohexane
	Dry	22.29	468.5	HFE7100
	Dry	34.39	487.3	R113
	Isentropic	36.68	456.7	R123
	Isentropic	37.96	425.1	n-Butane
	Isentropic	36.40	497.7	Isobutene
	Isentropic	44.08	471.2	R11

5. How does the thrust air bearing presented in Fig. 8.15 work?
6. Determine the minimum allowed condenser pressure for a micro ORC with different working fluids listed in Table 8.6. A permanent cooling water source with temperature of 20 °C is available. Assume the condenser operating temperature is at least 10 °C higher than the cooling water temperature.
7. MGT model of C65 (Fig. 8.17) manufactured by the Capstone company produces 65 kW of electricity with an electrical efficiency of 28%. The pressure ratio of the compressor is 3.5 and the turbine inlet temperature is 875 °C.The exhaust mass flow is 0.49 kg/ s. The natural gas HHV used in the MGT is 42,000 kJ/kg. Nominal full power performance is measured at ISO conditions of 15 °C, 101.325 kPa, 60% RH. The electric generator efficiency is 0.98, also only 0.97 of the total waste heat of the exhaust in the recuperator is absorbed by the compressor exit flow and the rest is dissipated to the environment through the MGT casing ($\eta_{recuperator} = 0.97$). Assume the process to be reversible in the compressor, and the pressure

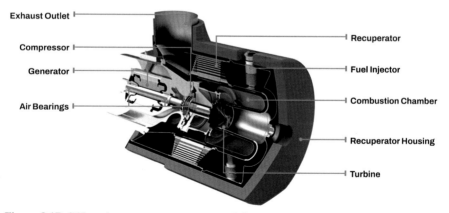

Figure 8.17 C65 engine components, courtesy of Capstone [1].

losses in the recuperator and the combustion chamber are neglected. The process in the turbine is real, and it has irreversibilities. Use the standard air assumptions wherever needed and determine:

(a) The mass flow of fuel consumption.

(b) The air to fuel mass flow ratio.

(c) The state point properties and the T-s of the MGT.

(d) The recuperator effectiveness.

(e) The adiabatic efficiency of the turbine section.

(f) The exergy destruction of the cycle and the exergy efficiency of the cycle.

8. In C65 MGT manufactured by Capstone, it produces net 65 kW of electricity, the turbine inlet temperature is 875 $°C$. The electrical efficiency is 28% and the heating value of natural gas as fuel is 42,000 $\frac{kJ}{kg}$. The pressure drop of the cold and hot streams in recuperator are 37 and 300 mbar. The pressure drop in the combustor is also 0.4 bar. The compressor pressure ratio is 3.5. The air flow through the compressor is 0.49 $\frac{kg}{s}$. The adiabatic efficiency of the turbine and compressor is 0.85 and 0.8265, respectively. It is assumed that the MGT is well insulated and no heat loss occur from the its body. The generator efficiency is also 0.98. Nominal full power performance is measured at ISO conditions of 15 $°C$, 101.325 kPa, and 60% RH. Determine:

(a) The fuel consumption mass flow rate.

(b) The air to fuel mass flow ratio.

(c) The state point properties and the T-s diagram.

(d) The turbine adiabatic efficiency.

(e) The exergy destruction and exergy efficiency.

(f) The recuperator effectiveness.

9. An ORC is supposed to be designed based on the Energent MST. The cycle is to be designed for three working fluids of ammonia, Cyclohexane and R11. The steam flow rate is 1.7 $\frac{kg}{s}$ and pressure at the inlet of the MST is 1034 kPa. The outlet pressure from the MST is supposed to be the minimum allowable pressure for cooling water source with temperature of 20 $°C$. The gearbox and the generator efficiencies are 97% and 96%. The cycle maximum temperature at the MST inlet is 250 $°C$. Assume the cycle is internally reversible.

(a) Compare the net power and thermal efficiency of the cycle for different working fluids.

(b) Draw the T-s diagram of the ORC for the three working fluids.

10. The ORC described in Example 8.3 is supposed to be coupled with the MGT model C200S which presented in Example 8.1 to build a micro scale combined cycle. The working fluid of the MST is supposed to be cyclohexane; the inlet/outlet pressure of the MST are still the same as the Example 8.3, but the mass flow rate is reduced to 0.6 kg/s, and the inlet temperature of the turbine depends on the amount and quality of the available waste heat. The exhaust gas of MGT enters the heat recovery steam generator (HRSG) as the only heat source, and it is cooled to 80 $°C$ at point 11 in Fig. 8.11. No supplementary fuel is used in the steam generator. The HRSG in this cycle is a kind of condensing steam generator that is protected against corrosion due to the condensation of water vapor in the exhaust gases. Determine:

(a) The inlet temperature of the turbine.

(b) The total net power generated by the combined cycle.

(c) The overall efficiency of the combined cycle.

References

[1] https://www.capstonegreenenergy.com/.
[2] http://www.energent.net/Technology/Microsteam-Turbine.html.
[3] Engineering ToolBox, Fuel gases—heating values [online] Available at: https://www. engineeringtoolbox.com/heating-values-fuel-gases-d_823.html, 2005 ([Accessed Day Month Year]).
[4] M. Ebrahimi, A. Keshavarz, Combined Cooling, Heating and Power, Decision-Making, Design and Optimization, Elsevier, 2014, ISBN 978-0-08-099985-2.
[5] https://www.celeramotion.com/westwind/support/technical-papers/what-is-an-air-bearing/.
[6] P. Welch, P. Boyle, New turbines to enable efficient geothermal power plants, GRC Transactions 33 (2009).

Reciprocating power generator engines

Chapter outline

Reciprocating engines (REs) are a well-matured technology and are used for different purposes including stationary power generation. Several REs can be paralleled for more power generation. They have a very quick startup time; therefore, they are perfect for peak shaving during peak hours. REs are available in different scales from about $30\,kW$ to about $20\,MW$ for power generation. They can use different fuel sources such as light and heavy liquid and gaseous fossil fuels, landfill, biogas, etc. The thermal efficiency of REs is usually higher than gas turbines in the same capacities and power output. They are more flexible than gas turbines for part load operation. In addition, the capital cost per kW of REs is less than any other heat engine including gas turbines, steam turbines, fuel cells, micro turbines, etc. The maintenance costs of REs are usually higher due to having more moving parts, specially due to the intake and exhaust valves. Their availability is higher than 95% for stationary power generation. They need very small auxiliary systems specially for startup (usually batteries or compressed air is enough); therefore, they are very suitable for emergency usages during the electricity outages.

Reciprocating engines basically are classified as spark ignition (SI) (Otto cycle for example) and compression ignition (CI) (Diesel-cycle for example). The main difference between SI and CI engines is their ignition method. The SI engines use light fuels such as gasoline or natural gas and combustion in each cycle starts by using an igniter. In these engines, air and fuel mixture enters the cylinder at the intake stroke.

Power Generation Technologies. https://doi.org/10.1016/B978-0-323-95370-2.00001-6
Copyright © 2023 Elsevier Inc. All rights reserved.

Compression ignition engines use heavy fuels such as diesel. In these engines, only air enters the cylinder in the intake stroke, and when the air temperature reaches near or above the self-ignition temperature of the fuel (due to compression), fuel is injected into the hot compressed air and combustion automatically starts.

The compression ratio of the SI engines is in the range of 9:1–12:1, while the diesel engines require compression ratio of 12:1–17:1 to increase the air temperature to the compression ignition condition.

REs are rated by different manufacturers according to how they are used throughout the year. Most manufacturers categorize them into *standby*, *prime*, and *baseload* (continuous) ratings.

Standby rated engines are emergency units, installed in places where the consumer is connected to the public grid, and the engines are switched on only for a short time during a power outage. Average power output is 70% of the standby rated *ekW* and typically operate about 200 h per year and with maximum usage of 500 h per year. Although output is available with varying load, but these engines can run at full load because they operate for short periods of time.

Prime-rated engines are generally the main generators of electricity and are installed in places where the consumer does not purchase electricity from the grid. These engines generate electricity continuously for an unlimited time with variable load, except during normal maintenance times and emergency shutdown. Average power output is 70% of the prime rated *ekW*. Typical peak demand is 100% of prime rated *ekW* with 10% overload for a maximum of 1 h in 12 h. Overload operation must stay under 25 h per year.

Baseload rated engines, also called continuous power engines, are used for non-varying load with unlimited operating hours except during normal maintenance times and emergency shutdowns. Average power output is 70%–100% of the continuous rated *ekW*. The peak demand is 100% of continuous rated *ekW* for all of the operating hours.

The main components of an RE include the cylinder, piston, connecting rod, crankshaft, intake valve, and exhaust valve (Fig. 9.1) [1]. Fuel burns with oxygen inside the cylinder and expands. This gas expansion pushes the piston and connecting rod to create the biggest possible volume between the inside walls of the cylinder and piston head (converting the chemical energy of fuel to a reciprocating motion). This causes the crankshaft to rotate and convert the reciprocating motion to rotational motion. Finally, the rotational motion of the crankshaft is converted to electricity using a generator.

Thermodynamic evaluation of power plants is essential for feasibility studies including technical, economic, and environmental evaluations. For this purpose, in the following, the thermodynamics of the RE is discussed from the first and second thermodynamics law. In addition, the components, auxiliary systems, etc. will be discussed.

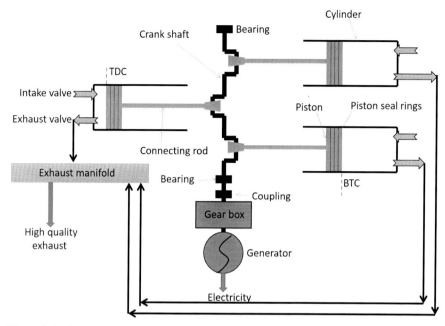

Figure 9.1 The common components of reciprocating engine power generators [1].

1. Thermodynamic analyses

1.1 The Otto cycle

Otto cycle is the internally reversible cycle of the spark ignition engines. Fig. 9.2 shows the $P-v$ and $T-s$ process diagrams of the Otto cycle [2].

For simplicity, the thermodynamic governing equations of the Otto cycle are derived using the **standard air** assumptions. These assumptions are as below:

1. The working fluid in the total process is pure air, which circulates in a close cycle.
2. Air behaves like an ideal gas.
3. Combustion is replaced by heat addition to air using an external heat source.
4. Exhaust processes is replaced by a heat rejection processes using an external heat sink.

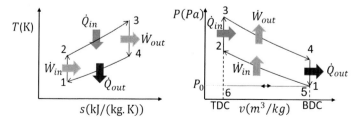

Figure 9.2 The internally reversible Otto cycle for SI reciprocating engines.

5. All of the processes are internally reversible.

In addition to the above five assumptions, if the specific heat is assumed constant at room temperature during the processes, these six assumptions are call called **cold-air-standard (CAS)** assumptions.

The description and governing equations of the processes under the cold-air-standard assumptions are presented below:

Process 6−1: constant pressure intake of air, which sucks in the air and fuel mixture. Intake valve is open, but exhaust valve is closed. This process consumes intake power as below:

$$P_1 = P_6 = P_0$$
$$\dot{W}_{intake} = \dot{m}_{intake} P_0 (v_1 - v_6)$$

(9.1)

in which v is the specific volume of air.

Process 1−2: isentropic compression ($s_1 = s_2$), in which, the piston moves from the bottom dead center (BDC) with the maximum volume in the cylinder to the top dead center (TDC) with the minimum volume in the cylinder. All the valves are closed. In this process, the fuel and air mixture are compressed to the pressure of P_2 (Stroke 2: compression). The required work for this compression is provided by other cylinders on the opposite side of the crank shaft and also the power stroke. The compression power or the inlet power can be determined as below:

$$\dot{W}_{compression} = \dot{W}_{in} = \dot{m}_{intake}(u_2 - u_1) = \dot{m}_{intake}\frac{P_1 v_1 - P_2 v_2}{1 - k} = \dot{m}_{intake} c_v (T_2 - T_1)$$

$$\frac{P_2}{P_1} = \left(\frac{v_1}{v_2}\right)^k = \left(\frac{V_1}{V_2}\right)^k = (r_c)^k = \left(\frac{T_2}{T_1}\right)^{\frac{k}{k-1}}$$

$$\frac{V_1}{V_2} = r_c = compression\ ratio$$

(9.2)

in which V is air volume, and k is the specific heat rartio of air.

Process 2−3: constant volume heat addition ($v_2 = v_3$), while piston is at the TDC, and all valves are closed, the spark plug ignites and a rapid combustion occurs, due to heat release, the gas pressure inside the cylinder increases at constant volume until it reaches P_3. The input heat rate due to the combustion process can be determined as below:

$$\dot{Q}_{2-3} = \dot{Q}_{in} = \dot{m}_{intake}(u_3 - u_2) = \dot{m}_{fuel} HV \eta_c = \dot{m}_{intake} c_v (T_3 - T_2)$$

$$\dot{m}_{intake} = \dot{m}_{air} + \dot{m}_{fuel} = (AF + 1)\dot{m}_{fuel}$$

$$T_3 = T_{max}\ \&\ P_3 = P_{max}$$

$$\frac{P_2 v_2}{T_2} = \frac{P_3 v_3}{T_3}, \quad v_2 = v_3 \rightarrow \frac{P_2}{T_2} = \frac{P_3}{T_3}$$

(9.3)

In which, AF is the air to fuel mass flow rate ratio. For SI engines, $12 : 1$ *(rich)* $\leq$ $AF \leq 20 : 1$ *(lean)*. The AF^1 value depends on operating parameters such as speed, pressure, allowable temperature, load, emission control systems, etc. In addition, η_c is the combustion efficiency.

Process 3—4: isentropic expansion ($s_3 = s_4$), the hot and high pressure gas trapped between the piston head, cylinder, and cylinder head, pushes the piston from the TDC to the BDC and generates power and rotation in the crankshaft. The power generated in this process can be determined as below:

$$\dot{W}_{expansion} = \dot{W}_{out} = \dot{m}_{intake}(u_3 - u_4) = \dot{m}_{intake}\frac{P_4v_4 - P_3v_3}{1 - k} = \dot{m}_{intake}c_v(T_3 - T_4)$$

$$\frac{P_4}{P_3} = \left(\frac{v_3}{v_4}\right)^k = \left(\frac{V_3}{V_4}\right)^k = \left(\frac{1}{r_c}\right)^k = \left(\frac{T_4}{T_3}\right)^{\frac{k}{k-1}}$$

(9.4)

Process 4—5: constant volume ($v_5 = v_4$) heat rejection (exhaust blowdown), after the expansion and generating power, the crankshaft keeps rotating, and it pushes the piston to the TDC, the combustion gases inside the cylinder are exhausted by opening the exhaust valve (while the intake valve remains closed), and while it keeps rotating, the piston goes back to the BDC and opens the intake valve to let the air and fuel mixture enters the cylinder (Stroke 1: intake).

As discussed above, the Otto engine is a four stroke engine with four main processes. The heat rejection can also be determined as below:

$$\dot{Q}_{out} = \dot{m}_{intake}(u_4 - u_5) = \dot{m}_{intake}c_v(T_4 - T_5)$$

$$v_1 = v_4$$

$$\frac{P_5v_5}{T_5} = \frac{P_4v_4}{T_4}, \ v_1 = v_4 = v_5 \rightarrow \frac{P_5}{T_5} = \frac{P_4}{T_4}$$

(9.5)

Process 5—6: constant pressure exhaust stroke, in this stroke, the exhaust valve is open and the intake valve is closed. The power consumed due to exhausting is determined as below:

$$P_5 = P_6 = P_0$$

$$\dot{W}_{exhaust} = \dot{m}_{intake}P_0(v_6 - v_5)$$

(9.6)

The exhaust power is a wasted power, unless it would be recovered by using a turbocharger for pressurizing the intake air.

---

[1] Stoichiometric ratio is the exact ratio of air to fuel that is required for complete combustion. No oxygen or fuel remain after stoichiometric combustion. If there is less or more air than the stoichiometric reaction the engine is called rich burn or lean burn.

Like other heat engines, using the first law of thermodynamics, the thermal efficiency can be determined as below:

$$\eta_{th} = \frac{\dot{W}_{net}}{\dot{Q}_{in}} = \frac{\dot{Q}_{in} - \dot{Q}_{out}}{\dot{Q}_{in}} = 1 - \frac{\dot{Q}_{out}}{\dot{Q}_{in}} \tag{9.7}$$

By applying the cold-air-standard assumptions, the thermal efficiency would be calculated as a function of compression ratio only as below:

$$\eta_{Otto-CAS} = 1 - \frac{1}{r_c^{k-1}} \tag{9.8}$$

The real spark ignition engines have some important differences with the Otto cycle under CAS assumptions. Below some of these differences are discussed.

1. The Otto cycle efficiency under CAS assumptions ignores the impact of temperature on the cycle efficiency, while in practice, the maximum temperature and pressure are both critical for engines. Therefore, for accurate designs, the impact of maximum temperature must be considered by rearranging the assumptions before calculations.

2. In the Otto cycle, specific heat is assumed constant, while in the SI engine, c_v is temperature dependent. This assumption can be modified by considering variable c_v in the cycle. This discussion is also valid for c_p, k and R. Some references use temperature average values of $c_v = 0.821 \frac{kJ}{kg.K}$, $c_p = 1.108 \frac{kJ}{kg.K}$, $k = 1.35$ and $R = 0.287\ 108 \frac{kJ}{kg.K}$ for the state points during compression, combustion, expansion, and exhaust. For the air properties in the intake process, it is recommended to use $c_v = 0.718 \frac{kJ}{kg.K}$, $c_p = 1.005 \frac{kJ}{kg.K}$, $k = 1.4$ and $R = 0.287\ 108 \frac{kJ}{kg.K}$. Using these assumptions make the calculations more accurate.

3. Another critical assumption is assuming internally reversible processes that ignores the irreversibilities due to friction, combustion, mixing, heat transfer, etc. This assumption can be corrected by using adiabatic compression and expansion efficiencies for the processes.

4. In the real SI cycles, the working fluid composition changes from the intake to the exhaust; at the intake, there is a mixture of air and fuel, but at the exhaust, there will be combustion products that may even change slightly in each cycle. Under CAS assumptions, the working fluid is pure air throughout the cycle. The thermodynamics of the real working fluid would be different from the pure air.

5. In the real cycle, all of the fuel energy is not consumed for temperature elevation, but some of it would be wasted trough the engine cooling system, oil, and engine radiation and convection to the surrounding. Hence, the maximum temperature would be lower than that is predicted by the CAS assumptions.

6. In the real cycle, the maximum pressure would be lower than that is predicted by the CAS assumptions because a significant amount of fuel energy would be wasted.

7. The combustion process is a continuous not an instantaneous process, that means it happens during a time not exactly at the TDC. This would consequently result in a continuous pressure elevation that is lower than that predicted by the CAS assumptions.

8. In the real cycle, the exhaust process does not happen in constant volume and the exhaust valve opens sooner and some power would be lost.

9. In the Otto cycle, there is no mention of fuel energy conversion components. The fuel energy is supposed to be converted to useful power, but a great portion of it would be wasted through

engine cooling system (jacketing), exhaust gases, engine body radiation and convection, oil cooling, intercooling (air cooling before entering the cylinder), exhaust power, intake power, and other losses. Hence, the energy balance can be written as below:

$$\dot{Q}_{in} = \dot{Q}_{fuel} = \dot{W}_{gross} + \dot{W}_{exhaust} + \dot{W}_{intake} + \dot{Q}_{jackting} + \dot{Q}_{exhasut} + \dot{Q}_{oil\ cooler}$$
$$+ \dot{Q}_{inintercooler} + \dot{Q}_{other\ losses}$$

$$(9.9)$$

A typical real spark ignition cycle is shown in Fig. 9.3. It consists of an upper and a lower loop. The area of the upper loop is the expansion power minus compression power, which is called the gross power. The lower loop is the summation of power consumed for intake and exhaust processes and is called pump power.

$$\dot{W}_{gross} = \dot{W}_{expansion} - \dot{W}_{compression}$$
$$\dot{W}_{pump} = \dot{W}_{intake} + \dot{W}_{exhaust}$$

$$(9.10)$$

For an engine without using the supercharger or turbocharger, the net power would be:

$$\dot{W}_{net} = \dot{W}_{gross} - \dot{W}_{pump}$$

$$(9.11)$$

And for an engine using a or turbocharger, the net power would be:

$$\dot{W}_{net} = \dot{W}_{gross} + \dot{W}_{pump}$$

$$(9.12)$$

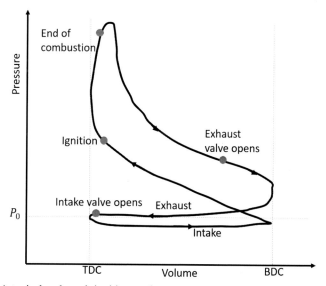

Figure 9.3 A typical real spark ignition cycle.

Because a turbocharger uses the exhaust stream to run a small compressor for pressurizing intake air before entering the engine. For such engines the lower loop in Fig. 9.3 would be above the P_0 line.

1.2 The diesel cycle

The Diesel cycle is the internally reversible cycle of the compression ignition engines. For simplicity, the thermodynamics of the Diesel cycle is discussed under CAS assumptions. Fig. 9.4 shows the $P - v$ and $T - s$ process diagrams of the diesel cycle [2,3].

The four processes of the Diesel cycle, along with the thermodynamically governing equations, are as below:

Process 6−1: The intake process, which sucks in the air and fuel mixture, consumes some power as below:

$$P_1 = P_6 = P_0$$
$$\dot{W}_{intake} = \dot{m}_{intake} P_0 (v_1 - v_6)$$

(9.13)

In this stroke, the intake valve is open and the exhaust valve is closed.

Process 1−2: isentropic compression ($s_1 = s_2$), in which the piston moves from the bottom dead center (BDC) with the maximum volume in the cylinder to the top dead center (TDC) with the minimum volume. In this process, pure air is compressed to the pressure of P_2 (Stroke 2: compression). The required work for the compression is provided by other cylinders on the opposite side of the crank shaft and also the power stroke. The compression power or the inlet power can be determined as below:

$$\dot{W}_{compression} = \dot{m}_{intake}(u_2 - u_1) = \dot{m}_{intake} \frac{P_1 v_1 - P_2 v_2}{1 - k} = \dot{m}_{intake} c_v (T_2 - T_1)$$

$$\frac{P_2}{P_1} = \left(\frac{v_1}{v_2}\right)^k = \left(\frac{V_1}{V_2}\right)^k = (r_c)^k = \left(\frac{T_2}{T_1}\right)^{\frac{k}{k-1}}$$

(9.14)

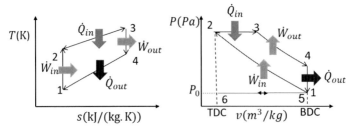

Figure 9.4 The internally reversible Diesel cycle for CI reciprocating engines.

Process 2–3: constant pressure heat addition ($P_2 = P_3$), while piston is at the TDC, fuel is injected to the hot compressed air in the cylinder and a rapid combustion occurs. Due to the heat release gas volume inside the cylinder increases at constant pressure until it reaches v_3. The input heat rate due to the combustion process can be determined as below:

$$\dot{Q}_{in} = \dot{m}_{intake}(h_3 - h_2) = \dot{m}_{fuel}HV\eta_c = \dot{m}_{intake}c_p(T_3 - T_2)$$

$$\dot{m}_{intake} = \dot{m}_{air} + \dot{m}_{fuel} = (AF + 1)\dot{m}_{fuel}$$

$$T_3 = T_{max} \ \& \ P_3 = P_{max}$$

$$\frac{P_2 v_2}{T_2} = \frac{P_3 v_3}{T_3}, \ P_2 = P_3 \rightarrow \beta = \frac{v_3}{v_2} = \frac{T_3}{T_2}$$

(9.15)

In which AF is the air to fuel mass flow rate ratio. For CI engines, $18 : 1$ (*rich*) $\leq$ $AF \leq 70 : 1$ (*lean*). The AF value depends on operating parameters such as speed, pressure, allowable temperature, load, emission control systems, etc. In addition, β is called **cutoff ratio** and shows the volume change during the combustion process. Volume change in constant pressure corresponds to power generation during the combustion process and is determined by writing the energy balance for this process as below:

$$\dot{W}_{2-3} = \dot{Q}_{in} - \dot{m}_{intake}(u_3 - u_2) = \dot{m}_{intake}P_2(v_3 - v_2)$$

(9.16)

Process 3–4: isentropic expansion ($s_3 = s_4$), the hot- and high-pressure gas trapped between the piston head, cylinder, and cylinder head, pushes the piston from the point 3 to the BDC at point 4 and generates power and rotation in the crankshaft. The power generated in this process can be determined as below:

$$\dot{W}_{expansion} = \dot{W}_{3-4} = \dot{m}_{intake}(u_3 - u_4) = \dot{m}_{intake}\frac{P_4 v_4 - P_3 v_3}{1 - k}$$

$$= \dot{m}_{intake}c_v(T_3 - T_4)$$

(9.17)

$$\frac{P_4}{P_3} = \left(\frac{v_3}{v_4}\right)^k = \left(\frac{V_3}{V_4}\right)^k = \left(\frac{T_4}{T_3}\right)^{\frac{k}{k-1}}$$

Process 4–5: constant volume ($v_5 = v_4$) heat rejection (exhaust blowdown), after the expansion and generating power, the crankshaft keeps rotating, and it pushes the piston to the TDC; the combustion gases inside the cylinder are exhausted from the exhaust valve (Stroke 4: exhaust), and while it keeps rotating the piston goes back to the BDC and opens the intake valve to let the air to enter the cylinder (stroke 1: intake).

As discussed above, the diesel engine is a four stroke engine with four main processes discussed above. The heat rejection can also be determined as below:

$$\dot{Q}_{out} = \dot{m}_{intake}(u_4 - u_5) = \dot{m}_{intake}c_v(T_4 - T_5)$$

$$v_5 = v_1 = v_4 \tag{9.18}$$

$$\frac{P_1 v_1}{T_1} = \frac{P_4 v_4}{T_4}, \ v_1 = v_4 = v_5 \rightarrow \frac{P_1}{T_1} = \frac{P_4}{T_4}$$

Process 5–6: constant pressure exhaust at P_0. In this stroke, the exhaust valve is open and intake valve is closed. The power wasted due to exhaust process is determined as below:

$$P_5 = P_6 = P_0$$

$$\dot{W}_{exhaust} = \dot{m}_{intake}P_0(v_6 - v_5) \tag{9.19}$$

The exhausted power can be recovered by using turbocharger for intake air compression.

Similar to other heat engines, and using the first law of thermodynamics, the thermal efficiency of the Diesel cycle under CAS assumptions can be determined as below:

$$\eta_{Diesel-CAS} = \frac{\dot{W}_{net}}{\dot{Q}_{in}} = \frac{\dot{Q}_{in} - \dot{Q}_{out}}{\dot{Q}_{in}} = 1 - \frac{\dot{Q}_{out}}{\dot{Q}_{in}} = 1 - \frac{1}{r_c^{k-1}}\left[\frac{\beta^k - 1}{k(\beta - 1)}\right] \tag{9.20}$$

The internally reversible Diesel cycle ignores many details in the six processes which were discussed. Fig. 9.5 shows a typical real compression ignition engine. It shows that fuel injection happens nearly at the end of compression stroke. Because

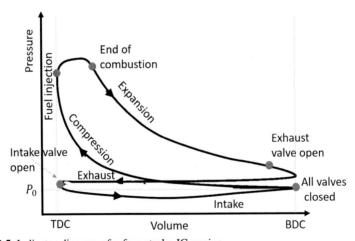

Figure 9.5 Indicator diagram of a four stroke IC engine.

of delay in ignition and finite time of fuel injection, the combustion continues to the expansion stroke and results in a constant pressure combustion.

Some important differences between the real CI engine and internally reversible Diesel engine are discussed below:

1. The Diesel cycle efficiency under CAS assumptions is calculated only by compression ratio and cutoff ratio, and it ignores the impact of temperature on the cycle efficiency, while in practice, the maximum temperature and pressure are both critical for Diesel engines.
2. In the Diesel cycle, specific heat is assumed constant, while in the real CI engine, it is temperature dependent. This assumption can be modified by considering variable c_v and c_p in the cycle. This discussion is also valid for k and R. Some references use temperature average values of $c_v = 0.821 \frac{kJ}{kg.K}$, $c_p = 1.108 \frac{kJ}{kg.K}$, $k = 1.35$ and $R = 0.287\ 108 \frac{kJ}{kg.K}$ for the state points during compression, combustion, expansion, and exhaust strokes. For the air properties in the intake process, it is recommended to use $c_v = 0.718 \frac{kJ}{kg.K}$, $c_p = 1.005 \frac{kJ}{kg.K}$, $k = 1.4$ and $R = 0.287\ 108 \frac{kJ}{kg.K}$. Using these assumptions makes the calculations more accurate.
3. Another critical assumption is assuming internally reversible processes that ignore the irreversibilities due to friction, combustion, mixing, heat transfer, etc. This assumption can be corrected by using adiabatic compression and expansion efficiencies for the processes.
4. In the real CI cycles, the working fluid composition changes from the intake to the exhaust, at the intake there is only air, but at the exhaust, there will be combustion products that may even change slightly in each cycle. Under CAS assumptions, the working fluid is pure air throughout the cycle. The thermodynamics of the real working fluid would be different from the pure air.
5. In the real cycle, all of the fuel energy is not consumed for temperature elevation, but some of it would be wasted trough the engine cooling system, oil, and engine radiation and convection to the surrounding. Hence, the maximum temperature and pressure would be lower than that is predicted by the CAS assumptions.
6. The combustion process is a continuous not an instantaneous process, that means it happens during the expansion stroke as well.
7. In the real cycle, the exhaust process does not happen in constant volume and the exhaust valve opens sooner and some power will be lost. This power can be used by turbocharger (if existed) for air compression in the intake stroke.
8. In the Diesel cycle, there is no mention of fuel energy conversion components. The fuel energy is supposed to be converted to useful power, but a great portion of it would be wasted through the engine cooling system (jacketing), exhaust gases, engine body radiation and convection, oil cooling, intercooling (air cooling before entering the cylinder), exhaust power, intake power, and other losses. Hence, the energy balance can be written as below:

$$\dot{Q}_{in} = \dot{Q}_{fuel} = \dot{W}_{gross} + \dot{W}_{exhaust} + \dot{W}_{intake} + \dot{Q}_{jackting} + \dot{Q}_{exhasut} + \dot{Q}_{oil\ cooler}$$
$$+ \dot{Q}_{intercooler} + \dot{Q}_{other\ losses}$$

$$(9.21)$$

Second law analysis is also required for the cycles evaluations to determine the cycle exergy efficiency and also identifying the exergy destructive processes for possible corrections and optimizations. The exergy efficiency of the heat engines can also be determined as below:

$$\eta_{II} = \frac{\dot{X}_{recovered}}{\dot{X}_{supplied}} = 1 - \frac{\dot{X}_{destroyed}}{\dot{X}_{supplied}} \qquad (9.22)$$

In which for Otto and Diesel engines, the recovered exergy is actually the net power output from the engine as bellow:

$$\dot{X}_{recovered} = \dot{W}_{net} \qquad (9.23)$$

The supplied exergy can also be determined as below:

$$\dot{X}_{supplied} = \dot{Q}_{in}\left(1 - \frac{T_L}{T_H}\right) = \eta_{carnot}\dot{Q}_{in} \qquad (9.24)$$

In which T_L is the heat sink average temperature. For these engines, the heat sink is the environment, so that $T_L = T_0$. In addition, T_H is the average temperature of heat addition process, that means average temperature of process 2–3. T_H can be determined as below:

$$\dot{Q}_{in} = \dot{m}_{intake}(h_3 - h_2) = \dot{m}_{intake}.T_H(s_3 - s_2) \rightarrow T_H = \frac{h_3 - h_2}{s_3 - s_2} \qquad (9.25)$$

In addition, the destroyed exergy in Otto and Diesel are as below:

$$\dot{X}_{destroyed} = \dot{X}_{supplied} - \dot{X}_{recovered} = \dot{Q}_{in}\left(1 - \frac{T_L}{T_H}\right) - \dot{W}_{net}$$

$$= \dot{Q}_{in}\left(1 - \frac{T_L}{T_H}\right) - \dot{Q}_{in} + \dot{Q}_{out}$$

$$= \dot{Q}_{out} - \dot{Q}_{in}\frac{T_L}{T_H} = T_L\left(\frac{\dot{Q}_{out}}{T_L} - \frac{\dot{Q}_{in}}{T_H}\right) = T_0\dot{S}_{gen}$$

Hence:

$$\dot{X}_{destroyed} = \dot{Q}_{out} - \dot{Q}_{in}\frac{T_L}{T_H} \qquad (9.26)$$

The above equation can also be written as below:

$$\dot{X}_{destroyed} = \dot{Q}_{in}(\eta_{carnot} - \eta_{th}) \qquad (9.27)$$

Hence, the second law efficiency also can be rewritten as below:

$$\eta_{II} = 1 - \frac{\dot{Q}_{in}\left(\eta_{carnot} - \eta_{th}\right)}{\eta_{carnot}\dot{Q}_{in}} = \frac{\eta_{th}}{\eta_{carnot}}$$

2. Reciprocating engine characteristics

For an RE, there are some operating characteristics which determine the engine capacity for power production. These characteristics limit the engine speed, the air consumption in each cycle, the fuel consumption, compression ratio, pressure ratio, etc. Although designing these characteristics is not the concern of this book, being familiar with these specifications is vital when working with REs power plants. Fig. 9.6 shows the main specifications of an RE.

According to the circular path of the crank shaft.

$$S = 2a \tag{9.28}$$

With the engine speed of N (RPM), and engine stroke of S (m) the piston average speed $\overline{U}_p \left(\frac{m}{s}\right)$ is:

$$\overline{U}_p = N(2S)\left(\frac{1}{60}\right) \tag{9.29}$$

The safe range of average piston speed is usually between 5 and 15 m/s. Unlike the dynamic flow equipment such as gas or steam turbines, the REs apply repeating and cyclic forces on the elements. In this engine, in each revelution of crankshaft, the piston twice reaches the maximum acceleration from the stop and then back to stop again. This repeating acceleration applies repeating force that can cause fatigue and life reduction of the components. It is clear that higher piston speed means bigger acceleration and bigger fatigue force. Therefore, to avoid the danger of components failure the piston speed is limited. According to the relation between the piston speed, engine rotational speed, and stroke length, there would be correlations between the engine size and piston speed. It is clear that for a specific piston speed, by increasing N, the stroke reduces. For example, an engine operating with piston speed of 15 m/s, and rotational speed of 3000 RPM, its stroke length is just 2.38 cm. Today with advances in material science and metallurgy, piston speeds of more than 25 m/s is achieved. Another parameter that limits the piston speed is the air flow in the intake and exhaust processes. The piston speed dictates the flow rate through the intake and exhaust valves. To avoid any pressure loss and restriction in these processes, the maximum valve size should be used for easy intake and exhaust.

The bore size (B), which is actually the cylinder diameter, is usually compared by the stroke length. B/S is called the bore to stroke ratio, and when it is equal to 1, the

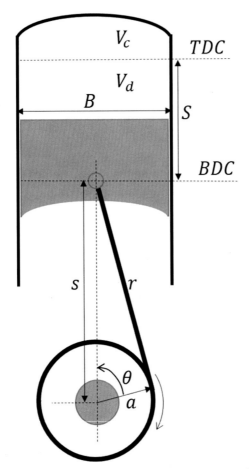

Figure 9.6 The geometrical specifications of the cylinder, piston, crankshaft, and connecting rod. B: Bore size, r: Connecting rod length, a: Crank offset, S:stroke, s: Piston position, θ: Crank angle, V_d: Displacement volume, V_c: Clearance volume, TDC: Top dead center, BDC: Bottom dead center.

engine is called *square* engine, is B/S is smaller than one it is called *under square* engine, and for B/S bigger than one it is called *over square* engine. Very big engines have stroke bigger than bore size, and they are under square always. The small engines usually have B/S of 0.8−1.2.

The instantaneous speed of piston can be determined by writing the equation for the instantaneous position of piston and the deviating it with respect to time. According to Fig. 9.6, the piston position can be formulated as a function of θ, a, and r as below:

$$s = a \cos \theta + \sqrt{r^2 - \sin^2 \theta} \tag{9.30}$$

Therefore:

$$\frac{U_p}{\overline{U}_p} = \frac{\pi}{2} \sin\theta \left[1 + \frac{\cos\theta}{\sqrt{R - \sin^2\theta}} \right] \tag{9.31}$$

The ratio of $R = \frac{r}{a}$ for small engines ranges from 3 to 4, and for large engines, it ranges from 5 to 10.

The displacement and clearance volume on cylinder of an engine are related to the bore and stroke size as below:

$$V_d = \frac{\pi}{4} B^2 S = V_{BDC} - V_{TDC}$$

$$V_c = V_{TDC} \tag{9.32}$$

$$V_{BDC} = V_d + V_c$$

The total displacement volume is usually reported based on the number of cylinders. For an engine with N_c cylinders, the total displacement volume is:

$$V_{d,total} = N_c \frac{\pi}{4} B^2 S \tag{9.33}$$

Displacement volume may differ from 0.1 cm^3 to 8000 cm^3 depending on the engine size and applications.

The cylinder volume at different crank angles can be written as below:

$$\frac{V}{V_c} = 1 + \frac{1}{2}(r_c - 1)\left[R + 1 - \cos\theta - \sqrt{R^2 - \sin^2\theta}\right] \tag{9.34}$$

And the combustion chamber surface area of a cylinder is equal to:

$$A = A_{ch} + A_p + \frac{\pi BS}{2}\left[R + 1 - \cos\theta - \sqrt{R^2 - \sin^2\theta}\right] \tag{9.35}$$

In which, A_{ch} is the cylinder heat surface area, A_p top piston area, and for the flat pistons, it is equal to $A_p = \frac{\pi}{4}B^2$. In some engines today, they manufacture nonflat piston heads to increase the combustion area.

The mean effective pressure of an engine is another parameter that is usually reported by the manufacturers and can be used for comparing engines. It is determined as the ratio of the work of a cycle divided by the displacement volume.

$$mep = \frac{w}{v_{max} - v_{min}} = \frac{w}{v_d} = \frac{W}{V_d} \tag{9.36}$$

The work and the power that can be generated by an engine is determined according to the speed and the torque generated by the engine. The torque that is required to stop the crankshaft from rotating is called the *braking torque*, and accordingly, the power and work are called *brake power* and *brake work*. It is clear that if brake work is used to determine the mep, it will be called *brake mean effective pressure* (bmep).

For a four stroke cycle engine, the torque, T, can be determined according to the displacement volume and bmep as below:

$$T = \text{bmep} \frac{V_d}{4\pi} \tag{9.37}$$

And the total power can be determined as follow:

$$\dot{W} = NTN_c = \text{mep} \frac{NV_d}{4} N_c = \text{mep} \frac{NA_p S}{4} N_c = \text{mep} \frac{\overline{U}_p A_p}{4} N_c \tag{9.38}$$

In which N_c is the number of cylinders in the engine, and N is the engine speed in RPM.

Example 9.1. In an Otto cycle with compression ratio of 12, after compression, air and fuel mixture is heated to the maximum temperature of $2700\,°C$ and then is exhausted to ambient. Determine the input heat, heat loss, and thermal efficiency by considering, (a) cold air standard assumptions and (b) air standard assumptions.

Solution. The cold air standard assumes constant specific heat, while the standard air assumptions consider it as temperature dependent. According to this point, the solution is provided below:

a:

Point 1:

$$T_1 = 25\,°C, \quad P_1 = 100\,kPa \rightarrow c_v = 0.7197 \frac{kJ}{kgK}, \quad k = 1.399$$

Point 2:

$$\frac{T_2}{T_1} = r_c^{k-1} \rightarrow \frac{T_2}{25 + 273.15} = 12^{1.399-1} \rightarrow T_2 = 530.1\,°C$$

$$\frac{P_2}{P_1} = r_c^k \rightarrow \frac{P_2}{100} = 12^{1.399} \rightarrow P_2 = 3233\,kPa$$

Point 3:

$$T_3 = 2700\,°C, \quad \frac{P_3}{P_2} = \frac{T_3}{T_2} \rightarrow \frac{P_3}{3233} = \frac{2700 + 273.15}{530.1 + 273.15} \rightarrow P_3 = 11966\,kPa$$

Point 4:

$$\frac{P_3}{P_4} = r_c^k \rightarrow \frac{11966}{P_4} = 12^{1.399} \rightarrow P_4 = 370.1 \; kPa$$

$$\frac{T_3}{T_4} = r_c^{k-1} \rightarrow \frac{2700 + 273.15}{T_4} = 12^{1.399-1} \rightarrow T_4 = 830.4°C$$

$$q_{in} = c_v(T_3 - T_2) = 0.7197(2700 - 530.1) = 1562 \frac{kJ}{kg}$$

$$q_{out} = c_v(T_4 - T_1) = 0.7197(830.4 - 25) = 579.6 \frac{kJ}{kg}$$

$$\eta_{th} = 1 - \frac{1}{r_c^{k-1}} = 1 - \frac{1}{12^{1.399-1}} = 0.6288$$

Or by using the following equation:

$$\eta_{th} = 1 - \frac{q_{out}}{q_{in}} = 1 - \frac{579.6}{1562} = 0.6288$$

The same result is obtained.

b:

Point 1:

$$T_1 = 25\,°C, \; P_1 = 100 \; kPa \rightarrow s_1 = 5.699 \frac{kJ}{kgK}, \; v_1 = 0.8558 \frac{m^3}{kg}, \; u_1 = 213 \frac{kJ}{kg}$$

Point 2:

$$s_2 = s_1 = 5.699 \frac{kJ}{kgK}, \; v_2 = \frac{v_1}{r_c} = \frac{0.8558}{12} = 0.07131 \frac{m^3}{kg} \rightarrow T_2 = 502.2°C, \; P_2$$
$$= 3121 kPa, \; u_2 = 572.5 \frac{kJ}{kg}$$

Point 3:

$$T_3 = 2700\,°C, \; v_3 = v_2 = 0.07131 \frac{m^3}{kg} \rightarrow P_3 = 11966 \; kPa, \; u_3 = 2636 \frac{kJ}{kg}, \; s_3$$
$$= 6.932 \frac{kJ}{kgK}$$

Point 4:

$$s_3 = s_4 = 6.932\,\frac{kJ}{kgK}, \quad \frac{v_4}{v_3} = r_c \rightarrow \frac{v_4}{0.07131} = 12 \rightarrow v_4 = 0.8558\,\frac{m^3}{kg} \rightarrow T_4$$

$$= 1145\,^{\circ}C, \quad P_4 = 475.5\,kPa, \quad u_4 = 1129\,\frac{kJ}{kg}$$

$$q_{in} = u_3 - u_2 = 2636 - 572.5 = 2063\,\frac{kJ}{kg}$$

$$q_{out} = u_4 - u_1 = 1129 - 213 = 916.2\,\frac{kJ}{kg}$$

$$\eta_{th} = 1 - \frac{q_{out}}{q_{in}} = 1 - \frac{916.2}{2063} = 0.5559$$

Comparing the results of the CAS with the standard air assumptions shows that the CAS is over estimating the thermal efficiency about $(0.6288-0.5559)/0.5559 = 13.11\%$.

More discussions:
Using constant specific heat assumption in the thermodynamics calculation of the power cycles especially those with highly variable temperature in the cycle, such as REs, gas turbines, steam turbines, and some fuel cells is not recommended for design purposes because it over estimates important evaluation criteria such as the thermal efficiency and may lead to wrong decision-making. In addition, the internally reversible assumption can be modified by considering the adiabatic efficiency for the compression and expansion processes.

The code generated in the EES for the Example 9.1 is given below in **Code 9.1**:

Code 9.1. The code generated in the EES for the Example 9.1

```
"Constant specific heat"
r_c=12
T[1]=25
P[1]=100
T[3]=2700
T0=273.15
c_v=CV(Air,T=T[1])
c_p=CP(AIR, T=T[1])
k=c_p/c_v
(T[2]+T0)/(T[1]+T0)=r_c^(k-1)
T[4]+T0=(T[3]+T0)/r_c^(k-1)
P[2]/P[1]=r_c^k
P[3]/P[2]=(T[3]+T0)/(T[2]+T0)
P[4]/P[3]=(1/r_c)^k
q_in=c_v*(T[3]-T[2])
q_out=c_v*(T[4]-T[1])
eta=1-1/r_c^(k-1)
eta1=1-q_out/q_in
```

```
"Variable specific heat"
r_c=12
T[1]=25
P[1]=100
T[3]=2700
s[1]=entropy(air, T=T[1], P=P[1])
v[1]=volume(air, T=T[1], P=P[1])
u[1]=intenergy(air, T=T[1])
s[2]=s[1]
v[2]=v[1]/r_c
T[2]=TEMPERATURE(Air,s=s[2],v=v[2])
P[2]=Pressure(air, v=v[2], T=T[2])
u[2]=intenergy(air, T=T[2])
v[3]=v[2]
P[3]=Pressure(air, v=v[3], T=T[3])
s[3]=entropy(air, T=T[3], P=P[3])
u[3]=intenergy(air, T=T[3])
S[4]=S[3]
V[4]/V[3]=r_c
T[4]=TEMPERATURE(Air,s=s[4],v=v[4])
P[4]=Pressure(air, v=v[4], T=T[4])
u[4]=intenergy(air, T=T[4])
q_in=(u[3]-u[2])
q_out=(u[4]-u[1])
eta=1-q_out/q_in
```

Example 9.2. The SGE-42HM (Fig. 9.7) manufactured by Siemens [4] is a gas engine capable of working with both natural gas and biogas. The engine power output is 1040 kW, and when connected to the generator, it produces 1006 kW of electricity. The engine speed is 1800 rpm, and its weight is 6250 kg. This engine is equipped for cogeneration of heat and power by heat recovery from the exhaust, engine jacket,

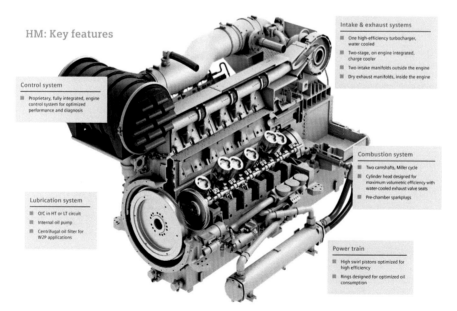

Figure 9.7 SGE-42HM genset, an SI gas engine (Siemens courtesy) [4].

intercooler, and lube oil cooling. The compression ratio of the engine is 11.9:1. The engine is turbocharged (Fig. 9.8), and the exhaust temperature after leaving the turbocharger is 455 °C and then is cooled to 120 °C in the exhaust heat recovery system to achieve 538 kW of heat. Ambient pressure and temperature are 100 kPa and 25 °C. The air temperature after being compressed in the turbocharger increases that will be cooled to 36.5 °C by the intercooler at the engine inlet. Assume the combustion efficiency is 100%. The combustion air and the exhaust mass flow rates are 4840 and 5020 kg/h, and the total displacement volume of the cylinders is also 42.2 L.

The energy balance of engine is as below:

Fuel energy	Electricity	Jacket	Exhaust recovery	Intercooler
2446 kW	1006 kW	586 kW	538 kW	60 kW
Engine radiation	Generator heat	Exhaust loss	Other losses	Power
70 kW	34 kW	?	18.5 kW	1040 kW

Determine:

A. The thermal and electrical efficiency of the engine. Why the electrical efficiency is smaller than thermal efficiency? How the power is lost?
B. The exhaust loss after heat recovery.
C. The air to fuel ratio, and the fuel heating value.
D. By using the air standard assumptions, determine all of the state point properties and then plot the $T - s$ and $P - v$ diagrams.
E. The overall efficiency of the engine.
F. The exergy efficiency and the exergy destruction of the engine.

Assume the standard air assumptions in the solution.

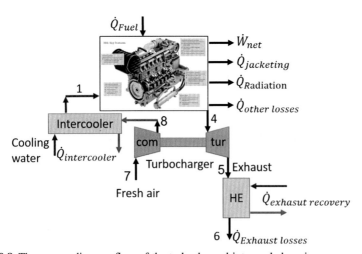

Figure 9.8 The energy diagram flow of the turbocharged intercooled engine.

Solution. A:

The thermal and electrical efficiency of the engine can be determined according to the given fuel energy and the net output power and electricity of the engine as below:

$$\eta_{th} = \frac{Power}{Fuel\ energy} = \frac{\dot{W}_{net}}{\dot{Q}_{in}} = \frac{1040}{2446} = 42.52\%$$

$$\eta_e = \frac{Electricity}{Fuel\ energy} = \frac{E_{net}}{\dot{Q}_{in}} = \frac{1006}{2446} = 41.13\%$$

The generator converts the rotational motion (power) of the crankshaft to electricity, and due to this rotation, the heat generated due to rotation of the generator's rotor some power will be lost. The energy balance is as below:

$$\dot{W}_{net} = E_{net} + \dot{Q}_{generator} \rightarrow \dot{Q}_{generator} = \dot{W}_{net} - E_{net} = 1040 - 1006 = 34\ kW$$

Which is also reported in the energy balance of the engine.

The generator efficiency can also be determined as below:

$$\eta_{generator} = \frac{E_{net}}{\dot{W}_{net}} = 1 - \frac{\dot{Q}_{generator}}{\dot{W}_{net}} = 1 - \frac{34}{1040} = 96.73\%$$

In addition, the following equation is valid for the efficiencies:

$$\eta_{th} \times \eta_{generator} = \eta_e \rightarrow \eta_{generator} = \frac{\eta_e}{\eta_{th}} = \frac{41.13}{42.52} = 96.73\%$$

B:

The ambient temperature can be determined according to the energy balance given for the exhaust energy loss after the heat recovery. The exhaust gas is cooled to $120\ ^{\circ}C$ by using the heat recovery system and then released to atmosphere. To do this, the exhaust mass flow rate must be determined.

$$T_6 = 120\ ^{\circ}C \rightarrow h_6 = 394.3 \frac{kJ}{kg.K}$$

$$T_7 = 25\ ^{\circ}C \rightarrow h_7 = 298.6 \frac{kJ}{kg.K}$$

$$\dot{Q}_{exhaust\ loss} = \dot{m}_{ex}(h_6 - h_7) \rightarrow \dot{Q}_{exhaust\ loss} = \frac{5020}{3600}(394.3 - 298.6) \rightarrow \dot{Q}_{exhaust\ loss}$$
$$= 133.5\ kW \rightarrow$$

$$\dot{Q}_{exhaust\ loss} = 133.5\ kW$$

C:

To determine the AF, the fuel consumption must be determined, according to the input fuel energy:

$$\dot{m}_{ex} = \dot{m}_{air} + \dot{m}_{fuel} \rightarrow 5020 = 4840 + \dot{m}_{fuel} \rightarrow \dot{m}_{fuel} = 180\frac{kg}{h} = 0.05\frac{kg}{s} \rightarrow AF$$

$$= \frac{\dot{m}_{air}}{\dot{m}_{fuel}} = \frac{4840}{180} = 26.8 \rightarrow AF = 26.8$$

$$HV = \frac{\dot{Q}_{in}}{\dot{m}_{fuel}\eta_{combustion}} = \frac{2446\ kW}{0.05\frac{kg}{s} \times 1} = 48920\frac{kJ}{kg}$$

D:

Point 1:
The specific displacement volume and compression ratio are related to v_1 and v_2 as below:

$$v_1 - v_2 = v_d = \frac{V_d}{m_{1\ cycle}} = \frac{V_d\left(m^3\right)}{\dfrac{\dot{m}_{ex}\left(\frac{kg}{s}\right) \times 60\left(\frac{s}{min}\right)}{\dfrac{N}{2}\left(\frac{rev}{min}\right)}} = \frac{\dfrac{42.2}{1000}}{\dfrac{\frac{5020}{3600} \times 60}{\frac{1800}{2}}} = 0.4539\frac{m^3}{kg}$$

$$v_2 = \frac{v_1}{r_c} = \frac{v_1}{11.9}$$

In which $\frac{\dot{m}_{ex} \times 60}{\frac{N}{2}}$ is the mass of air/fuel mixture in the displacement volume of cylinder. Every 2 rotations of engine the cylinders are filled and exhausted once. By solving these two equations, the specific volumes at points 1 and 2 would be determined:

$$v_2 = 0.04165\frac{m^3}{kg},\ v_1 = 0.4956\frac{m^3}{kg}$$

Since $T_1 = 36.5\ °C$, therefore:

$$P_1 = 179.5\ kPa,\ s_1 = 5.569\frac{kJ}{kgK},\ h_1 = 310.1\frac{kJ}{kgK}$$

Point 2:

$$v_2 = 0.04165\frac{m^3}{kg},\ s_2 = s_1 = 5.569\frac{kJ}{kgK} \rightarrow P_2 = 5509\ kPa,\ T_2 = 526.2\ °C$$

Point 4:

According to the data given for the exhaust temperature (point 5), intercooler heat recovery, and the turbocharger the following equations are valid:

$$T_5 = 455\,°C \rightarrow h_5 = 743.7\frac{kJ}{kgK}$$

$$\dot{Q}_{intercooler} = \dot{m}_{ex}(h_8 - h_1) \rightarrow 60 = \frac{5020}{3600}(h_8 - 310.1) \rightarrow h_8 = 353.2\frac{kJ}{kgK}$$

by using the power transfer in the turbocharger:

$$\dot{W}_{tur} = \dot{W}_{comp} \rightarrow \dot{m}_{ex}(h_8 - h_0) = \dot{m}_{ex}(h_4 - h_5) \rightarrow 353.2 - 298.6 = h_4 - 743.7$$

$$\rightarrow h_4 = 798.3\frac{kJ}{kgK}$$

$$\rightarrow T_4 = 505.2\,°C, \ T_8 = 79.22\,°C$$

$$v_4 = v_1 = 0.4956\frac{m^3}{kg}$$

Therefore:

$$P_4 = 450.8\ kPa, \ s_4 = 6.259\frac{kJ}{kgK}$$

Point 3:

$$v_3 = v_2 = 0.04165\frac{m^3}{kg}$$

$$s_3 = s_4 = 6.259\frac{kJ}{kgK}$$

Therefore:

$$P_3 = 12039\ kPa, \ T_3 = 1474\,°C$$

By using the state point properties and the EES code generated for the problem (Code 9.2), the $T - s$ and $P - v$ diagrams can be plotted as Fig. 9.9.

Attention must be paid that the pressure drops in the exhaust heat recovery and intercooler are neglected and they are supposed to be well insulated.

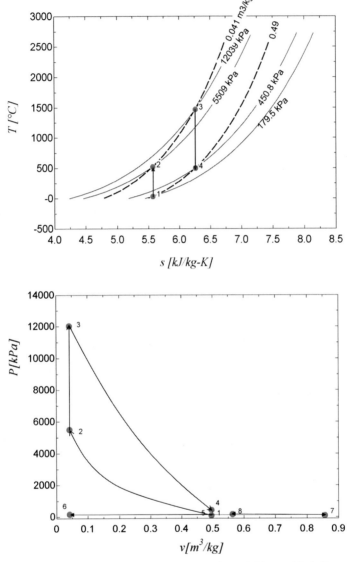

Figure 9.9 The T-s and P-v diagram of the Otto cycle with variable specific in Example 8.2 for SGE-42HM

E:

The overall efficiency of the cycle should be evaluated according to the recovered heat and the generated electricity be the engine, hence:

$$\eta_{overall} = \frac{E_{net} + Q_{recovered}}{Q_{in}} = \frac{E_{net} + \dot{Q}_{jacket} + \dot{Q}_{intercooler} + \dot{Q}_{exhasut\ recovery}}{\dot{Q}_{Fuel}}$$

$$= \frac{1006 + 586 + 60 + 538}{2446} = 89.53\%$$

F:

To determine the exergy efficiency of the engine, the average heating temperature, T_{H-mean} must be determined as below:

$$T_{H-mean} = \frac{u_3 - u_2}{s_3 - s_2} = \frac{1382 - 571.2}{6.385 - 5.7} = 1223.95 \ K = 950.8 \ ^\circ C$$

And the sink temperature which is ambient temperature is:

$$T_L = T_7 = 25 \ ^\circ C = 298.15 \ K$$

Therefore, the Carnot cycle efficiency can be determined:

$$\eta_{carnot} = 1 - \frac{T_L}{T_{H-mean}} = 1 - \frac{298.15}{1223.95} = 74.64\%$$

And the exergy efficiency or the second law efficiency is:

$$\eta_{II} = \frac{\eta_{th}}{\eta_{carnot}} = \frac{42.52}{74.64} = 56.21\%$$

The exergy destruction:

$$\underbrace{\left(\dot{X}_{in} - \dot{X}_{out} \right)}_{net\ rate\ of\ entropy\ transfer} - \dot{X}_{destroyed} = \frac{dX_{system}}{dt}$$

for steady-state operation $\frac{dX_{system}}{dt} = 0$, therefore:

$$\dot{X}_{destroyed} = \dot{X}_{in} - \dot{X}_{out} = \dot{Q}_{Fuel}\left(1 - \frac{T_L}{T_{H-mean}} \right) - \dot{W}_{net}$$

$$= 2455\left(1 - \frac{298.15}{1223.95} \right) - 1040 = 810.2 \ kW$$

The exergy destruction can also be determined as below:

$$\dot{X}_{destroyed} = \dot{Q}_{fuel}(\eta_{carnot} - \eta_{th}) = 2446(0.764 - 0.4252) = 810.2 \ kW$$

The code generated for this example is given in Code 9.2. Students are advised to use the code and evaluate parametric studies. For example, it can be used to investigate the impact of compression ratio on the engine performance.

Code 9.2. The code generated for the Example 8.2

```
"SGE-42HM "
Q_fuel=2446
Q_jacket=586
Q_Radiation=70
Q_exhaust_recovery=538
Q_intercooler=60
Q_other_loss=18.5
W_net=1040
E_net=1006
r_c=11.9
V_DISPLACEMENT=42.2/1000
N_C=12
m_ex=5020/3600
m_air=4840/3600
N=1800
T[1]=36.5
T[7]=25
T[5]=455
P[7]=100
k=1.4

"A"
eta_th=W_net/Q_fuel
eta_e=E_net/Q_fuel
Q_generator=W_net-E_net
eta_generator=E_net/W_net

"B"
Q_exhaust_loss=m_ex*(h[6]-h[7])
T[6]=120
h[6]=enthalpy(air, T=T[6])
h[7]=enthalpy(air, T=T[7])

"C"
m_fuel=m_ex-m_air
AF=m_air/m_fuel
HV_fue=Q_fuel/m_fuel

"D"

"TURBO CHARGER AND INTERCOOLER"
m_1_cycle=m_ex*60/(N/2)
v_d=V_DISPLACEMENT/m_1_cycle
v[1]-v[2]=v_d
v[1]=r_c*v[2]
h[5]=enthalpy(air, T=T[5])
h[1]=enthalpy(air, T=T[1])
m_ex*(h[8]-h[7])=m_ex*(h[4]-h[5])
```

```
Q_intercooler=m_ex*(h[8]-h[1])
T[4]=TEMPERATURE(air, h=h[4])
T[8]=TEMPERATURE(air, h=h[8])
P[1]/P[7]=((T[8]+273.15)/(T[7]+273.15))^(k/(k-1))
P[8]=P[1]
v1[1]=volume(air, t=t[1], p=p[1])
s[1]=entropy(air, T=T[1], P=P[1])
S[2]=S[1]
T[2]=temperature(air, s=s[2], v=v[2])
P[2]=pressure(air, t=t[2], v=v[2])
u[2]=intenergy(air, t=t[2])
v[3]=v[2]
s[3]=s[4]
P[3]=pressure(air, s=s[3], V=V[3])
T[3]=TEMPERATURE(air,P=P[3], V=V[3])
u[3]=intenergy(air, T=T[3])
v[4]=v[1]
s[4]=entropy(air, v=v[4], t=t[4])
P[4]=pressure(air, s=s[4], V=V[4])
P[5]=P[7]
P[6]=P[7]
V[5]=V[1]
V[6]=V[2]
V[7]=VOLUME(AIR, P=P[7], T=T[7])
V[8]=VOLUME(AIR, P=P[8], T=T[8])
s[5]=entropy(air, p=p[5], v=v[5])
s[6]=entropy(air, p=p[6], v=v[6])
s[7]=entropy(air, p=p[7], v=v[7])
s[8]=entropy(air, p=p[8], v=v[8])


"E"
eta_overall=(E_net+Q_exhaust_recovery+Q_jacket+Q_interc
ooler)/Q_fuel


"F"
T0=273.15
T_H+T0=(u[3]-u[2])/(S[3]-S[2])
T_L=T[7]
eta_carnot=1-(T_L+T0)/(T_H+T0)
eta_II=eta_th/eta_carnot
X_destroyed=Q_fuel*(1-(T_L+T0)/(T_H+T0))-W_net
X_destroyed2=Q_fuel*(eta_carnot-eta_th)
```

More discussions:

The engine discussed in the Example 8.2 utilizes turbocharger and also intercooler to elevate the engine thermal efficiency. The turbocharger increases the air inlet pressure by expanding the exhaust gases and turning the compressor section of the turbocharger. In fact, the turbocharger is a very small micro gas turbine, while its combustion chamber is replaced by the engine exhaust (Fig. 9.10). Inlet air compression increases the air mass flow rate and also the final operating pressure of the engine, which will eventually result in better performance.

During the air compression in the turbocharger, its temperature increases as well. Feeding hot air into the cylinder will result in more compression work. To avoid this problem and reducing compression power in the compression stroke, air is cooled in the intercooler by using cooling water before passing through the input valve. Some manufacturers call the air-cooler, *intercooler,* and some call it *aftercooler,* whatever its name is, its duty is to cool air before the engine input valve. It cools air after the turbocharger (called aftercooler), or it cools air between two consecutive compressions by the turbocharge and the engine (it is called intercooler). One may ask, why the oil cooling heat is not mentioned in the energy balance of the engine. It should be mentioned that the engine has two cooling water circuits, the main, and the auxiliary circuit. The cooling water in the main circuit, first cools the lube oil, and then enters the cylinder heads and sleeves. In the second circuit, water cools the inlet air in the intercooler. Fig. 9.9 shows these two circuits..

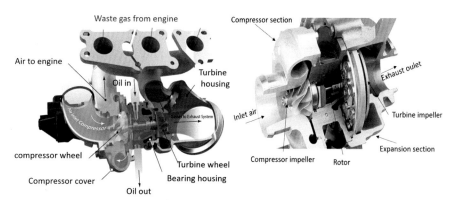

Figure 9.10 The turbocharger, and the rotor [5]. Courtesy of Mitsubishi heavy industry.

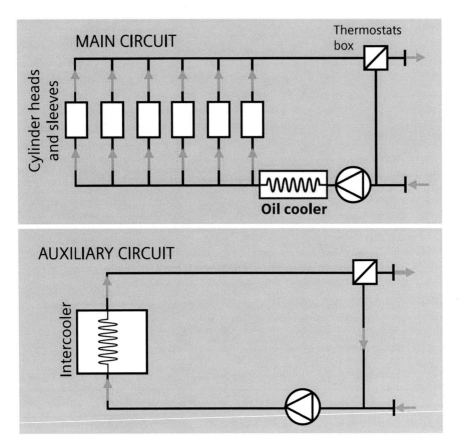

Figure 9.11 The main and auxiliary circuits of heat recovery from the oil, jacket, and inlet air in the SGE-42HM (Siemens courtesy) [4].

Example 9.3. The engine discussed in Example 9.2 (SGE-42HM) is a V12 engine (a V type with two groups of 6 cylinders on each side of the V frame). The piston mean speed is 10.5 m/s, the bore size is 160 mm, and the stroke length is 175 mm. The mean effective pressure is 16.4 bars, and the engine speed is 1800 RPM. Determine the net power of this engine and validate the answer with that reported in Example 9.2.

Solution. We are supposed to determine the power generated by the engine with the given characteristics as below:

$$N_c = 12, \ \overline{U}_p = 10.5 \, m/s, \ B = 0.16 \, m, \ S = 0.175 \, m, \ N = 1800 \, RPM, \ \text{mep}$$
$$= 16.4 \, bars$$

Hence:
According to Eq. (9.38), the power can be calculated as below:

$$\dot{W} = \text{mep}\frac{\overline{U}_p A_p}{4} N_c = 16.4 \times 100 \frac{10.5 \times \pi \times \frac{0.16^2}{4}}{4} 12 = 1039 \ kW$$

Which is very close to the 1040 ,kW which was reported in the Example 9.2. The EES code of Example 9.3 is given in the Code 9.3.

Code 9.3. The EEs code for the Example 9.3

```
"SGE-42HM"

U_bar_p=10.5
B=0.160
S=0.175
V_d=42.2/1000
mep=16.4*100
N=1800
N_c=12
W=mep*U_bar_p*pi*B^2/4* N_c /4
```

Example 9.4. The gas engine SGE-56HM is a V16 engine type with 16 cylinders on a V frame with 8 cylinders on each side manufactured by Siemens that produces 1350 kW of power. The engine bore size is 160 mm with the stroke length of 175 mm. Determine the mean effective pressure and mean speed of piston when the engine speed is 1800 RPM.

Solution. It is supposed to determine the mep of engine and the piston mean speed with the given characteristics as below:

$$N_c = 16, \ B = 0.16 \ m, \ S = 0.175 \ m, \ N = 1800 \ RPM, \ \dot{W} = 1350 \ kW$$

Hence:

$$\overline{U}_p = \frac{(2S)N}{60} = \frac{2 \times 0.175 \times 1800}{60} = 10.5 \frac{m}{s}$$

$$\dot{W} = \text{mep}\frac{\overline{U}_p A_p}{4} N_c \rightarrow 1350 = \text{mep} \times 100 \frac{10.5 \times \pi \times \frac{0.16^2}{4}}{4} 16 \rightarrow \text{mep}$$

$$= 1599 \ kPa \cong 16 \ Bars$$

The calculated values for $\overline{U}_p$ and mep are exactly the same as the values reported in the engine catalog published by the manufacturer [4]. The EES code of Example 9.4 is given in the Code 9.4.

Code 9.4.: The EEs code for the Example 9.3

```
"SGE-56HM"

B=0.160
S=0.175
"V_d=42.2/1000"
"mep=16.4*100"
N=1800
N_c=16
W=1350
W=mep*U_bar_p*pi*B^2/4*N_c/4
U_bar_p=N*S*2/60
```

Example 9.4. The four stroke diesel engine model of 6CM20C manufactured by the Caterpillar [6] is used for electricity generation in offshore applications (Fig. 9.12). This engine is turbocharged and aftercools the intake air after being compressed by the turbocharger. The turbocharger receives ambient air at standard conditions of 25 °C, 1 atm, and increases its pressure to the 3.4 bars as the engine charge pressure. The aftercooler cools down the turbocharger output to the engine charge temperature which is 45 °C. The intercooling is usually done by a Separate Circuit After-Cooled (SCAC) working with water as cooling medium. This engine has the following characteristics as reported in its published catalog.

$N_{@50HZ} = 1000\ RPM$, $S = 300\ mm$, $B = 200\ mm$, $V_d = 56\ L$, mep $= 24.2$ bars, and the air demand of engine is $\dot{V}_{air@25\ °C,\ 1\ atm} = 6790\ \frac{m^3}{h}$. The energy terms

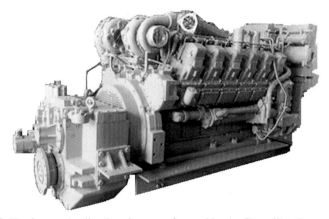

Figure 9.12 The four stroke diesel engine manufactured by the Caterpillar, Personal gallery of author.

including the net power $\dot{W}_{net}$, the heat rate losses from engine cooling ($\dot{Q}_{jacketing}$), aftercooler ($\dot{Q}_{SCAC}$), radiative and convective heat loss from the engine body ($\dot{Q}_{R\&C}$) of the engine are tabulated as below:

$\dot{W}_{net}$	$\dot{Q}_{R\&C}$	$\dot{Q}_{jacketing}$	$\dot{Q}_{oil\ cooler}$	$\dot{Q}_{SCAC}$	E_{net}	$\dot{Q}_{other\ losses}$	$\dot{Q}_{exhasut}$
1140 kW	52 kW	174 kW	158 kW	464 kW	1094 kWe	?	?

The engine is fueled with diesel oil, heavy fuel oil, and crude oil, and its specific oil fuel consumption (SFC) is 190 $\frac{g}{kWh}$. The heating value of the fuel is 42,700 $\frac{kJ}{kg}$, and the exhaust mass flow is $\dot{m}_{exhasut} = 8395 \frac{kg}{h}$. In addition, the firing pressure of the engine is 185 bars, and exhaust temperature is also 345 °C Determine:

A. The other losses and the exhaust heat loss.
B. The air to fuel mass flow rate ratio
C. Thermal and electrical efficiencies of the engine.
D. Mean speed of piston
E. Generator efficiency and the heat loss in this component.
F. Number of cylinders of the engine
G. The state points properties of the cycle and draw the $T - s$ and $P - v$.
H. The exergy efficiency and the exergy destruction of the engine.

Solution.   A:

To solve the problem, an energy flow diagram should be drawn as Fig. 9.13. So the energy balance of the engine is written as below:

$$\dot{Q}_{Fuel} = \dot{W}_{net} + \dot{Q}_{jacketing} + \dot{Q}_{R\&C} + \dot{Q}_{SCAC} + \dot{Q}_{exhasut} + \dot{Q}_{oil\ cooler} + \dot{Q}_{other\ losses}$$

The exhaust heat loss can be written as below:

$$\dot{Q}_{exhasut} = \dot{m}_{exhasut}(h_5 - h_6)$$

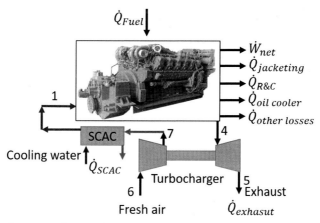

Figure 9.13 The energy diagram flow of the turbocharged after-cooled engine, photo from personal gallery.

In which:

$$T_6 = 25\,^{\circ}C \rightarrow h_6 = 298.6\,\frac{kJ}{kgK}$$

$$T_5 = 345\,^{\circ}C \rightarrow h_5 = 626.2\,\frac{kJ}{kgK}$$

Therefore:

$$\dot{Q}_{exhasut} = \frac{8395}{3600}(626.2 - 298.6) = 763.9\,kW$$

$\dot{Q}_{Fuel}$ is not explicitly given. Therefore, it must be calculated according to the given data.

$$\dot{Q}_{Fuel} = \dot{m}_{fuel}HV\eta_C$$

In which, $\eta_C = 1$ is assumed for the combustion efficiency. By using the SFC which is given, the $\dot{m}_{fuel}$ can be estimated as below:

$$SFC = \frac{m_{fuel}(kg)}{Work(kWh)} = 3600\left(\frac{s}{h}\right)\frac{\dot{m}_{fuel}\left(\frac{kg}{s}\right)}{\dot{W}_{net}\,(kW)} \rightarrow \dot{m}_{fuel} = 0.19 \times \frac{1140}{3600}$$

$$= 0.06017\,\frac{kg}{s} \rightarrow \dot{Q}_{Fuel} = 0.06017 \times 42700 \times 1 \rightarrow \dot{Q}_{Fuel} = 2569\,kW$$

From the energy balance equation:

$$2569 = 1140 + 174 + 52 + 464 + 549.7 + 158 + \dot{Q}_{other\;losses}$$

$$\rightarrow \dot{Q}_{other\;losses} = 31.38\,kW$$

B:

$$AF = \frac{\dot{m}_{exhasut} - \dot{m}_{fuel}}{\dot{m}_{fuel}} = \frac{\dfrac{8395}{3600} - 0.06017}{0.06017} = 37.76$$

Which is in the range previously discussed.

C:

$$\eta_{th} = \frac{\dot{W}_{net}}{\dot{Q}_{Fuel}} = \frac{1140}{2569} = 44.37\,\%$$

$$\eta_e = \frac{E_{net}}{\dot{Q}_{Fuel}} = \frac{1094}{2569} = 42.58\ \%$$

D:

$$\overline{U}_p = \frac{(2S)N}{60} = \frac{2 \times 0.3 \times 1000}{60} = 10\frac{m}{s}$$

E:

The generator efficiency is determined according to the net power and net electricity generated by the engine as below:

$$\eta_{generator} = \frac{E_{net}}{\dot{W}_{net}} = \frac{1094}{1140} = 95.96\ \%$$

The energy balance of the generator is as below:

$$\dot{W}_{net} = E_{net} + \dot{Q}_{generator} \rightarrow \dot{Q}_{generator} = 1140 - 1094 = 46\ kW$$

$$\rightarrow \dot{Q}_{generator} = 46\ kW$$

F:

The number of cylinders can be calculated according to the relation between the engine power and its characteristics is as below:

$$\dot{W}_{net} = mep\frac{\overline{U}_p A_p}{4}N_c \rightarrow 1140 = 24.2 \times 100 \times \frac{10 \times \frac{\pi \times 0.2^2}{4}}{4}N_c \rightarrow$$

$$N_c = 6$$

G:

Starting from point 1 at the engine charge:

$$T_1 = 45\ ^\circ C,\ P_1 = 340\ kPa \rightarrow v_1 = 0.2686\frac{m^3}{kg},\ s_1 = 5.413\frac{kJ}{kgK},\ h_1 = 318.7\frac{kJ}{kg}$$

Point 2:

The firing pressure is the maximum pressure of the cycle during the combustion:

$$P_2 = 18500\ kPa$$

And assuming the isentropic compression:

$$s_2 = s_1 = 5.409 \frac{kJ}{kgK}$$

Hence:

$$T_2 = 676.1\,^\circ C \;,\; v_2 = 0.01473 \frac{m^3}{kg}, \quad h_2 = 988.5 \frac{kJ}{kgK}$$

In other words, the compression ratio is:

$$r_c = \frac{v_1}{v_2} = \frac{0.2686}{0.01473} = 18.24$$

Point 4:

From the given data, the exhaust temperature is $T_5 = 345\,^\circ C$; this is the temperature at the turbocharger exit, and certainly, the inlet exhaust temperature to the turbocharger, which is T_4, must be higher. To calculate the T_4, an energy balance is written for the turbocharger and the intercooler as below:

$$\dot{Q}_{SCAC} = \dot{m}_{exhasut}(h_7 - h_1)$$

Hence, according to the atmospheric conditions:

$$464 = \frac{8395}{3600}(h_7 - 318.7) \rightarrow h_7 = 517.7 \frac{kJ}{kg}, \quad T_7 = 241\,^\circ C$$

In the turbocharger:

$$T_6 = 25\,^\circ C \rightarrow h_6 = 298.6 \frac{kJ}{kg}$$

according to the power transfer from the turbine to the compressor in the turbocharger:

$$\dot{m}_{exhasut}(h_7 - h_6) = \dot{m}_{exhasut}(h_4 - h_5)$$

Therefore:

$$T_5 = 345\,^\circ C \rightarrow h_5 = h_{exhaust} = 626.2 \frac{kJ}{kg}$$

$$517.7 - 298.6 = h_4 - 626.2 \rightarrow h_4 = 845.3 \frac{kJ}{kg} \rightarrow$$

$T_4 = 548\,^{\circ}C$

And from the constant volume exhaust process: $v_4 = v_1 = 0.2686\,\frac{m^3}{kg}$; therefore:

$$P_4 = 877.5\,kPa,\ s_4 = 6.122\frac{kJ}{kgK}$$

Point 3:

$$P_3 = P_2 = 18500\,kPa,\ s_3 = s_4 = 6.126\frac{kJ}{kgK}$$

$$\rightarrow h_3 = 1921\frac{kJ}{kgK} \rightarrow T_3 = 1461\,^{\circ}C,\ v_3 = 0.0269\frac{m^3}{kg}$$

According to the state point properties, the T-s and P-v diagrams are drawn as Fig. 9.14.

H:

To determine the exergy efficiency of the engine, the average heating temperature, T_{H-mean}, must be determined as below:

$$T_{H-mean} = \frac{h_3 - h_2}{s_3 - s_2} = \frac{1921 - 988.5}{6.126 - 5.413} = 1308.15\,K = 1035\,^{\circ}C$$

And the sink temperature, which is the ambient temperature, is:

$$T_L = T_6 = 25\,^{\circ}C = 298.15\,K$$

Therefore, the Carnot cycle efficiency can be determined:

$$\eta_{carnot} = 1 - \frac{T_L}{T_{H-mean}} = 1 - \frac{298.15}{1308.15} = 77.21\%$$

And the exergy efficiency or the second law efficiency is:

$$\eta_{II} = \frac{\eta_{th}}{\eta_{carnot}} = \frac{44.37}{77.21} = 57.47\%$$

The exergy destruction:

$$\underbrace{\left(\dot{X}_{in} - \dot{X}_{out} \right)}_{net\ rate\ of\ entropy\ transfer} - \dot{X}_{destroyed} = \frac{dX_{system}}{dt}$$

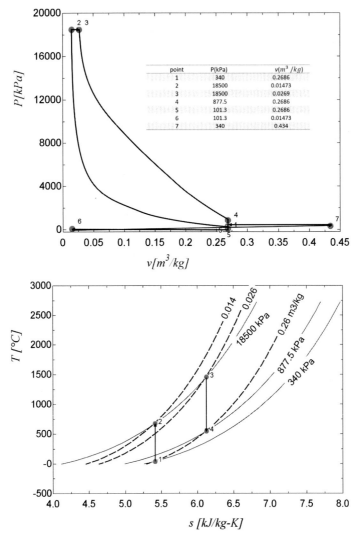

Figure 9.14 The T-s and P-v diagrams of the four stroke Diesel engine model of 6CM20C manufactured by the Caterpillar.

for steady-state operation $\frac{dX_{system}}{dt} = 0$, therefore:

$$\dot{X}_{destroyed} = \dot{X}_{in} - \dot{X}_{out} = \dot{Q}_{Fuel}\left(1 - \frac{T_L}{T_{H-mean}}\right) - \dot{W}_{net}$$

$$= 2569\left(1 - \frac{298.15}{1308.15}\right) - 1140 = 843.5 \; kW$$

The exergy destruction can also be determined as below:

$$\dot{X}_{destroyed} = \dot{Q}_{fuel}(\eta_{carnot} - \eta_{th}) = 2569(0.7721 - 0.4437) = 843.5 \, kW$$

The code generated for this example is given in Code 9.5. Students are advised to use the code and evaluate parametric studies. For example, it can be used to investigate the impact of compression ratio on the engine performance.

Code 9.5. The code used to simulate the Diesel engine of Example 9.5 for the 6CM25E

```
"6CM25E"
W_net=1140
E_net=1094
mep=24.2*100
SFC=190 "g/kWh"
V_d=56/1000 "m^3"
HV=42700 "kJ/kg"
N=1000
S=0.3
B=0.2
Q_jacketing=174
Q_R&C=52
Q_SCAC=464
Q_oil_cooler=158
m_exhasut=8395/3600
T[1]=45
P[1]=3.4*100
eta_c=1
T[5]=345
T[6]=25
P[6]=101.325
Q_SCAC=m_exhasut*(h[7]-h[1])
T[7]=temperature(air, h=h[7])
P[7]=P[1]
m_exhasut*(h[7]-h[6])=m_exhasut*(h[4]-h[5])
T[4]=temperature(air, h=h[4])
h[5]=enthalpy(air, T=T[5])
u[1]=intenergy(air, T=T[1])
h[1]=enthalpy(air, T=T[1])
u[4]=intenergy(air, T=T[4])
u[6]=intenergy(air, T=T[6])
h[6]=enthalpy(air, T=T[6])
u[5]=intenergy(air, T=T[5])
eta_th=W_net/Q_Fuel
eta_e=E_net/Q_Fuel
Q_Fuel=m_fuel*HV*eta_c
SFC=1000*3600*m_fuel/W_net

Q_exhasut=m_exhasut*(h[5]-h[6])
Q_Fuel=W_net+Q_jacketing+Q_R&C+Q_SCAC+Q_exhasut+
Q_other_losses+Q_oil_cooler
AF=(m_exhasut-m_fuel)/m_fuel
U_bar_p=2*S*N/60
W_net=mep*U_bar_p*pi*B^2/4*N_c/4
eta_generator=E_net/W_net
Q_generator=W_net-E_net
v[1]=volume(air, T=T[1], P=P[1])
S[1]=entropy(air, T=T[1], P=P[1])
s[2]=s[1]
p[2]=185*100
T[2]=TEMPERATURE(AIR, P=P[2], S=S[2])
v[2]=volume(air, t=t[2], p=p[2])
r_c=v[1]/v[2]
h[2]=ENTHALPY(Air,T=T[2])
S[4]=ENTROPY(air, t=t[4], V=V[4])
v[4]=v[1]
P[4]=PRESSURE(AIR, T=T[4], v=v[4])
P[3]=P[2]
s[3]=s[4]
T[3]=temperature(air, p=p[3], s=s[3])
V[3]=VOLUME(AIR, T=T[3], P=P[3])
h[3]=enthalpy(air, T=T[3])
p[5]=p[6]
v[5]=v[1]
V[6]=V[2]
V[7]=VOLUME(AIR, T=T[7], P=P[7])
T0=273.15
T_H+T0=(h[3]-h[2])/(S[3]-S[2])
T_L=T[6]
eta_carnot=1-(T_L+T0)/(T_H+T0)
eta_II=eta_th/eta_carnot
X_destroyed=Q_fuel*(1-(T_L+T0)/(T_H+T0))-W_net
X_destroyed2=Q_fuel*(eta_carnot-eta_th)
```

3. Efficiency improvement techniques

Efficiency improvement is a never ending process. The attempts to improve the efficiency of the REs need advances in different technologies, which are utilized for power generation. For instance, metallurgical improvement, thermodynamical optimization of the cycle, adding external components such as turbocharger, supercharger, and inlet air cooler can improve the cycle performance and its thermal efficiency. Furthermore, improvements in the engine body cooling technologies, combustion methods, and lubrication systems can increase the power output and thermal efficiency. There is still a high potential for exergy destruction reduction in the reciprocating power generators that need to be explored by the manufacturers, researchers and hardworking talented students.

According to the Carnot cycle principal, the thermal efficiency can be evaluated as below:

$$\eta_{carnot} = 1 - \frac{T_L}{T_H} \tag{9.39}$$

That means if you are supposed to seek ways to increase the cycle efficiency, those techniques must result in increasing T_H or decreasing T_L or both. In the following, the efficiency improvement methods of reciprocating internal combustion engines are reviewed.

3.1 Increasing the maximum temperature

Melting point temperature of stainless steel ranges from 1371 $°C$ to 1540 $°C$ depending on the alloy type; therefore, the cylinder body must not reach this temperature. For aluminum, the melting point temperature is 660 $°C$. In addition, the combustion temperature inside the cylinder may reach 1500 $°C$ to 2000 $°C$ that is well above the melting point temperature of the cylinder body. Increasing the $T_{max} = T_3$ in the Otto or Diesel engines can be done by decreasing the air to fuel ratio or increasing fuel consumption for the same air mass flow rate. Increasing the temperature is not very hard, but keeping the engine safe due to metallurgical limitations is realy hard. At higher temperatures, it may require to improve and change the cooling jacket of the engine or change the material used in the engine body specifically piston head, cylinder head, valve components, and the cylinder. As illustrated in Fig. 9.15 increasing the T_3 will increase the mean temperature of heat addition (T_H), and as a result, the efficiency will improve. This method increases the fuel consumption ($\dot{Q}_{in}$), the heat loss ($\dot{Q}_{out}$), and the net power ($\dot{W}_{net}$). Attention must be paid that increasing the maximum temperature and keeping the expansion ratio the same as the compression ratio will result in efficiency reduction. That is due to high temperature and pressure exhaust, which would be lost. To avoid this problem and make use of the benefits of higher maximum temperature, the gas must be expanded more than it was compressed from point 1 to 2. This will result in a cycle, which has an expansion ratio

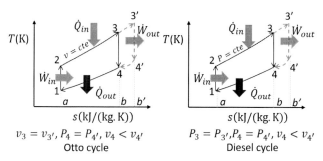

$$v_3 = v_{3'}, P_4 = P_{4'}, v_4 < v_{4'}$$
Otto cycle

$$P_3 = P_{3'}, P_4 = P_{4'}, v_4 < v_{4'}$$
Diesel cycle

Figure 9.15 Impact of maximum temperature elevation on the standard air Otto and Diesel cycles.

bigger than its compression ratio. Hence, increasing the maximum temperature can improve the efficiency if the gas could be expanded more such as $P_{4'} = P_4$ (and as a result $v_{4'} > v_4$). The bigger expansion ratio can be accomplished by late starting the compression process. Today engines operating based on the Miller cycle operate with expansion ratio bigger than compression ratio.

According to Fig. 9.15, the changes of these parameters can be written as below:

$$
\begin{aligned}
d\dot{W}_{net} &= A_{3-3'-4'-4-3} > 0 \\
d\dot{Q}_{in} &= A_{3-3'-b'-b-3} > 0 \\
d\dot{Q}_{out} &= A_{4-4'-b'-b-4} > 0
\end{aligned}
\tag{9.40}
$$

To determine the conditions that cause efficiency improvement by increasing the maximum temperature, the thermal efficiency of an Otto cycle is investigated as below:

The thermal efficiency is $\eta_{th} = 1 - \frac{\dot{Q}_{out}}{\dot{Q}_{in}}$, and it can be written as:

$$\eta_{th} = 1 - \frac{u_4 - u_1}{u_3 - u_2}$$

Under the cold air standard conditions:

$$\eta_{th} = 1 - \frac{T_4 - T_1}{T_3 - T_2} = 1 - \left(\frac{T_1}{T_2}\right)\frac{\frac{T_4}{T_1} - 1}{\frac{T_3}{T_2} - 1}$$

In an Otto cycle under the CAS assumptions and the same compression and expansion ratios:

$$\frac{T_4}{T_1} = \frac{T_3}{T_2}$$

Hence, the thermal efficiency would not change even if the maximum temperature is increased continuously.

To improve the efficiency, by increasing the maximum temperature, the following inequality must stand:

$$\frac{\dfrac{T_4}{T_1} - 1}{\dfrac{T_3}{T_2} - 1} < 1$$

Or:

$$\frac{T_4}{T_1} < \frac{T_3}{T_2}$$

Which means:

$$\frac{T_3}{T_4} > \frac{T_2}{T_1} \rightarrow r_{expansion}^{k-1} > r_{compression}^{k-1}$$

Or:

$$r_{expansion} > r_{compression} \tag{9.41}$$

Which means to increase the thermal efficiency by increasing the maximum temperature, the expansion ratio must be bigger than the compression ratio.

3.2   Higher compression ratio

Increasing the compression ratio will increase the engine efficiency. The compression ratio, $r_c = \frac{v_{BDC}}{v_{TDC}}$, which is in fact the volume ratio at BDC to the TDC can be increased by reducing v_{TDC}, or increasing v_{BDC} or both. Reducing v_{TDC} means reducing the clearance volume (V_c) that may cause the danger of touching cylinder head or valves by the piston at higher speeds. Such strike can bend the valve stems and failure of the engine valves. Therefore, depending on the piston speed, there will be a limitation for the minimum allowable space between the piston head and cylinder head at the TDC position. The clearance volume can be reduced by shaving the cylinder head or the engine block. Sometimes, a thinner gasket between the engine block and cylinder head may be used to reduce the clearance volume slightly. However, increasing the v_{BDC} means bigger stroke. The bigger stroke also will increase the piston speed for constant rotational speed of the crankshaft (N). Changing the stroke length is very challenging and makes many changes in the engine characteristics. The v_{BDC} can also be slightly increased by domed piston such as those showed in Figs. 9.7 and 9.16. The piston showed in Fig. 9.16 beside increasing the compression ratio, it also increases the turbulence inside the cylinder and improving combustion efficiency.

Figure 9.16 Triflow domed piston that increases both compression ratio and turbulence (courtesy of Siemens) [4].

Increasing compression ratio changes, the T-s diagram (Fig. 9.17), and results in changing the net power, decreasing the input heat rate and heat rate loss, and improving the thermal efficiency. Attention must be paid that the maximum temperature is kept constant. By comparing the average temperature in the process 2'-3' with 2-3, it is clear that the average temperature of heat addition (T_H) has increased due to higher compression ratio. In addition the mean temperature of heat rejection (T_L) has decreased, hence the efficiency improvement is guaranteed. However, this approach can be combined with increasing the maximum temperature to achieve higher rate of power and efficiency. According to Fig. 9.17:

$$d\dot{Q}_{in} = A_{2'-3'-b'-a-2'} - A_{2-3-b-a-2} < 0$$
$$d\dot{Q}_{out} = A_{1-4'-b'-a-1} - A_{1-4-b-a-1} < 0$$

$$(9.42)$$

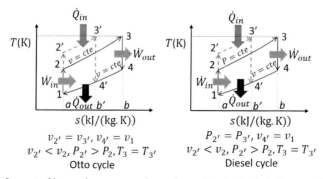

$$v_{2'} = v_{3'}, v_{4'} = v_1$$
$$v_{2'} < v_2, P_{2'} > P_2, T_3 = T_{3'}$$
Otto cycle

$$P_{2'} = P_{3'}, v_{4'} = v_1$$
$$v_{2'} < v_2, P_{2'} > P_2, T_3 = T_{3'}$$
Diesel cycle

Figure 9.17 Impact of increasing compression ratio on the standard air Otto and Diesel cycles.

3.3 Over expansion

Over expansion is required when the amount of pressure inside the cylinder at the moment before opening the discharge valve is significantly bigger than atmospheric pressure. In this case, further expansion of the combustion products can lead to the production of more work per cycle. If the expansion continues to a lower pressure, the exergy loss would reduce. The result is an increase in the thermal efficiency of the cycle. As a result, the heat loss is reduced, the fuel consumption remains unchanged, and more net power would be generated. Fig. 9.18 shows the impact of over expansion on the standard air Otto and Diesel cycles. Point 1 in the cycle is the atmospheric condition and does not change. Due to over expansion, the pressure and temperature would reduce to point $P_{4'}$ and $T_{4'}$, while the specific volume is the same as point 1. The changes in the input heat rate, heat rate loss, and net power are as below:

$$d\dot{Q}_{in} = 0$$
$$d\dot{Q}_{out} = A_{1-4'-b-a-1} - A_{1-4-b-a-1} < 0 \tag{9.43}$$
$$d\dot{W}_{net} = A_{1-4-4'-1} > 0$$

At the end of the expansion stork, when the piston is at the BDC, the pressure inside the cylinder is still considerably high (usually 3−5 atm) and is a good potential for more power generation. In addition, while the mean temperature of heat addition (T_H) is remained the same, the mean temperature of heat rejection (T_L) in the process $4'$-1 is lower than that of 4-1, hence the efficiency improvement is guaranteed. In 1985, the *Atkinson cycle*, or *over-expanded cycle*, was designed to make use of the pressure before opening the discharge valve, but failed to become commercialized mainly due to the complexity of the designed mechanisms.

The idea of the Atkinson cycle can also be implemented by reducing the compression ratio relative to the expansion ratio without adding complex mechanisms. The *Miller cycle* (Fig. 9.19), which was actually based on the idea of the Atkinson cycle, does this by closing the inlet valve before (point 7) the piston reaches the bottom dead center. Intake air in the Miller cycle is unthrottled. In this method, the pressure inside

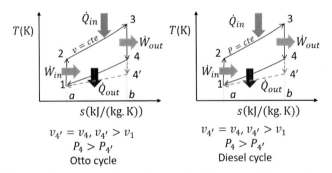

Figure 9.18 Impact of over expansion on the standard air otto and diesel cycles.

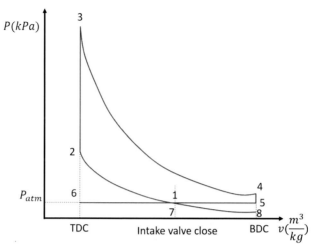

Figure 9.19 The Miller cycle for an SI engine.

the cylinder is reduced to point 8 due to the early closure of the inlet valve. In the return path (compression stroke), when the pressure starts to increase, the piston will spend some of the stroke length on increasing the reduced pressure inside the cylinder (point 8—7), and in practice, the compression stroke will decrease to point 1 to point 2 instead of point 8 to point 2.

In another design of the Miller cycle, the intake valve remains open until when the piston travels a portion of the compression stroke and then closes at point 1. In other words, the inlet valve closes a little later. Leaving the valve open causes some of the inlet air leave the cylinder and return to the intake manifold due to the upward movement of the piston from point 5 to point 1, which again reduces the compression ratio due to the late closure of the inlet valve. The compression stroke is from point 1 to point 2 again. The point at which the inlet valve is closed (point 1) plays a major role in the Miller cycle, and this point will also change at load and speed variations. For this purpose, a variable valve timing is now used. In a Miller spark ignition engine, the compression ratio may be 8:1, but the expansion ratio may be 10:1. The expansion stroke is from point 3 to point 4, which is larger than the compression stroke. Hence:

$$r_{compression} = \frac{v_1}{v_2} < \frac{v_5}{v_3} = r_{expansion} \tag{9.44}$$

In Fig. 9.19, the cycle 6-7-8-1-2-3-4-5-7-6 shows the operation of the Miller cycle in the event of an early closing of the intake valve, and the cycle 6-7-5-1-2-3-4-5-7-6 shows the Miller cycle in the event of a late closure of the intake valve.

3.4 Inlet air cooling

Inlet air temperature can alter the engine efficiency; inlet hot air will consume more power in the compression process. In addition, due to smaller density, the inlet air mass flow will reduce. The purpose of inlet air cooling is to cool air during the hot days of operation by different methods and hence to reduce compression work, and also let more air mass flow rate through the engine (Fig. 9.20).

The cooling process is a constant pressure process, and pressure loss due to standard air assumption is neglected. Due to cooling, the air density increases that means smaller specific volume and $v_{1'} < v_1$. Since the compression ratio is constant, therefore:

$$r_c = \frac{v_{1'}}{v_{2'}} = \frac{v_1}{v_2} \tag{9.45}$$

Therefore:

$$v_{2'} = \frac{v_{1'}}{v_1} v_2$$

since $\frac{v_{1'}}{v_1} < 1$ therefore $v_{2'} < v_2$. In addition, the maximum temperature is kept constant not to violate the metallurgical limitations.

Air can be cooled by passing it through an air cooler heat exchanger, which operates with cooling water. In this method, cooling water flows through finned tubes and the hot air flows around the tubes and gets cooled. For this purpose, a close circuit of water is required to cool the heated water for reusing. Hence, it needs pump, filters, cooling tower, fan, etc. This method is expensive because it has many equipment, and also the cooling potential is limited due to the low convective heat transfer coefficient on the air side. In addition, the operation costs of this system include electricity consumption by the pumps, and fans, the water consumption (which is much more for wet cooling towers), and annual maintenance costs of the cooling system. This method is nowadays used especially when the engine is turbocharged. In the turbocharged engines, air is cooled by an intercooler after the compressor and before the engine inlet. Fig. 9.21 shows an indirect air cooler heat exchanger, which uses a water close circuit for cooling purpose.

In another method which is not reported to be used in the REs, air can be cooled by a fog system or water spraying into the hot air before passing through the inlet

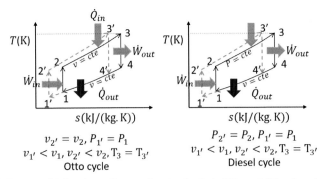

Figure 9.20 Impact of inlet air cooling on the standard air Otto and Diesel cycles.

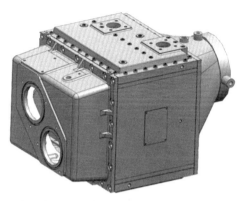

Figure 9.21 Inlet air cooler of a gas engine (Siemens courtesy) [4].

manifold. This approach has many advantages over the conventional indirect heat exchanger air coolers. For instance, it does not need cooling tower, fans, lengthy piping, heat exchanger, fan, etc. It only needs a demineralized water source, a pump, some nozzles, and a mist eliminator. Applying this method to the REs needs further researches. A schematic of this system is shown in Fig. 9.22.

3.5 Supercharging and turbocharging

Turbocharger is a small turbocompressor with radial impeller that is fed with the exhaust gases. It has no combustion chamber. It recovers a portion of the potential and thermal energy of the exhaust gases and use it in the compressor section to increase the inlet flow pressure to the engine.

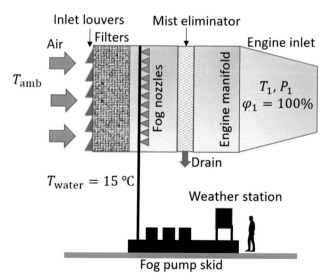

Figure 9.22 The fog system for inlet air cooling in the reciprocating engines.

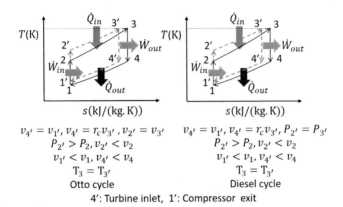

$v_{4'} = v_{1'}, v_{4'} = r_c v_{3'}, v_{2'} = v_{3'}$
$P_{2'} > P_2, v_{2'} < v_2$
$v_{1'} < v_1, v_{4'} < v_4$
$T_3 = T_{3'}$
Otto cycle

$v_{4'} = v_{1'}, v_{4'} = r_c v_{3'}, P_{2'} = P_{3'}$
$P_{2'} > P_2, v_{2'} < v_2$
$v_{1'} < v_1, v_{4'} < v_4$
$T_3 = T_{3'}$
Diesel cycle

4': Turbine inlet, 1': Compressor exit

Figure 9.23 Impact of turbocharging on the standard air Otto and Diesel cycles.

A supercharger also increases the inlet air pressure to the engine, but it uses the crankshaft power instead of exhaust power recovery. The compressor in this case can be radial type, lobe type, rotary vane type, etc. Fig. 9.23 shows the impact of a supercharger on the standard air Otto and Diesel cycles.

Due to turbocharger compression, the inlet air density and temperature increase that means specific volume decreases and $v_{1'} < v_1$. Since the compression ratio is kept constant, therefore:

$$r_c = \frac{v_{1'}}{v_{2'}} = \frac{v_1}{v_2} \tag{9.46}$$

Therefore:

$$v_{2'} = \frac{v_{1'}}{v_1} v_2$$

since $\frac{v_{1'}}{v_1} < 1$ therefore $v_{2'} < v_2$. In addition, the maximum temperature is kept constant not to violate the metallurgical limitations. The temperature rise due to compression in the turbocharger increases the compression power. Using an intercooler between the engine and compressor outlet can improve the turbocharger benefits.

3.6 Turbocharging and intercooling

Due to the air compression by the turbocharger, its temperature also increases. The higher the inlet temperature, the bigger the compression power consumption. To reduce the compression work, air should be cooled after the compressor and before the engine inlet. This air cooler is called intercooler. The simultaneous impact of intercooler and turbocharger on the standard air Otto and Diesel cycles is shown in Fig. 9.24.

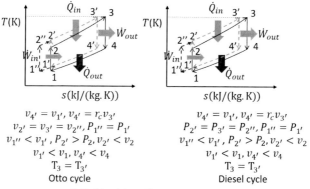

$$v_{4'} = v_{1'}, v_{4'} = r_c v_{3'}$$
$$v_{2'} = v_{3'} = v_{2''}, P_{1''} = P_{1'}$$
$$v_{1''} < v_{1'}, P_{2'} > P_2, v_{2'} < v_2$$
$$v_{1'} < v_1, v_{4'} < v_4$$
$$T_3 = T_{3'}$$
Otto cycle

$$v_{4'} = v_{1'}, v_{4'} = r_c v_{3'}$$
$$P_{2'} = P_{3'} = P_{2''}, P_{1''} = P_{1'}$$
$$v_{1''} < v_{1'}, P_{2'} > P_2, v_{2'} < v_2$$
$$v_{1'} < v_1, v_{4'} < v_4$$
$$T_3 = T_{3'}$$
Diesel cycle

4′: Turbine inlet, 1′: Compressor exit

Figure 9.24 Impact of turbocharging and intercooling on the standard air Otto and Diesel cycles.

3.7 Dual cycle

Dual cycle, which is shown in Fig. 9.25, is a combination of the standard air Otto and Diesel cycles. The purpose is to add the advantages of the Otto cycle to the Diesel cycle. In this cycle, the heat addition process is done partially at constant volume and partially constant pressure. This can be done by early fuel injection in the compression process; due to higher compression ratio of the Diesel engine, as the pressure increases, the temperature increases as well and fuel starts burning at the end of compression process. The pressure is further increased due to the combustion at constant volume at TDC until pressure reaches its maximum value at point x, since the fuel is not completely burnt yet, the piston pushes back and combustion continues at constant pressure until the fuel is fully burnt at point 3. After that combustion products expands to the point 4, generates power, and then exhausted.

When it comes to compare the dual cycle with the Otto and Diesel cycles, the best way for comparing them is to compare their thermal efficiency under the same input conditions at point 1 and also the same maximum operating design temperature and pressure. Under such conditions, the $T - s$ and $P - v$ diagrams of these three cycles are drawn as Fig. 9.26.

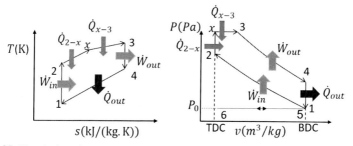

Figure 9.25 The dual cycle, a combination of the Otto and Diesel cycles.

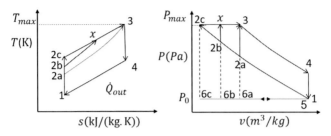

Otto:1-2a-3-4-1, Diesel:1-2c-x-3-4-1, Dual:1-2b-x-3-4-1

Figure 9.26 Comparison of otto, diesel, and dual cycles under the same input conditions and maximum operating design temperature and pressure.

Under such conditions, the thermal efficiencies would be as below:

$$\eta_{Diesel} > \eta_{Dual} > \eta_{Otto} \tag{9.47}$$

Example 9.6. In an Otto cycle with compression ratio of 12, air and fuel mixture is combusted and the maximum temperature of combustion products is 1500 °C. After expansion, combustion gases is exhausted to the ambient. Inlet air conditions are $T_1 = 25\ °C$ and $P_1 = 100\ kPa$. Determine:

a. The impact of changing compression ratio from 8 to 14 on the cycle efficiency
b. The impact of maximum operating temperature when it changes from 1200 °C to 1700 °C for the compression ratio of 12.
c. If the engine could expand the combustion gases to the pressure of $P_4 = 264.9\ kPa$ what would be the impact of increasing maximum temperature from 1200 °C to 1700 °C.

Use standard air assumptions.

Solution.   a:

To determine the cycle efficiency, the problem will be solved for the compression ratio of 12 and maximum temperature of 1500 °C, then by using the parametric table analysis in the EES software the impact of changing compression ratio and maximum temperature would be studied. For this purpose, the state point properties would be determined as below:

Point 1:

$$T_1 = 25\ °C,\ \ P_1 = 100\ kPa \rightarrow s_1 = 5.699 \frac{kJ}{kgK},\ v_1 = 0.8558 \frac{m^3}{kg},\ u_1 = 213 \frac{kJ}{kg}$$

Point 2:

$$s_2 = s_1 = 5.699 \frac{kJ}{kgK},\ \ v_2 = \frac{v_1}{r_c} = \frac{0.8558}{12} = 0.07131 \frac{m^3}{kg} \rightarrow T_2 = 502.2°C,\ P_2$$

$$= 3121kPa,\ \ u_2 = 572.5 \frac{kJ}{kg}$$

Point 3:

$$T_3 = 1500\,^\circ C, \quad v_3 = v_2 = 0.07131\frac{m^3}{kg} \rightarrow P_3 = 7137\ kPa, \quad u_3 = 1460\frac{kJ}{kg}, \quad s_3$$

$$= 6.427\frac{kJ}{kgK}$$

Point 4:

$$s_3 = s_4 = 6.427\frac{kJ}{kgK}, \quad \frac{v_4}{v_3} = r_c \rightarrow \frac{v_4}{0.07131} = 12 \rightarrow v_4 = 0.8558\frac{m^3}{kg} \rightarrow T_4$$

$$= 516\,^\circ C, \quad P_4 = 264.9\ kPa, \quad u_4 = 584.1\frac{kJ}{kg}$$

$$q_{in} = u_3 - u_2 = 1460 - 572.5 = 887.9\frac{kJ}{kg}$$

$$q_{out} = u_4 - u_1 = 584.1 - 213 = 371.1\frac{kJ}{kg}$$

$$w_{net} = q_{in} - q_{out} = 887.9 - 371.1 = 516.8\frac{kJ}{kg}$$

$$\eta_{th} = 1 - \frac{q_{out}}{q_{in}} = 1 - \frac{371.1}{887.9} = 0.5821$$

By using the parametric table analysis in the EES and commenting the compression ratio in the code, ("r_c") and setting the variation range for the r_c from 8 to 14 the impact of this variations on the specific fuel consumption, specific heat loss, and net specific work can be sought. The results are shown in Fig. 9.27. The figure shows that specific heat loss and fuel consumption decrease, while the net specific power increases slightly. As a result, the thermal efficiency has increased from 51.65% for the compression ratio of 8 to 60.46% for the compression ratio of 14.

b:

In order to determine the impact of bigger maximum temperature, by using the code provided for the part a, the maximum temperature (T_3) is commented in the code and its variation range is set from 1200 °C to 1700 °C. In this case, the expansion and compression ratios are equal. Due to this assumption, most of the power generation potentials are lost through the exhaust gases with high pressure and temperature as it is shown in Fig. 9.28.

As a result of exhaust high temperature and pressure, the specific net power increases slower than the input heat that results in efficiency reduction (Fig. 9.29).

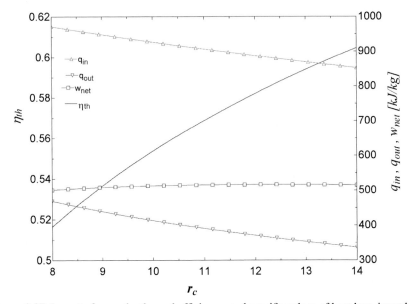

Figure 9.27 Impact of r_c on the thermal efficiency, and specific values of heat loss, input heat, and net work.

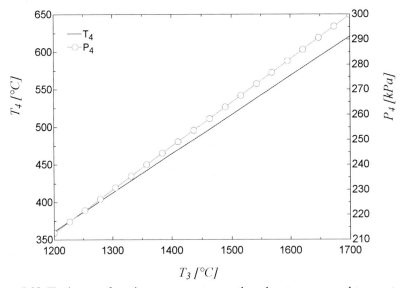

Figure 9.28 The impact of maximum temperature on the exhaust pressure and temperature.

c:

As discussed previously, to improve the thermal efficiency by increasing the maximum temperature, it is necessary to expand the exhaust gases more. In this part, the code generated for the part a and b is a little changed by assuming the

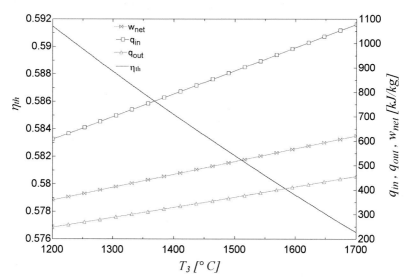

Figure 9.29 The impact of maximum temperature on the thermal efficiency, and specific values of heat loss, input heat, and net work under equal compression and expansion ratios of 12.

$P_4 = 264.9\ kPa$ constant, that means bigger expansion ratio. As a result, the exhaust temperature and pressure for example at $T_{max} = 1700\ °C$ have decreased from $621\ °C$ and $299.9\ kPa$ to $592.9\ °C$ and $264.9\ kPa$ (Fig. 9.30). Due to this improvement, the expansion ratio has increased with T_3, and accordingly the efficiency would increase as well. As it can be seen in Fig. 9.31, the efficiency change percentage would be positive when the expansion ratio and maximum temperature are in the highlighted box.

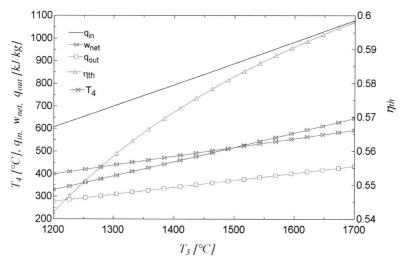

Figure 9.30 The impact of maximum temperature on the thermal efficiency, and specific values of heat loss, input heat, and net work, exhaust temperature under constant exhaust pressure of $P_4 = 264.9\ kPa$ and $r_c = 12$ assumptions.

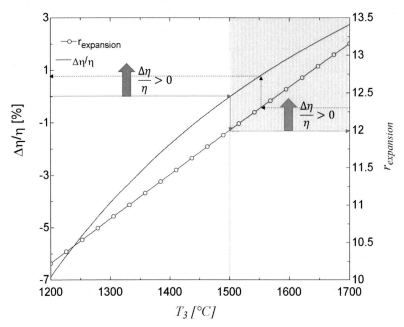

Figure 9.31 The impact of maximum temperature on the thermal efficiency change percentage and expansion ratio under constant exhaust pressure of $P_4 = 264.9\ kPa$, *and* $r_c = 12$ assumptions.

As discussed previously, late compression can result in over expansion and higher expansion ratio.

Code 9.6. the code generated for Example 9.6

```
"parts a and b"                            "Part c"
r_c=12                                     r_c=12
T[1]=25                                    T[1]=25
P[1]=100                                   P[1]=100
T[3]=1500                                  "T[3]=1500"

s[1]=entropy(air, T=T[1], P=P[1])          s[1]=entropy(air, T=T[1], P=P[1])
v[1]=volume(air, T=T[1], P=P[1])           v[1]=volume(air, T=T[1], P=P[1])
u[1]=intenergy(air, T=T[1])                u[1]=intenergy(air, T=T[1])
s[2]=s[1]                                  s[2]=s[1]
v[2]=v[1]/r_c                              v[2]=v[1]/r_c
T[2]=TEMPERATURE(Air,s=s[2],v=v[2])        T[2]=TEMPERATURE(Air,s=s[2],v=v[2])
P[2]=Pressure(air, v=v[2], T=T[2])         P[2]=Pressure(air, v=v[2], T=T[2])
u[2]=intenergy(air, T=T[2])                u[2]=intenergy(air, T=T[2])
v[3]=v[2]                                  v[3]=v[2]
P[3]=Pressure(air, v=v[3], T=T[3])         P[3]=Pressure(air, v=v[3], T=T[3])
s[3]=entropy(air, T=T[3], P=P[3])          s[3]=entropy(air, T=T[3], P=P[3])
u[3]=intenergy(air, T=T[3])                u[3]=intenergy(air, T=T[3])
S[4]=S[3]                                  S[4]=S[3]
v[4]=v[1]                                  T[4]=TEMPERATURE(Air,s=s[4],P=P[4])
T[4]=TEMPERATURE(Air,s=s[4],v=v[4])        P[4]=264.9
P[4]=Pressure(air,S=S[4], T=T[4])          v[4]=volume(air, s=s[4], P=P[4])
u[4]=intenergy(air, T=T[4])                r_e=V[4]/V[3]
q_in=(u[3]-u[2])                           u[4]=intenergy(air, T=T[4])
q_out=(u[4]-u[1])                          q_in=(u[3]-u[2])
w_net=q_in-q_out                           q_out=(u[4]-u[1])
eta=1-q_out/q_in                           eta=1-q_out/q_in
                                           w_net=q_in-q_out
                                           DELTAeta=(eta-0.5821)/0.5821*100
```

Example 9.7. The engine introduced in Example 9.6 is operating in a hot summer day with $T_1 = 50\,^\circ C$ and $P_1 = 100\,kPa$.

a. How much the efficiency will drop due to this hot temperature?
b. If a fog system is supposed to be used for inlet air cooling from $50\,^\circ C$ to $20\,^\circ C$, how much water would be consumed in the fog system?
c. Determine the impact of inlet air cooling from $50\,^\circ C$ to $20\,^\circ C$ on the thermal efficiency by a fog system.
d. How much the net power will change in percent due to using the fog system for every $5\,^\circ C$ of air cooling?

Solution. point 1:
    To determine the thermal efficiency of the cycle with the new ambient conditions, the same procedure which was taken in Example 9.6 part a would be followed as below:

$$T_1 = 50\,^\circ C, \quad P_1 = 100\,kPa \rightarrow s_1 = 5.78\,\frac{kJ}{kgK}, \quad v_1 = 0.9275\,\frac{m^3}{kg}, \quad u_1 = 231\,\frac{kJ}{kg}$$

Point 2:

$$s_2 = s_1 = 5.78\,\frac{kJ}{kgK}, \quad v_2 = \frac{v_1}{r_c} = \frac{0.9275}{12} = 0.07729\,\frac{m^3}{kg} \rightarrow T_2 = 559.5\,^\circ C, \quad P_2$$

$$= 3092\,kPa, \quad u_2 = 619.1\,\frac{kJ}{kg}$$

Point 3:

$$T_3 = 1500\,^\circ C, \quad v_3 = v_2 = 0.07729\,\frac{m^3}{kg} \rightarrow P_3 = 6584\,kPa, \quad u_3 = 1460\,\frac{kJ}{kg}, \quad s_3$$

$$= 6.45\,\frac{kJ}{kgK}$$

Point 4:

$$s_3 = s_4 = 6.45\,\frac{kJ}{kgK}, \quad \frac{v_4}{v_3} = r_c \rightarrow \frac{v_4}{0.07729} = 12 \rightarrow v_4 = 0.9275\,\frac{m^3}{kg} \rightarrow T_4$$

$$= 516.6\,^\circ C, \quad P_4 = 244.4\,kPa, \quad u_4 = 584.1\,\frac{kJ}{kg}$$

$$q_{in} = u_3 - u_2 = 1460 - 619.1 = 841.3\,\frac{kJ}{kg}$$

$$q_{out} = u_4 - u_1 = 584.1 - 231 = 353.1\,\frac{kJ}{kg}$$

$$w_{net} = q_{in} - q_{out} = 841.3 - 353.1 = 488.2 \frac{kJ}{kg}$$

$$\eta_{th} = 1 - \frac{q_{out}}{q_{in}} = 1 - \frac{353.1}{841.3} = 0.5803$$

In comparison with the Example 9.6 part a, the thermal efficiency has dropped from 58.21% to 58.03% that means 0.18% of efficiency reduction for 25 °C of warmer temperature, which comes to 0.036% efficiency drop for every 5 °C of warmer air. In addition, the net specific work has reduced from 516.8 $\frac{kJ}{kg}$ to 488.2 $\frac{kJ}{kg}$, which corresponds to 28.6 $\frac{kJ}{kg}$ of specific work loss. In other words, it is equivalent to 5.5% of loss for 25 °C of ambient air temperature increase that means 1.1% of specific work loss for every 5 °C.

b

To determine the water consumption rate, the specific humidity at the engine inlet should be determined. To reach the temperature of 20 °C the air must be saturated with the fog system operating with 20 °C demineralized water. Hence:

$$T_1 = T_{fog} = 20 \,°C \rightarrow P_{sat,1} = 2.339 \; kPa \rightarrow w_1 = \frac{0.622 \phi_1 P_{sat,1}}{P_1 - \phi_1 P_{sat,1}} \rightarrow$$

$$w_1 = \frac{0.622 \times 1 \times 2.339}{100 - 1 \times 2.339} = 0.0149 \; \frac{kg \; water}{kg \; dry \; air}$$

$$w_1 = \frac{\dot{m}_{fog}}{\dot{m}_{air}} \rightarrow \dot{m}_{fog} = 0.01065 \times 1 = 0.01065 \frac{kg \; water}{s}$$

Hence:

$$\dot{m}_{fog} = 0.01065 \frac{kg \; water}{s}$$

That means for every 1 kg of dry air, 0.01,065 kg of water would be added.

c:

point 1:
To determine the thermal efficiency of the cycle with the new ambient conditions, the same procedure which was taken in part a would be followed as below:

$$T_1 = 20 \,°C, \; P_1 = 100 \; kPa \rightarrow s_1 = 5.682 \frac{kJ}{kgK}, \; v_1 = 0.8414 \frac{m^3}{kg}, \; u_1 = 209.4 \frac{kJ}{kg}$$

Point 2:

$$s_2 = s_1 = 5.682 \frac{kJ}{kgK}, \quad v_2 = \frac{v_1}{r_c} = \frac{0.8414}{12} = 0.07012 \frac{m^3}{kg} \rightarrow T_2 = 490.6\,^\circ C, \quad P_2$$

$$= 3126\,kPa, \quad u_2 = 563.1 \frac{kJ}{kg}$$

Point 3:

$$T_3 = 1500\,^\circ C, \quad v_3 = v_2 = 0.07012 \frac{m^3}{kg} \rightarrow P_3 = 7258\,kPa, \quad u_3 = 1460 \frac{kJ}{kg}, \quad s_3$$

$$= 6.422 \frac{kJ}{kgK}$$

Point 4:

$$s_3 = s_4 = 6.422 \frac{kJ}{kgK}, \quad \frac{v_4}{v_3} = r_c \rightarrow \frac{v_4}{0.07012} = 12 \rightarrow v_4 = 0.8414 \frac{m^3}{kg} \rightarrow T_4$$

$$= 516.6\,^\circ C, \quad P_4 = 269.4\,kPa, \quad u_4 = 584.1 \frac{kJ}{kg}$$

$$q_{in} = \left(1 + m_{fog}\right)(u_3 - u_2) = (1 + 0.01065)(1460 - 563.1) = 910.7 \frac{kJ}{kg}$$

$$q_{out} = \left(1 + m_{fog}\right)(u_4 - u_1) = (1 + 0.01065)(584.1 - 209.4) = 380.3 \frac{kJ}{kg}$$

$$w_{net} = q_{in} - q_{out} = 910.7 - 380.3 = 530.4 \frac{kJ}{kg}$$

$$\eta_{th} = 1 - \frac{q_{out}}{q_{in}} = 1 - \frac{380.3}{910.7} = 0.5824$$

The efficiency at ambient temperature of 25 $^\circ C$ was 0.5821 which has increased to 0.5824 at 20 $^\circ C$ that corresponds to 0.03% of efficiency improvement for 5 $^\circ C$ of inlet air cooling.

d:

According to the Example 9.6, the net specific work generated at 25 $^\circ C$ was 516.8 $\frac{kJ}{kg}$, and due to fog cooling, its temperature dropped to 20 $^\circ C$ and the net specific work increased to 530.4 $\frac{kJ}{kg}$. This shows about 2.63% work improvement per 5 $^\circ C$ temperature reduction by using the fog system.

Simultaneous impact of air humidifying and cooling resulted in more power improvement than using the heat exchanger air coolers alone. It might be interesting that the impact of cooling was 1.1% in the work improvement, but the share of the added fog was 1.53% improvement per 5 °C temperature reduction. Hence, it can be concluded that for every 5 °C inlet air cooling:

$$\left(\frac{\Delta w}{w}\right)_{heat\ exchanger} \cong 1\%$$

$$\left(\frac{\Delta w}{w}\right)_{fog\ system} \cong 2.5\%$$

The code generated for the part a was exactly the same as the code written for the Example 9.6, part a. Only the ambient temperature was adjusted according to the part a of the present example.

For the part b, c, and d, the code was more developed accordingly that is reported in the Code 9.7.

Code 9.7. The code written for Example 9.7 parts b, c, and d

```
"parts  b, c, d"
r_c=12
T_amb=50
P_amb=100
T[1]=25
P_sat[1]=P_SAT(Water,T=T[1])
RH[1]=1
w[1]=0.622*RH[1]*P_sat[1]/(P[1]-RH[1]*P_sat[1])
m_fog=w[1]*1
P[1]=100
T[3]=1500

s[1]=entropy(air, T=T[1], P=P[1])
v[1]=volume(air, T=T[1], P=P[1])
u[1]=intenergy(air, T=T[1])
s[2]=s[1]
v[2]=v[1]/r_c
T[2]=TEMPERATURE(Air,s=s[2],v=v[2])
P[2]=Pressure(air, v=v[2], T=T[2])
u[2]=intenergy(air, T=T[2])
v[3]=v[2]
P[3]=Pressure(air, v=v[3], T=T[3])
s[3]=entropy(air, T=T[3], P=P[3])
u[3]=intenergy(air, T=T[3])
S[4]=S[3]
```

4. Stirling engine

Stirling engine is an external combustion RE, which can be fed by various types of fuels including gaseous, solid, liquid, and renewable energies such as solar and geothermal. It operates quietly and produces less air pollution. Stirling engine is classified in three main configurations of α type, β type, and γ type and shown in Figs. 9.32−9.34 [1]. In the α type configuration, the engine has two separate pistons and cylinders. The β type engine has a piston and a displacer in one single cylinder, and in the γ type, a piston and a displacer reciprocate in separate cylinders. In all of the three types, there are cooler and heater to contract and expand the working gas. The working gas may be air, nitrogen, helium, hydrogen, etc. As it shows the main components are cylinders, pistons, cooler, heater, crankshaft, gearbox, and generator .

The principal of operation of α type stirling engine can be described as below:

1. The gas is heated by the heater, which results in gas expansion to the maximum volume, and pressure reduction (to maintain the maximum temperature constant). This gas expansion pushes the expansion piston back to rotate the crankshaft.
2. As the crankshaft rotates, it pushes the expansion piston forward and pushing the gas into the regenerator, transferring its thermal energy to the regenerator to be used later.
3. Then, gas is cooled in the cooler to lose its thermal energy and make compression easier and consume less compression power. As the expansion piston moves forward, the compression piston moves back to maintain the available volume for the gas constant. At this stage, the crankshaft mechanism rotates and pushes the compression piston forward to compress the gas and pushing it back to the regenerator.

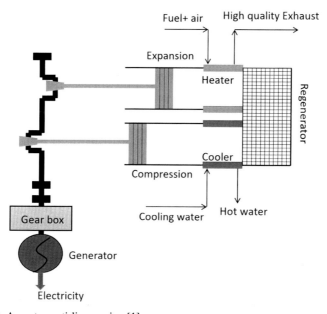

Figure 9.32 An α type stirling engine [1].

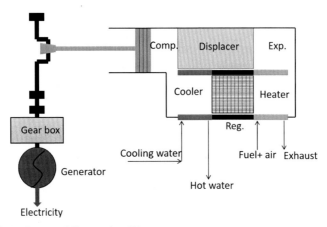

Figure 9.33 An β type stirling engine [1].

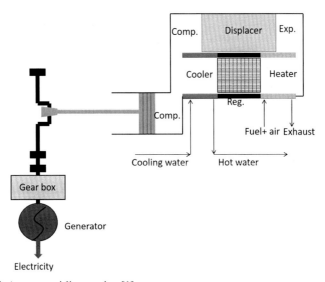

Figure 9.34 A γ type stirling engine [1].

4. As the gas flows through the regenerator, it captures the thermal energy stored in the regenerator in the step 2, its temperature increases and reaches the maximum. The pressure increases as well. This high-pressure gas pushes the expansion piston back, the gas pressure reduces, and this causes the temperature to decrease as well. To avoid more temperature reduction, gas must be heated to maintain the exit temperature from the regenerator. This completes the cycle and next cycle starts and ends the same.

As it can be seen, the Stirling engine described above can be operated in a close cycle. Sealing is very crucial for this engine to avoid leakage of working fluid to outside. However, it has no valve such as those can be found in the internal combustion engines. This will result in much less down time and maintenance costs. In addition, due to external combustion, no combustion product touches the internal parts of engine such piston and sealing system. This also helps to have a clean process inside the engine, and pressurized gases from the engine will not be exhausted into the ambient. This results in less corrosion, oxidation, and energy loss .

The $T - s$ and $P - v$ diagram of the reversible Stirling engine are shown in Fig. 9.35. As it can be seen, it receives heat and rejects it at constant temperatures of T_H and T_L. The thermal efficiency of the reversible Stirling engine is the same as the Carnot cycle and is equal to:

$$\eta_{Stirling} = 1 - \frac{T_H}{T_L} \tag{9.48}$$

The four reversible processes that make the Stirling cycle are as below:

1−2 constant temperature heat addition expansion from an external heat source.

2−3 constant volume internal heat transfer from the working fluid to the regenerator.

3−4 constant temperature compression heat rejection to the external sink.

4−1 constant volume internal heat transfer from the regenerator back to the working fluid.

5. Problems

1. Prove that for the Otto cycle, under CAS conditions, the thermal efficiency is only a function of compression ratio as below:

$$\eta_{th} = 1 - \frac{1}{r_c^{k-1}}$$

2. Prove that for the Diesel cycle, under CAS conditions, the thermal efficiency can be determined as below:

$$\eta_{Diesel-CAS} = 1 - \frac{1}{r_c^{k-1}} \left[\frac{\beta^k - 1}{k(\beta - 1)} \right]$$

3. Prove that the instantaneous speed of piston to the average piston ratio can be determined by using the following equation.

$$\frac{U_p}{\overline{U}_p} = \frac{\pi}{2} \sin \theta \left[1 + \frac{\cos \theta}{\sqrt{R - \sin^2 \theta}} \right]$$

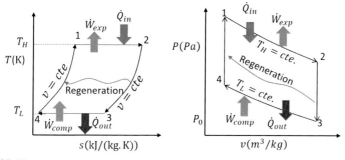

Figure 9.35 The P-v and T-s diagram of the stirling engine.

4. Plot the speed ratio $\frac{U_p}{U_p}$ vs. $0 \leq \theta \leq 180\,°$ for different values of $R = 4, 6, 8,$ and 10.

5. In an Otto cycle with compression ratio of 10, air and fuel mixture is combusted to the maximum temperature of 2500 $°C$ and then is exhausted to ambient. Determine the input heat, heat loss, and thermal efficiency by considering, (a) cold air standard assumptions and (b) air standard assumptions The inlet air is in standard atmospheric condition.

6. In problem 5, by using EES program, determine the impact of maximum temperature on the cycle efficiency when it changes from 2000 $°C$ to 3000 $°C$ under (a) cold air standard assumptions and (b) air standard assumptions. The inlet air is in standard atmospheric condition.

7. In problem 5, by using EES program, determine the impact of compression ratio on the cycle efficiency when it changes from 4 to 12 under (a) cold air standard assumptions and (b) air standard assumptions. The inlet air is in standard atmospheric condition.

8. Determine the impact of ambient air temperature when it changes for 20 to -20 $°C$ at $P_{atm} = 101.325\ kPa$ on the results of problem 5 under (a) cold air standard assumptions and (b) air standard assumptions.

9. Determine the impact of ambient pressure changes from 33.5 kPa on the Mount Everest to the pressure at the sea level 101.325 kPa on the results of problem 5 under (a) cold air standard assumptions and (b) air standard assumptions. The inlet air temperature is 25 $°C$.

10. The SGE-56HM manufactured by Siemens is a gas engine capable of working with both natural gas and biogas. The engine power output is 1350 kW, and when connected to the generator, it produces 1307 kW of electricity. The engine speed is 1800 rpm, and its weight is 7500 kg. This engine is equipped for cogeneration of heat and power by heat recovery from the exhaust, engine jacket, intercooler, and lube oil cooling. The compression ratio of the engine is 11.9:1. The engine is turbocharged, and the exhaust temperature after leaving the turbocharger is 470 $°C$ and then is cooled to 120 $°C$ in the exhaust heat recovery system to achieve 734 kW of heat. Ambient pressure and temperature are 100 kPa and 25 $°C$. The air temperature after being compressed in the turbocharger increases that will be cooled to 36.5 $°C$ by the intercooler at the engine inlet. Assume the combustion efficiency is 100%. The combustion air and the exhaust mass flow rates are 6330 and 5650 kg/h. The bore size, and the stroke length of the engine are 160 and 175 mm, the cylinder arrangement is V16, with 8 cylinders on each side of the V frame. The energy balance of the engine is as below:

Fuel energy	Electricity	Jacket	Exhaust recovery	Intercooler
3168 kW	1307 kW	717 kW	734 kW	83 kW
Engine radiation	Generator heat	Exhaust loss	Other losses	Power
84 kW	?	?	?	1350 kW

Determine:

A. The thermal and electrical efficiency of the engine. Why the electrical efficiency is smaller than the thermal efficiency? how the power is lost? Determine the energy loss in the generator.

B. The exhaust loss after heat recovery and the other losses term.

C. The air to fuel ratio, and the fuel heating value.

D. By using the air standard assumptions, determine all of the state point properties and then plot the $T - s$ and $P - v$ diagrams.

E. The overall efficiency of the engine.

F. The exergy efficiency, and the exergy destruction of the engine.

11. The engine discussed in problem 10 (SGE-56HM) is a V16 engine (a V type with two groups of 8 cylinders on each side of the V frame). The piston mean speed is 10.5 m/s, the bore size is 160 mm, and the stroke length is 175 mm. The mean effective pressure is 16 bars and the engine speed 1800 RPM. Determine the net power of this engine and validate the answer with that reported in problem 10.

12. The gas engine SGE-42HM is a V12 engine type with 12 cylinders on a V frame with 6 cylinders on each side manufactured by Siemens that produces 1040 kW of power. The engine bore size is 160 mm with the stroke length of 175 mm. Determine the mean effective pressure and mean speed of piston when the engine speed is 1800 RPM.

13. The four stroke diesel engine model of 9CM20C manufactured by the Caterpillar is used for electricity generation in offshore applications. This engine is turbocharged with turbocharger type of KBB HPR5000 and aftercools the intake air after being compressed by the turbocharger. The turbocharger receives ambient air at standard conditions of 25 °C, 1 atm and increases its pressure to the 3.4 bars as the engine charge pressure. The aftercooler, cools down the turbocharger output to the engine charge temperature which is 45 °C. The intercooling is usually done by a Separate Circuit After-Cooled (SCAC) working with water as cooling medium. This engine has the following characteristics as reported in its published catalog.

$N_{@50HZ} = 1000\ RPM$, $S = 300\ mm$, $B = 200\ mm$, $V_d = 84.6\ L$, mep $= 24.2$ bars, and the air demand of engine at 20 °C is $\dot{V}_{air@25\ °C,\ 1\ atm} = 10663\ \frac{m^3}{h}$. The energy terms including the net power $\dot{W}_{net}$, the heat rate losses from engine cooling ($\dot{Q}_{jacketing}$), aftercooler ($\dot{Q}_{SCAC}$), radiative and convective heat loss from the engine body ($\dot{Q}_{R\&C}$) of the engine are tabulated as below:

$\dot{W}_{net}$	$\dot{Q}_{R\&C}$	$\dot{Q}_{jacketing}$	$\dot{Q}_{oil\ cooler}$	$\dot{Q}_{SCAC}$	E_{net}	$\dot{Q}_{other\ losses}$	$\dot{Q}_{exhasut}$
1710 kW	78 kW	261 kW	238 kW	728 kW	1641 kWe	?	?

The engine is fueled with diesel oil, heavy fuel oil, and crude oil, and its specific oil fuel consumption (SFC) is 190 $\frac{g}{kWh}$. The heating value of the fuel is 42,700 $\frac{kJ}{kg}$ and the exhaust mass flow is $\dot{m}_{exhasut} = 13180\ \frac{kg}{h}$. In addition, the firing pressure of the engine is 185 bars and exhaust temperature is also 337 °C Determine:

A. The other losses and the exhaust heat loss.
B. The air to fuel mass flow rate ratio.
C. Thermal, and electrical efficiencies of the engine.
D. Mean speed of piston.
E. Generator efficiency, and the heat loss in this component.
F. Number of cylinders of the engine.
G. The state points properties of the cycle and draw the $T - s$ and $P - v$.
H. The exergy efficiency, and the exergy destruction of the engine.

14. Prove that under cold air standard conditions the thermal efficiency of a dual cycle would be as be below:

$$\eta_{Dual-CAS} = 1 - \frac{1}{r_c^{k-1}}\left[\frac{\alpha\beta^k - 1}{k\alpha(\beta - 1) + \alpha - 1}\right]$$

In which:

$$r_c = compression\ ratio = \frac{v_1}{v_2} = \frac{v_1}{v_x} = \frac{v_4}{v_x}$$

$$\alpha = pressure\ ratio = \frac{P_x}{P_2} = \frac{P_3}{P_2} = \frac{T_x}{T_2} = \frac{1}{r_c^k}\frac{P_3}{P_1}$$

$$\beta = cutoff\ ratio = \frac{v_3}{v_x} = \frac{v_3}{v_2} = \frac{T_3}{T_x}$$

15. Prove that for the same inlet conditions and the same maximum operating design temperature and pressure:

$$\eta_{Diesel} > \eta_{Dual} > \eta_{Otto}$$

16. Prove that for the same inlet conditions and the same compression ratio:

$$\eta_{Otto} > \eta_{Dual} > \eta_{Diesel}$$

17. In an Otto cycle with compression ratio of 12, after compression, air and fuel mixture is combusted to the maximum temperature of 1500 $°C$ and then is exhausted to ambient. Inlet air conditions are $T_1 = 50\ °C$ and $P_1 = 85\ kPa$. Determine:
    A. How much he efficiency will drop due to this hot temperature?
    B. If a fog system is supposed to be used for inlet air cooling from 50 $°C$ to 20 $°C$, how much water would be consumed in the fog system?
    C. The impact of inlet air cooling from 50 $°C$ to 20 $°C$ on the thermal efficiency by a fog system.
    D. How much the net power will change in percent due to using fog system for every 5 $°C$ of air cooling?
18. In problem 17, if you are supposed to use a supercharger to increase the inlet pressure to the engine from 85 kPa to 150 kPa, determine:
    A. The thermal efficiency of the cycle.
    B. The exhaust temperature leaving the turbocharger.
    C. The thermal efficiency if an intercooler heat exchanger could be used to cool the inlet air to the 35 $°C$.

References

[1] M. Ebrahimi, A. Keshavarz, Combined Cooling, Heating and Power, Decision-Making, Design and Optimization, first ed., Elsevier, 2014.
[2] Willard W. Pulkrabek, Engineering Fundamentals of the Internal Combustion Engine, Prentice Hall, Upper Saddle River, New Jersey 07458.
[3] Y.A. CENGEL, M.A. Boles, Thermodynamics, an Engineering Approach, fifth ed., McGraw-Hill, 2008.
[4] www.siemens-energy.com.
[5] https://www.mhi.com/group/mtee/company, https://www.turbocharger.mtee.eu/turbo-knowledge/turbo-explanation/.
[6] https://www.cat.com/.

Solar power plants

10

Chapter outline

Solar electricity is one of the main players for reducing CO_2 to achieve **Net Zero Emissions by 2050** scenario. The capacity of solar energy to generate electricity, heating, and cooling is extraordinary high. Earth is continuously receiving 172 TW (10^{12} W) of energy from sun, which is more than 10,000 times of the world total energy demand. According to the reports of International Energy Agency (IEA) [1], to reach the **Net Zero Emissions by 2050** scenario, 6970 TWh of photovoltaic (PV) power must be installed until 2030. This means that PV farms must increase with an average annual rate of 24% during 2020−2030.

Solar energy can be directly transformed to electricity by using PV panels or indirectly by using solar thermal power (STP) plants. In the STP plants, the solar radiation is concentrated at a point or along a line to increase the temperature of a working fluid; therefore, these kinds of power plants are mostly called concentrated solar power (CSP) plant. The concentrated heat can run a steam turbine power plant (SPP) by generating

Power Generation Technologies. https://doi.org/10.1016/B978-0-323-95370-2.00010-7

Copyright © 2023 Elsevier Inc. All rights reserved.

superheated steam or run an external combustion reciprocating Stirling engine by transferring heat to its working fluid or run a gas turbine working with a supercritical CO_2 (sCO_2) gas turbine closed cycle. Solar radiation can be concentrated by using different collectors mainly parabolic troughs, parabolic dishes, and solar towers. In general, the electrical efficiency of heat engines is higher than PV systems; hence, a CSP may be more efficient than a PV for electricity generation. An important parameter that may limit the use of solar heat for steam-based power plants is the long startup time of these power plants. The large-scale steam power plants require several hours for startup and warm up before injecting steam into the turbine. Hence, there will be a loss of energy and time during the start up. To overcome this problem, and avoid daily startup and shutdown, heat storage system can be integrated with the CSP for the nights or sunless/cloudy times, or they can be integrated with auxiliary fossil fuel system for continuous operation of the steam generator during the night. The CSP is not yet well matured due to technological gaps, which can be seen in the working fluid of the receiver in the solar tower, heat exchangers used for CSP, and the thermal energy storage system. They still need more investigations and attempts to overcome these problems. The U.S. Department of Energy, Solar Energy Technologies Office (SETO) is working to reduce the investment cost of CSP from 0.91 $\frac{\$}{kWh}$ in 2020 to 0.05 $\frac{\$}{kWh}$ in 2030 for at least 12 h of thermal energy storage to make it more affordable to be used in base load operation.

In this chapter, the concepts of photovoltaic system and also different types of CSP will be introduced.

1. Photovoltaic electricity

Solid materials can be classified into insulator, semiconductor, and conductor depending on their electrical conduction property. The conduction property of materials depends on the magnitude of energy band gap (E_g) between the valence band and the conduction band with respect to the photon energy of solar radiation in visible wavelengths. The amount of energy carried by a single proton is the photon energy and is linearly proportional to the proton's electromagnetic frequency.

The photon energy can be calculated as:

$$E_{ph} = hf = h\frac{c}{\lambda} \tag{10.1}$$

In which $h = 6.62607015 \times 10^{-34} \frac{m^2\ kg}{s}$ is the Plank's constant and f is the frequency. In addition, $c = 2.998 \times 10^8 \frac{m}{s}$ is the speed of light, and λ is the wavelength.

Example 10.1. Determine the energy of a red photon with a frequency of 4.57×10^{14} Hz. How much is the wavelength?

Solution.

$$E_{ph} = hf = 6.63 \times 10^{-34} \times 4.57 \times 10^{14} = 3.03 \times 10^{-19} J$$

$$f = \frac{c}{\lambda} \rightarrow 4.57 \times 10^{14} = \frac{2.998 \times 10^8}{\lambda} \rightarrow \lambda = 0.656 \times 10^{-6} m = 656\ nm$$

For insulators, the energy band gap is very large and $E_{ph} < E_g$; thus electrons can not reach the conduction band from the valence band. That means no conduction of current. For semiconductors, the energy band gap is smaller than the photon energy $(E_{ph} > E_g)$, and as a result, the electrons can reach the conduction band when they receive thermal or light energy. For the conductors $E_g \approx 0$ and no gap exist, that means electron can move to the conduction band easily. According to the discussion, semiconductors can be used to generate electricity by using the energy of sunlight. The energy band gap changes with temperature of the material, and the material properties as below:

$$E_g = E_g(0) - \frac{aT^2}{T + b} \tag{10.2}$$

in which T is the temperature in Kelvin, a and b depend on the material. It is evident that at zero absolute temperature $(T = 0$ K$)$, materials act like insulators.

For example, for silicon (Si) $a = 7 \times 10^{-4} \left(\text{eVK}^{-1}\right)$, $b = 1100$ (K), $E_g(0) = 1.166$ (eV), for gallium arsenide (GaAs) $a = 5.8 \times 10^{-4} \left(\text{eVK}^{-1}\right)$, $b = 204$ (K), $E_g(0) = 1.519$ (eV), and for germanium (Ge) $a = 4.77 \times 10^{-4} \left(\text{eVK}^{-1}\right)$, $b = 235$ (K), $E_g(0) = 0.7437$ (eV) [2].

It is worth mentioning that 1 eV is the work required to be done to cross an electric field with voltage of 1 V by an electron with the charge of 1.6×10^{-19} coulomb. Hence, 1 eV $= 1.6 \times 10^{-19}$ (C) $\times 1$(V) $= 1.6 \times 10^{-19}$ J.

In 1839, Edmund Becquerel discovered the phenomenon of converting sunlight into electricity by a solid material, and later, scientists discovered the photovoltaic process and used this phenomenon to generate electricity in semiconductor materials such as silicon.

The conductivity of a semiconductor in the absence of impurities (or very small impurity) is called *intrinsic conductivity*. The intrinsic conductivity can be increased by adding controlled amount of specific impurity ions. Such material is called *extrinsic semiconductor*. The process of adding impurity ions to the semiconductors is called *doping*.

Depending on the valency of the impurity ions, the extrinsic semiconductor may become an n-type or p-type material. If the valency of the impurity ions is less than that of the semiconductor, it produces holes, which accept and trap free electrons. These holes are positively charged and produce current through the material. Materials having holes more than electrons are called p-type material. If the valency of the impurity ions is more than that of the semiconductor, the electrons become the majority charge carriers and holes are the minority charge carrier. Such material is known as n-type material. Fig. 10.1 shows the classification of materials according to their conduction property.

According to the discussions, heterogeneous distribution of holes (acceptors) and electrons is vital to convert sunlight energy to electricity. For this purpose, a p-n junction is made by putting an n-type semiconductor in contact with a p-type

$$
Materials \begin{cases} semiconductors \\ (E_g < E_{ph}) \\ coductor \\ (E_g \approx E_{ph}) \\ insultor \\ (E_g > E_{ph}) \end{cases} \begin{cases} intrinsic\ (n_e = n_h) \\ extrinsic \begin{cases} n-type\ (n_e > n_h) \\ p-type\ (n_e < n_h) \end{cases} \end{cases}
$$

n_e: Concentration of electrons n_p: Concentration of holes

Figure 10.1 The classification of materials according to the conductivity property.

semiconductor. The total current density in the lightless condition for a p-n junction can be evaluated as below:

$$
I_D = I_0 \left[\exp\left(\frac{eV}{kT}\right) - 1 \right] \tag{10.3}
$$

This current is also called *dark current*. In which, I_0 is the saturation current density and can be determined as below for semiconductors:

$$
I_0 = AT^3 \exp\left(-\frac{E_g}{kT}\right) \tag{10.4}
$$

where A is the nonideality factor, and it is equal to 1 when the semiconductor follows the assumptions made for deriving the diode equation (Eq. 10.3). In addition, k is the Boltzmann's constant and is equal to $1.380649 \times 10^{-23} \frac{J}{K}$.

Example 10.2. Determine the dark current density for Si at the temperature of 323K and voltage of 0.75 V. In addition, plot the total current versus the voltage range of 0.5−0.8 V. What will happen for the current when the voltage approaches zero? Assume the nonideality factor to be 1.

Solution. The calculations for the voltage of 0.75 V is given below, but for the voltage range, the solution is coded in the EES program, and a parametric table is used to determine the current density in different voltages.

First of all, the energy band gap for Si at $T = 323$ K would be determined.

$$
E_g = E_g(0) - \frac{aT^2}{T+b} = 1.166\ eV - \frac{7 \times 10^{-4}\left(eVK^{-1}\right)323^2}{323 + 1100} = 1.11\ eV
$$
$$
= 1.11 \times 1.6 \times 10^{-19}J = 1.783 \times 10^{-19}J
$$

Hence, by considering an nonideality factor of $A = 1$, the saturation current would be:

$$I_0 = AT^3 \exp\left(-\frac{E_g}{kT}\right) = 1 \times 323^3 \exp\left(-\frac{1.783 \times 10^{-19}}{1.38 \times 10^{-23} \times 323}\right)$$

$$= 1.415 \times 10^{-10} \frac{A}{m^2} = 1.415 \times 10^{-14} \frac{A}{cm^2}$$

And the dark current density for the voltage of 0.75 V would be:

$$I_D = I_0 \left[\exp\left(\frac{eV}{kT}\right) - 1\right] = 1.415 \times 10^{-14} \left[\exp\left(\frac{1.6 \times 10^{-19} \times 0.75}{1.38 \times 10^{-23} \times 323}\right) - 1\right]$$

$$= 0.006959 \frac{A}{cm^2}$$

For plotting dark current against voltage, the **Code 10.1** is written, and the results are calculated by a parametric table and depicted in Fig. 10.2, as it shows when voltage approaches zero, the dark current also becomes zero.

Code 10.1. The code generated for the Example 10.2.

```
V=0.75
T=273+50
a=7*10^(-4)    "[eV/K]"
AA=1
b=1100    "[K]"
k=1.38*10^(-23)
E_g0=1.166    "[eV]"
e=1.6*10^(-19) "[Coulomb]"
E_g=(E_g0-a*T^2/(b+T))*e    "[J]"
I_0=AA*T^3*exp(-E_g/(k*T))  "[A/m^2]"
I_D=I_0*(EXP(e*V/(k*T))-1)*0.0001   "[A/cm^2]"
```

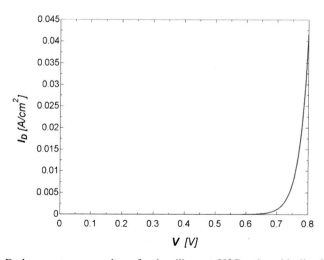

Figure 10.2 Dark current versus voltage for the silicon at 50°C and nonideality factor of 1.

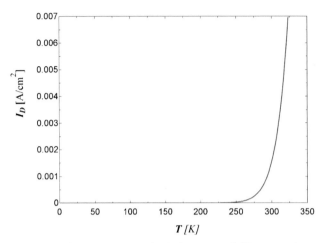

Figure 10.3 The impact of temperature on the dark current of silicon at voltage of 0.75 V.

Example 10.3. Plot the dark current versus temperature for Si in the temperature range of 1K to 323K and voltage of 0.75 V. Assume the nonideality factor to be 1.

Solution. The **Code 10.1** can be used by commenting the temperature as unknown and setting voltage at 0.75 V. By using a parametric table and changing temperature from 1 to 323K, the following plot would be resulted. As it can be seen at very low temperatures, the Si is behaving like an insulator (Fig. 10.3).

1.1   PV generator formulation

As discussed, the electronic asymmetric is vital for conversion of sunlight into electricity. In a p-n junction of semiconductors, the majority charge carriers flow in the opposite direction creates a negative charge on the p-region and a positive charge on the n-region. When the charge carriers flow, the *recombination*[1] process creates *depletion region* without mobile charge (Fig. 10.4).

Due to the solar radiation, electron hole pairs would be generated in the depletion region. The internal electric field drives these electron/hole pairs, and a photocurrent of I_L would be generated. This current is in the opposite direction of dark current. Absorption of more light creates more electron/hole pairs. This effect is called *photovoltaic effect,* and the corresponding p-n junction of semiconductor that produces this effect is called as *solar cell*. Solar cell is the smallest unit of a PV panel. Since the photocurrent is in the opposite direction of the dark current, the overall current of the solar cell would be:

---

[1] If "excess" charge carriers are created in a semiconductor, either by the absorption of light or by other means, the thermal equilibrium is disturbed, then these excess charge carriers must be annihilated after the source has been "switched off". This process is called recombination. The most important mechanisms for recombination are radiative recombination and recombination via defect levels [3].

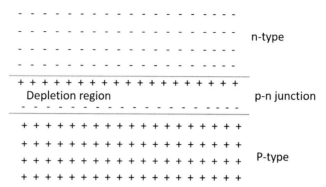

Figure 10.4 The p-n junction and depletion region.

$$I = I_L - I_D = I_L - I_0 \left[\exp\left(\frac{eV}{kT}\right) - 1 \right] \tag{10.5}$$

Different layers of a solar cell including the p-type material, n-type material, antireflection coating, metal contact, etc. produce a resistance, which is called series resistance (R_s). This reduces the effective voltage of the solar cell. Hence, the corrected current for a nonideal junction would be:

$$I = I_L - I_0 \left[\exp\left(\frac{e(V - IR_s)}{AkT}\right) - 1 \right] - \frac{(V - IR_s)}{R_p} \tag{10.6}$$

Due to improper junction or surface effect, current may find other ways to flow. This path creates *shunt resistance* of R_p to the current. The last term in the above equation is the effect of current passing through the shunt path. A is the nonideality factor of the junction, and usually is assumed 1.

The *short circuit current (I_{SC})* is the light induced current in the absence of load. In such condition, the positive and negative terminals of the solar cell are connected together. The *open-circuit voltage* is the voltage when the overall current is zero. This is the voltage of the maximum load in the circuit and can be found as below:

$$I_L - I_0 \left[\exp\left(\frac{eV}{kT}\right) - 1 \right] = 0 \rightarrow V_{OC} = \frac{kT}{e} \ln\left(\frac{I_L}{I_0} + 1\right) \tag{10.7}$$

The characteristics curve of current and power versus voltage can be drawn by using Eq. (10.6). In commercial solar cells, the shunt resistance is very large, while the series resistance is kept minimum. Hence, the overall current can be estimated by the following equation:

$$I = I_L - I_0 \left[\exp\left(\frac{e(V - IR_s)}{AkT}\right) - 1 \right] \tag{10.8}$$

Example 10.4. Draw the characteristics curves of (I−V) and (Power-V) for a solar cell which:

(a) Has an area of 1 cm^2, saturation current of $1.415 \times 10^{-14} \frac{A}{cm^2}$, light-induced current of 0.5 $\frac{A}{cm^2}$ and temperature of 300K. The series resistance is zero.
(b) Consider the series resistance to be 0.1 Ω.cm^2.
(c) Determine the maximum power for both cases in parts a and b.

Solution. The voltage changes from zero to the maximum value, which is the open-circuit voltage. Hence, the open-circuit voltage would be:

$$V_{OC} = \frac{kT}{e} \ln\left(\frac{I_L}{I_0} + 1\right) = \frac{1.38 \times 10^{-23} \times 323}{1.6 \times 10^{-19}} \ln\left(\frac{0.5}{1.415 \times 10^{-14}} + 1\right)$$
$$= 0.8072 \text{ V}$$

In addition, the short circuit current would be determined by using Eq. (10.5) at $V = 0$ as below:

$$I_{SC} = I_L = 0.5 \frac{A}{cm^2}$$

(a) By changing voltage and using Eq. (10.5), the ideal overall current would be determined. To determine the ideal power, voltage must be multiplied by the current. Hence:

$$I = I_L - I_0\left[\exp\left(\frac{eV}{AkT}\right) - 1\right] = 0.5 - 1.415$$
$$\times 10^{-14}\left[\exp\left(\frac{1.6 \times 10^{-19}V}{1 \times 1.38 \times 10^{-23} \times 323}\right) - 1\right]$$

And:

$$P = VI = V\left(0.5 - 1.415 \times 10^{-14}\left[\exp\left(\frac{1.6 \times 10^{-19}V}{1 \times 1.38 \times 10^{-23} \times 323}\right) - 1\right]\right)$$

By changing voltage from 0 to 0.8072 V, the characteristics curves would be drawn as Fig. 10.5. As it shows, there is a maximum condition for the power and the corresponding voltage and current are shown by $V_{P_{max}}$ and $I_{P_{max}}$.

(b) To plot the characteristics curves for the solar cell with a series resistance of 0.1 Ω.cm^2, Eq. (10.8) should be utilized as below:

$$I_c = I_L - I_0\left[\exp\left(\frac{e(V - IR_s)}{AkT}\right) - 1\right] = 0.5 - 1.415$$
$$\times 10^{-14}\left[\exp\left(\frac{1.6 \times 10^{-19}(V - 0.1 \times I)}{1 \times 1.38 \times 10^{-23} \times 323}\right) - 1\right]$$

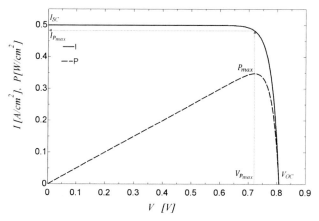

Figure 10.5 The characteristics curves for current and power vs. voltage for neglected series resistance.

In which, I_c is the corrected current due to the series resistance. The corrected power would be:

$$P_c = VI_c = V\left(0.5 - 1.415 \times 10^{-14}\left[\exp\left(\frac{1.6 \times 10^{-19}(V - 0.1 \times I)}{1 \times 1.38 \times 10^{-23} \times 323}\right) - 1\right]\right)$$

By changing V and determining I_c and P_c, the characteristics curves are presented in Fig. 10.6. The results are also compared with those in the part a. It shows that the series resistance has decreased the maximum power generated by the solar cell.

(c) According to the graphs in part b, the maximum power in the ideal conditions would be:

$$P_{max} = 0.3475 \text{ W/cm}^2$$

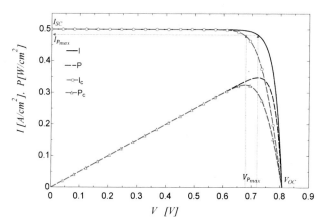

Figure 10.6 The comparison of the characteristics curves for current and power vs. voltage for two cases of with and without series resistance.

And for the real condition in the presence of series resistance:

$$P_{c, \, max} = 0.3236 \text{ W/cm}^2$$

That means the series resistance has decreased the maximum power achieved from the solar cell.

The code generated for the Example 10.4 is presented in the **Code 10.2**:

Code 10.2. The code generated for the Example 10.4.

```
"V=0.5"
T=300
AA=1
k=1.38*10^(-23)
e=1.6*10^(-19) "[Coulomb]"
I_0=1.415*10^(-14)
I_L=0.5
Rs=0.1
I=(I_L-I_0*(EXP(e*V/(AA*k*T))-1)) "[A/cm^2]"
I_C=(I_L-I_0*(EXP(e*(V+I_C*Rs)/(AA*k*T))-1)) "[A/cm^2]"
V_OC=k*T/e*ln(I_L/I_0+1)
P=I*V
P_C=I_C*V
```

As the previous example showed, there is a maximum value for the power generated by the solar cell, and the corresponding voltage and current are shown by $V_{P_{max}}$ and $I_{P_{max}}$, in which:

$$P_{max} = V_{P_{max}} \times I_{P_{max}} \tag{10.9}$$

And the optimum load resistance to get the maximum power from the solar cell would be:

$$R_{P_{max}} = \frac{V_{P_{max}}}{I_{P_{max}}} \tag{10.10}$$

The solar cell efficiency of a solar cell can be estimated as below:

$$\eta = \frac{P}{\phi} = \frac{V \times I}{I_T A} \tag{10.11}$$

In which P is the generated power by the PV system, I_T is the solar intensity, and ϕ is the solar radiation received by the PV system.

The ratio of the maximum achievable power from a PV system for a given short circuit current of (I_{SC}) and open-circuit voltage (V_{OC}) is called *fill factor (FF)*. It shows the deviation of the real solar cell from the ideal one. The deviation can be due to different faults or contact resistance and is determined as below:

$$FF = \frac{P_{max}}{V_{OC} \times I_{SC}} = \frac{V_{P_{max}} \times I_{P_{max}}}{V_{OC} \times I_{SC}} = \left(\frac{V_{P_{max}}}{V_{OC}} \right) \left(\frac{I_{P_{max}}}{I_{SC}} \right) \tag{10.12}$$

In other words:

$$P_{max} = FF \times V_{OC} \times I_{SC} \qquad (10.13)$$

And the solar cell maximum electrical efficiency (η_e) would be:

$$\eta_{max} = \frac{P_{max}}{P_{in}} = \frac{FF \times V_{OC} \times I_{SC}}{I_T A} \qquad (10.14)$$

For a silicon solar cell, the maximum value of FF is 0.88.

Example 10.5. Determine the FF for the Si solar cell in the Example 10.4.

Solution. According to the Example 10.4, the following parameters were determined for two cases of ideal cell and real cell with series resistance:
Ideal case:

$$P_{max} = 0.3475 \text{ W/cm}^2, \quad V_{OC} = 0.8072 \text{ V}, \quad I_{SC} = 0.5 \frac{\text{A}}{\text{cm}^2}$$

Hence:

$$FF = \frac{P_{max}}{V_{OC} \times I_{SC}} = \frac{0.3475}{0.8072 \times 0.5} = 0.86$$

And for the real conditions:

$$P_{c, \, max} = 0.3236 \text{ W/cm}^2$$

$$FF = \frac{P_{max}}{V_{OC} \times I_{SC}} = \frac{0.3236}{0.8072 \times 0.5} = 0.80$$

Example 10.6. Under standard test conditions of cell temperature of $T_c = 25°C$, and $I_T = 1000 \frac{\text{W}}{\text{m}^2}$ and maximum power output of 0.1 W determine the cell maximum electrical efficiency. The PV aperture area is 8 cm^2.

Solution. (a)

$$\eta_{max} = \frac{P_{max}}{P_{in}} = \frac{0.1}{1000 \times 8 \times 10^{-4}} = \frac{1}{8} = 12.5\%$$

1.2 Impact of cell temperature

All of the solar radiation falling on the solar cell will be converted to electricity and heat. A portion of the generated heat increases the cell temperature, and as a result, the electrical efficiency of the cell falls. The energy balance for a unit area of PV cell is as below:

$$\alpha \tau I_T = \eta_e I_T + U_L(T_c - T_a) \qquad (10.15)$$

In which: α and τ are the absorptivity and transmittivity of the solar cell, U_L is the overall heat loss coefficient, and T_c and T_a are the cell and ambient temperature.

The *nominal operating cell temperature* $(T_{c,NOCT})$ is the cell temperature under ambient temperature of $T_a = 20°C$, solar radiation of $I_{T,NOCT} = 800 \frac{W}{m^2}$, wind speed of $1 \frac{m}{s}$ and zero load. In such conditions by using Eq. (10.14):

$$\frac{\alpha\tau}{U_L} = \frac{T_{c,NOCT} - T_a}{I_{T,NOCT}} \tag{10.16}$$

By substituting Eq. (10.16) into (10.15) and solving for T_c, the cell temperature is determined as below:

$$T_c = T_a + \left(\frac{T_{c,NOCT} - T_a}{I_{T,NOCT}}\right)\left(1 - \frac{\eta_e}{\alpha\tau}\right)I_T \tag{10.17}$$

The electrical efficiency of solar cell as a function of cell temperature is given by Tiwari [2]:

$$\eta_e = \eta_0[1 - \beta_0(T_c - 298)] \tag{10.18}$$

In which, η_0 is the reference electrical efficiency under standard test conditions (ambient temperature of $T_a = 20\,°C$, and $I_T = 1000\frac{W}{m^2}$, $T_c = 25\,°C$), $\beta_0 = 0.0045\ K^{-1}$ or $0.0064\ K^{-1}$ is the silicon efficiency temperature coefficient.

Example 10.7. The reference electrical efficiency of a silicon solar cell is 0.16. If the cell temperature changes from -10 to $50°C$, how the electrical efficiency will change? Show it on a graph. Assume $\beta_0 = 0.0045\ K^{-1}$.

Solution. By using Eq. (10.18), and changing the cell temperature from 263 to 323K, the electrical efficiency versus cell temperature is plotted in Fig. 10.7. In addition, the reference efficiency is plotted to observe the impact of cell temperatures above and below 298K, which can be assumed as heating or cooling of the cell. As the results show, cooling of solar cell increases the electrical efficiency but heating decreases its efficiency.

$$\eta_e = \eta_0[1 - \beta_0[T_c - 298]] = 0.16[1 - 0.0045(T_c - 298)]$$

Important conclusion

Cooling of the solar cell increases the electrical efficiency of the cell. This idea is used today to manufacture the photovoltaic thermal (PVT) units. These units use cooling mediums such as water or air to cool the solar cells and also provide hot water or warm air. It means while the electrical efficiency is improved, the solar heat is also absorbed for heating purposes.

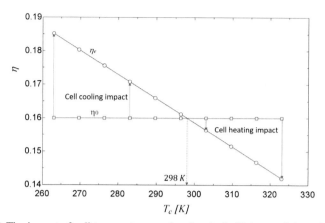

Figure 10.7 The impact of cell temperature on the electrical efficiency of the solar cell.

2. Solar cell, PV module, and PV array

The smallest unit of a PV generator is a solar cell. Solar cell converts sunlight into direct current (DC) power by using the photovoltaic effect. The amount of power generated by a single cell is very small and is not enough for daily demands by appliances and consumers. Fortunately, the solar cells can be coupled together in series and then sandwiched between top transparent cover and bottom opaque or transparent covers. The series coupled solar cells are known as *PV modules*. PV modules are available in different sizes depending on the electrical demand. The mostly used PV modules are made of glass-to-Tedlar PV modules made of 36 solar cells; each cell produces 0.5 V that results in max power voltage of 18 V. Fig. 10.8 shows a solar cell and a 20 W, PV panel, and its data sheet. A PV panel may have one or several modules.

A PV array is a combination of several PV panels coupled in series, parallel, or both and the number, dimensions, and capacity of panels used in an array depends on the geography of the location, electrical demand, load distribution during hours, site plan, and operation mode (on-grid operation or standalone). The DC power generated by PV arrays, if required, is converted to AC power by using solar inverters and then is fed into grid or load. When the PV arrays are supposed to operate stand alone, in addition to the inverter, it needs as well as charge controller and battery bank to store DC power for the sunless times. Fig. 10.9 shows some charge controller, inverter, and combined inverter-charge controller used with PV arrays [2].

Example 10.8. For the PV module presented in Fig. 10.8, determine the fill factor (FF) and the electrical efficiency under the standard test conditions reported in the data sheet of the module. What would be the electrical efficiency under cell temperature of -10 and $50°C$. Assume $\beta_0 = 0.0064$ K^{-1}.

Solution. According to the information given in the data sheet:

$$V_{OC} = 21.7 \text{ V}, I_{SC} = 1.23 \text{ A}, P_{max} = 20 \text{ W}, V_{P_{max}} = 18 \text{ V}, I_{P_{max}} = 1.12 \text{ A}, L$$
$$= 430 \text{ mm}, W = 350 \text{ mm}$$

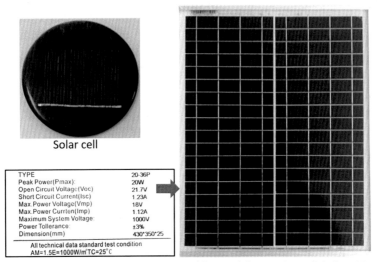

PV panel (4 modules, each module has 36 cell)

Figure 10.8 The solar cell, and a 20 W, and PV module (personal gallery).

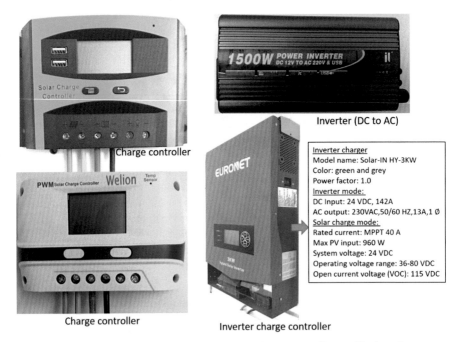

Figure 10.9 Charge controller, inverter, and inverter charge controller used in the solar systems (Personal gallery).

Hence:

$$FF = \frac{P_{max}}{V_{OC} \times I_{SC}} = \frac{20}{21.7 \times 1.23} = 0.7493$$

The standard electrical efficiency would be:

$$\eta_0 = \eta_e = \frac{P_{max}}{I_T \times A} = \frac{20}{1000 \times 0.430 \times 0.35} = 0.1329 = 13.29\%$$

At $T_c = -10\ ^\circ\text{C} = 263\ \text{K}$ the electrical efficiency is:

$$\eta_e = \eta_0[1 - \beta_0[T_c - 298]] = 0.1329[1 - 0.0064(263 - 298)] = 16.27\%$$

And at $T_c = 50\ ^\circ\text{C} = 323\ \text{K}$ the electrical efficiency is:

$$\eta_e = \eta_0[1 - \beta_0(T_c - 298)] = 0.1329[1 - 0.0064(323 - 298)] = 11.16\%$$

3. Material used in the solar cells

The materials used in the PV cells are semiconductor, they are called "semiconductor" because they conduct electricity better than insulators, but not as good as conductor materials such as copper or aluminum. When solar radiation falls on the PV cell, portions of it would be reflected (I_r), absorbed (I_a), or passed through (I_p) the solar cell. In other words:

$$I_T = I_r + I_a + I_p \tag{10.19}$$

The portion of the solar radiation absorbed by the cell is used to negatively charge electrons in the semiconductor. The absorbed energy let the electrons create an electric current in the material. This current is extracted by using metal contacts, which can be seen as grid lines on the solar panels. The extracted electricity can be used to supply electricity demands by the consumer. The efficiency of a solar cell, as its definition tells, depends on the amount of solar radiation shining on the solar cell as well as the amount of electricity generated by the semiconductor. The amount of electricity generated by the semiconductor depends on some important characteristics such as intensity and wavelength of the radiation and also bandgap of the semiconductor. The bandgap of a semiconductor specifies what wavelengths of the radiation can be absorbed by the semiconductor to generate electricity. That means choosing a material with a matched bandgap with the available wavelength of the radiation will increase the cell efficiency and electricity production [4,5].

The most common materials used for manufacturing of solar cells can be categorized as monocrystalline, polycrystalline, amorphous silicon, and compound thin film.

3.1 Silicon solar cells

Silicon is the mostly used material in the PV cells, and approximately 95% of the total solar cells sold in the market are made from silicon. Silicon, after oxygen is the most available material on earth, it is also used in computer chips. Silicon-made solar cells are today highly efficient; they can operate for about 25 years, their price has reduced considerably, and their production processes are optimizing day to day. Silicon on earth is usually found in the form of SiO_2. Another important reason for silicon popularity is that its electrical characteristics can be precisely controlled during the doping process and adding impurity to the intrinsic silicon. Silicon wafer is a thin slice of pure crystalized silicon that can be found today in many electronic applications including PV systems. A silicon wafer doped with boron impurities creates a p-type wafer, and if it is doped with arsenic or phosphorous, it creates n-type wafer. The manufacturing process of silicon wafer starts with melting SiO_2 or quartzite in an electrode arc furnace in the presence of carbon (coke, coal). The molten silicon in this stage is 98% pure, and it has minor impurities of aluminum and iron. The chemical process is as below:

$$SiC + SiO_2 \rightarrow Si + SiO + CO \qquad (10.20)$$

In the next step of purification, silicon reacts with hydrochloric gas at 300°C, as below:

$$Si + 3HCl \xrightarrow{300°C} SiHCl_3 + H_2 \qquad 2SiHCl_3 + 2H_2 \xrightarrow{300°C} 2Si + 6HCl \qquad (10.21)$$

The above process may be repeated many times to reach highly pure silicon. The silicon produced until this step is mostly a polycrystalline silicon.

The mostly used method for monocrystalline production is the *Czochralski growth method*. The created monocrystalline silicon needs some after processing such as shaping, cutting by industrial diamond, grinding, and chemical etching to remove damages and contaminants. Fig. 10.10 shows some monocrystalline rode and silicon wafers.

3.2 Thin-film solar cells

By depositing one or more layers of a PV material on one or two sides of a module surface a thin-film solar cell would be made [4]. The most common thin film semiconductors in the market include cadmium telluride (CdTe) and copper indium gallium diselenide (CIGS). The module surface which in fact a supporting plate can be made of glass, plastic, metal, etc.

Figure 10.10 The monocrystalline silicon rod and silicon wafer.
Drawing by Amir Hossain Pajooh, Mechanical Engineering Student, University of Kurdistan.

In comparison with silicon, thin-film PV cells made of CdTe are less expensive, but their efficiency is lower as well. CdTe is the second mostly used semiconductor used in the PV cells after silicon. CIGS, which is a combination of four materials, has high lab efficiency, but the complexity of the combination four materials has made it difficult to find its way to be mass manufactural and marketable [4]. Fig. 10.11 shows silicon solar cell materials including thin-film, monocrystalline, and polycrystalline.

3.3 Perovskite solar cells

Perovskite solar cells are a thin-film solar cell and are made up of layers of materials that are printed, coated, or vacuum deposited on an underlying backing layer known as the substrate. They are usually easy to assemble and can achieve the same performance as crystalline silicon. In the laboratory, the efficiency of perovskite solar cells has reached more than 25%. Perovskite solar cells must be stable enough for outdoor operation of more than 20 years [4], and for this purpose, they need more investigation to become mass manufactural and marketable. Figs. 10.12 and 10.13 show the production line of inkjet printing solar cell and a lightweight and semi-transparent perovskite solar module made by SAULE technology.

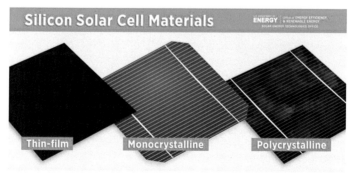

Figure 10.11 Silicon solar cell materials [4].
Credit: U.S. Department of Energy.

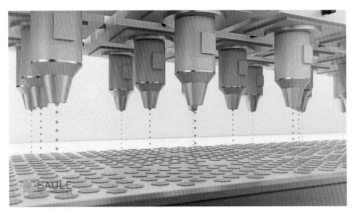

Figure 10.12 Inkjet printing solar cell production line [6].
SAULE technology, used with permission.

Figure 10.13 Lightweight and semi-transparent perovskite solar module [6].
SAULE technology, used with permission.

3.4   Organic solar cells

Organic solar cells (Fig. 10.14) are made up of carbon-rich (organic) compounds and can be designed to improve specific characteristics of a solar cell such as bandgap, transparency, or color. The efficiency of organic solar cells is currently only half of the crystalline silicon cells and have a shorter lifespan. The production cost may reduce in mass production. They can also be applied to a variety of backing materials such as flexible plastics, allowing organic solar cells to offer a variety of applications [4]. Organic solar cells are categorized in two groups of small-molecule solar cells and polymer-based solar cells [4].

3.5   Multijunction solar cells

Multijunction solar cells, unlike single junction cells, are made of several layers of different semiconductor materials. The radiation that passes through the first layer is absorbed by the subsequent layers and thus can absorb more light per unit area and generate more electricity. This makes multijunction solar cells more efficient than single junction cells. The reason that light passing through the first layer is absorbed by the subsequent layers is due to having different bandgaps in different layers. Multijunction solar cells that combine the semiconductors of columns III and V in the periodic table are called III-V multijunction solar cells. The efficiency of multijunction cells has reached 45%, but they are highly expensive, and they are only used for special applications such as space explorations, drones, and some military applications [4].

3.6   Quantum dots solar cells

Quantum dot solar cells are actually made up of particles of different semiconductor materials that are nanometers in size. Because it is very difficult to make electrical connections between these particles, the efficiency of these cells is still low. These solar cells can be used to absorb light that is not absorbed by other solar cells and generate

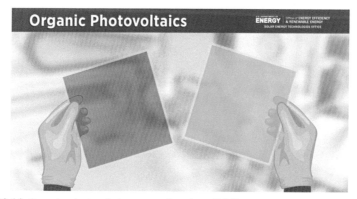

Figure 10.14 Organic photovoltaic or organic solar cell [4].
Credit: U.S. Department of Energy.

electricity. These cells can be made on a substrate by various methods such as spin-coat, printing, spraying, etc. These cells can be coupled with Perovskite, and multi-junction solar cells for capturing hardly captureable sunlight. Fig. 10.15 shows a schematic of the quantum dot solar cell production [4].

3.7 Concentration photovoltaic

Another way to increase the efficiency of a solar cell is to increase the intensity of radiation in the same area of the cell with the same amount of semiconductor material. This can be done by focusing sunlight using a mirror or lens. Concentration photovoltaic (CPV) has the highest overall efficiency but requires more expensive materials, a solar tracking system, and very sophisticated fabrication techniques. Because of these challenges, they have not yet been mass-produced [4].

4.   Concentrated solar thermal power plant

CSP generators use solar heat to run heat engines. Solar radiation is concentrated at a point or along a pipeline to generate high-temperature working fluid (generally above 150°C). Different types of concentrators are shown in Fig. 10.16A. They can produce different temperatures and can be used in different thermal power cycles such as organic Rankine cycle, steam power plants, open and closed cycle gas turbines, Stirling engines, etc. However, among the concentrators, the parabolic trough, parabolic dish, linear Fresnel reflectors, and solar tower with reflective flat mirrors are the most popular. The characteristics of a solar parabolic trough is shown in Fig. 10.16B.

4.1 General requirements for CSP

Rocky and stony lands with pebbles and gravels that have little vegetation and are not suitable for profitable economic activities can be used to build a CSP. Therefore, this advantage causes the cost of these power plants to be significantly reduced. The

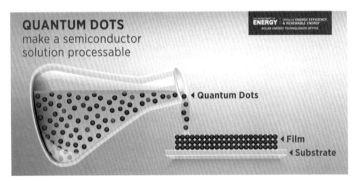

Figure 10.15 The schematic of quantum dots solar cell production [4].
Credit: U.S. Department of Energy.

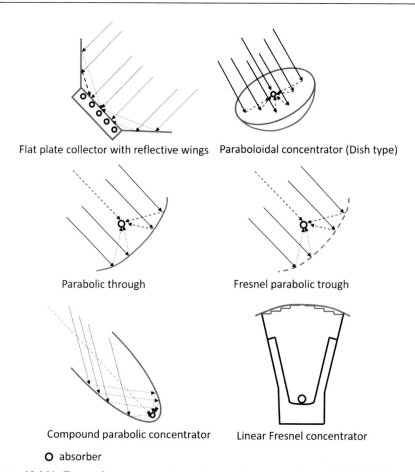

Flat plate collector with reflective wings   Paraboloidal concentrator (Dish type)

Parabolic through   Fresnel parabolic trough

Compound parabolic concentrator   Linear Fresnel concentrator

O absorber

Figure 10.16A Types of concentrator heat collectors, Regenerated from ASHRAE [7].

ground required for parabolic through should be as flat as possible, this is because of the working fluid circuit, which connects the parabolic mirrors together. The amount of area required for mirrors is large; for example, the land area used for the through parabolic CSP Andasol 3 with capacity of 50 MW is about 200 ha. The ground used for solar towers does not have to be flat because each mirror is independent. Another parameter that can change the price of these power plants is the existing infrastructure such as access to the road network, transmission network with different voltages, as well as water resources. In the absence of any of these infrastructures, additional investment and obtaining the necessary permits will be required. Natural phenomena such as storm and earthquake at the solar farm site should be considered in the design. If the land in question is high risk in terms of natural phenomena, special measures should be taken in the design, which means additional costs. Other parameters that encourage investment in CSP and reduce investment risk, include the existence of long-term guaranteed purchase contracts for electricity, scheduled payments for the

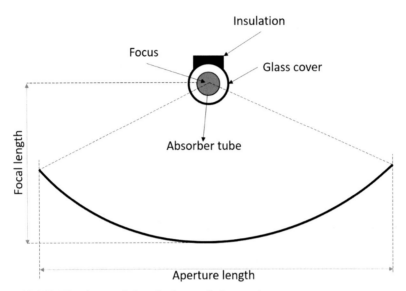

Figure 10.16B The characteristics of solar parabolic trough.

purchased electricity, and tax exemptions. As mentioned, the CSP exists in three main configurations of parabolic trough, parabolic dish, and solar tower. In the following, these three CSPs are presented [8].

4.2 Heat gain by the concentrated collectors

The general equation used for estimation of heat gained by a collector per unit area of the aperture can be estimated as below [7]:

$$q_u = I_{t\theta}(\tau\alpha)_\theta - U_L(t_p - t_a) = \frac{\dot{m}c_p(t_{fe} - t_{fi})}{A_{ap}} \tag{10.22a}$$

$$q_u = F_R\left[I_{t\theta}(\tau\alpha)_\theta - U_L(t_{fi} - t_a)\right] \tag{10.22b}$$

The above equation can be modified for the concentrating collectors as below:

$$q_u = \left[I_{DN}(\tau\alpha)_\theta(\rho\Gamma) - U_L\frac{A_{abs}}{A_{ap}}(t_{abs} - t_a)\right]$$
$$= F_R\left[I_{DN}(\tau\alpha)_\theta(\rho\Gamma) - U_L\frac{A_{abs}}{A_{ap}}(t_{fi} - t_a)\right] \tag{10.23}$$

In which:

$q_u \left(\frac{W}{m^2}\right)$ = useful heat gained by collector per unit area of the aperture.

$I_{t\theta}\left(\frac{W}{m^2}\right)$ = total irradiation of collector.

$I_{DN}\left(\frac{W}{m^2}\right)$ = direct normal irradiation.

$(\tau\alpha)_\theta$ = transmittance τ of cover times absorptance α of plate at incident angle of θ.

$U_L\left(\frac{W}{m^2 K}\right)$ = upward heat loss coefficient.

$t_p(^\circ C)$ = temperature of the absorber plate.

$t_a(^\circ C)$ = temperature of atmosphere.

$t_{abs}(^\circ C)$ = temperature of absorber.

$\dot{m}\left(\frac{kg}{s}\right)$ = fluid mass flow rate.

$c_p\left(\frac{kJ}{kg.K}\right)$ = specific heat of fluid.

$t_{fe}(^\circ C)$ = exit fluid temperature.

$t_{fi}(^\circ C)$ = entering fluid temperature.

$\rho\Gamma$ = reflectance of the concentrator surface times fractions of reflected or refracted radiation that reaches absorber.

$A_{abs}\,(m^2)$ = area of the absorber that admits radiation.

$A_{ap}\,(m^2)$ = area of the aperture that admits radiation.

F_R = collector heat removal factor that is the ratio of the actual heat transfer to the maximum possible heat transfer through the collector. This term is suggested by ASHRAE 1977 to allow using fluid inlet temperature in Eq. (10.22a).

The ratio $C_r = \frac{A_{ap}}{A_{abs}}$ is the concentration ratio of the collector. The collector efficiency can be determined by dividing the heat gain to the total irradiation of collector.

$$\eta_c = \frac{q_u}{I_{t\theta}} \tag{10.24}$$

The total solar irradiation $I_{t\theta}$ for unit area in any direction, tilting angle of Σ and incidence angle of θ can be calculated as Eq. (10.25):

$$I_{t\theta} = I_{DN}\cos\theta + I_{d\theta} + I_{re} \tag{10.25}$$

In which $I_{DN}\cos\theta$ is the direct irradiation component, $I_{d\theta}$ is the diffusion coming from the sky, and I_{re} is the reflected shortwave irradiation from the foreground that possibly reaches the collector. To estimate the diffuse component, dimensionless parameter C is defined that depends on dust and moisture content of atmosphere and changes through the year. It is defined by Eq. (10.26):

$$C = \frac{I_{dH}}{I_{DN}} \tag{10.26}$$

In which, I_{dH} is the diffuse component on a horizontal surface in a cloudless day. Eq. (10.27) is used to estimate the diffuse irradiation on a collector with tilting angle of Σ:

$$I_{d\theta} = C \times I_{DN} \times F_{ss}$$

$$F_{ss} = \frac{1 + \cos\Sigma}{2} \tag{10.27}$$

$$C = a_1 \exp\left[-\left(\frac{N - b_1}{c_1}\right)^2\right] + a_2 \exp\left[-\left(\frac{N - b_2}{c_2}\right)^2\right] \tag{10.27a}$$

$$a_1 = 0.08041; \ b_1 = 187.4; c_1 = 85.71; a_2 = 0.05888; b_2 = 179.3; c_2 = 589.6$$

In which, C is curve fitted by using the data presented in Ref. [8]. The reflected component can be calculated by Eq. (10.28):

$$I_{re} = I_{tH} \times \rho_g \times F_{sg}$$

$$F_{sg} = \frac{1 - \cos\Sigma}{2} \tag{10.28}$$

In which, ρ_g is the reflectance coefficient and I_{tH} is the total radiation on a horizontal surface. Bituminous surfaces reflect less than 10% of the total solar irradiation [8].

The direct solar radiation on the earth surface in a cloudless and clean sky is calculated by the following equation:

$$I_{DN} = A \times \exp\left(-\frac{B}{\sin\sigma}\right) \tag{10.29}$$

In which, A and B are apparent extraterrestrial irradiation and atmospheric extinction coefficients, respectively, and are functions of the date and take into account the seasonal variation of the earth-sun distance and the air's water vapor content. The magnitudes of A and B are curve fitted by Ebrahimi and Keshavarz [9] according to the data presented in Ref. [8] as follow:

$$A = a_3 \exp\left[-\left(\frac{N - b_3}{c_3}\right)^2\right] + a_4 \exp\left[-\left(\frac{N - b_4}{c_4}\right)^2\right]$$

$$a_3 = 1187; b_3 = 371.9; c_3 = 225.9; a_4 = 1144; b_4 = -6.401; c_4 = 206.2 \tag{10.30}$$

$$B = a_5 \ exp\left[-\left(\frac{N-b_5}{c_5}\right)^2\right] + a_6 \ exp\left[-\left(\frac{N-b_6}{c_6}\right)^2\right] + a_7 \ exp\left[-\left(\frac{N-b_7}{c_7}\right)^2\right]$$

$a_5 = 0.007325; b_5 = 235; c_5 = 22.82; a_6 = 0.07095; b_6 = 184.90;$

$c_6 = 99.03; a_7 = 0.1409; b_7 = 783.3; c_7 = 3624$

(10.31)

The axis which earth spins about is tilted 23.45 degrees. In order to find the collector heat gain, the declination angle δ must be calculated for each day of year. This angle is shown in Fig. 10.17. As it can be seen, it is the angle between the sun arrays and equator and is calculated as below:

$$\delta = 23.45 \sin\left(360° \times \frac{284+N}{365}\right)$$

(10.32)

In which, N is the day number starting from $N = 1$ for the Jan. 1st and ending with $N = 365$ for the Dec. 31st, and annual changes of δ is negligible. The sun coordination is determined by solar azimuth angle az_s in the horizontal plane (angle HOS), and the solar altitude σ (angle HOQ in Fig. 10.18).

Solar angular hour H is also calculated as below:

$$H = (\text{Number of hours from solar noon}) \times 15 \text{ degrees}$$
$$= |12 - TIH| \times 15 \text{ degrees}$$

(10.33)

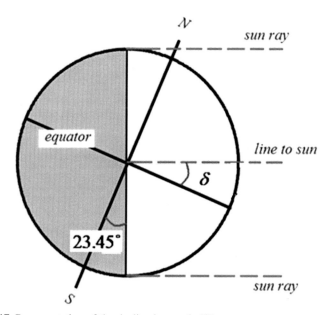

Figure 10.17 Representation of the declination angle [9].

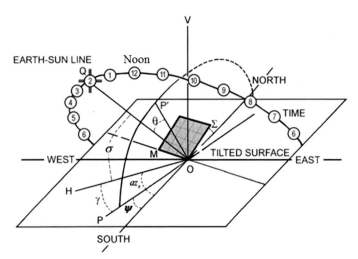

Figure 10.18 Sun coordination with respect to the collector aperture at point O on earth ($\theta, az_s, \sigma, \psi$, and γ are the incident, solar azimuth, solar altitude, surface azimuth, and surface-solar azimuth angles, respectively) [9].

In which, *TIH* is the time in hour in 24 h scale, for example 04:00p.m. is 16 in the above equation. Therefore, σ and az_s are calculated as below:

$$\sin \sigma = \cos LAT \cos \delta \cos H + \sin LAT \sin \delta \tag{10.34}$$

$$\sin az_s = \frac{\cos \delta \sin H}{\cos \sigma} \tag{10.35}$$

$$\cos az_s = \frac{\sin \sigma \sin LAT - \sin \delta}{\cos \sigma \cos LAT} \tag{10.36}$$

The angle between the sun array falling on earth at point O (OQ) and the perpendicular axis to the collector aperture (OP′) is called incident angle θ and is calculated as below:

$$\cos \theta = \cos \sigma \cos \gamma \sin\Sigma - \sin \sigma \cos\Sigma \tag{10.37}$$

In which Σ is the tilting angle of the collector (QOP′) and γ is the collector-sun azimuth angle. When the collector is facing east of south γ is as below.

$$\gamma = \begin{cases} az_s - \psi; \; morning \\ az_s + \psi; afternoon \end{cases} \tag{10.38}$$

In which, ψ is the collector azimuth angle, if the collector faces west of south γ is as below:

$$\gamma = \begin{cases} az_s + \psi; \ morning \\ az_s - \psi; afternoon \end{cases} \tag{10.39}$$

For calculation of the total solar heat gained during each day of a year, sunrise time, sunset time, and day length are required and are calculated as below:

$$DL = \frac{2}{15}\arccos(-\tan LAT \tan \delta)$$

$$sunrise = 12 - 0.5DL \tag{10.40}$$

$$sunset = sunrise + DL$$

Example 10.9. A linear parabolic trough solar concentrator is used to heat water with inlet temperature of 200°C. It rotates horizontally on an east-west axis facing south and reflects solar radiation on a 60 m long heat receiver. The absorber creates a shadow on the reflector, which its width is 0.2 m. The collector aperture is 2.2 m and focal length is 1 m. The direct normal radiation to the aperture is 970 W/m², $\alpha\tau = 0.77$, $\rho\Gamma = 0.85$, ambient air temperature is 20°C, the heat removal factor is 0.85, and the heat loss coefficient is 13.5 $\frac{W}{m^2K}$. In addition, the top side of the absorber, which account for one sixth of the absorber area, is insulated to avoid upward heat loss. The absorber diameter is 20 cm. Determine (a) the useful heat rate gain by the absorber (b) the exit water temperature from the absorber if the operating pressure is 2000 kPa, and water flow rate is 0.02 $\frac{kg}{s}$ (c) concentration ratio.

Solution. (a) According to the data given:

$$F_R = 0.85$$

The effective reflector area can be determined by subtracting the shadow of the absorber as below:

$$A_{ap} = (aperture \ length - shadow) \times absorber \ length = (2.2 - 0.2) \times 60$$
$$= 120 \ m^2$$

And the effective area of the absorber would be:

$$A_{abs} = d_{abs}l_{abs} = 0.2 \times 60 = 12 \text{ m}^2$$

$$\dot{Q}_u = F_R A_{ap}\left[I_{DN}(\tau\alpha)_\theta(\rho\Gamma) - U_L\frac{A_{abs}}{A_{ap}}(t_{fi} - t_a)\right]$$

$$= 0.85 \times 120\left[970 \times 0.77 \times 0.85 - 13.5\frac{12}{120}(200 - 20)\right]$$

$$= 39970W = 39.97 \text{ } kW$$

(b) Since due to high operating temperature of CSP, phase change of water may happen, instead of constant specific heat and temperature difference, it would be more accurate to use enthalpy difference. The exit fluid temperature can be determined by using the enthalpy and operating pressure.

$$\dot{Q}_u = \dot{m}c_p(t_{fe} - t_{fi}) = \dot{m}(h_{fe} - h_{fi})$$

Since $T_{fi} = 200\ ^\circ C$ and $P_{fi} = 2000$ kPa, $P_{fe} = 2000$ kPa then:

$$h_{fi} = 852.6\frac{kJ}{kg}$$

$$44.1 \text{ kW} = 0.02(h_{fe} - 852.6)$$

$$\rightarrow h_{fe} = 2851\frac{kJ}{kg}$$

$$\rightarrow t_{fi} = 230.8\ ^\circ C$$

which means the exit water is superheated steam.

(c) the concentration ratio is determined as the ratio of absorber area to the aperture area:

$$C_r = \frac{A_{ap}}{A_{abs}} = \frac{120}{12} = 10$$

Important discussion

As it can be seen a 60 m long absorber placed at the focus point of a parabolic trough concentrator with the same length, is able to produce superheated steam that can be used in a steam turbine power plant for power generation. To increase the capacity, similar arrays must be installed in parallel.

It is worth mentioning that most of thermal heat is consumed for phase change of water during the boiling process. After superheating adding only 5 m to the length of the absorber would increase the exit temperature from 230.8 to 297.8°C.

The code written for Example 10.9 is presented in the **Code. 10.3**.

Code 10.3. The code generated for Example 10.9, a parabolic trough concentrator.

```
T_atm=20
T_fi=200
p[1]=2000
t[1]=T_fi
p[2]=p[1]
m_dot=0.02
F_R=0.85
tawalpha=0.77
rogama=0.85
L_aperture=2.2
w_shadaw=0.2
L_collector=60
d_abs=0.2
A_ap=(L_aperture-w_shadaw)*L_collector
A_abs=d_abs*L_collector
I_DN=970
U=13.5
Q_u=F_R*A_ap*(I_DN*tawalpha*rogama-U*A_abs*(T_fi-T_atm)/A_ap)/1000
H[1]=ENTHALPY(WATER, T=T[1], P=P[1])
Q_u=m_dot*(H[2]-H[1])
T[2]=TEMPERATURE(WATER, H=H[2], P=P[2])
S[2]=ENTROPY(WATER, H=H[2], P=P[2])
S[1]=ENTROPY(WATER, H=H[1], P=P[1])
C_r=A_ap/A_abs
```

4.3 Thermal energy storage

Concentrated solar power plants use solar heat to generate power; however, the solar radiation has two essential problems for power generation:

1 It is changeable during daytime; in general, it starts from zero solar heat in the sunrise, then it reaches a peak point, and after that it declines until sunset when it becomes zero again. As is can be seen in Fig. 10.19 depending on the capacity of the power generator unit (PGU) and also the magnitude of solar heat gain, the PGU may operate in full load, part load, or even switched off.

2 At night, there is no solar heat, and as a result, the PGU is switched off if there is no back up plan.

The changeable solar heat results in frequently part load operation and shutdown at night. Repetitive startup, shutdown, and part load operation of PGU results in efficiency drop, wear in the bearings, repetitive thermal stresses, and finally lifetime reduction of PGU.

In addition, operating under the minimum operating load (low load operation) for a long time results in irreversible damages to the PGU, specially due to lower compression ratio, low temperature working fluid, poor lubrication, and possible excessive vibration. Hence, it is preferred to switch off the PGU when the available solar heat is less than the required heat for the minimum allowable operating load [10].

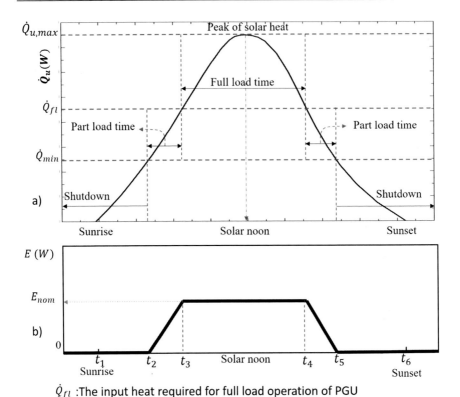

$\dot{Q}_{fl}$:The input heat required for full load operation of PGU
$\dot{Q}_{min}$:The input heat required for minimum allowable load operation of PGU

Figure 10.19 A typical daily solar heat gain, and operation mode of the PGU working with concentrated solar heat and without TES system.

For steam-based PGUs, the startup/shutdown procedure may take several hours; hence, it is impossible to run a steam power plant without coupling the solar system with an auxiliary burner or thermal energy storage (TES) system.

To avoid problems associated with startup/shutdown, part load/low load operation of the PGU, the CSP plants are strongly recommended to be equipped with a TES system.

Thermal storage systems remove heat from or add heat to a storage medium for use at another time [11]. They also can be used when there is a potential of heat loss in a process or equipment and the possibility of using this heat in another time for the process or heating purposes. In CSP systems, we need to store high-quality heat in the TES system for the night time or the time when sunlight is weak due to cloud, mist, dust, etc. Depending on the investor decision, the PGU of the CSP plant may be fully fed with solar heat or hybrid solar/fossil fuel. Heat-quality needed for thermal power plants is usually very high. For example, in a steam power plant, steam must be superheated, and the steam temperature is usually higher than 500°C. In a gas turbine, depending on the working fluid and cycle type (open or close cycle), the gas temperature may be even higher than 1000°C. This means that storing such high-quality heat needs a very special TES system and also a very large solar concentrator system. According to Fig. 10.19, heat can be stored only when the solar heat gain is bigger than the required heat for full load operation, the rest of time the TES system is discharging to omit the impact of solar heat fluctuation on the PGU operation.

If the PGU is supposed to operate continuously at full (*fl*) load mode, its electrical efficiency would be as Eq. (10.41).

$$\eta_{e,fl} = \frac{E_{nom}}{\dot{Q}_{fl}} \rightarrow \dot{Q}_{fl} = \frac{E_{nom}}{\eta_{e,fl}} \tag{10.41}$$

In which E_{nom} is the nominal electricity generation by the PGU at full load mode under standard test conditions specified by the manufacturer.

The PGU electrical efficiency at part load (*pl*) operation is less, hence:

$$\eta_{e,pl} = \frac{E}{\dot{Q}_{pl}} \rightarrow \dot{Q}_{pl} = \frac{E}{\eta_{e,pl}} \tag{10.42}$$

In which the part load efficiency is a function of load ratio (*lr*) as below:

$$lr = \frac{E}{E_{nom}}$$
$$\eta_{e,pl} = f(lr) \tag{10.43}$$

In addition if the minimum allowable part load operation is E_{min}, then the heat required for this condition would be:

$$\dot{Q}_{min} = \frac{E_{min}}{\eta_{e,pl}} = \frac{E_{min}}{f(lr_{min})} \tag{10.44}$$

In which $lr_{min} = \frac{E_{min}}{E_{nom}}$.

The heat gained by the solar concentrator, $\dot{Q}_u$, is very unstable. Since the TES is fed with the solar system, different conditions may occur for charging and discharging of the TES system. Since solar heat is changeable it can not be used directly in the PGU. It must be stored in the TES system to dampen the fluctuations and then use it with the stable and desired temperature demanded by the PGU manufacturer. In the following, according to Fig. 10.19, different cases are studied for better understanding of the CSP system.

Case 1: $\dot{Q}_u = 0$ that happens at nigh:

For full load and continuous operation of the PGU, the heat required by the PGU must be absorbed from the TES material; therefore, the TES system is discharging in this case; hence, the amount of high-quality heat that must be discharged (Q_{dch1}) from the TES can be determined as below:

$$Q_{dch1}(J) = \int_{sunset \ (t_6)}^{sunrise \ (t_1)} \dot{Q}_{fl}(W)dt \tag{10.45}$$

Case 2: $\dot{Q}_u < \dot{Q}_{min}$

Since the useful heat gain is less than $\dot{Q}_{fl}$, for full load and continuous operation of the PGU, the lack of input heat must be compensated from the TES system. In such cases, depending on the working fluid of the solar heat receiver, usually there are several units of TES system in which while one unit is charging with the available solar heat, another unit is discharging and providing high-quality heat for the PGU. In this case, the available heat stored in the TES is reduced (discharged) as below:

$$Q_{dch2}(J) = \int_{t_1}^{t_2} \left[\dot{Q}_{fl} - \dot{Q}_u \right] dt + \int_{t_5}^{t_6} \left[\dot{Q}_{fl} - \dot{Q}_u \right] dt \tag{10.46}$$

Case 3: $\dot{Q}_{min} < \dot{Q}_u \le \dot{Q}_{fl}$

Similar to the second case, the TES would be discharging since $\dot{Q}_u \le \dot{Q}_{fl}$. The amount of discharged heat would be:

$$Q_{dch3}(J) = \int_{t_2}^{t_3} \left[\dot{Q}_{fl} - \dot{Q}_u \right] dt + \int_{t_4}^{t_5} \left[\dot{Q}_{fl} - \dot{Q}_u \right] dt \tag{10.47}$$

Case 4: $\dot{Q}_{fl} \le \dot{Q}_u$

In this case, there is extra solar heat that can be used for charging the TES system. The amount of heat that can be stored in the TES system would be:

$$Q_{ch}(J) = \int_{t_3}^{t_4} \left[\dot{Q}_u - \dot{Q}_{fl} \right] dt \tag{10.48}$$

Hence, by writing the energy balance for the TES system for continuous operation of the PGU, and assuming zero heat loss from the TES and other auxiliary systems, the charge and discharge heat must be equal.

$$Q_{ch}(J) = Q_{dch}(J) \tag{10.49}$$

Therefore:

$$\int_{t_3}^{t_4}\left[\dot{Q}_u - \dot{Q}_{fl}\right]dt = \int_{sunset\ (t_6)}^{sunrise\ (t_1)} \dot{Q}_{fl}dt + \int_{t_1}^{t_3}\left[\dot{Q}_{fl} - \dot{Q}_u(t)\right]dt + \int_{t_4}^{t_6}\left[\dot{Q}_{fl} - \dot{Q}_u(t)\right]dt$$

$$\int_{t_3}^{t_4}\dot{Q}_u dt - t_{34}\dot{Q}_{fl} = t_{61}\dot{Q}_{fl} + t_{13}\dot{Q}_{fl} - \int_{t_1}^{t_3}\dot{Q}_u(t)dt + t_{46}\dot{Q}_{fl} - \int_{t_4}^{t_6}\dot{Q}_u(t)dt \rightarrow$$

$$\dot{Q}_{fl}(t_{13} + t_{34} + t_{46} + t_{61}) = \int_{t_1}^{t_6}\dot{Q}_u(t)dt$$

In which $t_{13} + t_{34} + t_{46} + t_{61} = 24$ h or 24 h $\times 3600\left(\frac{s}{h}\right) = 86400$ s, hence:

$$\dot{Q}_{fl} = \frac{\int_{t_1}^{t_6}\dot{Q}_u(t)dt}{86400} \rightarrow \frac{E_{nom}}{\eta_{e,fl}} = \frac{\int_{t_1}^{t_6}\dot{Q}_u(t)dt}{86400}$$

Therefore, the PGU nominal electrical capacity in the absence of heat loss would be:

$$E_{nom}(W) = \eta_{e,fl} \times \frac{\int_{t_1}^{t_6}\dot{Q}_u(t)dt}{86400} \qquad (10.50)$$

The design procedure presented above was based on the solar radiation curve during an arbitrary day. However, the solar radiation curve is different on different days of the year. The hottest and coldest days of the year are important for determining the size of the PGU, solar concentrator capacity, and the TES capacity. Fig. 10.20 shows a CSP

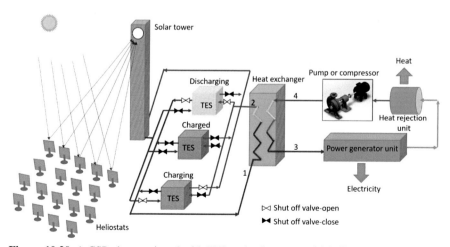

Figure 10.20 A CSP plant equipped with TES and solar tower with heliostats mirrors.

in which the solar system is a tower with heliostats, three TES units, and a PGU cycle. The working fluid of the PGU depending on the engine can be gas or steam. As it can be seen, there are three TES units (they can be even more depending on the design conditions and TES material), one of them is charging, one of them is discharging, and the third one is fully charged and left for the time when there is no or weak solar radiation. As it can be seen solar heat is not used directly in the PGU cycle, it is first stored in the TES system and then used to provide stable amount of heat with demanded quality by the PGU manufacturer. The heat transfer to the PGU cycle happens in a heat exchanger as it is shown in Fig. 10.20. Depending on the working fluid (steam or gas) of PGU cycle, the heat transfer inside the counter flow heat exchanger can be shown as Fig. 10.21.

If the highest allowable temperature of the working fluid at the inlet of PGU cycle is T_3 (Figs. 10.20 and 10.21), this temperature must be stable during the operation to avoid fluctuating thermal stresses, efficiency drop, and energy waste in the PGU cycle. Under such conditions the returning temperature to the discharging unit of TES is T_2. When the TES unit is fully discharged, its temperature would be T_2 and the re-charging process starts from this temperature until it approaches T_1 again. In the absence of heat loss from the exchanger body and piping:

$$\dot{m}_3(h_3 - h_4) = \dot{m}_1(h_1 - h_2) \tag{10.51}$$

Since the minimum return temperature to the TES unit is T_2 and the charging process starts from this temperature, the TES unit will have a dead capacity that stores heat with the temperature of T_2. This dead capacity must be considered in capacity calculations of the TES unit. In other words, the total heat capacity of the TES would be:

$$Q_{TES} = Q_{ch} + Q_{dead} = \int_{t_3}^{t_4} \left[\dot{Q}_u - \dot{Q}_{fl} \right] dt + m_{TES}\left(h_{@T_2} - h_0\right) \tag{10.52}$$

In which h_0 is the enthalpy of the TES medium at ambient conditions.

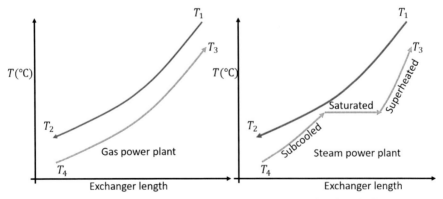

Figure 10.21 Heat transfer inside the counter flow heat exchanger in Fig. 10.20.

4.4   Popular CSPs

Most popular CSP systems include parabolic trough, parabolic dish, linear Fresnel reflector, and solar tower with heliostat and are shown in Fig. 10.22. Except for the parabolic dish, which mostly runs a Stirling engine, the others can run steam turbine power plant and gas turbine power plant. The parabolic dish generates less electricity than other CSP. It generally produces 3 to 25 kW of electricity per dish [4].

> The U.S. Department of Energy (DOE) SunShot Initiative is a collaborative national effort launched in 2011 that has planned to drive down the cost of solar electricity to 6 cents per kilowatt-hour, without incentives, by the year 2020 *[12]*.

The 6 cents/kWh cost includes solar field (1.7 cent/kWh), thermal storage (1 cent/kWh), power plant (1.8 cent/kWh), and receiver and heat transfer fluid (HTF) (1.7 cent/kWh).

Sunshot has specified technical goals for the CSP plants as shown in Fig. 10.23. It shows that the receiver thermal efficiency is more than 90%, and the HTF exit temperature from the receiver is more than 720°C. The optical error of the solar field is smaller than 3 mrad, with lifespan of more than 30 years. The HTF must be stable at temperatures higher than 800°C with specific heat capacity of more than 3 $\frac{J}{g.K}$, corrosion of less than 15 $\frac{\mu m}{year}$, and melting point temperature of less than 250°C. The thermal storage system must provide inlet temperature of more than 720°C at the inlet of power cycle; the energy and exergy efficiencies of the TES are more than 99% and 95%. The power plant efficiency is more than 50% and dry cooled to avoid water loss and water consumption. The sunshot proposes using sCO_2 that is a closed cycle gas turbine working with CO_2 that operates at supercritical temperature and pressure

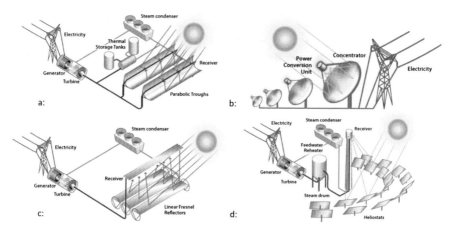

Figure 10.22 Most popular CSP systems, (A) Parabolic trough (B) Parabolic dish (C) Linear Fresnel reflectors (D) solar tower with heliostats [4].
Credit: U.S. Department of Energy.

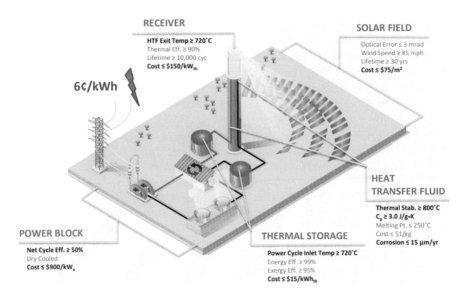

Figure 10.23 A solar tower CSP plant with the technical goals specified by sunshot [12]. Credit: U.S. Department of Energy.

where liquid and gas are not distinguishable. The turbomachinery footprint of a 300 MWe steam turbine is about 10 times of the footprint required by a sCO_2 for the same MW of electricity [12].

Depending on the receiver phase (gas, liquid, or solid particle) and the TES material, different CSP cycles can be proposed. In the following, some cycles which are investigated by U.S. Department of Energy/Sunshot are presented with some technical data [12].

4.4.1 Gas-phase receiver system with PCM storage

This CSP plant is shown in Fig. 10.24. The working fluid in the receiver is a gas such as helium, which is an inert gas and can be assumed as ideal gas. Other gases such as air or CO_2 may be considered as the receiver working fluid as well, but corrosion due to these gases must be evaluated. However, air is free and abundant, and CO_2 is very cheap in comparison with helium. The working pressure of the receiver is in the range of 60−120 bar and uses a gas circulator to move gas from the receiver to the multitank PCM²-TES system and return it to the receiver again. The gas circulator works at very high pressure of 50 bar and temperature of 550°C; therefore, it needs a very special design. This circulator must overcome the pressure drops in the pipes and storage tanks. The pipes also must be designed for tolerating high operating temperature and pressure of the working fluid. In this system, heat is first stored in the TES, and later used for power generation [12]. The TES system protects the power cycle from the fluctuating radiation

---

² Phase change materials that are used in the thermal energy storage systems.

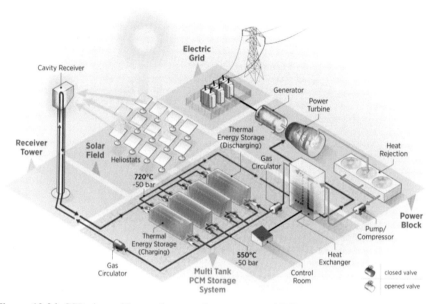

Figure 10.24 CSP plant with gas phase receiver system and PCM storage [13].
Credit: U.S. Department of Energy.

and can be designed to transfer steady heat to the power cycle for continuous and stable operation. The power cycle can be either gas turbine or steam turbine depending on the site plan and availability of water. However development of supercritical Brayton closed cycle CO_2 (or briefly sCO_2 cycle) can be a very good option for using with dry heat rejection system. The main challenge of the TES system is to maintain the temperature profile stable as it is discharging; this can be done by forming zones or layers of material that each zone would be discharged during a period of time to a specified temperature, and when the specified conditions happen, the next zone/layer would be discharged [13].

4.4.2 Gas-phase receiver system with particle storage

This CSP plant is shown in Fig. 10.25. Similar to the cycle presented in Fig. 10.24, the working fluid in the receiver can be helium, argon, air, CO_2, or other gases that are less corrosive, less expensive, and as inert as possible. The working pressure of the receiver is usually in the range of 60−120 bar and uses a gas circulator for flowing gas from the receiver to the gas-to-particle heat exchanger on top of the hot silo and returns it to the receiver again. The gas circulator works at very high pressure of about 50 bar and temperature of 550°C; hence, it needs a very special design. This circulator must overcome the pressure drops due to the pipes, fittings, and gas-to-particle heat exchanger. The pipes also must be designed for tolerating high operating temperature and pressure of the working fluid. In this system, heat is first stored in the solid particles of the hot silo; then, the hot particles are carried by a conveyer to the gas-to-particle heat exchanger on top of the cold silo. In the exchanger, heat is transferred to the coming

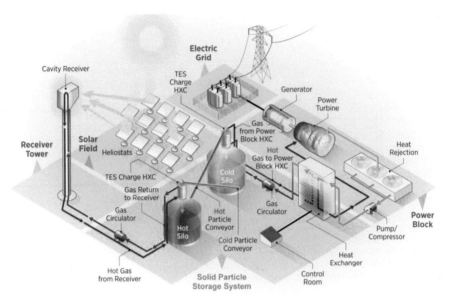

Figure 10.25 CSP plant with gas phase receiver system with particle heat storage [13]. Credit: U.S. Department of Energy.

gas from the heat exchanger between of the power cycle and the cold silo. Then, the cold particles drop into the cold silo. The cold particles would be returned to the gas-to-particle heat exchanger on top of the hot silo; they receive heat and drop into the hot silo again [12]. The hot silo of particles protects the power cycle from the fluctuating radiation and can be designed to transfer steady heat to the power cycle for continuous and stable operation. The number of silos can be more depending on the heat storage demand by the power cycle. The power cycle can be either gas turbine or steam turbine [13].

4.4.3   Liquid-phase receiver system with liquid storage

This CSP plant is shown in Fig. 10.26. The working fluid in the receiver of the solar tower is a liquid such as molten salt. Heat is stored in the hot salt tank. The hot liquid with temperature of about 720°C is pumped to the power cycle's heat exchanger. After exchanging heat to the working fluid of power cycle, it returns to the cold slat tank with the temperature of about 520°C. Then, it is pumped back to the receiver to be heated again. If the power cycle is a sCO_2, the carbon dioxide would be heated to about 700°C. The power cycle can also be a steam turbine cycle. An important challenge for using nitrate salts such as $Na - K/NO_3$ is that they decompose at temperatures above 600−620°C. Chlorides and carbonates are good candidates for using in the liquid phase receiver. Another challenge is designing and manufacturing of components such as heat exchangers, valves, fittings, etc. to operate in such high temperature with molten salts, while validated and published thermophysical properties of molten

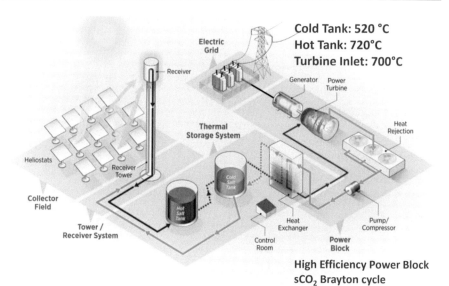

Figure 10.26 CSP plant with liquid phase receiver system and liquid salt heat storage[14]. Credit: U.S. Department of Energy.

salt at temperatures above 700 is scarce. In the following four salt mixtures are mentioned for using in the liquid phase receiver [14]:

- ($NaNO_3$, KNO_3), melting point 220°C, heat capacity of 1.5 $\frac{J}{kgK}$, and density of 1.7 $\frac{kg}{l}$.
- ($MgCl_2$, KCl), melting point 426°C, heat capacity of 1.15 $\frac{J}{kgK}$, and density of 1.66 $\frac{kg}{l}$.
- ($ZnCl_2$, KCl, NaCl), melting point 204°C, heat capacity of 0.81 $\frac{J}{kgK}$, and density of 2.4 $\frac{kg}{l}$.
- (Na_2CO_3, K_2CO_3, Li_2CO_3), melting point 398°C, heat capacity of 1.61 $\frac{J}{kgK}$, and density of 2 $\frac{kg}{l}$.

4.4.4 Solid-phase receiver system with solid particle storage

This CSP plant is shown in Fig. 10.27. Solid particles are directly used to absorb and store thermal heat in the receiver. Solid particles pour down look like a curtain (Fig. 10.28) in the receiver, and concentrated radiation enters from the aperture and heats up the solid particles to the temperature of 750°C. The hot particles store in a hot silo placed below the receiver. Below the hot silo, there is a particle-to-gas heat exchanger, which transfers heat to the working fluid of the power cycle. The power cycle is considered to be a supercritical closed cycle gas turbine working with 720°C carbon dioxide at the turbine inlet. Below the heat exchanger a cold silo is placed. The cold particles with temperature of 575°C are stored in this silo, and then they are continuously conveyed to the receiver by using a vertical elevator. The main advantage of the proposed system is the integration of the receiver loop, TES system and its piping, and the heat exchanger of the power block in one hot

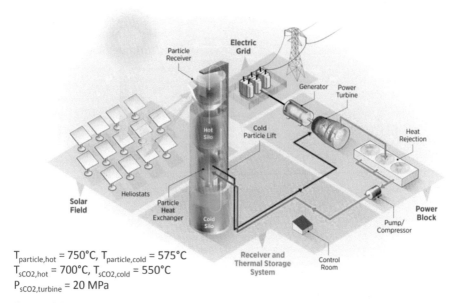

$T_{particle,hot} = 750°C$, $T_{particle,cold} = 575°C$
$T_{sCO2,hot} = 700°C$, $T_{sCO2,cold} = 550°C$
$P_{sCO2,turbine} = 20$ MPa

Figure 10.27 CSP plant with solid phase receiver system and solid particle storage [15]. Credit: U.S. Department of Energy.

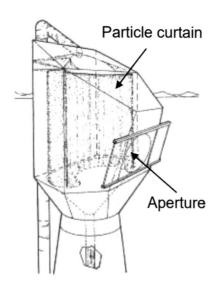

Figure 10.28 The receiver, aperture, and particle curtain [15]. Credit: U.S. Department of Energy.

silo and a particle-to-gas heat exchanger placed in the tower. In other words, heat is directly transferred to the power cycle, while in the liquid and gas phase receivers, the heat transfer was done indirectly. The particle material plays an important role in this system. The particle tolerance against high-temperature, continuous free falling, wear and impingement with other particles must be considered. In addition, they must have high absorptivity and conductivity to absorb and conduct solar heat. They must be stable and abundant in the nature. Some alternatives with their important characteristics for the particles are as below [12,15]:

Silica sand: with composition of SiO_2, density of 2610 $\frac{kg}{m^3}$, specific heat of 1000 $\frac{J}{kg.K}$, stable and abundant, low cost, and low solar absorptivity and conductivity.

Alumina: with composition of Al_2O_3, density of 3960 $\frac{kg}{m^3}$, specific heat of 1200 $\frac{J}{kg.K}$, stable, and low solar absorptivity.

Coal ash: with composition of $SiO_2 + Al_2O_3$ + minerals, density of 2100 $\frac{kg}{m^3}$, specific heat of 720 $\frac{J}{kg.K}$ at ambient temperature, stable, abundant, low or no cost, for good absorptivity and conductivity suitable ash must be identified.

Calcined flint clay: with composition of $SiO_2 + Al_2O_3 + TiO_2 + Fe_2O_3$, density of 2600 $\frac{kg}{m^3}$, specific heat of 1050 $\frac{J}{kg.K}$, mined abundant, and low absorptivity.

Ceramic proppants: with composition of 11% SiO_2 + 75% Al_2O_3 + 3% TiO_2 + 9% Fe_2O_3, density of 3300 $\frac{kg}{m^3}$, specific heat of 1200 $\frac{J}{kg.K}$ at 700°C, high solar absorptivity, stable, and high cost [15].

5. Problems

1. Determine the frequency and wavelength of a red photon with energy of 3.5×10^{-19} J.
2. Determine the energy and wavelength of a yellow photon with the frequency of 5.16×10^{14} Hz.
3. According to Fig. 10.29 approximate the energy of red, orange, yellow, green, blue, and violet photons in the visible light range. The wavelength is given on the figure. Which color has more energy?
4. Determine the total current versus voltage for silicon at the temperature of 310 K and voltage of 0.75 V. Plot the total current versus the voltage range of 0.5−0.8 V. What will

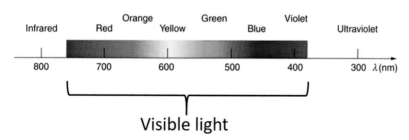

Figure 10.29 Wavelength of different colors.

happen for the current when the voltage approaches zero? Assume the nonideality factor to be 1.

5. Determine the total current versus temperature for Si in the temperature range of 273 K to 323 K and voltage of 0.7 V. Assume the non-ideality factor to be 1.

6. Draw the characteristics curves of (I−V) and (Power-V) for a solar cell which:

   A. Has an area of 10 cm^2, saturation current of $1.3 \times 10^{-14} \frac{A}{cm^2}$, light induced current of $0.45 \frac{A}{cm^2}$ and temperature of 305 K. The series resistance is zero.

   B. Consider the series resistance to be $0.15 \ \Omega.cm^2$.

   C. Determine the maximum power for both cases in parts a and b.

7. Determine the fill factor for the Si solar cell in the example problem 6. In both real and ideal conditions.

8. Under standard test conditions of cell temperature of $T_c = 25 \, °C$, and $I_T = 1000 \frac{W}{m^2}$ and maximum power output of 0.15 W determine the cell electrical efficiency. The PV aperture area is 10 cm^2.

9. The reference electrical efficiency of a Silicon solar cell is 0.15. If the cell temperature changes from -10 to 50°C, how the electrical efficiency will change? Show it on a graph. Assume $\beta_0 = 0.0064 \, K^{-1}$.

10. For the monocrystalline PV module presented in Fig. 10.30, determine the fill factor (FF) and the electrical efficiency under the standard test conditions reported in the data sheet of the module. What would be the electrical efficiency under cell temperature of -40 and 85°C. Assume $\beta_0 = 0.0045 \, K^{-1}$.

11. A linear parabolic trough solar concentrator is used to heat water with inlet temperature of 180°C. It rotates horizontally on an east-west axis facing south and reflects solar radiation on a heat receiver with length of 60 m. The absorber creates a shadow on the reflector which its width is 0.21 m. The collector aperture is 2.25 m and focal length is 1m. The direct

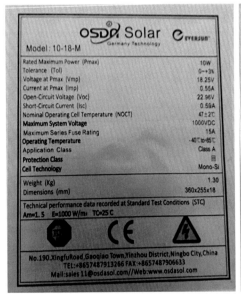
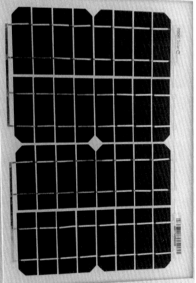

Figure 10.30 The 10 W solar panel and it technical data, (Personal gallery).

normal radiation to the aperture is $990\ \text{W}/\text{m}^2$, $\alpha\tau = 0.79$, $\rho\Gamma = 0.84$, ambient air temperature is $20°\text{C}$, and heat removal factor is 0.88 and the heat loss coefficient is 13.5 $\frac{\text{W}}{\text{m}^2\text{K}}$. In addition, the top side of the absorber which account for one sixth of the absorber area is insulated to avoid heat loss. The absorber diameter is 20 cm. Determine (a) the useful heat rate absorbed by the absorber (b) the exit water temperature from the absorber if the operating pressure is 3000 kPa, and water flow rate is $0.02\ \frac{\text{kg}}{\text{s}}$ (c) concentration ratio (d) plot the impact of absorber/collector length on the exit temperature if it changes from 30 to 90 m, (e) how many arrays must be installed in parallel to absorb 1 MW of thermal heat with superheated steam at temperature of $280°$ C.

References

[1] International Energy Agency (IEA).
[2] G.N. Tiwari, A.T. Shyam, Handbook of Solar Energy Theory, Analysis and Applications, Springer, 2016.
[3] A. Goetzberger, V.U. Hoffmann, Photovoltaic Solar Energy Generation, Springer, 2005.
[4] U.S. Department of Energy. https://www.energy.gov/eere/solar/solar-photovoltaic-cell-basics. (Available on 4 April 2022).
[5] https://www.sumcosi.com/english/. (Available on 4 April 2022).
[6] https://sauletech.com/press/. (Available on 4 April 2022).
[7] Solar Energy Use, (Chapter 33), ASHRAE Handbook 2007 - HVAC applications SI.
[8] Solar Thermal Power Plants Heat, Electricity and Fuels from Concentrated Solar Power, German Aerospace Center (DLR), Cologne, May 2021.
[9] M. Ebrahimi, A. Keshavarz, Combined Cooling Heating and Power, Decision-Making, Design Optimization, Elsevier, 2014, ISBN 978-0-08-099985-2.
[10] Espen Dalsøren Tufte, Impacts of Low Load Operation of Modern Four-Stroke Diesel Engines in Generator Configuration, Norwegian University of Science and Technology, Department of Marine Technology, 2014.
[11] ASHRAE Handbook 2007 - HVAC Applications SI, (Chapter 34), thermal storage.
[12] U.S. Department of Energy, Sunshot. https://www.energy.gov/sunshot.
[13] M. Wagner, Gas-phase Receiver Technology Pathway, U.S. Department of Energy/NREL, 2017.
[14] J. Gomez-Vidal, A. Kruizenga, Technology Pathway Molten Salt, U.S. Department of Energy/NREL and SNL, 2017.
[15] C. Ho, Technology Pathway − Particle Receivers, U.S. Department of Energy/SNL, 2017.

Wind power plants

11

Chapter outline

Wind power was discovered by our ancestors and used for various purposes. Windmills were powerful tools that worked around the clock to grind agricultural products such as wheat or pump water to higher elevations. Sailing ships used wind power to make long voyages possible for human. Fig. 11.1 shows a windmill. In recent decades, the use of wind power has taken on wider aspects. A variety of wind turbines are used around the world to generate electricity. Giant planes also take off by creating a wind current on their wings due to the engine thrust force. Air flowing over and under the wings generates a pressure difference between the top (low-pressure surface) and underneath surfaces (high-pressure surface) of wings that results in an upward lift force that must overcome the weight of plane for takeoff. Such a force in wind turbines causes the turbine and generator rotors to rotate, which ultimately leads to electricity generation.

Wind power, a renewable energy source, which is continuously generating on earth, can play a critical role in reducing CO_2 production. Wind power can be harvested by onshore, offshore, and distributed wind turbines. The wind turbines installed on land for large-scale power production are called onshore, those installed in the sea or ocean

Power Generation Technologies. https://doi.org/10.1016/B978-0-323-95370-2.00012-0
Copyright © 2023 Elsevier Inc. All rights reserved.

Nasrin Ebrahimi

Figure 11.1 A windmill used for grinding agricultural products like wheat.
This painting is drawn by my dear sister Nasrin.

are called offshore, and small-scale wind turbines utilized for providing electricity for residential or very small consumers are called distributed wind turbines. They are usually installed at the place of demand (Fig. 11.2).

International Energy Agency (IEA), as a part of its offshore wind outlook 2019, has assessed the technical potential of offshore wind by geospatial analysis, and the results showed that the best offshore wind sites close to shore could generate about 36,000 $\frac{TWh}{year}$, which is more than the global electricity demand in 2019. The IEA reports that to reach the **Net Zero Emissions by 2050** Scenario, 8000 TWh of wind power must be generated at 2030. This means that wind farms must increase with an average yearly rate of 18% during 2021−2030. Therefore, it is necessary to add onshore and offshore capacities annually by 310 GW and 80 GW respectively [2].

Land based (onshore) wind turbine   Offshore wind turbine

Distributed wind turbine

Figure 11.2 Onshore, offshore, and distributed wind turbines [1].
Credit: U.S. Department of Energy, NREL.

Unlike the fossil fuels, wind power is abundant and distributed around the world; the potential for using this energy source can be found everywhere. Hence, the market is growing fast and countries such as China and the United States are the leaders in this field. Manufacturers need to invest more in the research and developments of wind turbines to capture a bigger share of this market. Currently, the main wind turbine manufacturers around the world include Vesta in Denmark, Siemens Gamesa in Spain, Goldwind in China, GE in the United States, Envision in China, MingYang in China, Windey in China, and Nordex in Germany [3].

This chapter is devoted to make the reader familiar with the wind power and wind turbines. For this purpose, wind generation would be discussed, then the governing equations between wind power and wind turbines would be introduced and different types of turbines will be presented.

1. Wind generation

Wind is the air flowing from a high pressure region to a low pressure region. Larger pressure difference generates more powerful wind. Several factors may generate the pressure difference, among them uneven solar heating, Coriolis force and local geography are the most dominate factors.

1.1 Uneven solar heating

Among the impacting factors of wind generation, unequal solar radiation is the most important one. One of the important factors in creating unequal radiation in different parts of the earth is the spherical shape of the earth and its different angle in different places with the sun's rays. For example, the equator is perpendicular to the sun's rays and receives the most heat energy from the sun, while the poles are approximately parallel with the sun's rays, so they receive the least radiant energy. This difference in solar heating causes a temperature gradient from the equator to the poles. Therefore, the air in the equator becomes warmer and lighter and moves upwards, while at the poles, the air becomes heavier and moves downwards. This difference in air density creates a pressure gradient from the poles to the equator, which results in a wind flow on the ground from each pole to the equator.

Second reason for the uneven solar radiation on earth is the declined angle of earth with respect to the self-rotation axis. This angle is 23.45 degrees that is shown in Fig. 11.3. This angle causes uneven heating on earth surface and results in cyclic seasonal changes of weather.

Third reason for uneven solar heating on earth is that earth is covered with different materials and different colors and shapes. Some places are covered with water, some with ice, some with grass, sand, soil, trees, etc. Different materials behave differently for absorbing, reflecting, or passing sunlight. As a result, uneven heating happens on earth [5].

The fourth factor that causes uneven heating is the earth's topographic surface. Mountain ranges, valleys, hills, and deserts provide different potentials for receiving solar radiation. While one side of a mountain is receiving sunlight, the other side is in shadow and receives no direct sunlight. At the same time, desert is open to receive sunlight.

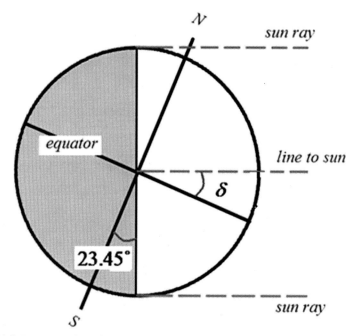

Figure 11.3 Representation of the declination angle [4].

1.2 Coriolis force

Earth's self-rotation while moving around the sun creates the Coriolis force that deflects the movement of atmosphere and wind patterns. This force changes with earth's latitude; hence, the Coriolis force would be zero at equator and maximum at the Poles. In the northern hemisphere wind is deflected to the right due to the earth rotation direction, and in the southern hemisphere, it is deflected to the left. The balance between the Coriolis force and pressure gradient, which is generated due to the air density gradient, determines the atmospheric circulation in different regions of earth. If the earth did not have self-rotation about its axis, due to the uneven heating, atmosphere would only circulate between the poles and the equator. As it can be seen in Fig. 11.4, due to this force balance, the single circulation pattern of atmosphere in each hemisphere (which was made due to uneven solar heating) is divided into 3 cells with different circulation patterns. These cells are called Hadley cell (from latitude of 0−30 degrees), Ferrel or mid-latitude cell (from latitude of 30−60 degrees), and polar cell (from latitude of 60−90 degrees). In the Hadley cell, wind direction is deflected to the northeast; while in the Ferrel cell, it is deflected to the southwest [5].

1.3 Local geography

Earth's roughness changes the wind speed profile on earth. The roughness may be natural like trees, mountains, hills, and valleys or manmade structures such as buildings,

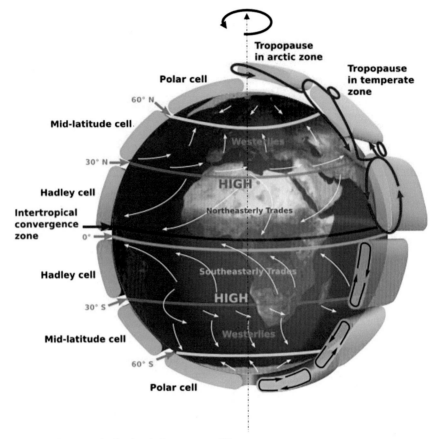

Figure 11.4 Atmospheric circulation patterns [6].
Free of copy right images by wpclipart.

towers, bridges, etc. Due to the viscosity of air (no slip condition), and roughness on the ground, wind loses part of its momentum in the layers close to ground, and its speed would be slower in this region. But out of this region (the boundary layer) where the roughness and viscosity effects diminish, wind speed increases. Wind speed increases at higher altitude where there is no obstacle. A reliable approximation says that by doubling the height, wind speed increases by 10% [7]. From this, it can be concluded that tall tower turbines result in more power generation than short tower turbines.

2. Wind power and wind turbine characteristics

Wind power can be used for different purposes, such as sailing, water pumping, electricity generation, grinding, flying, or producing hydrogen. In this book, it is used to generate power and electricity.

2.1 Wind power

Wind, which is made of moving air molecules, contains kinetic energy. Every moving particle with mass of m and velocity of u would have a kinetic energy as Eq. (11.1).

$$E_k = \frac{1}{2} m u^2 \qquad (11.1)$$

Due to the turbulence effect, wind speed is unsteady. It can be written as the summation of a mean term $(\bar{u})$ and a fluctuating term (u') as Eq. (11.2).

$$u(t) = \bar{u} + u'(t) \qquad (11.2)$$

where t is time. The strength of the turbulence is determined by the turbulence intensity, which is determined as the ratio of root mean square of the fluctuating terms to the mean term as below:

$$I = \frac{\sigma}{\bar{u}} = \frac{\sqrt{\frac{1}{3}\left(u_x'^2 + u_y'^2 + u_z'^2\right)}}{\sqrt{\bar{u}_x^2 + \bar{u}_y^2 + \bar{u}_z^2}} \qquad (11.3)$$

The bigger the turbulence intensity, the bigger the dynamic force and fatigue forces will act on the turbine blades. Controlling different motion modes of a turbine at higher turbulence intensity is much more difficult. Hence, determining turbulence intensity for the proposed sites to erect a wind turbine farm is crucial.

The power of wind is estimated according to the mean wind speed and the air mass flow of $(\dot{m})$ as below:

$$P_w = \frac{1}{2} \dot{m} \bar{u}^2 \qquad (11.4)$$

The wind mass flow rate is also defined as below:

$$\dot{m} = \rho \bar{u} A \qquad (11.5)$$

where A is the effective swept area by the wind turbine blades, ρ is the air density that is variable with height and temperature. By combining Eqs. (11.4) and (11.5), the wind power would be determined by knowing the air density, swept area, and mean wind speed as Eq. (11.6):

$$P_w = \frac{1}{2}\rho A \bar{u}^3 \tag{11.6}$$

The swept area is the area that is the effective for power generation. For example, in a horizontal axis wind turbine as shown in Fig. 11.4, the swept area is as Eq. (11.7a).

$$A = \pi\left[(l+r)^2 - r^2\right] = \pi l(l+2r) \tag{11.7a}$$

where r is the hub radius, and l is the blade length.

In addition, the ratio of the total area of blades to the swept area of the turbines is called solidity, if the area of a blade is A_{blade} and the turbine has N blades, the solidity would be:

$$\text{Solidity} = \frac{NA_{blade}}{A} \tag{11.7b}$$

Turbines with low solidity (about 0.1) have lower torque but higher speed, and turbines with high solidity (about 0.8) have higher torque but lower speed (Fig. 11.5).

According to Eq. (11.5), doubling wind speed result in eight-fold increase of wind power. This shows why wind speed is very critical for power production. The swept area is proportional to the square length of the blade. It means, in large wind turbines, by neglecting the hub radius, the wind power will increase by four-fold when the blade length is doubled.

Wind power is directly proportional with the air density, and air density varies with temperature and pressure (altitude above the see level). The following equation can be used to approximate air density at altitude of Z(m) from sea level and temperature of T(K) [8, 9].

$$\rho = \frac{353.049}{T}\exp\left(-\frac{0.034Z}{T}\right) \tag{11.8}$$

For example at sea level ($Z = 0$) and temperature of 15°C the air density would be:

$$\rho = \frac{353.049}{273.15 + 15}\exp\left(-\frac{0.034 \times 0}{273.15 + 15}\right) = 1.22\frac{\text{kg}}{\text{m}^3}$$

In addition, by combining the hydrostatic and total pressure equations, perfect gas equation, and wind shear equation, for air with Eq. (11.8) a nonlinear equation would be derived for temperature and altitude.

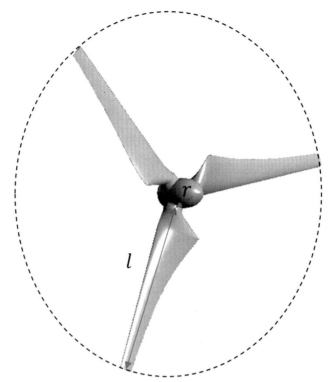

Figure 11.5 The swept area of a horizontal wind turbine.
Simulation by Omid Rasooli, M.Sc. Mechanical engineering student at University of Kurdistan.

$$P_{total}(Z) = P_{static}(Z) + \frac{1}{2}\rho\bar{u}(Z)^2 \tag{11.9}$$

$$P_{static}(Z) = P_{atm@Z_0} - \rho gZ \tag{11.10}$$

$$P_{total}(Z) = \rho RT \tag{11.11}$$

$$\bar{u}(Z) = \bar{u}(Z_0)\left(\frac{Z}{Z_0}\right)^a \tag{11.12}$$

Hence:

$$P_{atm@Z_0} - \frac{353.049}{T}\exp\left(-\frac{0.034Z}{T}\right)gZ + \frac{1}{2}\frac{353.049}{T}$$

$$\exp\left(-\frac{0.034Z}{T}\right)\left(\bar{u}(Z_0)\left(\frac{Z}{Z_0}\right)^a\right)^2 = \frac{353.049}{T}\exp\left(-\frac{0.034Z}{T}\right)RT \tag{11.13}$$

which shows a nonlinear equation between temperature and altitude. In the above equations, P_{total} is the total pressure and $\frac{1}{2}\rho\bar{u}^2$ is the dynamic pressure. In addition, Z_0 is the reference altitude in which $\bar{u}(Z_0)$ and $P_{atm@Z_0}$ are measured at this altitude. a is the wind shear coefficient that the empirical results show that it follows the one seventh law $\left(a = \frac{1}{7}\right)$.

2.2 Power density and power coefficient

The *wind power flux* or *wind power density* is a vital criterion to compare the power content in unit of swept area at the same altitude for different sites and can be used to classify wind power. The wind power density (W_{PD}) is determined as below:

$$W_{PD} = \frac{P_w}{A} = \frac{1}{2}\rho\bar{u}^3 \tag{11.14}$$

The wind power at altitudes of 10 and 50 m is classified according to the mean wind speed and power density in Table 11.1. For large-scale wind turbines, class rating of 4 or higher is preferred for wind farms [5].

It is clear that all of the wind power is not extractable by the turbine's blades. It depends on the design of the wind turbine, especially the blades. Blades convert the wind power to mechanical power that results in the rotation of the turbine shaft. The amount of extracted power (P) with respect to the total wind power is called the *power coefficient*, C_p. Hence:

$$C_p = \frac{P}{P_w} \tag{11.15}$$

Hence:

Table 11.1 Classification of wind power class according to wind speed and Refs. [5,10].

Wind power class	Wind potential	10 m height		50 m height	
		$W_{PD}\left(\frac{W}{m^2}\right)$	$\bar{u}\left(\frac{m}{s}\right)$	$W_{PD}\left(\frac{W}{m^2}\right)$	$\bar{u}\left(\frac{m}{s}\right)$
1	Poor	<100	<4.4	<200	<5.6
2	Marginal	100–150	4.4–5.1	200–300	5.6–6.4
3	Fair	150–200	5.1–5.6	300–400	6.4–7.0
4	Good	200–250	5.6–6.0	400–500	7.0–7.5
5	Excellent	250–300	6.0–6.4	500–600	7.5–8.0
6	Outstanding	300–350	6.4–7.0	600–700	8.0–8.8
7	Superb	>400	>7	>800	>8.8

$$P = \frac{1}{2}C_p\rho A\bar{u}^3 \tag{11.16}$$

Since there are many irreversibilities in the interaction of wind with turbine, the magnitude of the C_p never reaches its theoretical limit, which is called the Lanchester-Betz limit. According to Betz theory, [11] the maximum possible power coefficient, C_p, is 59%. In other words, the maximum theoretical efficiency of the wind turbine is never 100% and is always below the Betz limit. The reason is that if there was no irreversibility between wind and turbine, the wind speed just after the turbine must reach zero and full stop to give up all of its power. However, it is evident that there are many irreversibilities such as blade friction, tip losses, root losses, wake formation, etc. In addition, wind full stop will never happen by using wind turbine. Hence, the reversible efficiency of wind turbines is 0.59, but in reality, it is ranging from 30% to 45%.

The *turbine power density* (T_{PD}) can also be determined as below:

$$T_{PD} = \frac{P}{A} = \frac{1}{2}C_p\rho\bar{u}^3 = C_p W_{PD} \tag{11.17}$$

Example 11.1. The diameter of the Siemens SWT-3.6-120 onshore wind turbine is 120 m, and its power density is 318.6 $\frac{W}{m^2}$. The rated wind speed for this turbine is 12.5 $\frac{m}{s}$. Determine (a) the nominal (rated) power of this turbine (b) The power coefficient. Assume air temperature and pressure as 25°C and 101.3 Kpa.

Solution. (a) By neglecting the hub area, the swept area can be calculated as below:

$$A = \pi l^2 = \pi\left(\frac{120}{2}\right)^2 = 11310 \text{ m}^2$$

And using the turbine power density, the nominal power would be determined.

$$T_{PD} = \frac{P_{nom}}{A} \rightarrow 318.6 = \frac{P_{nom}}{11310} \rightarrow P_{nom} = 3.603 \text{ MW}$$

(b) The air density at 25°C and 101.3 Kpa is 1.184 $\frac{kg}{m^3}$, and the power coefficient at the rated power and the rated wind speed would be:

$$T_{PD} = \frac{1}{2}C_p\rho\bar{u}^3 \rightarrow 318.6 = \frac{1}{2}C_p \times 1.184 \times 12.5^3 \rightarrow C_p = 0.28$$

The code written in EES for the Example 11.1 is given below (**Code 11.1**):

Code 11.1. The code written in EES for the Example 11.1.

```
T_amb=25
P_amb=101.3
ro_air= DENSITY(Air,T=T_amb,P=P_amb)
D=120
l=D/2
p_D=318.6
u_bar=12.5
A=pi*l^2
P_nom=A*p_D
P_w=ro_air*A*u_bar^3/2
C_p=P_nom/P_w
```

Example 11.2. For the wind turbine presented in Example 11.1, determine the wind speed at the downstream of the turbine. How much is the $\frac{\bar{u}_{out}}{\bar{u}_{in}}$ at nominal operation of the wind turbine?

Solution. According to the first law of thermodynamics for the nominal operation conditions:

$$P_{w,in} = P_{w,out} + P_{nom} \rightarrow \frac{1}{2}\rho A \bar{u}_{in}^3 = \frac{1}{2}\rho A \bar{u}_{out}^3 + P_{nom} \rightarrow \frac{1}{2} 1.184 \times 11310 \times 12.5^3$$

$$= \frac{1}{2} 1.184 \times 11310 \times \bar{u}_{out}^3 + 3603000 \rightarrow \bar{u}_{out}$$

$$= 11.23 \frac{m}{s}$$

$$\alpha = \frac{\bar{u}_{out}}{\bar{u}_{in}} = \frac{11.23}{12.5} = 0.90$$

which means only 10% of the wind speed is used for power generation. This percent of wind speed corresponds to 28% of wind power ($C_p = 0.28$).

The code written in EES for the Example 11.2 is given below (**Code 11.2**):

Code 11.2. The code written in EES for the Example 11.2.

```
T_amb=25
P_amb=101.3
ro_air= DENSITY(Air,T=T_amb,P=P_amb)
D=120
l=D/2
p_D=318.6
u_bar_in=12.5
A=pi*l^2
P_nom=A*p_D
P_w=ro_air*A*u_bar_in^3/2
P_w=P_nom+ro_air*A*u_bar_out^3/2
alpha=u_bar_out/u_bar_in
```

Example 11.3. Determine the impact of temperature variation from -20 to $45°C$ on the power coefficient and wind power of the turbine presented in Example 11.1. Assume constant pressure at 101.3 kPa. Plot the results.

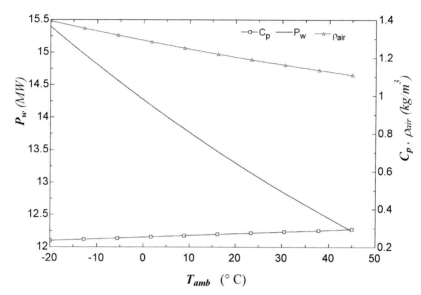

Figure 11.6 The ambient air temperature impact on turbine power coefficient, wind power, and air density.

Solution. To determine the impact of temperature variation on the turbine characteristics, the density variation should be determined. The code written for Example 11.1 is used in a parametric table by EES, and the results are shown in Fig. 11.6. To do the parametric analysis, temperature (T_amb), is commented as "T_amb" in the code. The results show that in the given temperature range, wind power changes 3.15 MW, which is very considerable. In addition, due to this variation, the power coefficient has increased from 0.234 in −20°C to 0.2941 in 45°C. Density has decreased from 1.394 in −20°C to 1.109 in 45°C. The impact of altitude is neglected in this problem.

Example 11.4. Determine the impact of pressure variation from 101.3 to 85 kPa on the power coefficient of the turbine and the wind power of the turbine presented in Example 11.1. Plot the results. Assume temperature constant at 25°C.

Solution. To determine the impact of pressure variation on the turbine characteristics, the density variation should be determined. The code written for Example 11.1 is used in a parametric table by EES, and the results are shown in Fig. 11.7. To do the parametric analysis, pressure (P_amb) is commented as "P_amb" in the code. The results show that in the proposed pressure range wind power changes 2.1 MW, which is considerable. In addition, due to this variation, the power coefficient has decreased from 0.2756 in 85 kPa to 0.3285 in 101.3 kPa. Density has decreased from 1.184 in 85 kPa to 0.9932 in 101.3 kPa.

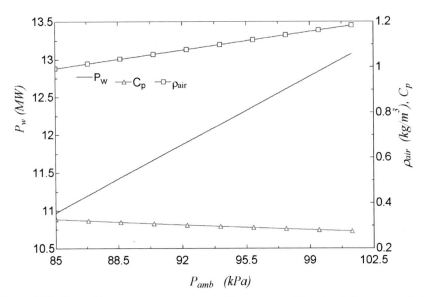

Figure 11.7 The ambient pressure impact on turbine power coefficient, wind power, and air density.

2.3 Lanchester-Betz limit

Example 11.5. By using the first law of thermodynamics and definition of power coefficient, prove that the reversible power coefficient (highest theoretical power coefficient) for wind turbines is 0.59.

Solution. The control volume for the wind turbine is drawn as Fig. 11.8, and the energy conservation can also be written as below:

$$\dot{E}_{in} - \dot{E}_{out} = \frac{dE_{cv}}{dt}$$

For the steady-state conditions, $\frac{dE_{CV}}{dt} = 0$ and:

$$\dot{E}_{in} = \dot{E}_{out}$$

$$\left(\dot{m}h + \frac{1}{2}\dot{m}\bar{u}^2 + \dot{m}gz \right)_{in} = \left(\dot{m}h + \frac{1}{2}\dot{m}\bar{u}^2 + \dot{m}gz \right)_{out} + P_{out} + \dot{Q}_{out}$$

By neglecting the heat transfer $\dot{Q}_{out} = 0$, and knowing that $z_{in} = z_{out}$, and assuming air as a perfect gas, $h = h(T)$, and since $T_{in} = T_{out}$ due to no heat transfer result in $h_{in} = h_{out}$, therefore:

Figure 11.8 The control volume for the wind turbine.

$$\left(\frac{1}{2}\dot{m}\bar{u}^2\right)_{in} = \left(\frac{1}{2}\dot{m}\bar{u}^2\right)_{out} + P_{out} \rightarrow P_{out} = \frac{1}{2}\dot{m}\left(\bar{u}_{in}^2 - \bar{u}_{out}^2\right)$$

Since $\dot{m} = \rho u A$, assume:

$$u = \frac{\bar{u}_{in} + \bar{u}_{out}}{2}$$

Hence:

$$\dot{m} = \rho A \frac{\bar{u}_{in} + \bar{u}_{out}}{2}$$

Therefore:

$$P_{out} = \frac{1}{2}\rho A \frac{\bar{u}_{in} + \bar{u}_{out}}{2}\left(\bar{u}_{in}^2 - \bar{u}_{out}^2\right)$$

The input power is the inlet power of wind:

$$P_{in} = \dot{E}_{in} = \frac{1}{2}\rho A \bar{u}_{in}^3$$

According to the definition of power coefficient:

$$C_p = \frac{P_{out}}{P_{in}} = \frac{\frac{1}{2}\rho A \dfrac{\bar{u}_{in} + \bar{u}_{out}}{2}\left(\bar{u}_{in}^2 - \bar{u}_{out}^2\right)}{\frac{1}{2}\rho A \bar{u}_{in}^3} = \frac{1}{2}\left(1 + \frac{\bar{u}_{out}}{\bar{u}_{in}}\right)\left(1 - \left(\frac{\bar{u}_{out}}{\bar{u}_{in}}\right)^2\right)$$

By assuming $\alpha = \frac{\bar{u}_{out}}{\bar{u}_{in}}$, the power coefficient can be maximized as below:

$$C_p = \frac{1}{2}(1+\alpha)(1-\alpha^2) \tag{11.18}$$

$$\rightarrow \frac{dC_p}{d\alpha} = 1 - 2\alpha - 3\alpha^2 = 0 \rightarrow \alpha = 0.33 \rightarrow$$

$$C_p^{max} = \frac{1}{2}(1+0.33)(1-(0.33)^2) = 0.59$$

Hence, the theoretical maximum of power coefficient would be:

$$C_p^{max} = 0.59 \tag{11.19}$$

which is the Lanchester-Betz limit for the wind turbine efficiency. In addition, the theoritical minimum velocity of wind after passing over the wind turbine would be:

$$\alpha = \frac{\bar{u}_{out}}{\bar{u}_{in}} = 0.33 \rightarrow \bar{u}_{out}^{min} = 0.33\bar{u}_{in} \tag{11.20}$$

2.4 Wind speed utilization coefficient

The α which is the ratio of the wind speed loss to the input wind speed can be named as the *wind speed loss coefficient* because it shows the ratio of the wind speed, which passes over the turbine without being used. Under reversible conditions and desired environment, $\alpha = \alpha_{min}$ shows the minimum speed loss of wind turbine. In such a manner, the *wind speed utilization coefficient* can be defined as:

$$C_u = \frac{wind\ speed\ utilized}{input\ wind\ speed} = \frac{\bar{u}_{in} - \bar{u}_{out}}{\bar{u}_{in}} = 1 - \frac{\bar{u}_{out}}{\bar{u}_{in}} = 1 - \alpha \tag{11.21}$$

The analogy between the definition of the Carnot cycle efficiency $\eta_{Carnot} = 1 - \frac{T_L}{T_H}$ for internal reversible heat engines and the wind speed utilization coefficient is very interesting. According to the definition, the maximum C_u happens under reversible conditions and desired environmental conditions that corresponds to $\alpha = \frac{1}{3}$, hence:

$$C_u^{max} = 1 - \alpha = 1 - \frac{1}{3} = \frac{2}{3} \tag{11.22}$$

That means under reversible (ideal) conditions, only two third of the wind speed would be utilized for power production. Under real conditions, the wind speed utilization coefficient is always less than $\frac{2}{3}$.

2.5 Thrust coefficient

Air passing through the swept area (from point 1 to point 2 in Fig. 11.9) of the wind turbine experiences a pressure drop, and due to this pressure drop, a thrust force will act on the turbine [8,12].

The thrust force is equal to:

$$T = (P_1 - P_2)A \tag{11.23}$$

To evaluate the $\Delta P = P_1 - P_2$, two Bernoulli equations can be written. Between the upstream and point 1:

$$\frac{1}{2}\rho \bar{u}_{in}^2 + P_{amb} = \frac{1}{2}\rho \bar{u}_1^2 + P_1 \tag{11.24}$$

and between point 2 and downstream:

$$\frac{1}{2}\rho \bar{u}_{out}^2 + P_{amb} = \frac{1}{2}\rho \bar{u}_2^2 + P_2 \tag{11.25}$$

Hence by subtracting Eq. (11.25) from (11.24):

$$P_1 - P_2 = \frac{1}{2}\rho\left(\bar{u}_{in}^2 - \bar{u}_{out}^2\right) + \frac{1}{2}\rho\left(\bar{u}_1^2 - \bar{u}_2^2\right)$$

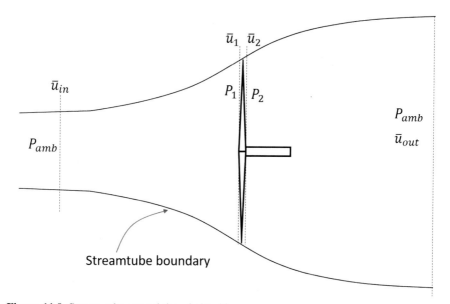

Figure 11.9 Stream tube around the wind turbine.

Since:

$$\dot{m}_1 = \dot{m}_2 \rightarrow \rho A \bar{u}_1 = \rho A \bar{u}_2 \rightarrow \bar{u}_1 = \bar{u}_2$$

Therefore:

$$P_1 - P_2 = \frac{1}{2} \rho \left(\bar{u}_{in}^2 - \bar{u}_{out}^2 \right) \tag{11.26}$$

And:

$$T = \frac{1}{2} \rho \left(\bar{u}_{in}^2 - \bar{u}_{out}^2 \right) A \tag{11.27}$$

The thrust coefficient which is a dimensionless parameter is defined as below:

$$C_T = \frac{T}{\frac{1}{2} \rho \bar{u}_{in}^2 A} = \frac{\frac{1}{2} \rho \left(\bar{u}_{in}^2 - \bar{u}_{out}^2 \right) A}{\frac{1}{2} \rho \bar{u}_{in}^2 A} = 1 - \frac{\bar{u}_{out}^2}{\bar{u}_{in}^2} = 1 - \alpha^2 \tag{11.28}$$

The value of C_T under desired environmental conditions in which the power coefficient became maximum would be:

$$C_T^{max} = 1 - \frac{1}{3^2} = \frac{8}{9} = 0.89 \tag{11.29}$$

2.6   Axial induction factor

The axial induction factor, a, is the fractional decrease of wind speed from the upstream to the turbine rotor. The induction factor can be adjusted by using the pitch angle and generator torque, which are standard inputs for utility scale wind turbine. The induction factor can be stated as below:

$$a = \frac{\bar{u}_{in} - \bar{u}_1}{\bar{u}_{in}} \tag{11.30}$$

Since $\bar{u}_1 = \bar{u}_2 = \frac{\bar{u}_{in} + \bar{u}_{out}}{2}$ therefore:

$$a = \frac{\bar{u}_{in} - \bar{u}_1}{\bar{u}_{in}} = a = \frac{\bar{u}_{in} - \frac{\bar{u}_{in} + \bar{u}_{out}}{2}}{\bar{u}_{in}} = \frac{\bar{u}_{in} - \bar{u}_{out}}{2\bar{u}_{in}} = \frac{1}{2}(1 - \alpha) = \frac{C_u}{2}$$

Or:

$$\alpha = 1 - 2a$$

$$C_u = 2a \tag{11.31}$$

The thrust coefficient can also be stated as a function of induction factor.

$$C_T = 1 - \alpha^2 = 1 - (1 - 2a)^2 = 4a(1 - a)$$

Hence:

$$C_T = 4a(1 - a) \tag{11.32}$$

And the power coefficient can also be stated as a function of induction factor:

$$C_p = \frac{1}{2}(1 + \alpha)(1 - \alpha^2) = \frac{1}{2}(1 + 1 - 2a)(1 - (1 - 2a)^2) = 4a(1 - a)^2$$

Hence:

$$C_p = 4a(1 - a)^2 \tag{11.33}$$

Example 11.6. Plot the wind speed coefficient, power coefficient, and thrust coefficient versus induction factor for $0 \leq a \leq 1$ by using parametric table in EES.

Solution. Since $\alpha = 1 - 2a$ and $0 \leq \alpha \leq 1$; therefore:

$$0 \leq 1 - 2a \leq 1 \rightarrow 0 \leq a \leq \frac{1}{2} \tag{11.34}$$

That means for the $a > \frac{1}{2}$ the momentum equation would not be satisfied. Therefore, the wind speed, thrust, and power coefficients are drawn in the induction factor range from 0 to 0.5 in Fig. 11.10.

2.7 Capacity factor

Wind turbine's nameplates show the nominal capacity of the turbine. It is shown by P_{nom} and is the maximum power that can be generated by the turbine under desired wind conditions and the generator specifications. Due to the variable nature of wind, turbine nominal capacity is not always reached. When there is no wind or slow wind, turbine remains still or small power will be generated. The ratio of the annual mean power generated by the turbine to the annual nominal power generation under ideal conditions for wind is called *Capacity factor (CF)*. In other words, *CF* is the ratio of the actual energy generated by the wind turbine to the energy, which could be potentially harvested from the wind under nominal environmental conditions.

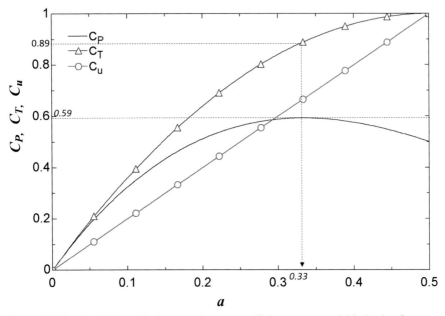

Figure 11.10 The wind speed, thrust, and power coefficients versus axial induction factor.

$$CF = \frac{E_{actual}}{E_{ideal}} = \frac{time \times \overline{P}}{time \times P_{nom}} \tag{11.35}$$

In addition, the CF can be defined as the ratio of the effective time of operation to the total time as below:

$$CF = \frac{E_{actual}}{E_{ideal}} = \frac{E_{actual}}{time \times P_{nom}} = \frac{\dfrac{E_{actual}}{P_{nom}}}{time} = \frac{time_{eff}}{time} \tag{11.36}$$

CF usually ranges from 0.26 to 0.5 depending on site and turbine's characteristics.

2.8 Characteristics of wind turbines

The characteristic curve of a typical wind turbine is shown in Fig. 11.11 as it shows wind turbines do not generate power for very slow wind speeds smaller than the cut-in speed. Operation at very low speeds may cause damage to the wind turbine specially for lubrication of the bearings. At the cut-in speed, the turbine rotates and starts generating power. Power generation increases as the wind blows faster. Power generation keeps increasing until it reaches the rated (nominal) power at the rated wind speed. Since the turbine is matched with the generator, its speed must be controlled to avoid damages to the generator and blades. For this purpose, the control

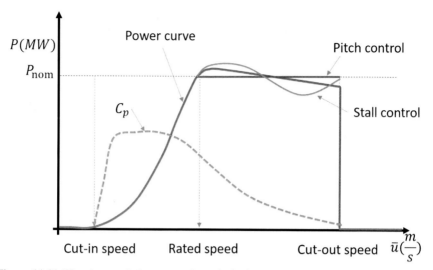

Figure 11.11 The characteristics curve of a typical wind turbine.

system controls the turbine rotational speed by monitoring power generation and adjusting the angle of blades with respect to wind direction. The control system is working until wind speed reaches the cut-out speed. At this speed, the control systems shutdown the turbine by braking to avoid any irreversible damage. In addition, survival wind speed is the maximum wind speed that a turbine is designed to tolerate (survive) before sustaining damage. Typical values of survival wind speed for commercial wind turbines range from $40 \frac{m}{s}$ to $72 \frac{m}{s}$, and mostly used survival wind speed is about 60 m/s.

2.8.1 Control of wind turbine

The wind turbine control system must be activated when the wind speed is too low or too high. If the wind speed is less than the cut-in speed, it will be shutdown to prevent damage to the turbine bearings and turbine generator. Also, if the wind speed exceeds the nominal (rated) speed, since the power output is proportional with the speed cube, applying too much power into the generator will cause it to overheat and damage it, so in this case, the control system should limit the rotational speed of rotor. If the wind speed exceeds the cut-out speed, the control system will turn off the turbine to prevent damage to the generator and turbine. Several control systems including pitch control, passive and active stall control, and yaw control that act like braking system exists in the wind turbines.

Pitch control is performed by changing the angle between the turbine blade and the wind flow. If the turbine blade is not completely aligned with the wind, the pressure difference between the two sides of the blade airfoil will reduce, resulting in a reduction in lift force acting on the blade. Reducing the lift force will also reduce the rotational speed of the turbine. The control system does this control by monitoring the output power of the generator. If the power generation tends to increase too much, the control system will reduce the rotor speed by increasing the angle between the blade and the wind.

Stall, which is the separation of the current from the surface of the airfoil, creates a vortex and turbulence on one side of airfoil. Occurrence of this phenomenon on the wings of an aircraft or axial compressor blades can cause the plane to fall and severe vibrations in the compressor and even break the compressor blades. In wind turbines, however, it is used in a controlled way to reduce the rotational speed of the blades. Passive stall uses fixed blades on the hub of the wind turbine. In this case, stall is generated by the blade's twists. As the wind speed increases, the flow separation tendency increases on the low pressure side of the airfoil, and as a result, the pressure increases on the low pressure side and balancing the high pressure side of the airfoil and reducing lift force and rotor rotational speed.

Active stall unlike the passive stall, uses blades with the variable and adjustable pitch angle. In this method, like the pitch control, the generated power is monitored by the control system, but instead of misaligning the blade with the wind flow, the pitch angle is adjusted for the occurrence of stall.

When the wind speed is less than the rated speed of the turbine, aligning the turbine blades with the wind direction will increase the turbine efficiency and more wind power will be captured. This is done by Yaw control. In small turbines, a small vane at the end of the rotor and behind the blades does this, but in large turbines, an anemometer installed on the rear of the hub measures the wind speed and direction, and the control system adjusts the rotor direction to maximize the power generation. The pitch and yaw rotation directions are shown in Fig. 11.12.

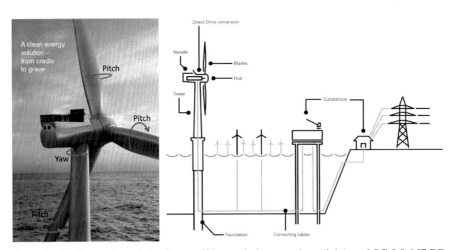

Figure 11.12 System boundaries for an offshore wind power plant (right), and SG 8.0-167 DD (Left) [13].
Courtesy of Siemens Gamesa Renewable Energy.

Example 11.7. The SG-8.0-167 DD offshore wind turbine is a Siemens Gamesa product (Fig. 11.12). This wind turbine is a benchmark solution for sites with all wind speeds. The rotor diameter is 167 m with nominal power of 8 MW, and the blade length is 81.4 m. This turbine produces 34,000 MWh of electrical energy for the grid annually. And its life time is estimated to be 25 years. Its power regulation is done by pitch control and variable speed. The tower height is site specific. This turbine uses direct drive technology to reduce the number of wear-prone and rating components that makes the turbine smaller and lighter, and as a result, the installation costs drop. In addition, it uses integral blade technology that casts the whole blade in one piece and eliminates weaker points that may be produced due to glued joints. The sensors installed on this turbine continuously transmit data to the Siemens Gamesa Diagnostic Center, enabling the early detection of anomalies and preventing potential failures. By analyzing vibration pattern, they can identify possible problems and troubleshoot the turbine to avoid serious damages. Most of turbines should be shutdown when wind speed is higher than 25 m/s. The turbines equipped with The High Wind Ride Through (HWRT) system that is a breakthrough in stabilizing energy output can continue operation. This system will slowly ramp down power output instead, enabling smoother production ramp-down and thereby a more reliable electrical grid. The characteristics of the SG-8.0-167 DD is reported in Table 11.2. Determine: (a) swept area of the turbine (b) potential wind power (c) capacity factor, (d) wind speed behind the turbine for inlet wind speed of 25 m/s (e) power coefficient, thrust coefficient, and wind speed coefficient if the wind turbine is supposed to be installed at 1000 m above the sea level and air temperature is $-10°C$.

Solution. (a) The swept area is determined by using the rotor diameter and blade length as below:

Table 11.2 The characteristics of SG 8.0-167 DD [13].

Product and system description	Main characteristics
Turbine	SG 8.0-167 DD
Number of turbines in wind power plant	80
Expected lifetime	25
Expected average wind speed	10 m/s
Distance to shore/shore to grid	50/km/22 km
Annual energy production to grid per turbine (wake, availability and electrical losses subtracted)	Approx. 34,000 MWh
Nacelle	8.0 MW DD (steel, iron, copper)
Blades	81 m (fiberglass, epoxy)
Tower	92 m (steel)
Foundation	925 t (steel)
Substations	12,700 t (steel, concrete)

$$A = \pi\left[(l+r)^2 - r^2\right] = \pi l(l+2r) = \pi\frac{167}{2}\left(\frac{167}{2} + 2\left(\frac{167}{2} - 81\right)\right)$$

$$= 21890 \text{ m}^2$$

(b) The wind power would be calculated at the nominal and expected average wind speeds, to do this the air density at the given altitude and temperature must be determined:

$$\rho = \frac{353.049}{T}\exp\left(-\frac{0.034Z}{T}\right) = \frac{353.049}{273.15 - 10}\exp\left(-\frac{0.034 \times 1000}{273.15 - 10}\right) = 1.179\frac{\text{kg}}{\text{m}^3}$$

The rated wind speed is half of the cut-out speed which is $\frac{25}{2} = 12.5\frac{\text{m}}{\text{s}}$

$$P_w = \frac{1}{2}\rho A\bar{u}^3 = \frac{1}{2}1.179 \times 21890 \times 12.5^3 = 25.2 \text{ MW}$$

And for the average wind speed of 10 m/s:

$$P_{w,avg} = \frac{1}{2}\rho A\bar{u}^3 = \frac{1}{2}1.179 \times 21890 \times 10^3 = 12.9 \text{ MW}$$

(c) Annual energy production for the electricity grid per turbine after subtracting wake, availability and electrical losses is 34,000 MWh per year, hence the real average power during a year would be:

$$E_{actual} = 34000\frac{\text{MWh}}{\text{year}} = 34000\frac{\dfrac{\text{MWh}}{\text{year}}}{24\dfrac{\text{h}}{\text{day}} \times 365\dfrac{\text{day}}{\text{year}}} = 3.88 \text{ MW}$$

which means by average the wind turbine is continuously producing 3.881 MW of electricity through a year. The capacity factor of this turbine would be:

$$CF = \frac{E_{actual}}{E_{ideal}} = \frac{3.88 \text{ MW}}{8 \text{ MW}} = 0.485$$

(d) to determine the wind velocity behind the turbine, the energy conservation equation for the actual conditions would be as below:

$$P_{w,avg} = E_{actual} + \frac{1}{2}\rho A\bar{u}_{out}^3 \rightarrow 12.9 = 3.881 + \frac{\frac{1}{2}1.179 \times 21890 \times \bar{u}_{out}^3}{10^6} \rightarrow \bar{u}_{out}$$

$$= 8.88\frac{\text{m}}{\text{s}}$$

In the above calculations the efficiency of the gear box, and generator are assumed to be 100%.

(e) to determine the wind speed coefficient, the wind speed loss coefficient is required, hence:

$$\alpha = \frac{\bar{u}_{out}}{\bar{u}_{in}} = 0.710$$

Hence:

$$C_u = 1 - \frac{\bar{u}_{out}}{\bar{u}_{in}} = 1 - \alpha = 1 - 0.71 = 0.290$$

And the induction coefficient would be:

$$a = \frac{1 - \alpha}{2} = \frac{0.29}{2} = 0.145$$

The thrust coefficient:

$$C_T = 4a(1 - a) = 4 \times 0.145 \, (1 - 0.145 \,) = 0.496$$

And the power coefficient:

$$C_p = 4a(1 - a)^2 = 4 \times 0.06 \, (1 - 0.06 \,)^2 = 0.424$$

The code written in the EES for this example is given in **Code 11.3**:

Code 11.3. The code written in the EES for the Example 11.7.

```
D=167
L=81.4
L+r=D/2
P_nom=8 "MW"
A=pi*((L+r)^2-r^2)
u_in=12.5
u_avg=10
T=273.15-10
Z=1000
ro_air=353.049*exp(-0.034*Z/T)/T
P_w=ro_air*A*u_in^3/2*10^(-6)
P_w_avg=ro_air*A*u_avg^3/2*10^(-6)
P=34000  "MWh"
E_actual=P/(24*365)
CF=E_actual/P_nom
P_w_avg=E_actual+ro_air*A*u_out^3/2*10^(-6)
alpha=u_out/u_in
alpha=1-2*a_induction
C_u=2*a_induction
C_T=4*a_induction*(1-a_induction)
C_p=4*a_induction*(1-a_induction)^2
```

3.   Wind turbine classifications

Depending on the axis of rotation, wind turbines are classified as horizontal axis wind turbines (HAWTs) and vertical axis wind turbines (VAWTs).

3.1   Horizontal axis wind turbines

The axis of rotation in these turbines is parallel to the ground surface and approximately the wind direction as well. Figs. 11.12 and 13 show horizontal axis wind turbines. Most of the todays' commercialized wind turbines fall in this class. The main components of a HAWTs are shown in Fig. 11.13. As previously explained, these turbines work based on lift force. Lift is generated due to pressure difference between two sides of the blade's airfoil. Lift force creates a torque, and it rotates the main shaft of the turbine. The main shaft is connected to a gearbox. The gearbox increases the rotational speed to adjust it with the required speed by the generator. The generator then

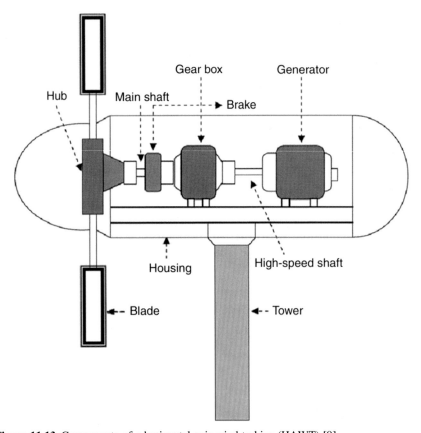

Figure 11.13 Components of a horizontal axis wind turbine (HAWT) [9].
Courtesy of Elsevier.

converts the rotation to electricity. In addition to these main components, the control systems for controlling yaw, pitch, power and safety are also placed in between. In addition, the brake system which can be activated by the control system is placed between the generator and hub. This brake operates when wind speed exceeds the cut-out speed. The power received by the blades would reduce at the outlet of the generator due to energy loss in the generator, gearbox, bearings, converter, switches, controls, and cables. Hence the total efficiency of a wind turbine would be as below:

$$\eta_t = C_p \eta_{gear} \eta_{gen} \eta_{ele} \tag{11.37}$$

And the effective power generated by the wind turbine would be as:

$$P_{eff} = C_p \eta_{gear} \eta_{gen} \eta_{ele} P_w \tag{11.38}$$

In which η_{gear}, η_{gen}, and η_{ele} are the gearbox, generator, and electrical elements efficiencies.

HAWTs may be single-blade, double-blade, triple-blade, or multi-blade depending on the application, design, and site specifications. More blades mean higher solidity, and higher solidity means higher torque and lower speed. Although single-blade rotors are cheaper, they have less drag, but they are not popular due to rotational imbalance, vibration, and visual unacceptance. The two-blade rotors have the same problems but less than the single bladed rotors. However, the three bladed rotors are the most popular rotor being used today because they have a relatively uniform load pattern. They are also visually more popular [9].

Higher number of blades are only required when the wind turbine is directly connected to an equipment such as pump, which they require a high torque at the beginning to overcome the initial inertia of the system. In such applications, the required torque may be 2 to 4 times the required torque during normal operation.

The HAWTs may be upwind or downwind (Fig. 11.14). In the downwind turbines, wind first flows over the tower and nacelle then it reaches the blades. Due to this problem, wind profile changes due to the wakes that would be generated behind the tower and nacelle, and as a result, the wind input energy reduces and wind turbine efficiency drops as well. This effect is called the tower shadow. However, the upwind turbines do no suffer from tower shadow. The tower shadow may cause imbalance and unwanted vibration and fatigue loads due to asymmetric turbulence and loading behind the tower and nacelle.

The upwind turbines require yaw control system to always the blades face the wind, but in the downwind turbines wind first passes over the nacelle and may not always require the yaw control mechanism. This makes the downwind turbines less expensive.

The main advantages of HAWTs are:

- *"They are the most stable and commercially accepted design. Today, most of the large − grid-integrated − commercial wind turbines work on three-bladed horizontal axis designs.*
- *They have a relatively lower cut-in wind velocity and higher power coefficient resulting in higher system efficiency and energy yield.*

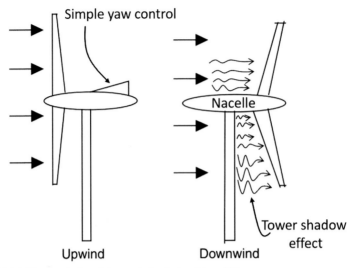

Figure 11.14 The upwind and downwind wind turbines [9].
Courtesy of Elsevier.

- *There are possibilities of using taller towers to tap the better wind potential available at higher elevations. This would be a distinct advantage at sites with strong wind shear where the velocity at higher levels could be significantly higher.*
- *There is greater control over the angle of attack, which can be optimized through variable blade pitching. This results in better system output under fluctuating wind regimes [9]"*
- They have the highest power coefficient among the wind turbines.
- In comparison with other wind turbines, more well-known and international companies are manufacturing HAWTs.

The main disadvantages of HAWTs are:

- *"HAWTs require yaw drives (or tail mechanism in case of small turbines) to orient the turbine toward wind.*
- *The heavy units of generator and gearbox are to be placed over the tall tower, which requires stronger structural support. This makes the HAWTs more complex and expensive.*
- *Taller towers make installation and maintenance more difficult and expensive [9]."*
- *Again, the taller mast height can make HAWT visible even from longer distances, which may aggravate problems related to the visual impact of wind farms."*

3.2 VAWTs

In the VAWTs the tower and the rotor are one piece or aligned. In fact, the blades are rotating about the tower (rotor). This has caused the VAWTs to accept wind from any direction. That means they do not need a yaw/pitch control system to adjust the blades angle or rotor angle with the wind direction. Since the rotor and tower are coaxial, the gear box and generator are usually placed on the ground/or near the ground. This makes them easily accessible especially for maintenance jobs. These advantages have made the VAWTs cheaper than HAWTs.

Unlike the HAWTs, the VAWTs are not generally self-starting (except for the turbines such as Savonius, which work based on drag force difference on the blades). That means they frequently need a small engine to start the turbine; hence, electricity access is inevitable. If grid electricity was not available, solar electricity and DC motors are good choices. Every advantage may bring a drawback. The short tower and closeness of generator and gearbox to the ground means low wind speed because according to the shear velocity law, wind speed increases with height. Furthermore, while a blade is passing through a favorable wind, the other blades are moving in a relatively dead zone, because favorable wind is not available in every direction at once. As a result, the overall performance is low. Among the VAWTs types, the Darrieus, and Savonius turbines are more popular. Fig. 11.15 shows the working principal of these two turbines [9].

The tip speed ratio (λ) of turbines is defined as the ratio of the rotor tip speed ($R\omega$) to the wind speed ($\bar{u}$):

$$\lambda = \frac{R\omega}{\bar{u}} = \frac{2\pi}{60} \frac{N}{\bar{u}} \tag{11.39}$$

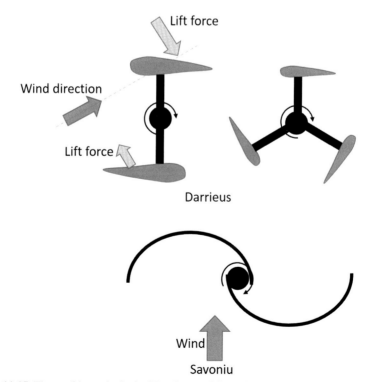

Figure 11.15 The working principal of Darrieus and Savonius.

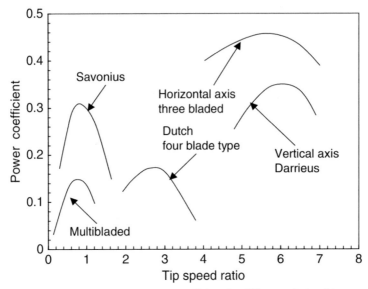

Figure 11.16 The comparison of the power coefficient for different wind turbines versus tip speed ratio [9].
Courtesy of Elsevier.

In which $\omega \left(\frac{rad}{s} \right)$ is the angular velocity, N (rpm) is the rotational speed, and R is the rotor radius. Fig. 11.16 shows the comparison of the power coefficient as the main parameter for evaluating the wind turbine's efficiency in term of tip speed ratio. As the graph shows, HAWTs are the best for the higher tip speed ratios ranging from 4 to 7. For the small tip speed ratio Savonius is brilliant.

4. Problems

1. By using the global wind atlas on the https://globalwindatlas.info/ find regions at your living place which is appropriate to erect a wind farm. This application gives the mean wind speed at 100 m above the ground and also the power density for the specified region by the user. You can also insert the technical data of your chosen turbines and compare them.
2. The diameter of the Siemens SG-2.9 -129 onshore wind turbine is 129 m, and its nominal power 2.9 MW. The blade length is 63.5 m, and it has three blades. The rated wind speed for this turbine is 12.5 $\frac{m}{s}$. The annual average wind speed is 8.5 m/s. Determine (a) the power density of this turbine (b) The power coefficient. (c) The capacity factor (d) and the annual energy production (AEP). Assume air temperature and pressure as 25°C and 101.3 kPa.
3. The diameter of the Siemens SG-4.7 -155 onshore wind turbine is 155 m and is specially designed for high wind speed. The blade length is 76 m, and it has three blades. Its power density is 249 $\frac{W}{m^2}$. The rated wind speed for this turbine is 12.5 $\frac{m}{s}$. The annual average wind speed is 8.5 m/s. Determine (a) the power density of this turbine (b) The power

coefficient. (c) and the capacity factor, d) and the annual energy production (AEP). The tower height is 120.5 m and the altitude from the see level is 2000 m. Assume the air temperature to be 15 °C.

4. In the previous example the air temperature may vary from −20 to 45°C during a year. The altitude and the tower height remain the same. For the rated wind speed of 12.5, plot the power coefficient, thrust coefficient, wind speed utilization factor versus ambient temperature.

5. The wind turbine presented in problem 2, is supposed to be installed at two places, on near the sea (altitude 100 m above the sea level) and the other at altitude of 2500 m above the sea level. If the ambient air temperature is 25°C, compare the power coefficient and power output for the rated wind speed of 8.5.

6. Determine the impact of pressure variation from 101.3 to 85 kPa on the power coefficient of the turbine and the wind power of the SG-2.9 -129 presented in problem 2. Plot the results. Assume temperature constant at 25°C.

7. The SWT-2.3-120 wind turbine is a Siemens product. It is supposed to be installed at 1000 m above the sea level and air temperature is − 10°C. This wind turbine is a benchmark solution for sites with low and medium wind speeds. The rotor diameter is 120 m; the blade length is 59 m with nominal power of 2.3 MW. This turbine produces 10,400 MWh of electrical energy for the grid annually (AEP) at average wind speed of 7.5 m/s. Its life span is estimated to be 25 years. Maximum operating temperature and elevation are 40°C and 2000 m. The rotor weight is 70 tons, and the nacelle weight is 88 tons. The power regulation is done by pitch control and variable speed. The tower height is 92.4 m. The cut-out speed is 25 m/s. Determine: (a) swept area of the turbine (b) potential wind power (c) capacity factor, (d) wind speed behind the turbine for inlet wind speed of 25 m/s e) power coefficient, thrust coefficient, and wind speed coefficient.

References

[1] https://www.energy.gov/eere/wind/how-do-wind-turbines-work#size.
[2] Offshore Wind Outlook, World Energy Outlook Special Report, International Energy Agency (IEA), 2019, www.iea.org.
[3] BloombergNEF. https://about.bnef.com/.
[4] M. Ebrahimi, A. Keshavarz, Combined Cooling Heating and Power, Decision-Making, Design Optimization, Elsevier, 2014, ISBN 978-0-08-099985-2.
[5] W. Tong, Fundamentals of wind energy, WIT Transactions on State of the Art in Science and Engineering, Vol 44. https://dx.doi.org/10.2495/978-1-84564-205-1/01.
[6] https://www.wpclipart.com/.
[7] P. Gipe, Wind Energy Comes of Age, John Wiley & Sons, 1995.
[8] T. Burton, N. Jenkins, D. Sharpe, E. Bossanyi, Wind Energy Handbook, second ed., John Wiley & Sons, Ltd, 2011.
[9] Ali Sayigh - Comprehensive Renewable Energy, Elsevier, 2012.
[10] www.nrel.gov.
[11] A. Betz, The maximum of the theoretically possible exploitation of wind by means of a wind motor, Wind Eng. 37 (4) (2013) 441−446.
[12] W. Tong, Fundamentals of wind energy, in: WIT Transactions on State of the Art in Science and Engineering, Vol. 44, © 2010 WIT Press. www.witpress.com. ISSN 1755-8336 (online).
[13] https://www.siemensgamesa.com/en-int/products-and-services/offshore/wind-turbine-sg-8-0-167-dd.

Hydro power plants

12

Chapter outline

Moving water has energy, and this energy is called hydropower that can be harnessed for different purposes. Our ancestors have used hydropower for sailing along rivers, and later they replaced the human or animal power with hydropower to turn huge and heavy millstones used to grind seeds like wheat. More than 2000 years ago, ancient Greeks used hydropower to turn wooden wheel (Fig. 12.1), which was connected to the millstone to crush grains for making bread.

Using hydropower for electricity generation is not very old, and it just started a short time after Thomas Edison invented electric power. Hydropower is the second source for electricity generation after the thermal steam power plants. Hydropower produces the biggest amount of renewable electricity among the renewable energy sources as well.

Today, hydropower is converted to electricity by using different turbo-generators. In general, hydro turbines can be classified according to their blade or impeller shapes as impulse or reaction. To make electricity out of hydropower, the water stream must strike or rush through the blades or vanes of the turbine. Momentum of moving water

Copyright © 2023 Elsevier Inc. All rights reserved.

Figure 12.1 A hydromill used for grinding seeds.
The painting is the artwork of my sister, Nasrin Ebrahimi.

applies to the blades or impeller's vanes causes the turbine's rotor to spin. The rotor
will also cause the generator's shaft to spin, and electricity will be generated. According to the Bernoulli's equation, pressure and velocity are interchangeable. Therefore,
the momentum required to turn the rotor of a turbine may be provided by the natural
flow of water in a river (which is called diversion) or by a column (head) of water
behind a dam (which is called impoundment).

In addition, artificial lakes may be created to store water at higher altitudes and utilize them to generate electricity when needed by establishing water flow downward
and placing some turbines and generators in the waterway. This type of technology
is called pump-storage hydropower. When electricity consumption is low, electricity
producers sell electricity cheaper due to lower demand and also to prevent frequent
shutdowns of their power plants. At these times, cheap electricity can be used to
pump water to the storage at higher altitude, which is a very big lake, and at peak
load times, when electricity is more expensive, they use it to generate electricity.

An important advantage of hydropower is that unlike solar or wind electricity, it is
easily, and cheaply storable. Pump-storage hydropower, in fact, is a huge electricity
storage system. This technology can be coupled with wind turbines, and photovoltaic
systems to pump water with the renewable electricity during the time when the wind
and solar energies are accessible, and reuse the stored water for electricity generation
during the time it is needed. According to the second law of thermodynamics; the electricity generation by the pump-storage hydropower would be less than the electricity
consumption for pumping water, but it is necessary for peak shaving.

Another unique advantage of hydropower is its renewability due to the *water cycle*.
As the water flows from higher elevations to the lower elevations, it loses its power.
The water cycle returns it to the higher elevations and makes it accessible for power
generation. It is interesting that we do not consume energy for this natural pumping

of water, sun does it. Water cycle, which is also called *hydrologic cycle* or the *hydro-logical cycle,* is the continuous process of water evaporation in the oceans (with zero elevation), lakes, streams, and transpiration from plants, forming cloud, and moving it by the wind to the higher elevations with lower temperature, precipitation as rain and snow on the ground and again flowing down in the rivers to the oceans. This pumping system is only fueled with solar energy (Fig. 12.2).

Another advantage of hydropower is its costs. It is the cheapest among all of the renewable energy resources. The reason is that the infrastructure has been in place for a long time, and it produces electricity consistently. The lifespan of a hydropower plant may exceed 60 years [2], which is more than twice the life of solar and wind power plants and more than three times the life of gas turbine power plants. Long life will split costs and increase the total positive cash flow of the project. In addition, the operation and maintenance costs are very low. Hydropower is a domestic source of energy for power generation and approximately distributed and accessible everywhere around the world. Different countries and states can use it without depending on fossil fuels which exist and produced only by some limited countries.

Impoundment technology provides some important opportunities for fishing, swimming, flood control, irrigation, providing drinking water, boating, etc. In addition, the infrastructure provided for the dams can be used by the rural people and industries.

Hydropower plants start up time from zero load to full load may only take some minutes. Hence, they would be very helpful especially during peak hours and emergency shut down of other plants.

Hydropower can by combined with other renewable and nonrenewable energy sources especially as a big energy storage (pump storage) system for the time when the other sources of energy are not accessible.

Hydropower is an essential source of carbon saving to reach the **Net Zero Emission (NZE) by 2050** scenario. In the NZE report released in 2021, International Energy

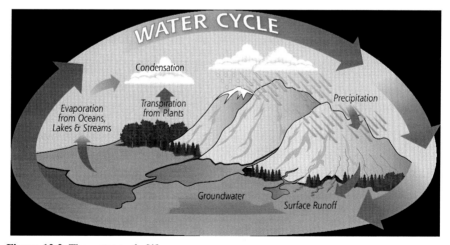

Figure 12.2 The water cycle [1].
Courtesy of NASA.

Agency (IEA) called for the global hydropower capacity to double by 2050 to keep global temperature rising below 1.5°C°C. In other words, in the next 30 years, hydropower plants should be launched as much as in the last 100 years, which is equivalent to 1300 GW [3].

In this chapter, the basic principal of operation of hydropower systems would be discussed, and classifications and applications of different hydro-turbines would be presented. The governing equations and characteristic curves of the hydro-turbines would be introduced. In addition, the reader would be able to choose the right turbine type for the available hydropower source in place.

1. Classification of hydropower plants

Hydropower plants can be classified depending on the stream that is used for electricity generation, the size of plant, or the type of turbine used. In the following, each of these classifications is discussed.

1.1 Classification by the stream type

Water can be stored behind a dam and then be utilized for power generation. This type of hydropower plant is called *impoundment*. If a portion of water from a river be separated by using a pipe or a canal and then be used for power generation, it is called *diversion*. This type of hydropower plant is also called run-of-river. Another type of power plant in this classification is the *pump-storage*, in which water head is created by pumping water to a storage at higher altitude during the time when excess electricity is available. Another type of hydropower plant is *power in pipe*. Water flowing in pipes to be used by consumers usually have extra pressure, placing small turbines in the pipe can be used to generate distributed small-scale power generation when water is supposed to flow due to water demand.

1.1.1 Impoundment

The most common type of hydropower plant is the impoundment facility. Dam stops rivers to flow naturally and control it to be used for different purposes such as electricity generation, flood control, irrigation, recreation, hydro sports, fish passage, and other environmental and water quality needs. To generate electricity, water is channeled through a *penstock* to the turbine inlet nozzles. A penstock is a closed canal that guides the flow of water from higher elevation to the lower elevation at the turbine inlet. The nozzles convert the static head behind the dam and above the turbine to velocity. Fig. 12.3 shows an impoundment hydropower plant with its main components.

1.1.2 Diversion (run-of-river)

A diversion hydropower plant channels a portion of a river water into a canal, penstock, or pipe to make use of the natural slopes of the region around the river to

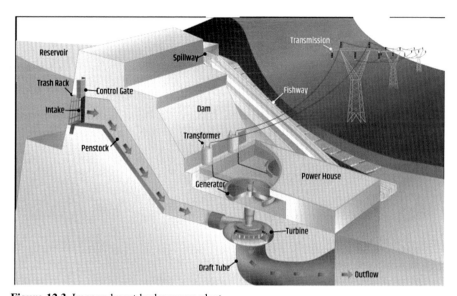

Figure 12.3 Impoundment hydropower plant.
Credit: U.S. Department of Energy, Office of Energy Efficiency & Renewable Energy.

produce electricity. Diversion does not need a dam to store water and stop the river. Hence, there will be no need to provide fishway, which fished use it to migrate upstream to feed and mate. Fig. 12.4 shows a diversion hydropower plant.

1.1.3 Pump storage

Pump storage power plants use cheap electricity or renewable electricity during the low demand time to pump water from the lower elevation to the higher elevation. Water stored in the reservoir flows downstream through the penstock to generate electricity as demand for electricity increases. The pump storage power plants can also be combined with wind or/and solar power plants to use their excess electricity to pump water when the sun is shining or the wind is blowing, and the energy stored in the water be used to generate electricity when the sun is not shining or wind is not blowing. The pump storage hydropower plants can be either open loop or close loop. The close loop does not disturb the natural water resources; hence, it is more popular from the environmental point of view. Fig. 12.5 shows the open and closed loop pump storage hydropower.

1.1.4 Hydropower in pipe

Water pipes coming down from hills or water pipes having extra pressure than it is required by the end users can be utilized for electricity production by planting vertical or horizontal axis turbines, which are basically very similar to the wind turbines. Fig. 12.6A and B show schematics of vertical axis and horizontal axis hydropower turbines. An important advantage of this type of hydropower system is that environmental impact is minimized because it does not require any change in the ecosystem as it does in the rivers or dams [4].

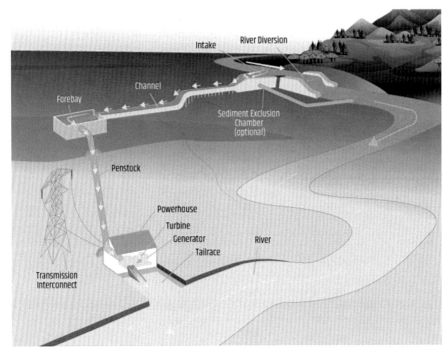

Figure 12.4 Diversion hydropower plant.
Credit: U.S. Department of Energy, Office of Energy Efficiency & Renewable Energy.

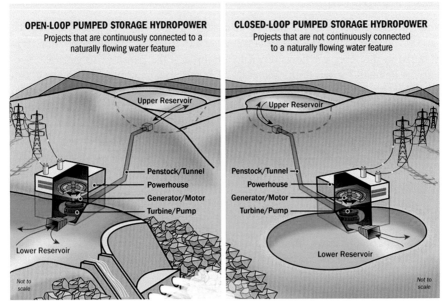

Figure 12.5 The close (right) and open loop (left) pump storage power plants.
Credit: U.S. Department of Energy, NREL.

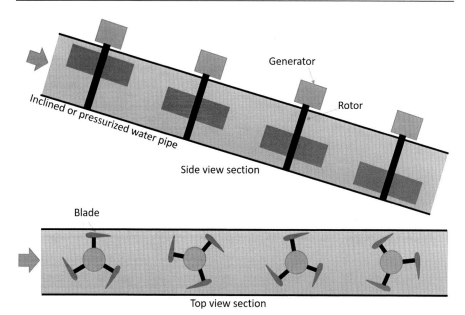

Figure 12.6A Vertical axis hydropower turbine in water pipes.

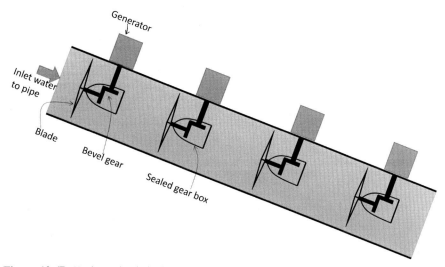

Figure 12.6B Horizontal axis hydropower turbine in water pipes.

1.2   Classification by turbine type

As water passes through the hydropower plant, its power content would be converted into rotational motion of the turbine rotor. The rotation would be transmitted to the generator through a coupling to convert the mechanical power to electrical power. In fact, the water energy is first received by the vanes, blades, or buckets of the *runner*. Runner is the rotating element containing the vanes, blades, or buckets (Fig. 12.7).

Depending on the shape and function of the blades, hydro-turbines can be classified as impulse, or reaction type.

1.2.1   Impulse turbines

Impulse turbines utilize the kinetic energy of water jet to provide mechanical power to rotate the runner. The impulse turbines have single- or double-bowel-shaped buckets that receive water jet tangentially through one or several nozzles. The nozzles convert pressure to velocity. Among the impulse turbines *Pelton* and *cross-flow* are the main types.

1.2.1.1   Pelton turbine

The Pelton turbine was invented by American inventor Lester Allan Pelton in 1870s. A Pelton turbine receives high velocity water jets impinging on the buckets. Draft tubes are not necessary for impulse turbines since the runner is located above the maximum tail water height to permit operation at atmospheric pressure. This turbine is used when high head and low flow rate is available (Figs. 12.8 and 12.9).

1.2.1.2   Cross flow turbine

Cross flow turbines (Fig. 12.10) are mostly developed to make use of water sources with lower head and higher flow rate. It has several curved vanes on a cylindrical runner. Inlet guide vanes (IGV) adjust the water flow direction against the vanes. The IGVs are placed inside a rectangular shrinking cross section outlet nozzle. As water passes through the rectangular nozzle, its velocity increases and its direction is adjusted to act against the vanes. This turbine was first invented by Anthony Michelll, an Austrian engineer, in the early 1900s.

Figure 12.7 The runner with fixed blades of a Francis turbine. Personal gallery.

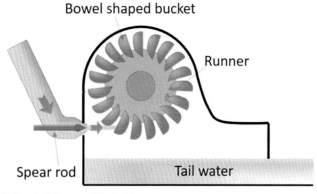

Figure 12.8 Pelton turbine, modified picture.
Credit: U.S. Department of Energy, Office of Energy Efficiency & Renewable Energy.

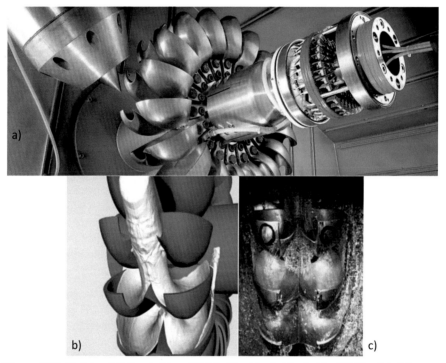

Figure 12.9 High head Pelton turbine, (A) Machined bucket backside, (B) Simulated flow through Pelton runner, (C) Actual flow through model [6].
Courtesy of Voith and Siemens Company.

1.2.2 Reaction turbines

Reaction turbines generate power based on the pressure difference on both sides of the blade. The runner is directly placed inside the water and water flows over all of the blades at the same time. The blades shape (airfoil) changes the pressure of water on

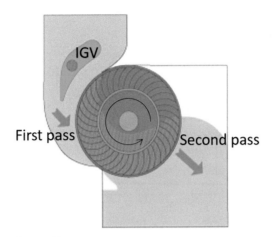

Figure 12.10 Cross flow turbine.
Credit: U.S. Department of Energy, Office of Energy Efficiency & Renewable Energy.

both sides of the blade that results in a reacting tangential force. These turbines are mostly used for water sources with lower head and higher flow rates. The three most common types of reaction turbines are propeller, Francis, and kinetic turbines.

1.2.2.1 Propeller turbines

Propeller turbines are very similar to propeller of boat if supposed working in a pipe. They usually have three to six blades, and all of the blades are in contact with water all of the times. The pitch angle of the blades may be fixed or adjustable according to the design. The main components of this hydro turbine include runner, wicket gate, draft tube, and a scroll case. The scroll case contains the generator, and shaft which connects the runner to the generator. Since the runner and scroll case are submerged in water, sealing is very important to protect the generator from leakage of water into the scroll case. Some types of the propeller turbines include Kaplan, bulb, Straflo (Straight flow), and tube turbine.

1.2.2.1.1 Bulb turbine It has a bulb-shaped casing that contains the generator and other auxiliary systems. The bulb-shaped casing also acts like a nozzle and increases the water velocity as it approaches the turbine's blades. The lubricating and cooling systems are mostly water operated. Fig. 12.11 shows the schematic of a bulb turbine with its components. Fig. 12.12 is a bulb turbine made by Voith and Siemens.

1.2.2.1.2 Straflo turbine In this turbine, instead of the bulb-shaped casing, a straight casing is used. The generator is directly connected to the turbine's blade peripherally, and the stator is outside the rotor. Sealing is very important. Fig. 12.13 shows the schematic of a Straflo turbine.

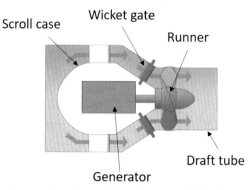

Figure 12.11 Bulb turbine, a kind of propeller turbine, modified picture.
Credit: U.S. Department of Energy, Office of Energy Efficiency & Renewable Energy.

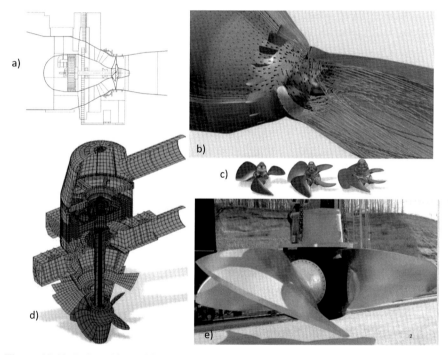

Figure 12.12 Bulb turbine, a kind of propeller (Kaplan) turbine, (A) Cross section of a bulb turbine and generator, (B) CFD illustration: Pressure and velocity distribution in a bulb turbine, (C) 3-, 4-, 5-blade bulb turbine runners, (D) 3D section of a vertical bulb unit, (E) a three-blade runner of a 75 MW bulb turbine, [6].
Courtesy of Voith and Siemens Company.

1.2.2.1.3 Tube turbine In this turbine, the generator is placed outside the tube. This can be done by using a bevel type gearbox inside the casing or bending the penstock before the turbine or bending the draft tube after the runner. Fig. 12.14A shows a tube turbine using the bevel gear to put the generator outside of the water flow. Fig. 12.14B

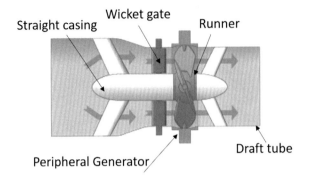

Figure 12.13 The schematic of a Straflo turbine modified picture.
Credit: U.S. Department of Energy, Office of Energy Efficiency & Renewable Energy.

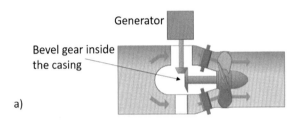

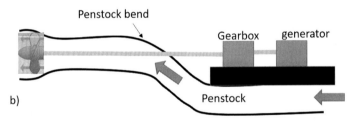

Figure 12.14 The tubular turbine A: used a bevel gear, B penstock bend before the runner modified picture.
Credit: U.S. Department of Energy, Office of Energy Efficiency & Renewable Energy.

shows a tube turbine using the penstock bend before the runner to let direct connection to the generator out of tube.

1.2.2.1.4 Kaplan turbine Kaplan (Fig. 12.15) is another propeller type of hydro turbine, which both the wicket gate and the blades of the runner are adjustable. Automatic adjusting the wicket gate and blades provide high efficiency over a wide range of head and flow rate of water. This turbine was first developed by Viktor Kaplan an Austrian professor in 1913. The head range for Kaplan turbines is usually from 10 to 70 m with an output power of 5–200 MW. The runner diameter is from 2 to 11 m, and the rotational speed may vary from 54 to 450 rpm [6].

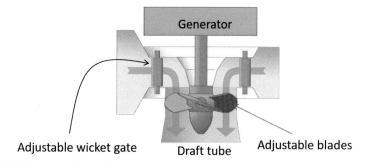

Figure 12.15 The Kaplan turbine, modified picture.
Credit: U.S. Department of Energy, Office of Energy Efficiency & Renewable Energy.

1.2.2.2 Francis turbine

This turbine has a runner with fixed blades but adjustable wicket gate. The number of blades may exceed 9 and receive flow peripherally. Penstock is connected to the scroll case of the turbine and guide the water to the wicket gates. Depending on the electrical demand, the wicket gate would be adjusted automatically to control the flow rate. After turning the runner, water goes through the draft tube. This turbine can be installed both vertically and horizontally. Its operating head and flow rate range is in the middle with respect to the Kaplan and Pelton. The recommended head is usually between 40 and 600 m. Fig. 12.16 shows a schematic of the Francis turbine and its main components. This turbine was invented by British-American engineer James Francis in 1849. Fig. 12.17 shows some photo from the Francis turbine installed in Azad dam, which was taken by the author during a visit with the students for the power plant course offered at the University of Kurdistan. The Scroll case, runner and the wicket gate are shown in this figure.

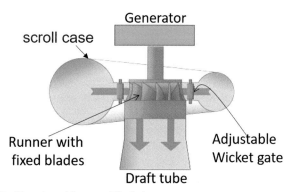

Figure 12.16 The Francis turbine, modified picture.
Credit: U.S. Department of Energy, Office of Energy Efficiency & Renewable Energy.

Scroll case

Adjustable wicket gate   Runner

Figure 12.17 The Francis turbine used in a dam, photo taken during the overhaul
Personal gallery.

1.2.2.3   Kinetic turbine

As its name implies, it is a kind of turbine that uses water velocity to produce power. It
means that velocity is preferred over the potential head for this turbine. It can be
installed in water transfer canals, rivers, ocean beds, tidal waters, and even pipes or
diversion power plant types. They do not need very much civil work because they
use the existing infrastructure. Fig. 12.18 shows a simulated kinetic turbine designed
to be used for water velocities more than 1 m/s in the free surface water transfer canals
from the Gavshan dam, near Kamyaran, Kurdistan. These turbines exist from about
1 kW to several MW.

   When comparing the hydro turbines, the main parameters to be considered are the
head and flow rate ranges, which they can operate. Initial decision-making about the
turbine type depends on the available head and flow rate. Different manufacturers offer
different graphs for the head and flow rate ranges, which the turbines can operate in.
Hence, for choosing appropriate type and size of turbine, the proposed charts by the
manufacturer is recommended to be used. Fig. 12.19 shows a head-flow rate diagram
proposed by Ref. [8] and reproduced by the author. For example it shows that the flow
rate and head ranges for Kaplan are $0.3-120 \frac{\text{m}^3}{\text{s}}$, and $2-30$ m, respectively.

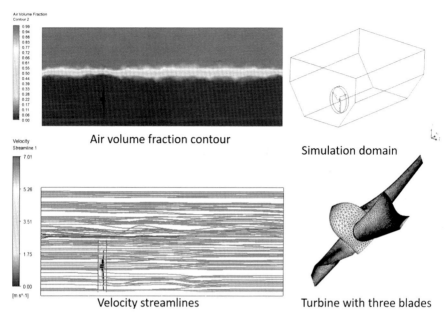

Figure 12.18 Hydro-kinetic turbine used to extract power from water canals [7].

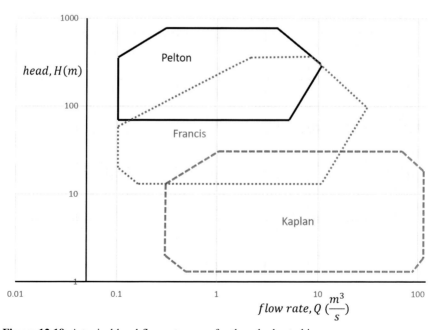

Figure 12.19 A typical head-flow rate range for three hydro turbines.
Reproduced chart from O. Rasooli, M. Ebrahimi, Design and Multi-Criteria Evaluation of
Combined Production of Cooling, Heating and Power for a Village Nearby the Gavshan Dam
Water Transfer Canal (M.Sc. thesis), University of Kurdistan, 2022.

1.3   Classification by size

Hydropower plants may be installed and operated from MW scale (*large scale*) to only several kilowatt scales. In the large scales usually, the electricity demand of a large number of consumers is provided, but in the *small or micro scales,* only the electricity demand of one or some limited consumers would be provided. The small and micro scale hydropower plants are best suited for villages with natural rivers and places near to the water transfer canals. The small or micro units are usually installed at the places close to the consumer. The Department of Energy in the U.S. has defined large-scale hydropower plants to have capacity of more than 30 MW. The small scale is defined as the hydropower plants, which produce 100 kW to 10 MW. The micro scale hydropower plants produce less than 100 kW.

2.   Hydraulic of hydropower plant

Water molecules entering the turbine at point 2 in Fig. 12.20 are forced by the pressure of water column behind them. The pressure at head water level (point 1) and tail water level (point 3) is atmospheric, but at point 2, it depends on the h and the losses in the penstock. Between point 2 and point 3, the turbine converts the kinetic and potential energy of water into power.

   If it could be assumed that the volume of water in the reservoir is infinite, and with the withdrawal of any amount of water from the dam, the same amount is added to the reservoir water at the same time, then the outflow of water from the penstock will not affect the water level behind the dam. In this case, each molecule of water that enters the reservoir to do a work on the turbine must travel through the point 1 to point 2 and then 3. To do the calculations, a differential element of water with differential volume

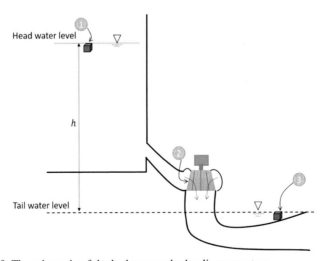

Figure 12.20 The schematic of the hydropower hydraulic parameters.

of dV is assumed to travel from point 1 to point 3. The theoretical work done by the weight force would be calculated as below:

$$dW_t = \rho g dV h \tag{12.1}$$

In which $\rho\left(\frac{kg}{m^3}\right)$ is the water density, $g\left(\frac{m}{s^2}\right)$ is the gravity acceleration, and $dV\left(m^3\right)$ is the differential volume of water, $h(m)$ is the vertical distance that the differential volume of water travels from the head water, to the tail water level, dW_t (J) is the differential theoretical work done by the weight force.

$$\dot{W}_t = \frac{dW_t}{dt} = \rho g h \frac{dV}{dt} = \rho g h Q \tag{12.2}$$

In which $Q\left(\frac{m^3}{s}\right)$ is the volume flow rate, and $\dot{W}_t(W)$ is the theoretical power stored in the moving element of water that passes it to the turbine as it moves from the head water to the tail water level.

The real power passing from the fluid particles to the turbine depends on the term η which is called the *overall efficiency*.

$$\dot{W} = \dot{W}_t\eta = \rho g h Q \eta = \gamma h Q \eta \tag{12.3}$$

In which η is responsible for the irreversibilities from the head water to the tail water level that includes the friction loss in the penstock, internal losses of the turbine such as friction, tip losses, leakages, flow separations, possible cavitation, gearbox irreversibilities, etc. In addition, $\gamma = \rho g$ is the specific weight of water.

2.1 Bernoulli equation

Since water is an incompressible fluid, and inviscid assumption is applicable in most of the regions from point 1 to point 3, (except for the turbine inside), the Bernoulli equation can be applied to the problem to calculate the power as following.

For the points 1 and 2 in Fig. 12.21:

$$\frac{V_1^2}{2g} + \frac{P_1}{\gamma} + Z_1 = \frac{V_2^2}{2g} + \frac{P_2}{\gamma} + Z_2 + h_f \tag{12.4}$$

In which h_f is the head loss from point 1 to point 2 especially due to friction in the penstock. For the points 2 and 3, the Bernoulli equation is as below:

$$\frac{V_2^2}{2g} + \frac{P_2}{\gamma} + Z_2 = \frac{V_3^2}{2g} + \frac{P_3}{\gamma} + Z_3 + h_{eff} \tag{12.5}$$

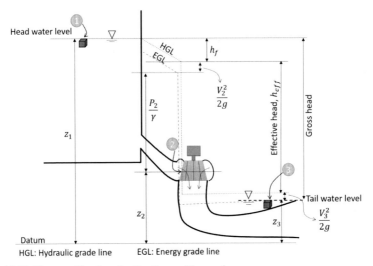

Figure 12.21 The specification for the Bernoulli equation.

In which h_{eff} is the effective head. Solving Eq. (12.4) for P_2 and substituting in Eq. (12.5):

$$\frac{V_2^2}{2g} + \frac{V_1^2}{2g} - \frac{V_2^2}{2g} + \frac{P_1}{\gamma} + Z_1 - Z_2 - h_f + Z_2 = \frac{V_3^2}{2g} + \frac{P_3}{\gamma} + Z_3 + h_{eff}$$

Since in practice, the $V_1 \cong 0$, $P_1 = P_3 = P_{atm}$, hence the effective head would be:

$$h_{eff} = Z_1 - Z_3 - h_f - \frac{V_3^2}{2g} = Z_1 - Z_3 - \frac{fL_P V_P^2}{2gD_P} - \frac{V_3^2}{2g} \tag{12.6}$$

In which the subscript P represents the characteristics of the penstock including, length, diameter, and water velocity. In addition, f is the friction factor and can be determined according to the Reynolds number and roughness of the penstock by using the Moody diagram.

Hence, the effective theoretical power would be:

$$\dot{W}_{t,eff} = \gamma h_{eff} Q \tag{12.7}$$

And the real power at turbine shaft would be calculated by considering the turbine efficiency, η_t:

$$\dot{W} = \eta_t \dot{W}_{t,eff} = \gamma h_{eff} Q \eta_t \tag{12.8}$$

According to the discussions, efficiency of a hydropower plant can be stated by different definitions.

The hydraulic efficiency, which is the most known and useful term for comparing different turbines, is defined as below [9]:

$$\eta_h = \frac{power\ deployed\ by\ the\ runner}{power\ supplied\ at\ the\ inlet\ to\ the\ turbine} = \frac{power\ deployed\ by\ the\ runner}{\gamma h_{eff} Q}$$

(12.9)

And the mechanical efficiency of hydro turbine is defined as below:

$$\eta_m = \frac{power\ available\ at\ the\ tubine\ shaft}{power\ deployed\ by\ the\ runner} = \frac{\dot{W}}{power\ deployed\ by\ the\ runner}$$

(12.10)

The volumetric efficiency of a turbine is defined as the actual flow of water striking the turbine with respect to the total water supplied to the turbine as below:

$$\eta_v = \frac{actual\ flow\ of\ water\ striking\ the\ turbine}{total\ water\ supplied\ to\ the\ turbine} = \frac{Q_{real}}{Q_{total}} = \frac{Q_{real}}{Q_{real} + Q_{loss}} \quad (12.11)$$

The total (overall) turbine efficiency is in fact the product of hydraulic and mechanical efficiencies of the turbine, as below:

$$\eta_t = \eta_h \eta_m = \frac{\dot{W}}{\gamma h_{eff} Q}$$

(12.12)

Fig. 12.22 compares different types of hydro-turbines according to their efficiency. As it shows, the efficiency is mostly over 80%. In addition, it shows that Pelton and cross-flow (impulse turbines) reach their maximum efficiency even at very low flow ratios of smaller than 0.2, but the propeller type, Kaplan, and Francis (reaction turbines) are very dependent on the flow rate ratio.

2.2 Kinetic theory

Depending on the runner's blade type (impulse or reaction), the kinetic theory can be applied to determine the force and the torque applied to the runner by water. As discussed previously, the impulse turbines mostly work on the jet velocity while reaction turbine work on pressure difference and kinetic energy. In the following, force and torque applied on the runner are formulated to make a better understanding on how the water energy is converted to power [11,12].

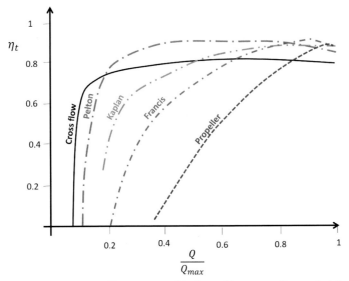

Figure 12.22 Comparison of different types of hydro-turbines according to their hydraulic efficiency [10].
Reproduced from ASCE standards.

2.2.1 Impulse turbine

According to the momentum theory the dynamic force imparted on a moving bucket of the impulse turbine can be determined according to Fig. 12.23 as below:

$$F = \dot{m} v_r (1 - c_{loss} \cos \theta) \tag{12.13}$$

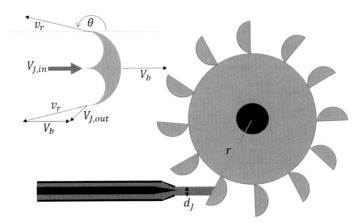

Figure 12.23 The velocity vectors and kinematic diagram of the bucket.

In which $\dot{m}\left(\frac{kg}{s}\right)$ is the water mass flow rate, $v_r\left(\frac{m}{s}\right)$ is the relative velocity between water jet (V_J) and linear velocity of moving bucket (V_b), $F(N)$ is the force applied to the bucket, θ is the deflection angle of water jet after striking the bucket, and c_{loss} is the loss coefficient of velocity as water moves across the bucket. The relative velocity is calculated as below:

$$v_r = V_J - V_b \tag{12.14}$$

The torque applied to the runner is the product of force and lever, hence:

$$T = \dot{m}v_r(1 - c_{loss}\cos\theta)r \tag{12.15}$$

In which $r(m)$ is the radius of runner, and $T(N.m)$ is the torque applied to the runner.

And the theoretical power of the moving runner would be:

$$\dot{W}_t = FV_b = \dot{m}v_r(1 - c_{loss}\cos\theta)V_b \tag{12.16}$$

Or:

$$\dot{W}_t = T\omega = \dot{m}v_r(1 - c_{loss}\cos\theta)r\omega \tag{12.17}$$

In which $\omega\left(\frac{rad}{s}\right)$ is the angular velocity in of the runner.

The *velocity coefficient* (also called *flow ratio*) C_v is defined as the ratio of the real jet velocity, V_J to the *ideal jet velocity*, which is also called the *spouting velocity* and is equal to $\sqrt{2gh}$, hence:

$$C_v = \frac{V_J}{\sqrt{2gh}} \tag{12.18}$$

The C_v is usually taken as 0.98.

The *speed ratio* is the ratio of the peripheral velocity to the spouting velocity as below:

$$K_u = \frac{V_b}{\sqrt{2gh}} \tag{12.19}$$

The peripheral velocity of a Pelton wheel at the moment of inlet and outlet flow is the same and is equal to:

$$V_b = \frac{2\pi}{60}rN \tag{12.20}$$

In which N is the rotational speed in RPM. And the ratio of bucket velocity to the jet velocity is called the *velocity ratio*:

$$\phi = \frac{V_b}{V_J} \tag{12.21}$$

Therefore, the relative velocity can be written in term of velocity ratio and jet velocity:

$$v_r = V_J(1 - \phi) \tag{12.22}$$

Therefore, the theoretical power of the runner would be:

$$\dot{W}_t = \dot{m}V_J^2(1 - \phi)\phi(1 - c_{loss}\cos\theta) \tag{12.23}$$

For a given $0° \leq \theta \leq 180°$, and c_{loss} the maximum power deployed by the runner would happen when $1 - \phi = \phi$, which means $\phi = 0.5$, we know that $0 \leq \phi \leq 1$ in which $\phi = 0$ stands for stationary wheel, and $\phi = 1$ stands for runaway speed. Hence:

$$\dot{W}_{t,max} = \frac{\dot{m}V_J^2(1 - c_{loss}\cos\theta)}{4} \tag{12.24}$$

The input power of the water jet is also:

$$\dot{W}_{in} = \frac{\dot{m}V_J^2}{2} \tag{12.25}$$

Therefore, the hydraulic efficiency would be:

$$\eta_h = \frac{\dot{m}V_J^2(1 - \phi)\phi(1 - c_{loss}\cos\theta)}{\frac{\dot{m}V_J^2}{2}} \rightarrow \tag{12.26}$$

$$\eta_h = 2(1 - \phi)\phi(1 - c_{loss}\cos\theta)$$

And the maximum hydraulic efficiency for a given θ and c_{loss} would be:

$$\eta_{h,max} = \frac{1 - c_{loss}\cos\theta}{2} \tag{12.27}$$

In addition, the jet diameter can be determined as below by using the flow rate equation:

$$Q = V_J A = C_v \sqrt{2gh} A = C_v \sqrt{2gh} \ \frac{\pi d_J^2}{4} \ \rightarrow$$

$$d_J = \left(\frac{4Q}{\pi C_v \sqrt{2gh}} \right)^{\frac{1}{2}}$$

$$(12.28)$$

By substituting $g = 9.8$, $C_v = 0.98$, the jet diameter equation would be simpler as below:

$$d_J = 0.5416 \frac{Q^{0.5}}{h^{0.25}} \tag{12.29}$$

And for an unknown C_v:

$$d_J = 0.5361 \frac{Q^{0.5}}{C_v^{0.5} h^{0.25}} \tag{12.30}$$

A rule of thumb, which is proposed by Ref. [13], states that the runner diameter in feet can be estimated as the jet diameter in inches. That means if the jet diameter is 2 in, the runner diameter can be estimated as 2 ft.

2.2.2 Reaction turbines

The torque applied to the runner according to the velocity vectors at the inlet and outlet of the runner's blades is given by the following equation:

$$T = \dot{m}(r_1 V_1 \cos \alpha_1 - r_2 V_2 \cos \alpha_2) \tag{12.31}$$

Fig. 12.24 shows the velocity diagram of a reaction turbine. The subscript 2 refers to the properties at the runner blade exit. In addition, the theoretical power transferred to the runner by the water is equal to:

$$\dot{W} = T\omega = \dot{m}(r_1 V_1 \cos \alpha_1 - r_2 V_2 \cos \alpha_2)\omega \tag{12.32}$$

In which, $\omega \left(\frac{\text{rad}}{\text{s}} \right)$ is the angular velocity of the runner.

The magnitude of the relative velocity can be found by using the sine and cosine laws as below:

$$v_1 = \sqrt{u_1^2 + V_1^2 - 2u_1 V_1 \cos \alpha_1}$$

$$V_1 \sin \alpha_1 = v_1 \sin(\pi - \beta_1)$$

$$(12.33)$$

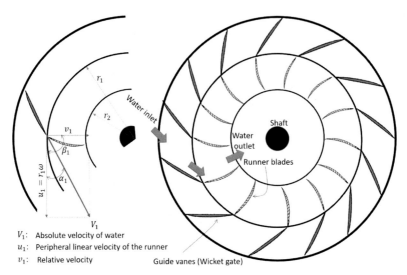

V_1: Absolute velocity of water
u_1: Peripheral linear velocity of the runner
v_1: Relative velocity

Figure 12.24 The velocity diagram of a reaction turbine.

The angle β_1 is usually bigger than 90 degrees and as great as 135 degrees. In addition, α_1 is usually in the range of 15–35 degrees.

The height of the passage of water, which is equal to the height of blade, can be determined by the following equation [13]:

$$B = \frac{A_1}{2\pi C_1 r_1 \sin \alpha_1} \tag{12.34}$$

In which, B is the height of water passage at the entrance of the runner, A_1 is the required cross section area of water passages in perpendicular direction of water flow, and C_1 is a constant coefficient and usually is about 0.95 [11].

For the best performance of a reaction turbine, the outlet water speed must be very small and the flow leaving the runner stay axial, which in practice, it is not possible to have completely axial flow. The absolute velocity at the inlet of the runner is equal to:

$$V_1 = C_v\sqrt{2gh} \tag{12.35}$$

In which, the velocity coefficient varies from 0.6 to 0.8 for the reaction turbines.

2.2.3 Hydrokinetic turbines

Hydrokinetic turbine (Fig. 12.25) looks like the horizontal axis wind turbines discussed in the previous chapter; hence, the design and governing equations are very similar. Due to this similarity, the hydrokinetic turbines need moving water to produce electricity. In other words, they convert the kinetic energy of water flow into mechanical power, and the mechanical power is converted to electricity through the generator.

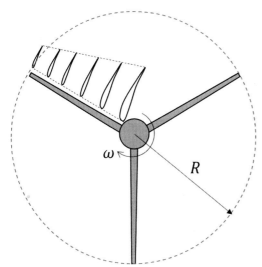

Figure 12.25 Hydrokinetic turbine.

The electricity generated (E) by a hydrokinetic turbine depends on the water density $\rho\left(\frac{\mathrm{kg}}{\mathrm{m}^3}\right)$, swept area of the turbine $A\,(\mathrm{m}^2)$, velocity of water, $V\left(\frac{\mathrm{m}}{\mathrm{s}}\right)$, drive train efficiency (η) (gearbox, generator, electrical, controls, etc.), and power coefficient (C_p). The η for small turbines is about 79% but for larger turbines is usually over 90%.

The theoretical limit (Betz limit) for the C_p as proved in the previous chapter is 59.3%. However, in general, it depends on the pitch angle (θ (rad)) and the tip speed ratio (λ) as below [14]:

$$\dot{W} = \frac{1}{2}C_p\rho A V^3$$

$$E = \eta P$$

$$C_p(\lambda,\ \theta) = 0.22\left(\frac{116}{\lambda_i} - 0.4\theta - 5\right)\exp\left(-\frac{12.5}{\lambda_i}\right) \tag{12.36}$$

$$\frac{1}{\lambda_i} = \frac{1}{\lambda + 0.08\theta} - \frac{0.035}{\theta^3 + 1}$$

$$\lambda = \frac{R\omega}{V}$$

In which R is the turbine radius, and ω is the angular velocity.

Example 12.1. The velocity coefficient of a Pelton turbine is 0.98, and a water source is available with flow rate of $1\ \frac{\mathrm{m}^3}{\mathrm{s}}$, and effective head of 200 m. The turbine bucket is somehow designed that deflection angle is 150 degrees. The radius of the runner is 1 m. Determine:

A: the maximum power of runner, if $c_{loss} = 1$
B: the maximum hydraulic efficiency, if $c_{loss} = 1$
C: the nozzle jet diameter

Solution. The velocity coefficient is defined as below:

$$C_v = \frac{V_J}{\sqrt{2gh}}$$

Therefore, the jet speed would be:

$$V_J = C_v\sqrt{2gh} = 0.98\sqrt{2 \times 9.81 \times 200} = 61.39\,\frac{m}{s}$$

Since it is asked to estimate the maximum power; therefore $\phi = \frac{1}{2}$, as a result, the bucket linear velocity would be:

$$\phi = \frac{V_b}{V_J} \rightarrow \frac{1}{2} = \frac{V_b}{61.39} \rightarrow V_b = 30.69\,\frac{m}{s}$$

The angular velocity of the runner would be:

$$V_b = r\omega \rightarrow \omega = \frac{30.69}{1} = 30.69\,\frac{rad}{s}$$

Hence the maximum power of the runner would be:

$$
\begin{aligned}
\dot{W}_{t,max} &= \frac{\dot{m}V_J^2(1 - c_{loss}\cos\theta)}{4} = \frac{\rho Q V_J^2(1 - c_{loss}\cos\theta)}{4} \\
&= \frac{1000 \times 1 \times 61.39^2 \times (1 - 1 \times \cos 150°)}{4} = 1.758 \times 10^6 W \\
&= 1.758\,MW \rightarrow \dot{W}_{t,max} = 1.758\,MW
\end{aligned}
$$

B: the maximum efficiency would be as below:

$$\eta_{h,max} = \frac{1 - c_{loss}\cos\theta}{2} = \frac{1 - 1 \times \cos 150°}{2} = 93.3\%$$

C: The jet diameter also is calculated by:

$$d_J = 0.5361\frac{Q^{0.5}}{C_v^{0.5}h^{0.25}} = 0.5361\frac{1^{0.5}}{0.98^{0.5}200^{0.25}} = 0.144m = 14.4\,cm$$

The code written in EES to solve the Example 12.1 is presented in **Code 12.1**.

Code 12.1. The code used to solve Example 12.1.

```
C_v=0.98
g=9.81
Q=1
h=200
phi=0.5
C_v=V_J/(2*g*h)^0.5
PHI=V_b/V_J
V_r=V_J-V_b
ro=1000
M_dot=Q*ro
V_b=r*omega
teta=150
r=1
T=M_dot*V_r*(1-cos(teta))*r
P=T*omega/10^6 "MW"
ETA_2=(1-c_loss*cos(teta))/2
d_J=0.5361*Q^0.5/(c_V^0.5*h^0.25)
```

Example 12.2. The velocity coefficient of a Pelton turbine is 0.97, and a water source is available with flow rate of 1.2 $\frac{m^3}{s}$ and effective head of 400 m. The radius of the runner is 1.5 m. Determine:

A: the impact of $0 \leq \phi \leq 1$ on the power of runner, if $c_{loss} = 1$, and $\theta = 180$degrees, plot the results.

B: The impact of deflection angle 0degree $\leq \theta \leq 180$degrees on the hydraulic efficiency, if $c_{loss} = 1$, and $\phi = \frac{1}{2}$, plot the results.

Solution. The velocity coefficient is defined as below:

$$C_v = \frac{V_J}{\sqrt{2gh}}$$

Therefore, the jet speed would be:

$$V_J = C_v\sqrt{2gh} = 0.97\sqrt{2 \times 9.81 \times 400} = 85.93 \frac{m}{s}$$

Therefore the power of runner in term of ϕ would be as below:

$$\dot{W}_t(MW) = \frac{\dot{m}V_J^2(1-\phi)\phi(1 - c_{loss}\cos\theta)}{10^6}$$

$$= 1000 \times 1.2 \times \frac{85.93^2(1-\phi)\phi(1 - 1 \times \cos 180°)}{10^6}$$

$$= 1000 \times 1.2 \times 85.93^2(1-\phi)\phi \times \frac{2}{10^6}$$

$$= 17.72 \times (1-\phi)\phi$$

By changing ϕ from 0 to 1 and plotting the data, Fig. 12.24 would be generated for the power in MW. It shows that the maximum power is achieved at $\phi = 0.5$. The parametric table of EES is used to do the calculations. And the code is given in **Code 12.2**.

To investigate the impact of ϕ on the hydraulic efficiency, the following equation is used for the given data:

$$\eta_h = 2(1 - \phi)\phi(1 - c_{loss} \cos \theta) = 2(1 - \phi)\phi(1 - 1 \times \cos 180°) = 4(1 - \phi)\phi$$

By changing ϕ from 0 to 1 and plotting the data, Fig. 12.26 would be generated for the hydraulic efficiency of the Pelton turbine in MW.

Code 12.2. The EES code written for Example 12.2.

```
C_v=0.97
g=9.81
Q=1.2
h=400
c_loss=1
"phi=0.5"
teta=180
C_v=V_J/(2*g*h)^0.5
phi=V_b/V_J
V_r=V_J-V_b
ro=1000
M_dot=Q*ro
V_b=r*omega
r=1.5
T=M_dot*V_r*(1-c_loss*cos(teta))*r
W_dot=T*omega/10^6
eta=W_dot/(0.5*M_dot*V_J^2)*10^6
d_J=0.5361*Q^0.5/(c_V^0.5*h^0.25)
```

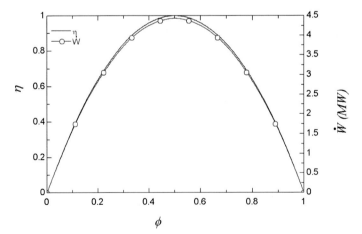

Figure 12.26 Hydraulic efficiency and runner power versus the bucket to jet speed ratio ϕ.

To evaluate the impact of deflection angle of water, the equation for power of runner is rewritten, and by inserting the given data, an equation will be generated for the power in term of deflection angle as below:

$$\dot{W}_t(MW) = \frac{\dot{m}V_J^2(1-\phi)\phi(1-c_{loss}\cos\theta)}{10^6}$$

$$= 1000 \times 1.2 \times \frac{85.93^2(1-0.5)0.5(1-1\times\cos\theta)}{10^6}$$

$$= 2.215(1-\cos\theta)$$

And for the hydraulic efficiency:

$$\eta_h = 2(1-\phi)\phi(1-c_{loss}\cos\theta) = 2(1-0.5)0.5(1-1\times\cos\theta) = \frac{(1-\cos\theta)}{2}$$

By changing θ from 0 to 180 degrees, the hydraulic efficiency and power of runner were evaluated and plotted in Fig. 12.27.

Example 12.3. For a Francis turbine which its water passage height is 25 cm, $r_1 = 70$ cm, $r_2 = 30$ cm, velocity coefficient is 0.75, the effective head of water source is 80 m, and flow rate through the penstock is $1 \frac{m^3}{s}$, and the exit velocity from the runner has no tangential velocity component.

The other given parameters include $\alpha_1 = 25$degrees, $\alpha_2 = 90$degrees, and $\beta_1 = 100$degrees. Determine:

A: The angular velocity and rotational speed of the runner.

B: the torque generated by the runner.

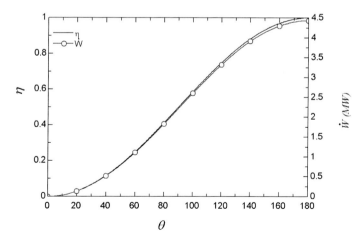

Figure 12.27 Hydraulic efficiency and runner power versus the deflection degree of water jet θ.

C: the power of the runner.

D: the hydraulic efficiency of the turbine.

Solution.   **a.** To determine the absolute inlet velocity to the runner:

$$V_1 = C_v \sqrt{2gh} = 0.75\sqrt{2 \times 9.81 \times 80} = 29.71 \frac{m}{s}$$

Using the sine law, the relative velocity of water at inlet would be determined:

$$V_1 \sin \alpha_1 = v_1 \sin(180° - \beta_1) \rightarrow 29.71 \times \sin 25° = v_1 \sin(180° - 100°)$$

$$\rightarrow v_1 = 12.75 \frac{m}{s}$$

And using the cosine law the tangential component of velocity is determined:

$$v_1 = \sqrt{u_1^2 + V_1^2 - 2u_1 V_1 \cos \alpha_1} \rightarrow$$

$$12.75 = \sqrt{u_1^2 + 29.71^2 - 2 \times u_1^2 \times 29.71 \times \cos 25°} \rightarrow u_1 = 24.72 \frac{m}{s}$$

Therefore, the angular velocity would be:

$$u_1 = r_1 \omega \rightarrow 24.72 = 0.7 \times \omega \rightarrow \omega = 35.31 \frac{rad}{s}$$

And the rotational speed is:

$$\omega = \frac{2\pi N}{60} \rightarrow 35.31 = \frac{2\pi N}{60} \rightarrow N = 337.36 \text{ rpm}$$

b. Assuming no tangential velocity at the outlet of the runner's blades, the flow rate can be written as below to determine the outlet absolute velocity from the runner's blade:

$$Q = 2\pi r_2 B V_2 \rightarrow 1 = 2 \times \pi \times 0.3 \times 0.25 \times V_2 \rightarrow V_2 = 2.122 \frac{m}{s}$$

The mass flow rate of water is:

$$\dot{m} = \rho Q = 1000 \times 1 = 1000 \frac{kg}{s}$$

And the torque of the runner that will be transmitted to the generator would be:

$$T = \dot{m}(r_1 V_1 \cos \alpha_1 - r_2 V_2 \cos \alpha_2) = 1000(0.7 \times 29.71 \times \cos 25°$$
$$- 0.3 \times 2.122 \times \cos 90°) \rightarrow T$$
$$= 18851 \text{ N.m}$$

a. The power is simply determined by multiplying torque with angular velocity.

$$\dot{W} = T\omega = 18851 \times 35.31 = 665580 \text{ W}$$

d. The hydraulic efficiency would be:

$$\eta_h = \frac{\dot{W}}{\rho g h Q} = \frac{665580}{1000 \times 9.81 \times 80 \times 1} = 84.81\%$$

The **Code 12.3** is the EES code used to solve the Example 12.3.

Code 12.3. The code used to solve Example 12.3.

```
B=0.25
r_1=0.7
r_2=0.3
h=80
g=9.81
Q=1
ro=1000
C_v=0.75
C_1=0.95
alpha_1=25
alpha_2=90
beta1=100
m_dot=ro*Q
V_1=C_v*(2*g*h)^0.5
Q=2*PI*r_2*B*V_2
V_1*sin(alpha_1)=v_r*sin(180-beta1)
v_r=(u1^2+V_1^2-2*u1*V_1*cos(alpha_1))^0.5
omega1=u1/r_1
T=m_dot*(r_1*V_1*cos(alpha_1)-r_2*V_2*cos(alpha_2))
P=m_dot*(r_1*V_1*cos(alpha_1)-r_2*V_2*cos(alpha_2))*omega1
ETA=P/(ro*g*h*Q)
```

Example 12.4. For the previous example determine:

A: The impact of the angle between the absolute and tangential velocity component at the inlet of runner on the turbine efficiency and output power if 15degrees $\leq \alpha_1 \leq$ 35degrees.

B: The impact of the angle between the absolute and tangential velocity component at the inlet of runner on the turbine efficiency, and output power if 45degrees $\leq \alpha_2 \leq$ 90degrees.

Solution.   **a.** To evaluate the impact of α_1 on the performance of the turbine, the **Code 12.3** written for the Example 12.3 is used in a parametric table of EES by changing α_1 from 15 to 35 degrees, and the plot in Fig. 12.28 would be provided.

The results show that the maximum efficiency happens when $\alpha_1 = 15$ degrees, with the assumption of no tangential velocity component at the exit of blades.

b. To evaluate the impact of α_2 on the performance of the turbine, the code written for the Example 12.3 is used in a parametric table of EES by changing α_2 from 45 to 90 degrees, and the plot in Fig. 12.29 would be provided.

The results show that the maximum efficiency happens when $\alpha_2 = 90$ degrees, which means no tangential velocity. The $\alpha_1 = 25$ degrees is assumed constant in this evaluation.

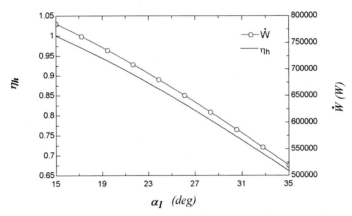

Figure 12.28 The impact of α_1 on the power and efficiency of a Francis turbine.

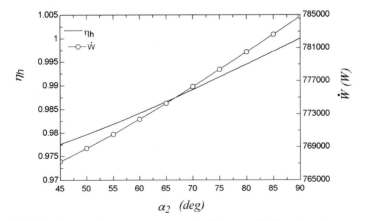

Figure 12.29 The impact of α_2 on the power and efficiency of a Francis turbine.

3. Turbine constants

Turbine constants are dimensionless numbers that can be calculated based on the principles of geometric, kinematic, and dynamic similarity by using the Buckingham's π theorem. Dimensionless numbers eliminate the need to examine all of the parameters governing the problem, and only by knowing dimensionless groups, we can make better decisions and designs. For example, in the fluid mechanics, it is very important to determine whether the flow is laminar or turbulent. It is possible to distinguish it in many cases by knowing the Reynolds number only (especially in smooth currents). The important point is that the Reynolds number itself is a function of viscosity, velocity, and length scale. In many cases, we will not need to examine the effects of all of the parameters separately on whether the flow is laminar or turbulent, only the Reynolds number will show us the way.

Turbine constants will help us in decision-making and designing of the proper turbine for the available water sources.

Hydro turbines main parameters that may alter its design, include head (h), flow rate (Q), runner diameter as the length scale (D), angular velocity of the runner (ω), water density (ρ), gravity (g), and output power ($\dot{W}$). By using the Buckingham's π theorem, the following dimensionless groups will be generated:

$$C_H = head\ coefficient = \frac{gh}{\omega^2 D^2} \tag{12.37}$$

$$C_Q = capacity\ coefficient = \frac{Q}{\omega D^3} \tag{12.38}$$

$$C_P = power\ coefficient = \frac{\dot{W}}{\rho \omega^3 D^5} \tag{12.39}$$

$$\eta_h = hydraulic\ efficiency = \frac{\dot{W}}{\rho g h Q} \tag{12.40}$$

Because the capacity and head coefficients are functions of power coefficient, the performance curve of turbines is usually plotted against power coefficient.

The affinity laws for the hydro turbines are as below.

$$\frac{Q_1}{Q_2} = \frac{\omega_1}{\omega_2}\left(\frac{D_1}{D_2}\right)^3 \tag{12.41}$$

$$\frac{h_1}{h_2} = \left(\frac{\omega_1}{\omega_2}\right)^2 \left(\frac{D_1}{D_2}\right)^2 \tag{12.42}$$

$$\frac{\dot{W}_1}{\dot{W}_2}=\frac{\rho_1}{\rho_2}\left(\frac{\omega_1}{\omega_2}\right)^3\left(\frac{D_1}{D_2}\right)^5 \tag{12.43}$$

The affinity laws are valid for a model and a prototype only if the Reynolds number for both of them is the same, and they have geometrical similarity. The geometrical similarity is very important to be established about every geometrical parameters including surface roughness, clearances, tolerances, and other main dimensions. Since most of time, it is not possible to satisfy all of the requirements for the geometrical similarity, or the Reynolds number the efficiency of the prototype is higher than the model. This is the main reason why smaller components, although they look similar in the appearance, have less efficiency. It is because of not satisfying the complete similarity rules that may be due to manufacturing limitations. Due to these limitations, some estimations have been made to evaluate the prototype efficiency (η_p) as a function of the model efficiency (η_m) and the scale down ratio. The following estimation is done by Refs. [15,16],

$$\eta_p = 1 - (1 - \eta_m)\left(\frac{D_m}{D_p}\right)^{\frac{1}{5}} \tag{12.44}$$

Specific speed or *power specific speed* is another important dimensionless parameter which can be used to choose the right turbine at its maximum efficiency for a given water source (Fig. 12.30). It is defined based on the head coefficient and power coefficient.

$$N_s = \frac{C_P^{\frac{1}{2}}}{C_H^{\frac{5}{4}}} = \frac{\omega \dot{W}^{\frac{1}{2}}}{\rho^{\frac{1}{2}}(gh)^{\frac{5}{4}}} \tag{12.45}$$

Fig. 12.28 shows that for:

- $N_s \leq 0.3$, use impulse turbines
- $0.3 \leq N_s \leq 2$ use Francis turbines
- And for $2 \leq N_s$ use Kaplan turbine.

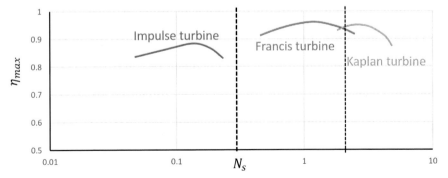

Figure 12.30 The maximum efficiency of hydro turbines versus the specific speed. Reproduced from [15].

Example 12.5. A water source is available with effective head of 200 m. It is supposed to produce 1.8 MW of electricity by using a turbine that its angular velocity is 30.7 $\frac{rad}{s}$. Specify the turbine type, which is proper for this water source.

Solution.

$$N_s = \frac{\omega \dot{W}^{\frac{1}{2}}}{\rho^{\frac{1}{2}}(gh)^{\frac{5}{4}}} = \frac{30.7 \times 1.8 \times 10^6}{1000^{\frac{1}{2}}(9.81 \times 200)^{\frac{5}{4}}} = 0.09975$$

Since $N_s \leq 0.3$, the proposed turbine would be a Pelton turbine.

Example 12.6. For the turbine proposed in the Example 12.5, if the demand for electricity is tripled, for the same angular velocity, what would be the new turbine's diameter?

Solution. By using the similarity laws:

$$\frac{\dot{W}_1}{\dot{W}_2} = \frac{\rho_1}{\rho_2}\left(\frac{\omega_1}{\omega_2}\right)^3\left(\frac{D_1}{D_2}\right)^5 \rightarrow$$

$$\frac{\dot{W}_1}{3\dot{W}_1} = \frac{\rho_1}{\rho_1}\left(\frac{\omega_1}{\omega_1}\right)^3\left(\frac{D_1}{D_2}\right)^5 \rightarrow \left(\frac{D_1}{D_2}\right)^5 = \frac{1}{3} \rightarrow \frac{D_2}{D_1} = 3^{\frac{1}{5}} = 1.246 \rightarrow$$

Hence:

$$D_2 = 1.246 D_1$$

4. Problems

1. Find the 5 top hydro turbine manufacturers around the world.
2. What is the surge tank in the hydropower installations? What is its main duty?
3. What is the breaking nozzle, how it works? When does it operate?
4. What are the main functions of the wicket gates in a Francis turbine?
5. In your country there are many dams or natural waterfalls or rivers. Make a list of the 5 main water sources and by searching through the website of main hydro turbine manufacturers choose the most suitable turbines for those sites.
6. The velocity coefficient of a Pelton turbine is 0.96 and a water source is available with flow rate of 2.5 $\frac{m^3}{s}$, and effective head of 500 m. The turbine bucket is somehow designed that deflection angle is 165 degrees. The radius of the runner is 1.5 m. Determine:
   a. the maximum power of runner, if $c_{loss} = 0.98$.
   b. the maximum hydraulic efficiency, if $c_{loss} = 0.98$.
   c. the nozzle jet diameter.
7. The velocity coefficient of a Pelton turbine is 0.98 and a water source is available with flow rate of 1.2 $\frac{m^3}{s}$, and effective head of 400 m. The radius of the Pelton runner is 1.5 m. Determine:

    a. the impact of $0 \leq \phi \leq 1$ on the power of runner, if $c_{loss} = 0.97$, and $\theta = 160$degrees, plot the results.

    b. The impact of deflection angle 0degree $\leq \theta \leq 180$degrees on the hydraulic efficiency, if $c_{loss} = 0.97$, and $\phi = \frac{3}{5}$, plot the results.

8. For a Francis turbine, which its water passage height is 20 cm, $r_1 = 80$ cm, $r_2 = 25$ cm, velocity coefficient is 0.77, the effective head of water source is 120 m, and flow rate through the penstock is 1.2 $\frac{m^3}{s}$. The other given parameters include $\alpha_1 = 25$degree, $\alpha_2 = 60$degree, and $\beta_1 = 115$degree. Determine:

    a. The angular velocity and rotational speed of the runner.

    b. the torque generated by the runner.

    c. the power of the runner.

    d. the hydraulic efficiency.

9. A water source is available with effective head of 75 m. It is supposed to produce 1 MW of electricity by using a turbine that its angular velocity is 40 $\frac{rad}{s}$. Specify the turbine type which is proper for this water source.

10. For a turbine if the demand for electricity is 8 folded, for the same angular velocity, what would be the new turbine's diameter?

References

[1] https://gpm.nasa.gov/education/.

[2] https://www.nrel.gov/research/hydropower.html.

[3] Offshore Wind Outlook, World Energy Outlook Special Report, International Energy Agency (IEA), 2019, www.iea.org.

[4] https://www.altenergymag.com/article/2016/04/what-is-hydropower-in-a-pipe/23404.

[5] A Voith and Siemens Company, https://voith.com/corp-en/Voith_t3366e_Bulb_Pit_S-turbines_screen.pdf, https://voith.com/cn/t3371e_Pelton_turbine_screen.pdf, https://www.siemens-energy.com/.

[6] https://www.energyencyclopedia.com/en/renewable-energy-sources/water-energy/choosing-a-turbine-turbine-selection-graph.

[7] O. Rasooli, M. Ebrahimi, Design and Multi-Criteria Evaluation of Combined Production of Cooling, Heating and Power for a Village Nearby the Gavshan Dam Water Transfer Canal, M.Sc. thesis, University of Kurdistan, 2022.

[8] https://jhd.co.jp/en/products/.

[9] https://www.linquip.com/blog/pelton-wheel/.

[10] https://ascelibrary.org/action/showPublications?pubType=standards&sortBy=Ppub.

[11] C.C. Warnick, Hydropower Engineering, Prentice-Hall. Inc, Englewood, New Jersey Cliff, NJ 07639, 1984.

[12] S. Kemal Furkan, Pelton Turbine Test, Experiment Sheet, Bursa Technical University, 2016.

[13] L.P. Brown, W.E. Whippen, Hydraulic Turbines, International Correspondence Schools, 1965.

[14] E. Chica, F. Perez, A. Rubio-Clemente, S. Agudelo, Design of a hydrokinetic turbine, energy and sustainability VI 137, in: WIT Transactions on Ecology and The Environment, 195, WIT Press, 2015.

[15] Y.A. Cengel, J.M. Cimbala, Fluid Mechanics, Fundamentals and Applications, McGraw-Hill, 2006.
[16] L.F. Moody, The propeller type turbine, ASCE Trans 89 (1926) 628.

Further reading

[1] https://www.energy.gov/eere/water/benefits-hydropower.

Fuel cell power plants

13

Chapter outline

Fuel cells are energy conversion equipment that use electrochemical reaction rather than combustion of fuel. As a result of electrochemical reactions, heat, electricity, and water would be generated as the main products. Fuel cells exist in different types, including but not limited to the solid oxide fuel cell (SOFC), molten carbonate fuel cell (MCFC), proton exchange membrane fuel cell (PEMFC), alkaline fuel cell (AFC), phosphoric acid fuel cell (PAFC), etc. They also use diverse fuels such as hydrogen, natural gas, biogas, alcohol, etc. to produce electricity. They operate in different

Power Generation Technologies. https://doi.org/10.1016/B978-0-323-95370-2.00013-2
Copyright © 2023 Elsevier Inc. All rights reserved.

temperatures ranging from about $60°C$ to more than $1000°C$ and exist in different capacities for different applications.

A fuel cell is usually made of several cells connected in series to increase the voltage. A single cell has three main components: anode, cathode, and an electrolyte, which is sandwiched between the anode and the cathode. Fuel enters at the anode, while air or oxygen enters at the cathode side. A catalyst at the anode converts hydrogen into protons and electrons; protons travel through the electrolyte, while electrons transfer from an external circuit and produce electrical current. The electrons and the protons meet the oxygen molecules at the cathode to produce water vapor at the exhaust.

Fuel cells have some auxiliary components depending on the fuel cell and fuel type and its thermodynamic properties. These components may include air compressor, fuel compressor, reformer (which can be internal, or external, wet or dry), after burner, preheater heat exchanger, humidifier, DC/AC inverter, etc.

Fuel cells have several advantages over the conventional combustion—based power generators. They produce less to zero environmental pollution, and in case of using hydrogen as the fuel, the only exhaust product is water vapor (zero pollution). Most of the fuel cells operate at temperatures less than the operating temperature of combustion-based heat engines; therefore, cheaper materials can be used for tolerating lower temperature.

Their efficiency can exceed 60%, which is considerably higher than that for most of the conventional thermal power generators.

They are silent due to having minimum moving parts; therefore, they are appropriate for using in residential buildings, hospitals, passenger cars, and applications, in which silence is critical.

Sometimes, fuel cells are mistaken by batteries, batteries run out of electricity after a while, but fuel cells keep generating electricity as long as they receive fuel.

In the present chapter, the basics of the fuel cells are presented, and the reader would be familiar with the components and structure of the fuel cells. In addition, the thermodynamics of the fuel cells and governing equations are discussed thoroughly to prepare the reader for simulation of the fuel cells. Several solved examples with their EES codes are presented that can be extended more to investigate the impact of operating parameters.

1. Basics of fuel cells

Electrolysis is the splitting of water molecules into hydrogen and oxygen gases by using electric current. The water electrolyzer comprises an electrolyte (which can be alkaline or acidic water), two electrodes as anode and cathode, and an external electricity source. Fig. 13.1 shows the electrolysis of water with a diluted acidic electrolyte. The electric current on the anode converts water molecules to hydrogen cation, oxygen molecules, and electrons as below:

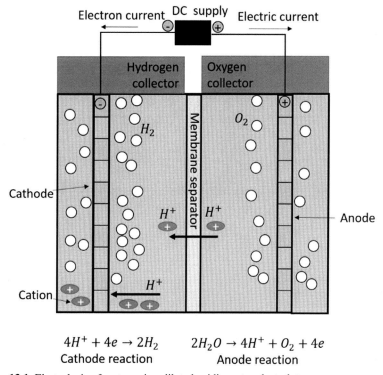

$$4H^+ + 4e \rightarrow 2H_2 \qquad 2H_2O \rightarrow 4H^+ + O_2 + 4e$$

Cathode reaction   Anode reaction

Figure 13.1 Electrolysis of water using diluted acidic water electrolyte.

$$2H_2O \rightarrow 4H^+ + O_2 + 4e^- \qquad (13.1)$$

The electrons travel through the external circuit due to the voltage difference created by the DC supply. The cation penetrates through the porous membrane and reaches the cathode, while oxygen bubbles leave the electrolyzer and get collected by the oxygen collector. On the other side, the hydrogen cations combine with the electrons again on the cathode and hydrogen bubbles form that can be collected by the hydrogen collector.

$$4H^+ + 4e^- \rightarrow 2H_2 \qquad (13.2)$$

This process keeps going until when the electric current and water are supplied.

While electrolyzer is an electricity consumer equipment, fuel cells which do the same process, but in the reverse direction, are electricity generator equipment. Fuel cells consume hydrogen with oxygen to produce electricity, heat, and water. As long as the hydrogen and oxygen are fed into the fuel cell, electricity, heat, and water would be generated. Fuel cells were first demonstrated by William Robert Grove (1811−96) (Fig. 13.2). The "*Grove cell*" used a platinum electrode immersed in nitric

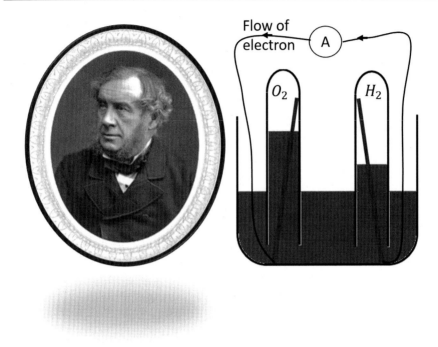

Figure 13.2 William Robert Grove inventor of fuel cells and the schematic of his setup.

acid (HNO_3) and a zinc electrode in zinc sulfate ($ZnSO_4$) to generate about 12 amps of current at about 1.8 V.

Grove used two platinum electrodes as the anode and cathode. One end of each electrode was immersed in a container of sulfuric acid (H_2SO_4) and the other end separately sealed in containers of oxygen and hydrogen gases.

Fig. 13.2 shows the schematic of the Grove setup. As a result of this experiment, a constant current flowed between the anode and the cathode. The sealed containers held water as well as the hydrogen and oxygen, and Grove reported that the water level increases in both containers as the current flowed. This in fact was the result of producing water as well as electricity during the electrochemical reaction.

According to the setup presented by Grove, the amount of electricity generation is very small. This has two main reasons; first, the rods have very small contact area; second, they are far apart, and the electrolyte between them creates a resistance against flow of electrons. To solve these problems, electrodes are made as a flat plate with a very thin layer of electrolyte between them. The flat electrodes create a considerable contact area with hydrogen or oxygen, and this electrolyte omits the unnecessary distance between the electrodes. Fig. 13.3A and B show schematic of such cell with both acid or alkaline electrolyte and the related reactions on the electrodes.

In an acid electrolyte fuel cell, hydrogen gas gets ionized on the anode as H^+, releases electrons and heat as below:

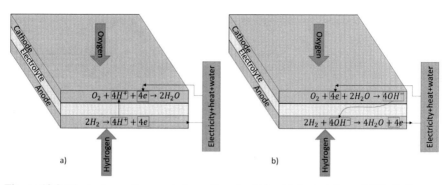

Figure 13.3 Electrode reactions and charge flow for (A) acid electrolyte, and (B) alkaline electrolyte fuel cells, the electrons flow from anode to cathode while the electric current is from cathode to anode.

$$2H_2 \rightarrow 4H^+ + 4e^- + heat \tag{13.3}$$

On the cathode, oxygen reacts with hydrogen cation and electron to generate water and completing the reactions as below:

$$O_2 + 4e^- + 4H^+ \rightarrow 2H_2O \tag{13.4}$$

While these processes are happening simultaneously, electrons flow from the anode to the cathode through an external circuit, while hydrogen cations pass through the electrolyte from the anode to the cathode.

The overall reaction of the acid electrolyte fuel cell would be as simple as combining hydrogen and oxygen to make water as below:

$$O_2 + 2H_2 \rightarrow 2H_2O \tag{13.5}$$

It must be emphasized that the electrons must not pass through the electrolyte, if not there would be less or no electricity.

In an alkaline electrolyte fuel cell, mobile ions of hydroxyl (OH^-) are available. They react with hydrogen on the anode, and as a result, electron, water, and heat would be released. Therefore, the anode reaction can be written as below:

$$2H_2 + 4OH^- \rightarrow 4H_2O + 4e^- + heat \tag{13.6}$$

Oxygen reacts with the electrons coming from the anode through the external circuit and water in the electrolyte forming hydroxyl anions as below:

$$O_2 + 4e^- + 2H_2O \rightarrow 4OH^- \tag{13.7}$$

The overall reaction in an alkaline electrolyte fuel cell is the same as the acid electrolyte fuel cell as Eq. (13.5).

It is clear that using only 1 cell may produce a very small amount of electricity, but using several cells in series can increase the amount of electricity production. Several cells together are called *cell stack* (Fig. 13.4).

Every reaction needs enough activation energy to continue. If the activation energy is not provided, the reaction slows down, and electricity generation reduces as well. The energy released in the electrochemical reaction may not be high enough to provide the activation energy. To avoid dropping the electricity generation, three main methods, that are, using catalysts, increasing temperature, and increasing the electrode contact area are recommended. The third method is specially used for fuel cells, while the other two methods can be used to every chemical reaction. It should be mentioned that in addition to using flat electrodes with bigger contact area, they are also porous, which increases the effective contact area significantly. In addition to increasing the contact area of the electrodes, they have to tolerate high operating temperature and include catalysts in themselves.

In order to make effective contact between the gas and the electrodes, the gas is directed through manifolds to the designed channels in the vicinity of the electrodes. Fig. 13.5 schematically shows the simple gas distribution grooves in the vicinity of the electrodes.

Clearly, sealing is very important to prevent hydrogen from entering the cathode and oxygen or air from entering the anode.

The plates on which the gas grooves are created are called bipolar plates, and in addition to distributing the gas on the electrodes, they transfer electrons between

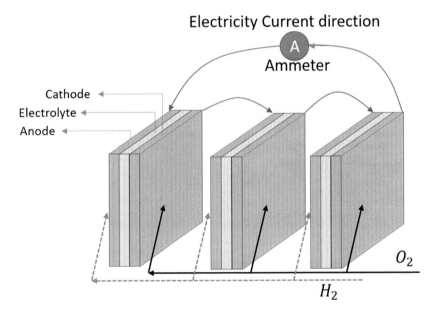

Figure 13.4 Three-cell edge connected in series.

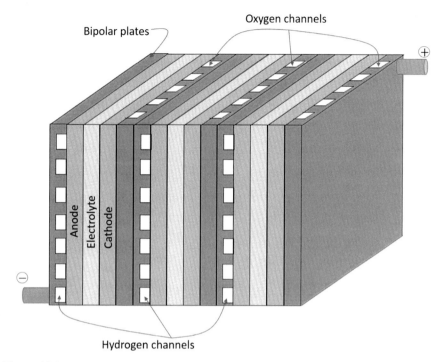

Figure 13.5 Three-cell stack with gas channels between cells.

cells in series. Bipolar plates are a very important part of a fuel cell stack due to their important functions. The material may be graphite, stainless steel, ceramic, etc. depending on the fuel cell characteristics such as corrosivity, operating temperature, etc.

The arrangement used in Fig. 13.5 for gas distribution is called external manifold since an *external manifold* will cover the stack sides with the gas grooves. If the manifold and grooves could be designed internally with only one input and one output ports on the bipolar plates, it is called the *internal manifold*.

Fig. 13.6 shows an internal manifold with inlet and exit ports. These ports are installed on all bipolar plates and holding plates of the electrodes, so that the gas can be introduced from one point, divided equally between the electrodes at the same time and finally discharged from the electrode.

2. Fuel cell types

Fuel cells are silent, efficient, and environmentally friendly. However, their cost is still high due to manufacturing challenges. As discussed in the previous section, hydrogen is the preferred fuel for the fuel cells, but unfortunately, it is scarce in the nature. Due to this problem, there are two options, producing hydrogen by using different methods

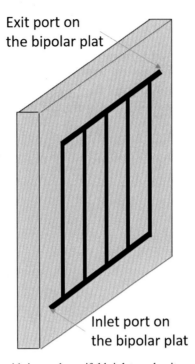

Figure 13.6 Bipolar plate with internal manifold, inlet, and exit ports.

such as hydrolysis, reforming, etc. or developing fuel cells to use other fuels such as natural gas or biogas without poisoning the fuel cell. In addition, slow reaction is another challenging problem for the fuel cells, which as previously discussed; three methods can be used to solve this problem, using catalysts, raising operational temperature, and increasing contact area. These two important challenges have led the researchers and manufacturers to use different electrolytes, electrodes, bipolar plates, and other auxiliary systems to create different fuel cells. Important fuel cells can be classified according to their main characteristics as in Table 13.1. In the following, important fuel cells are introduced briefly.

2.1 PEMFC

Starting with PEMFC, it has the lowest operating temperature that makes it very suitable for residential applications and electronic appliances; however, the low operating temperature causes a slow reaction rate that must be compensated by using expensive and complicated platinum catalysts and electrodes. The electrolyte is solid, which makes it very suitable for packaging and portable applications. This cell requires pure hydrogen to avoid poisoning of cell with CO, sulfur, and NH_3; therefore, internal reforming is not allowed to be used. It is available in small sizes of 50–200 kW, which makes it very suitable for small demands in building sector, portable power, electric

Fuel cell type	PEMFC	AFC	PAFC	MCFC	SOFC
Charge carrier	H^+ ions	OH^- ions	H^+ ions	$CO_3^=$ ions	$O^=$ ions
Electrolyte type	Polymeric membrane	Aqueous potassium, KOH	Phosphoric acid solutions	Phosphoric acid (immobilized liquid), molten alkali carbonates	Stabilized zirconia ceramic matrix with free oxide ions
Electrode material	Porous carbon	Metals	Porous carbon or graphite	Nickel or nickel oxide	Ceramic
Typical construction	Plastic, metal, or carbon	Plastic, metal	Carbon porous ceramic	High temperature metal, porous ceramic	High temperature metal, ceramic
Catalyst	Platinum	Platinum, nickel, metal oxides, or noble metals	Platinum	Nickel or nickel oxide	Parasites, ceramic
Interconnections material for bipolar plates	Carbone or metal	Metal	Graphite	Nickel or stainless steel	Nickel, ceramic, or stainless steel
Prime cell components	Carbone	Carbone	Graphite	Stainless steel	Ceramic
Oxidant	Air or O_2	Purified air or O_2	Air or O_2	Air	Air
Fuel	Hydrocarbons or methanol	Clean hydrogen, or hydrazine	Hydrocarbons or alcohols	Clean hydrogen, or natural gas, propane, diesel	Natural gas or propane
Operational temperature	50–100°C	50–220°C	100–200°C	600–700°C	500–1000°C
Size range	3–250 kW	10–200 kW	100–200 kW	250 kW–5MW	1–10 MW
Electrical efficiency[a]	30%–50%	32%–70%	40%–55%	55%–57%	50%–60%
Primary contaminants	CO, sulfur, and NH_3	CO, CO_2, H_2S, CH_4	CO>1%, H_2S	H_2S	H_2S
Internal fuel reforming	No	No	No	Yes	Yes
Applications	Vehicles, mobile, CHPs and CCHPs	Space vehicles	Medium-scale stationary, CHPs, and CCHPs	Large and average size power stations, hybrid, CHPs and CCHP	All sizes of power stations, CHP, CCHP

[a]The efficiencies are based on values for hydrogen fuel, and do not include the electricity for hydrogen reforming.

cars, etc. The low temperature of this fuel cell shorten its startup time and makes it safe for portable applications such as cars.

2.2 AFC

Alkaline fuel cell that exchanges hydroxyl anions through the electrolyte operate in the temperature range of $50-220°C$ and is used for space applications such as Apollo and Shuttle Orbiter craft to produce electricity and water. The slow reaction rate due to the low operation temperature is solved by using platinum catalysts and porous electrodes. Pure hydrogen and pure oxygen must be used on the anode and cathode to avoid cell poisoning.

2.3 PAFC

A phosphoric acid fuel cell, which exchanges hydrogen cations through the phosphoric acid solutions, uses porous carbon or graphite electrodes. The interconnections (bipolar plates) are made of graphite and can use other hydrocarbon fuels and alcohol. Internal reforming is not allowed and gets poisoned by CO more than 1%. PAFC is the first commercialized fuel cell, which was made in capacities of 200 kW by the International Fuel Cell Corporation, operating in $220°C$ and was installed in the United States and Europe. They are mostly maintenance free and can operate for a year with minimum shutdown and repair.

2.4 MCFC

MCFC operates at high temperatures therefore the slow reaction rate is resolved; as a result, expensive catalysts are not required. The electrolyte does not remain solid at high temperatures and it melts. It needs CO_2 in the air to work and works with different hydrocarbons by using internal reformer.

2.5 SOFC

Solid oxide fuel cell operates at high temperatures, and the electrolyte remains solid; therefore, low rate of reaction is not a problem for this fuel cell. The high temperature exhasut of this fuel cell can be used in a gas turbine to generate more power, and then recovering heat from the gas turbine exhaust can boost the overall efficiency. In addition, the electrical efficiency of this fuel cell is higher than PEMFC, AFC, and PAFC, which have lower operating temperature. The electrolyte is solid in this cell, which makes it suitable for packaging and sealing is easier. It is available in different scales and capacities. Different hydrocarbons including natural gas and propane can be used as fuel. In addition, internal reforming is possible for this cell, and it does not get poisoned with CO. The high operating temperature of SOFC result in slow start up and makes it unsuitable for portable applications. But, it is highly suitable for large scale utility applications.

The reactions happening on the anode and cathode of different fuel cells, and the cell reactions are presented in Table 13.2 [3]. As it shows for high-temperature fuel cells such as MCFC and SOFC, pure hydrogen or pure oxygen is not required. In addition, for the MCFC, on the cathode, CO_2 should be used with air to complete the reactions.

From the environmental point of view, since high temperature fuel cells have higher efficiency they produce less CO and CO_2 per MWh of electricity production. However, due to the high temperature, nitrogen reacts with oxygen and produce more NO_x in comparison with the low temperature fuel cells. The environmental information about different fuel cells for some specific capacities can be seen in Table 13.3.

2.6   RFC

Another type of fuel cells is the *regenerative* or *reversible fuel cell* (RFC). At the beginning of this chapter, it was mentioned that fuel cell and electrolysis do the same reactions but in the opposite directions. A fuel cell that is able to use electricity to produce and store hydrogen and also is able to use hydrogen to generate electricity is a reversible fuel cell. These fuel cells use the electricity produced by the renewable energies such as solar or wind at the time when demand for electricity is low to generate hydrogen and store it for the time when electricity demand is high. When demand for electricity increases the fuel cell uses the stored hydrogen to generate electricity [4].

For instance, SOFC is a good candidate for RFC due to its high operating temperature and high electrical efficiency. Due to the high operating temperature, less electricity is required for the electrolysis process in solid oxide reversible fuel cells (SORFCs).

3.   Fuel cell subsystems

The main components of a fuel cell depend on the fuel type, the fuel cell type, and the application. The basic components that can be found in many fuel cells include the fuel cell stack, the fuel processor, power conditioners, compressors, and humidifiers [4].

3.1   Fuel cell stack

Stack is the heart of a fuel cell system. It converts fuel and oxidizer into electricity, heat, and water as the main products. A stack may comprise of hundreds of single cells. Single cells produce electricity with voltage of less than 1 V. To increase the voltage and the power, many cells must be used in series as a stack. The amount of power produced by a stack depends on the number of cells, the fuel type, fuel and air pressure, operating temperature, and the cell size [4].

Table 13.2 The reactions on the anode and cathode of different fuel cells [3].

	Anode reaction	Cathode reaction	Cell reaction
PEMFC	$2H_2 \rightarrow 4H^+ + 4e^-$	$O_2 + 4H^+ + 4e^- \rightarrow 2H_2O$	$H_2 + \frac{1}{2}O_2 \rightarrow H_2O$
AFC	$H_2 + 2OH^- \rightarrow 2H_2O + 2e^-$	$\frac{1}{2}O_2 + H_2O + 2e^- \rightarrow 2OH^-$	$H_2 + \frac{1}{2}O_2 \rightarrow H_2O$
PAFC	$H_2 \rightarrow 2H^+ + 2e^-$	$\frac{1}{2}O_2 + 2H^+ + 2e^- \rightarrow H_2O$	$H_2 + \frac{1}{2}O_2 \rightarrow H_2O$
MCFC	$H_2 + CO_3^= \rightarrow H_2O + CO_2 + 2e^-$ $CO + CO_3^= \rightarrow 2CO_2 + 2e^-$	$\frac{1}{2}O_2 + CO_2 + 2e^- \rightarrow CO_3^=$	$H_2 + \frac{1}{2}O_2 + (CO_2)_{ca} \rightarrow H_2O + (CO_2)_{an}$ $CO + \frac{1}{2}O_2 + (CO_2)_{ca} \rightarrow 2(CO_2)_{an}$
SOFC	$H_2 + O^= \rightarrow H_2O + 2e^-$ $CO + O^= \rightarrow CO_2 + 2e^-$ $CH_4 + 4O^= \rightarrow 2H_2O + CO_2 + 8e^-$	$O_2 + 4e^- \rightarrow 2O^=$	$H_2 + \frac{1}{2}O_2 \rightarrow H_2O$ $CO + \frac{1}{2}O_2 \rightarrow CO_2$ $CH_4 + 2O_2 \rightarrow 2H_2O + CO_2$

Table 13.3 Comparison of different types of some fuel cells from environmental point of views [2].

Fuel cell type	PEMFC	PEMFC	PAFC	MCFC	SOFC
E_{nom} (kWe)	10	200	200	250	100
η_e (%) HHV	30	35	36	46	45
NO_x (ppmv at 15% O_2)	1.8	1.8	1.0	2.0	2.0
NO_x(lb/MWh)	0.06	0.06	0.03	0.06	0.05
CO (ppmv at 15% O_2)	2.8	2.8	2.0	2.0	2.0
CO (lb/MWh)	0.07	0.07	0.05	0.04	0.04
CO_2 (lb/MWh)	1360	1170	1135	950	910
Carbon (lb/MWh)	370	315	310	260	245
Unburnt hydrocarbons (ppmv at 15% O_2)	0.4	0.4	0.7	0.5	1.0
Unburnt hydrocarbons (lb/MWh)	0.01	0.01	0.01	0.01	0.01

3.2 Fuel processor

Different fuel cells demand fuel with different qualities. Providing the favorable fuel for the fuel cell is the main duty of a fuel processor. Depending on the fuel cell type and its requirements, the fuel processor may be a simple sorbent bed that removes impurities or a complex of some reactors and sorbents.

If the fuel cell demands hydrogen rich, fuels such natural gas, gasoline, methanol, or gasified coal should be reformed first to convert it into hydrogen and carbon compounds called "reformate." Reformate may need to be used in a set of reactors that converts CO to CO_2 and use a sorbent bed to remove other impurities and avoiding poisoning of the fuel cell catalysts. Poisoning reduces fuel cell efficiency, reaction rate, power production, and lifespan of the fuel cell. Such fuel cells require *external reformer* and sorbent beds.

Some fuel cells such as MCFC and SOFC can reform the fuel internally. This is called *internal reforming*. In the internal reforming, there may be some unused fuel that may reach the cells. To avoid this problem, some traps are needed to remove impurities [4].

3.3 Power conditioner

Power conditioning includes controlling voltage, frequency, ampere (current), etc. to provide the electrical characteristics that different applications need to operate. Fuel cells generate direct current (DC) electricity that means electrons move only in one direction. Electricity consumers and appliances around us may consume DC or alternating current (AC). In an AC electricity, electrons flow in both directions on alternating cycles. Both AC and DC currents must be conditioned and controlled before being consumed. If your application needs AC power, the power of the fuel cell must be converted to AC power by a current inverter. Power loss in the inverter and conditioner may reach 2%–6% [4].

3.4 Fuel/air compressor

Higher inlet gas pressure increases the fuel cell efficiency; for this reason, usually the inlet air/fuel pressure is about 2—4 atm. The compressor can be centrifugal or rotary to provide continuous flow. Its adiabatic efficiency usually is above 75%, and the high pressure of hot exhaust gases can be recovered by using an expander such as a micro-gas turbine or a turbocharger to generate more power. To reduce the compression power of compressor at the higher pressures, the compression may be done in two stages and an intercooler may be utilized between the two compression stages.

3.5 Humidifiers

Polymer membrane electrolytes used in fuel cells such as PEMFC operate well when they are humidified. For this reason, many fuel cell systems need an air humidifier to provide the required humidity. The water content of exhaust can be recovered in some cases for the humidifying [4].

3.6 Other auxiliary systems

A *heat recovery unit,* which is usually a *heat exchanger,* may be used to recover the exhaust heat for combined heat and power generation (CHP) or for preheating of the inlet air and fuel. In some systems, a *preheater* is used to heat the inlet gases to the fuel cell. In addition to the heat recovery system, a *cooling system,* which is a heat exchanger, may be used to keep the fuel cell body temperature constant. *Control valves* as well as *pressure regulators* are usually needed. A *controller* is also needed to control the fuel and air injection specially during the start-up and shutdown. Controlling these two processes (start-up and shutdown) in most of power generators is very critical [5].

4. Thermodynamics of fuel cells

An acid electrolyte fuel cell receives hydrogen and oxygen as inputs and produces electricity, heat, and water as the outputs (Fig. 13.3A). The power generated by the fuel cell is the product of voltage and ampere.

$$P = VI \tag{13.8}$$

Everything starts with the electrochemical reactions inside the fuel cell. The amount of energy released due to a reaction can be determined by subtracting the Gibbs free energy of formation of reactants from the products:

$$\Delta G_f = \sum_{products} G_f - \sum_{reactants} G_f \tag{13.9}$$

In which, G_f represents the Gibbs free energy of formation of the component in the reactants or products. The subscript f stands for formation. Eq. (13.9) can be written in the molar form as below:

$$\Delta \bar{g}_f \left(\frac{kJ}{mol} \right) = \sum_{products} \bar{g}_f - \sum_{reactants} \bar{g}_f \tag{13.10}$$

The Gibbs free energy is defined as a function of enthalpy, entropy, and temperature of the substance as below:

$$\bar{g}_f \left(\frac{kJ}{mol} \right) = \bar{h} \left(\frac{kJ}{mol} \right) - T(K)\bar{s} \left(\frac{kJ}{mol.K} \right)$$

$$\bar{h} \left(\frac{kJ}{mol} \right) = \bar{h}_f + \bar{h}(T, P) - \bar{h}(T_0, P_0) \tag{13.11}$$

In which, $\bar{h}_f$ is the formation enthalpy of the species.

Example 13.1. (a) Determine the $\Delta \bar{g}_f$ for the overall reaction of an acid electrolyte fuel cell if the cell temperature is 300K and pressure is 1 bar. Assume the species as ideal gas and the formation enthalpy of hydrogen, and oxygen is zero, for steam water it is -242 $\frac{kJ}{mol}$ (b) If the temperature is supposed to change from 300K to 1300K, how the $\Delta \bar{g}_f$ will change? Plot the changes. (c) Keep the temperature constant at 1000 K but change the pressure from 1 to 5 bar and plot the changes of $\Delta \bar{g}_f$. (c) Determine the impact of the pressure and temperature for the similar ranges as parts a and b on the Gibbs energy of the reactants.

Solution. **a.** The overall reaction of an acid electrolyte fuel cell is as below:

$$O_2 + 2H_2 \rightarrow 2H_2O(g)$$

Or for 1 mole of water:

$$\frac{1}{2}O_2 + H_2 \rightarrow H_2O(g)$$

Hence:

$$\Delta \bar{g}_f = \bar{g}_{f,H_2O} - \bar{g}_{f,H_2} - \frac{1}{2}\bar{g}_{f,O_2}$$

In addition:

$$\bar{g}_{f,H_2O} = \bar{h}_{f,H_2O} + \bar{h}_{H_2O}(T, P) - \bar{h}_{H_2O}(T_0, P_0) - T \times s_{H_2O}(T, P)$$

$$\overline{g}_{f,H_2} = \overline{h}_{f,H_2} + \overline{h}_{H_2}(T, P) - \overline{h}_{H_2}(T_0, P_0) - T \times s_{H_2}(T, P)$$

$$\overline{g}_{f,O_2} = \overline{h}_{f,O_2} + \overline{h}_{O_2}(T, P) - \overline{h}_{O_2}(T_0, P_0) - T \times s_{O_2}(T, P)$$

Since all of the gases are assumed as perfect gas, then:

$$\overline{g}_{f,H_2O} = \overline{h}_{f,H_2O} + \overline{h}_{H_2O}(T) - \overline{h}_{H_2O}(T_0) - T \times s_{H_2O}(T, P)$$

$$\overline{g}_{f,H_2} = \overline{h}_{f,H_2} + \overline{h}_{H_2}(T) - \overline{h}_{H_2}(T_0) - T \times s_{H_2}(T, P)$$

$$\overline{g}_{f,O_2} = \overline{h}_{f,O_2} + \overline{h}_{O_2}(T) - \overline{h}_{O_2}(T_0) - T \times s_{O_2}(T, P)$$

The thermodynamic properties are as below:

$$\overline{h}_{f,H_2O} = -242000 \frac{kJ}{kmol}, \overline{h}_{f,H_2} = 0, \overline{h}_{f,O_2} = 0$$

At $T = 300K \rightarrow \overline{h}_{H_2O} = -241758 \frac{kJ}{kmol}, \quad \overline{h}_{H_2} = 52.97 \frac{kJ}{kmol}, \quad \overline{h}_{O_2} = 54.56 \frac{kJ}{kmol}$

At $T_0 = 298.15K \rightarrow \overline{h}_{H_2O} = -241820 \frac{kJ}{kmol}, \quad \overline{h}_{H_2} = 0 \frac{kJ}{kmol}, \quad \overline{h}_{O_2} = 0 \frac{kJ}{kmol}$

At $T = 300K, andP = 1bar \rightarrow s_{H_2O} = 188.9 \frac{kJ}{kmol.K}, \quad s_{H_2} = 130.9 \frac{kJ}{kmol.K}, \quad s_{O_2} = 205.2 \frac{kJ}{kmol.K}$

By substituting these values in the Gibbs functions, the change in the Gibbs free energy would be determined as below:

$$\overline{g}_{f,H_2O} = -242000 - 241758 + 241820 - 300 \times 188.9 = -298616 \frac{kJ}{kmol}$$

$$\overline{g}_{f,H_2} = 0 + 52.97 - 0 - 300 \times 130.9 = -39205 \frac{kJ}{kmol}$$

$$\overline{g}_{f,O_2} = 0 + 54.56 - 0 - 300 \times 205.2 = -61510$$

Therefore, the Gibbs free energy change of the reaction would be:

$$\Delta \overline{g}_f = -298616 + 39205 + \frac{1}{2} \times 61510 = -228656 \frac{kJ}{kmol}$$

In fact, $\Delta \overline{g}_f$ is the available (reversible) energy to do work or produce electricity.

(b) to determine the impact of temperature on the $\Delta \overline{g}_f$, the above calculations are coded in the EES as shown in Code 13.1, and a parametric table is used to change the temperature from 300 to 1300 K. Do not forget to set the *unit system* from options > unit system menu for the units, which are used in this example. The units should be like pressure (bar), specific properties (molar base), and temperature (Kelvin).

Code 13.1. The code written to determine the Gibbs free energy variation with temperature in Example 13.1

```
"T1=300"
P1=1
T0=298.15
h_f_WATER=-242000
h_f_H2=0
h_f_O2=0
h_water=h_f_water+h_water_T-H0_WATER
h_water_T=enthalpy(h2o, t=T1)
H0_WATER=enthalpy(h2o, t=T0)
S_WATER=ENTROPY(H2O, T=T1, P=P1)
G_water=h_water-(t1)*s_WATER
h_H2=h_f_H2+h_H2_T-H0_H2
h_H2_T=enthalpy(h2, t=T1)
H0_H2=enthalpy(h2, t=T0)
S_H2=ENTROPY(H2, T=T1, P=P1)
G_H2=h_H2-(t1)*s_H2
h_O2=h_f_O2+h_O2_T-H0_O2
h_O2_T=enthalpy(O2, t=T1)
H0_O2=enthalpy(O2, t=T0)
S_O2=ENTROPY(O2, T=T1, P=P1)
G_O2=h_O2-(t1)*s_O2
DELTA_g_bar=G_water-G_H2-0.5*G_O2
```

By using the Code 13.1, it can be concluded that increasing temperature would increase the Gibbs free energy change. The results are plotted in Fig. 13.7. The negative sign of the numbers shows the release of energy.

By using the Code 13.1, keeping the temperature constant at 1000 K, commenting pressure (P1) and changing it from 1 to 5 bars in another parametric table the impact of pressure on the Gibbs free energy variation would be determined as Fig. 13.8. It shows that increasing pressure decreases the Gibbs free energy change for the overall reaction of an acid electrolyte fuel cell.

(C) To determine the impact of temperature and pressure on the reactants and products of the acid electrolyte fuel cell the Code 13.1 can be used again with the parametric tables used in parts a and b of the present example. The results are shown in Figs. 13.9 and 13.10.

4.1   Open circuit voltage

If there was no irreversibility, all of the change in the Gibbs free energy in the electrochemical reaction must be converted to electricity. However, it does not happen in practice, some of the energy would be lost as heat, and some of it would be lost due to electrical and internal resistances or other polarization losses.

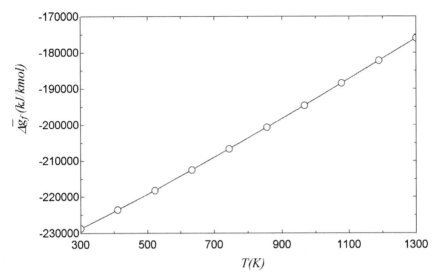

Figure 13.7 Changes of Gibbs free energy with temperature for the $O_2 + 2H_2 \rightarrow 2H_2O(g)$ under $P = 1$ bar.

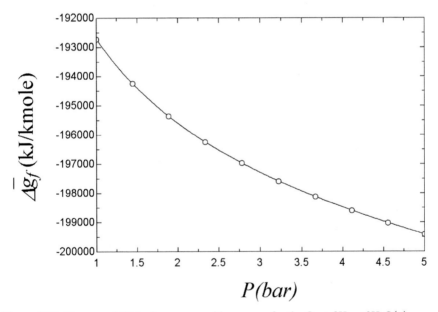

Figure 13.8 Changes of Gibbs free energy with pressure for the $O_2 + 2H_2 \rightarrow 2H_2O(g)$ under.$T = 1000K$.

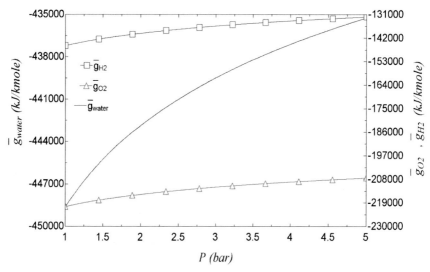

Figure 13.9 Changes of Gibbs free energy with pressure for the reactants and products of $O_2 + 2H_2 \rightarrow 2H_2O(g)$ under. $T = 1000K$.

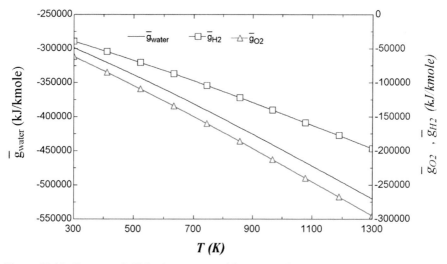

Figure 13.10 Changes of Gibbs free energy with pressure for the reactants and products of $O_2 + 2H_2 \rightarrow 2H_2O(g)$ under. $P = 1$ bar.

For the hydrogen fuel cell, as discussed for every molecule of hydrogen consumed, two electrons flow in the external circuit. Therefore, for every mole of hydrogen, $2N$ electron would flow in the external circuit in which $N = 6.02214076 \times 10^{23}$ is the

Avogadro's number. Therefore, since $-e$[1] is the charge on one electron, the charge of electrons flowing due to consuming of 1 mole of hydrogen in an acid electrolyte fuel cell would be:

$$-2Ne = -2F \qquad (13.12a)$$

And the electrical current due to consuming z mole of hydrogen per second in a fuel cell, which transfers n_e electrons for one molecule of hydrogen would be as blow:

$$I(A) = n_e zF \qquad (13.12b)$$

According to the Eq. 13.6, for consuming 1 molecule of hydrogen in the hydrogen fuel cell, 2 electrons would be released, hence $n_e = 2$.

In which $F = 96485 \frac{coulomb}{mole}$ is the *faraday constant* and is the absolute electric charge of 1 mole of electrons. If the voltage of the fuel cell is the V, then the reversible work done by the current of the electrons would be:

$$Electrical work = \Delta\bar{g}_f = -2F.V \qquad (13.13)$$

Or:

$$V_{OCV} = \frac{-\Delta\bar{g}_f}{2F} \qquad (13.14)$$

In which V_{OCV} is the open circuit voltage (OCV) or electromotive force (EMF) of the reversible hydrogen fuel cell. Attention must be paid that the number 2 in the Eq. 13.14 shows the number of electrons for each molecule of hydrogen consumed in the acid electrolyte fuel cell. If the number of electrons changed for example to n_e, this number must be updated, and in general the V_{OCV} would be as below:

$$V_{OCV} = \frac{-\Delta\bar{g}_f}{n_e F} \qquad (13.15)$$

Example 13.2. Determine the OCV of a hydrogen fuel cell for the temperature range of 300–1300 K.

Solution. By using the Code 13.1, the value of the Gibbs free energy change is determined as it was shown in Fig. 13.7. The voltage of the reversible fuel cell would be determined by adding Eq. (13.14) to the code as below (**Code 13.2**) and calculating this value in a parametric table.

Code 13.2. The code used to determine the OCV in different temperatures.

---

[1] Electron charge or electric charge of one electron shown by e is equal to 1.60217663 x 10^{-19} coulomb.

```
T1=473
P1=1
T0=298.15
h_f_WATER=-242000
h_f_H2=0
h_f_O2=0
h_water=h_f_water+h_water_T-H0_WATER
h_water_T=enthalpy(h2o, t=T1)
H0_WATER=enthalpy(h2o, t=T0)
S_WATER=ENTROPY(H2O, T=T1, P=P1)
G_water=h_water-(t1)*s_WATER
h_H2=h_f_H2+h_H2_T-H0_H2
h_H2_T=enthalpy(h2, t=T1)
H0_H2=enthalpy(h2, t=T0)
S_H2=ENTROPY(H2, T=T1, P=P1)
G_H2=h_H2-(t1)*s_H2
h_O2=h_f_O2+h_O2_T-H0_O2
h_O2_T=enthalpy(O2, t=T1)
H0_O2=enthalpy(O2, t=T0)
S_O2=ENTROPY(O2, T=T1, P=P1)
G_O2=h_O2-(t1)*s_O2
DELTA_g_bar=G_water-G_H2-0.5*G_O2

F=96485
V_ocv=-DELTA_g_bar/2/F
```

And the results are plotted in Fig. 13.11. It shows that increasing temperature reduces the voltage.

4.2 Efficiency of a fuel cell

Efficiency is always limited by the real conditions and irreversibilities. As discussed in the previous chapters, for a thermal power plant operating between a high temperature energy source with temperature of T_H and a low temperature heat sink with temperature of T_L, the efficiency would be limited by the Carnot efficiency defined as:

$$\eta_{Carnot} = 1 - \frac{T_L}{T_H} \tag{13.16}$$

For horizontal axis wind turbine and hydro turbine, it was limited by the Betz limit. For a fuel cell, there will be a limit as well. Since the total energy of a fuel will be released during the electrochemical process the input energy to the fuel cell would be the change in the formation enthalpy $(\Delta \bar{h}_f)$. The maximum electricity received from a reversible fuel cell would be the Gibbs free energy change. Hence, the maximum possible theoretical efficiency or the efficiency limit of fuel cells would be as below:

$$\eta_{max} = \frac{\Delta \bar{g}_f}{\Delta \bar{h}_f} = \frac{\Delta \bar{g}_f}{HHV_{fuel}} \tag{13.17}$$

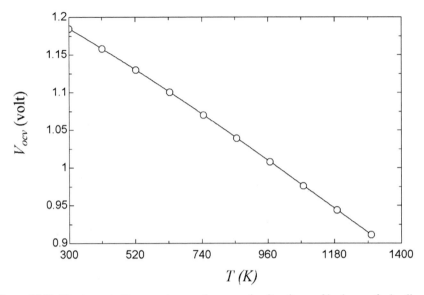

Figure 13.11 The impact of temperature on the open circuit voltage of hydrogen fuel cell.

In reality due to irreversibilities, the electrical energy received from the fuel cell would be smaller than $\Delta \bar{g}_f$; therefore, the real efficiency would be:

$$\eta = \frac{-\Delta \bar{g}_f + T\Delta s}{-\Delta \bar{h}_f} = \frac{E}{-\Delta \bar{h}_f} = 1 - \frac{T\Delta s}{\Delta \bar{h}_f} \tag{13.18}$$

In which, Δs is the change in entropy of the electrochemical reaction due to irreversibilities, and E is the actual electrical work produced by the fuel cell.

The energy of fuel can be calculated according to the state of H_2O in the combustion products. If combustion gases are cooled until steam water is condensed, some energy related to vaporization enthalpy would be added; in this case, the change in the enthalpy of formation is called higher heating value (HHV). If H_2O remains as steam in the products, the change of formation enthalpy is smaller and is called lower heating value (LHV). In other words:

$$HHV = LHV + h_{fg} \tag{13.19}$$

In which h_{fg} is the vaporization enthalpy (or latent heat of vaporization). For hydrogen reacting with oxygen, $HHV = -285.84 \frac{kJ}{mole}$, and $LHV = -241.83 \frac{kJ}{mole}$.

Example 13.3. Determine the maximum thermal efficiency of a hydrogen fuel cell by assuming (a) water as liquid in the products for the temperature range of 200 K to 1300 K. (b) Water as steam in products for the same temperature range. Plot both parts on the same graph.

Solution. As previously defined the maximum efficiency would be:

$$\eta_{max} = \frac{\Delta \bar{g}_f}{HHV_{fuel}}$$

That means the Gibbs free energy change should be calculated once for water as liquid (Part a) and the other time as steam (Part b).

By using the Code 13.1, the value of the Gibbs free energy changes is determined as it was shown in Fig. 13.7. The fuel cell efficiency of the reversible fuel cell would be determined by adding Eq. (13.13) to the code as below (**Code 13.3**) and calculating this value in the parametric table.

Code 13.3. The code used to determine the efficiency limit of hydrogen fuel cell in different temperatures.

```
"T1=473.15"
P1=1
T0=298.15
h_f_WATER_HHV=-285840
h_f_WATER_LHV=-241830
h_f_H2=0
h_f_O2=0

h_water_HHV=h_f_WATER_HHV+h_water_T-H0_WATER
h_water_T=enthalpy(h2o, t=T1)
H0_WATER=enthalpy(h2o, t=T0)
S_WATER=ENTROPY(H2O, T=T1, P=P1)
G_water_HHV=h_water_HHV-(t1)*s_WATER
h_water_LHV=h_f_WATER_LHV+h_water_T-H0_WATER
G_water_LHV=h_water_LHV-(t1)*s_WATER
h_H2=h_f_H2+h_H2_T-H0_H2
h_H2_T=enthalpy(h2, t=T1)
H0_H2=enthalpy(h2, t=T0)
S_H2=ENTROPY(H2, T=T1, P=P1)
G_H2=h_H2-(t1)*s_H2
h_O2=h_f_O2+h_O2_T-H0_O2
h_O2_T=enthalpy(O2, t=T1)
H0_O2=enthalpy(O2, t=T0)
S_O2=ENTROPY(O2, T=T1, P=P1)
G_O2=h_O2-(t1)*s_O2
DELTA_g_bar_HHV=G_water_HHV-G_H2-0.5*G_O2
DELTA_g_bar_LHV=G_water_LHV-G_H2-0.5*G_O2

DELTA_h_f_bar_HHV=h_f_WATER_HHV-h_f_H2-0.5*h_f_O2

eta_max_HHV=DELTA_g_bar_HHV/DELTA_h_f_bar_HHV
eta_max_LHV=DELTA_g_bar_LHV/DELTA_h_f_bar_HHV
```

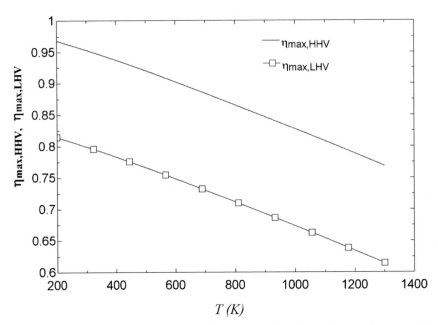

Figure 13.12 The impact of temperature on the hydrogen fuel cell maximum efficiency for two cases of water as liquid or as steam in the products.

And the results are plotted in Fig. 13.12. It shows that by increasing temperature, the efficiency limit reduces. In addition, the results show that the maximum efficiency for hydrogen fuel cell is higher when the water content in the products is condensed.

4.3 The Nernst equation

As presented in the previous section, the Gibbs free energy changes with temperature. The impact of pressure and concentration of the reactants and products are not discussed yet. The Gibbs free energy change also depends on the partial pressure of the components in the reaction. For example, for a hydrogen fuel cell with acid electrolyte, the Gibbs free energy would be as below:

$$\Delta \bar{g}_f = \Delta \bar{g}_f^0 - RT \ln \left(\frac{P_{H_2} . P_{O_2}^{\frac{1}{2}}}{P_{H_2O}} \right) \tag{13.20}$$

In which, $R = 8.3145 \frac{J}{mol.K}$ is the universal gas constant and $\Delta \bar{g}_f^0$ is the Gibbs free energy change under standard conditions ($P = 1atm, T = 25°C$). The molar fraction for the ideal gases can also be stated as below:

$$y_{spicies} = \frac{N_{species}}{N_{total}} = \frac{V_{species}}{V_{total}} = \frac{P_{species}}{P_{total}} \tag{13.21}$$

In which $N_{species}$, $V_{species}$, and $P_{species}$ are the number of moles, partial volume, and partial pressure of the species. And the ratio is the molar fraction, volume fraction, and pressure fraction of the species. By assuming the total pressure as the cell pressure (P_{cell}) that can be assumed equal on both anode and cathode and substituting Eq. (13.21) into Eq. (13.20):

$$\Delta \bar{g}_f = \Delta \bar{g}_f^0 - RT\ln\left(\frac{y_{H_2} \cdot y_{O_2}^{\frac{1}{2}} P_{cell}^{\frac{1}{2}}}{y_{H_2O}}\right) = \Delta \bar{g}_f^0 - RT\ln\left(\frac{y_{H_2} \cdot y_{O_2}^{\frac{1}{2}}}{y_{H_2O}}\right) - \frac{1}{2}RT\ln(P_{cell})$$

Hence, according to Eq. (13.15), the reversible voltage of the hydrogen fuel cell would be:

$$V_{rev} = V^0 + \frac{RT}{2F}\ln\left(\frac{y_{H_2} \cdot y_{O_2}^{\frac{1}{2}}}{y_{H_2O}}\right) + \frac{1}{4F}RT\ln(P_{cell}) \tag{13.22}$$

In which V^0 is the standard voltage.

Example 13.4. Determine (a) the voltage change of a hydrogen fuel cell if the cell pressure is changed, while the molar fraction of the reactants and products and the operating temperature stay the same, (b) the voltage change for the PEMFC with operating temperature of 60°C, (c) the voltage change for the PAFC with operating temperature of 200°C, and (d) the voltage change for the SOFC with operating temperature of 1100°C,

Solution. (a) According to Eq. (13.21), the voltage change would be:

$$V_2 - V_1 = \frac{1}{4F}RT\ln(P_{cell_2}) - \frac{1}{4F}RT\ln(P_{cell_1}) = \frac{1}{4F}RT\ln\left(\frac{P_{cell_2}}{P_{cell_1}}\right) \rightarrow$$

$$\Delta V = \frac{1}{4F}RT\ln\left(\frac{P_{cell_2}}{P_{cell_1}}\right)$$

(b) for the PEMFC

$$\Delta V_{PEMFC} = \frac{1}{4F}RT\ln\left(\frac{P_{cell_2}}{P_{cell_1}}\right) = \frac{8.3145 \times 333.15}{4 \times 96485}\ln\left(\frac{P_{cell_2}}{P_{cell_1}}\right)$$

$$= 0.0072\ln\left(\frac{P_{cell_2}}{P_{cell_1}}\right)$$

(c) for the PAFC:

$$\Delta V_{PAFC} = \frac{1}{4F}RT\ln\left(\frac{P_{cell_2}}{P_{cell_1}}\right) = \frac{8.3145 \times 473.15}{4 \times 96485}\ln\left(\frac{P_{cell_2}}{P_{cell_1}}\right) = 0.0102\ln\left(\frac{P_{cell_2}}{P_{cell_1}}\right)$$

(d) for the SOFC:

$$\Delta V_{SOFC} = \frac{1}{4F} RT\ln\left(\frac{P_{cell_2}}{P_{cell_1}}\right) = \frac{8.3145 \times 1373.15}{4 \times 96485}\ln\left(\frac{P_{cell_2}}{P_{cell_1}}\right)$$

$$= 0.0296\ln\left(\frac{P_{cell_2}}{P_{cell_1}}\right)$$

As it can be seen, voltage would increase with increasing pressure and decreases with decreasing pressure. In addition, the change in voltage for the fuel cells operating at higher temperature is bigger.

Example 13.5. In a fuel cell operating with air, it is decided to use pure oxygen instead of air. What would be the change in voltage for (a) SOFC with operating temperature of 1000°C (b) PEMFC with operating temperature of 80°C. The other factors remain constant.

Solution. The oxygen percentage in air is 21, while in the pure oxygen, it is 100; in other words, the molar fraction of oxygen has changed from 21% to 100%, and as a result, the partial pressure would change accordingly. Since:

$$V_{rev} = V^0 + \frac{RT}{2F}\ln\left(\frac{y_{H_2} \cdot y_{O_2}^{\frac{1}{2}}}{y_{H_2O}}\right) + \frac{1}{4F} RT\ln(P_{cell})$$

Therefore:

$$V_{rev} = V^0 + \frac{RT}{2F}\ln\left(\frac{y_{H_2} \cdot y_{O_2}^{\frac{1}{2}}}{y_{H_2O}}\right) + \frac{1}{4F} RT\ln(P_{cell})$$

$$= V^0 + \frac{RT}{2F}\ln\left(y_{H_2}\right) + \frac{RT}{2F}\ln\left(y_{O_2}^{\frac{1}{2}}\right) - \frac{RT}{2F}\ln\left(y_{H_2O}\right)\frac{1}{4F} RT\ln(P_{cell})$$

Since except for oxygen, the other parameters remain unchanged:

$$\Delta V = \frac{RT}{2F}\left(\ln\left(y_{O_2}^{\frac{1}{2}}\right)_2 - \ln\left(y_{O_2}^{\frac{1}{2}}\right)_1\right) = \frac{RT}{4F}\ln\frac{(y_{O_2})_2}{(y_{O_2})_1} = \frac{RT}{4F}\ln\left(\frac{1}{0.21}\right)$$

$$= 1.561\frac{RT}{4F}$$

(a) For the SOFC:

$$\Delta V_{SOFC} = 1.561\frac{RT}{4F} = 1.516\frac{8.3145 \times 1373.15}{4 \times 96485} = 0.04281\,V$$

(b) for the PEMFC:

$$\Delta V_{PEMFC} = 1.561 \frac{RT}{4F} = 1.516 \frac{8.3145 \times 353.15}{4 \times 96485} = 0.012V$$

As it can be seen, using pure oxygen increases the voltage, and its impact for the high operating temperature fuel cells is bigger.

4.4 Irreversibilities in the fuel cells

The voltage calculated by Eq. (13.22) is the reversible voltage generated by the fuel cell. In practice, there are several reasons for voltage drop. The four main causes of voltage drop include *activation loss,* which is mainly due to slow reaction rate on the electrodes. Part of generated voltage is lost for driving the reactions. *Ohmic loss* is the voltage loss due to the resistances against flowing of electrons and ions. *Mass transport or concentration loss* is due to the concentration changes of the reactants on the electrodes as the fuel is consumed. *Fuel crossover and internal currents* are due to passing some of the fuel molecules through the electrolyte without participating in the electrochemical reactions. In addition, some electrons may not flow through the external circuit and pass through the electrolyte. This will also reduce the voltage.

4.4.1 Activation polarization

For the activation loss bigger than 50−100 mV, the activation loss is usually approximated by using the *Tafel equation* as below:

$$\Delta V_{act} = A \ln \left(\frac{i}{i_0} \right) \tag{13.23}$$

The constant A depends on the reaction rate of the reactants. For slow reactions, it is bigger, and for faster reactions, it is smaller. Inversely, the faster the reaction rate, the bigger the constant current density i_0. i_0 is called the *exchange current density* and must stay smaller than current density of the fuel cell (i) to satisfy the Tafel equation. For low-temperature fuel cells i_0 is about $0.1 \frac{mA}{cm^2}$ and for high-temperature fuel cells such as 800−1000°C, the exchange current density i_0 is about $10 \frac{mA}{cm^2}$. The constant A is defined as below:

$$A = \frac{RT}{\alpha n_e F} \tag{13.24}$$

In which, α is the *electron transfer coefficient* that depends on the reactions and the material of the electrodes. Its value ranges from 0 to 1, but for hydrogen fuel cells, its value is 0.5 for many materials.

The voltage drop due to activation polarization is bigger for low- to medium-temperature fuel cells due to slow reaction rates. For high-temperature fuel cells and also high operating pressure, the activation voltage drop is smaller and less important. For hydrogen fuel cells, the activation loss occurs mostly on the cathode, but for the fuel cells using other fuels, the activation loss happens on both electrodes and can be determined by considering both anode and cathode characteristics as Eq. (13.25).

$$\Delta V_{act} = A \ln\left(\frac{i}{b}\right)$$

$$A = A_a + A_c \tag{13.25}$$

$$b = i_{0a}^{\frac{A_a}{A}} + i_{0c}^{\frac{A_c}{A}}$$

In which, the subscripts a and c refer to anode and cathode.

The activation loss reduces by increasing temperature and pressure, using rough electrodes that increase the contact area, increasing concentration of reactants on the electrodes, and using more effective catalysts.

According to the discussions by assuming the activation voltage loss, the new voltage would be as Eq. (13.26).

$$V = V_{rev} - \Delta V_{act} \tag{13.26}$$

Example 13.6. Plot the voltage of an acid electrolyte hydrogen fuel cell with operating temperature of 80°C vs. the current density variation from 0.1 $\frac{mA}{cm^2}$ to 1000 $\frac{mA}{cm^2}$ (a) as a reversible fuel cell (b) by considering the impact of activation loss on the reversible voltage. The exchange current density is $i_0 = 0.1 \frac{mA}{cm^2}$. In addition, the electron transfer coefficient is $\alpha = 0.5$, and pure hydrogen and oxygen are fed into the anode and cathode. The water in the products appears as steam. The formation enthalpy of steam water is $-241,830 \frac{kJ}{kmol}$.

Solution. The voltage by considering the activation loss for an acid electrolyte hydrogen fuel cell follows Eqs. 13.23 and 13.26. Hence, first of all, the OCV must be calculated. This can be done by using the Code 13.2 but for the given operating temperature in the present example. In addition, the equations of 13.23 and 13.26 should be included in the code to calculate the real voltage by considering the activation polarization. The modified code is given in the Code 13.4 as below:

Code 13.4. The code written to calculate the voltage loss due to activation polarization

```
T1=353.15
P1=1
T0=298.15
h_f_WATER_LHV=-241830
h_f_H2=0
h_f_O2=0
F=96485
i_0=0.1
"i=200"
ALPHA=0.5
n_e=2
R=8.3145

h_water_T=enthalpy(h2o, t=T1)
H0_WATER=enthalpy(h2o, t=T0)
S_WATER=ENTROPY(H2O, T=T1, P=P1)
h_water_LHV=h_f_WATER_LHV+h_water_T-H0_WATER
G_water_LHV=h_water_LHV-(t1)*s_WATER
h_H2=h_f_H2+h_H2_T-H0_H2
h_H2_T=enthalpy(h2, t=T1)
H0_H2=enthalpy(h2, t=T0)
S_H2=ENTROPY(H2, T=T1, P=P1)
G_H2=h_H2-(t1)*s_H2
h_O2=h_f_O2+h_O2_T-H0_O2
h_O2_T=enthalpy(O2, t=T1)
H0_O2=enthalpy(O2, t=T0)
S_O2=ENTROPY(O2, T=T1, P=P1)
G_O2=h_O2-(t1)*s_O2
DELTA_g_bar_LHV=G_water_LHV-G_H2-0.5*G_O2
V_OCV=-DELTA_g_bar_LHV/(n_e*F)

DV_act=A*ln(i/i_0)
A=R*T1/(ALPHA*n_e*F)
Voltage=V_OCV-DV_act
```

By using the code and the parametric table for changing the current density in the given range, the OCV and the corrected voltage by considering the activation loss can be plotted as Fig. 13.13. As the graph shows, the activation polarization is more important at smaller current densities. It causes a sharp drop at very small current densities, and then it decreases slowly.

4.4.2   Fuel crossover and internal currents

The fuel crossover is passing fuel through the electrolyte, reaching cathode, and having chemical reaction with the oxidant instead of electrochemical reaction. In other words, due to the fuel crossover, only heat is released on the cathode, and no electron is freed to flow through the external circuit. That means the voltage loss due to fuel crossover can be modeled by a loss of current. The internal currents are due to flowing electrons from the passages other than the external circuit such as electrolyte or interconnections. Hence, the internal current can also be modeled as a loss of current. Due to the similar impact of fuel crossover and the internal currents, they both can be modeled as internal currents, and the voltage equation can be stated by modifying the Tafel equation for both activation and internal currents losses as below:

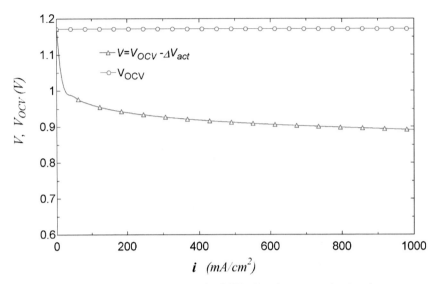

Figure 13.13 Impact of activation loss on the OCV when the current density changes.

$$V = V_{rev} - A\ln\left(\frac{i + i_n}{i_0}\right) \tag{13.27}$$

In which, i_n is the internal current density and for a typical low-temperature cell, it is about $3\frac{mA}{cm^2}$. By using the Code 13.4 and including the impact of internal current and plotting the voltage for two cases of with and without the internal currents, it can be seen that the impact of internal current is very small. As the cell current density increases, the impact of internal current vanishes. Fig. 13.14 shows the impact of considering the internal current on the voltage.

4.4.3 Ohmic polarization

Ohmic loss occurs due to the resistance in the electrodes, electrolyte, and the interconnections. This resistance is against the flow of ions and electrons and can be simply modeled by:

$$\Delta V_{ohm} = IR = i(AR) = ir \tag{13.28}$$

In which, I is the current and R is the total resistance, $i\left(\frac{A}{cm^2}\right)$ is the current density, and $r(cm^2.\Omega)$ is the area-specific resistance.

The ohmic loss is very important for all of the fuel cells and can be reduced by using electrodes with high conductivity, using low resistance materials for the interconnections and thinning the electrolyte as much as the technical designs permits. There are many experimental or theoretical equations in the literature, which let us calculate the ohmic losses. For example, the following equation is proposed to be used in SOFC [6].

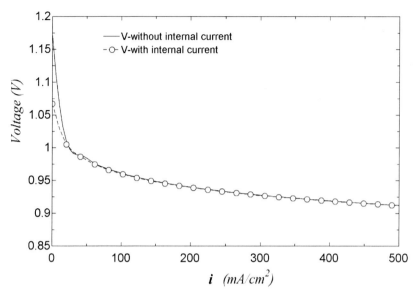

Figure 13.14 Impact of internal currents on the voltage loss.

$$\Delta V_{ohm} = n_e Fz \sum_{N=1}^{4} \delta_N \rho_N = n_e Fz \sum_{N=1}^{4} \frac{\delta_N A_N}{A_{cell}} \exp\left(\frac{B_N}{T}\right) \tag{13.29}$$

In addition, $\rho_N = A_N \exp\left(\frac{B_N}{T}\right)$ is the corresponding material-specific resistivity and δ_N, A_N, and B_N are constants that depend on components materials, in which an example is given in Table 13.4. In addition, $z\left(\frac{mol}{s}\right)$ is the molar consumption rate of the fuel. Furthermore, in the above equation, the resistance is $R = \sum_{N=1}^{4} \frac{\delta_N A_N}{A_{cell}} \exp\left(\frac{B_N}{T}\right)$, the current is $I = n_e Fz$ and A_{cell} is the area of the electrodes, electrolyte, or the interconnectors in 1 cell. Besides, δ_N is the thickness of the components and T is the operating temperature of the cell in Kelvin.

Table 13.4 Ohmic loss constant parameters [7].

Components	N	$A_N(\Omega - m)$	$B_N(K)$	$\delta_N(m)$
Cathode (ca)	1	8.11E-5	6.0E2	2.2E-3
Anode (an)	2	2.98E-5	−13.92E2	1.0E-3
Electrolyte (el)	3	2.94E-5	10.35E3	4.0E-4
Interconnector (in)	4	1.2E-3	4.69E3	8.5E-5

Example 13.7. Plot the voltage of an acid electrolyte hydrogen fuel cell with operating temperature of 1000°C vs. the current density variation from 10 $\frac{mA}{cm^2}$ to 800 $\frac{mA}{cm^2}$ (a) as a reversible fuel cell (b) by considering the impact of activation loss, internal current loss, and (c) considering activation loss, internal current loss, and ohmic loss on the same graph. The exchange current density is 10 $\frac{mA}{cm^2}$. In addition, the electron transfer coefficient is 0.5, and pure hydrogen and oxygen are fed into the anode and cathode. The water in the products appear as steam. The formation enthalpy of steam water is $-241,830 \frac{kJ}{kmol}$. Compare the results with that presented in Example 13.6. Use the data given in Table 13.4 for the properties of the components. Assume the internal current density to be 3 $\frac{mA}{cm^2}$.

Solution. The code written for the Example 13.6 could calculate the voltage by considering the activation and internal current losses. The voltage loss due to the ohmic polarization can be stated as below and added to the Code 13.4. The Code 13.5 shows the new code, which considers the activation, internal current, and ohmic losses.

$$\Delta V_{ohm} = n_e F z \sum_{N=1}^{4} \frac{\delta_N A_N}{A_{cell}} \exp\left(\frac{B_N}{T}\right)$$

Since the current density is changed over a range from 10 to 1000 $\frac{mA}{cm^2}$, so that he above equation can be used as below:

$$\Delta V_{ohm} = I \sum_{N=1}^{4} \frac{\delta_N A_N}{A_{cell}} \exp\left(\frac{B_N}{T}\right) = \frac{I}{A_{cell}} \sum_{N=1}^{4} \delta_N A_N \exp\left(\frac{B_N}{T}\right)$$

$$= i \sum_{N=1}^{4} \delta_N A_N \exp\left(\frac{B_N}{T}\right) = i\left(\delta_{an} A_{an} \exp\left(\frac{B_{an}}{T}\right) + \delta_{ca} A_{ca} \exp\left(\frac{B_{ca}}{T}\right)\right.$$

$$\left. + \delta_{el} A_{el} \exp\left(\frac{B_{el}}{T}\right) + \delta_{in} A_{in} \exp\left(\frac{B_{in}}{T}\right)\right)$$

By substituting the data from Table 13.4 into the above equation and multiplying with current density in $(\frac{A}{m^2})$, the voltage loss due to the ohmic polarization would be calculated.

$$\Delta V_{ohm} = i\left(10^{-3} \times 2.98 \times 10^{-5} \exp\left(\frac{-1392}{T}\right)\right.$$

$$+ 2.2 \times 10^{-3} \times 8.11 \times 10^{-5} \exp\left(\frac{600}{T}\right)$$

$$+ 4 \times 10^{-4} \times 2.94 \times 10^{-5} \exp\left(\frac{10350}{T}\right)$$

$$\left. + 8.5 \times 10^{-5} \times 1.2 \times 10^{-3} \exp\left(\frac{4690}{T}\right)\right)$$

As it can be seen, the ohmic loss is a function of temperature and the current density. In the present example, the temperature is constant and equal to $1000\,°C$ therefore $T = 1273.15K$ in the above equation. Care must be taken that the current density in the above equation must be in $\frac{A}{m^2}$. The voltage now can be calculated by using the following equation:

$$V = V_{rev} - A\ln\left(\frac{i + i_n}{i_0}\right) - \Delta V_{ohm} \tag{13.30}$$

Code 13.5. the code written to calculate the voltage loss due to ohmic polarization

```
T1=1273.15
P1=1
T0=298.15
h_f_WATER_LHV=-241830
h_f_H2=0
h_f_O2=0
F=96485
i_0=10
"i=200"
ALPHA=0.5
n_e=2
R=8.3145
i_n=3
delta_ca=0.0022
delta_an=0.0001
delta_el=0.0004
delta_in=0.000085
A_an=0.0000298
A_ca=0.0000811
A_el=0.0000294
A_in=0.001256
B_an=600
B_ca=-1392
B_el=10350
B_in=4690
z=1

h_water_T=enthalpy(h2o, t=T1)
H0_WATER=enthalpy(h2o, t=T0)
S_WATER=ENTROPY(H2O, T=T1, P=P1)
h_water_LHV=h_f_WATER_LHV+h_water_T-H0_WATER
G_water_LHV=h_water_LHV-(t1)*s_WATER
h_H2=h_f_H2+h_H2_T-H0_H2
h_H2_T=enthalpy(h2, t=T1)
H0_H2=enthalpy(h2, t=T0)
S_H2=ENTROPY(H2, T=T1, P=P1)
G_H2=h_H2-(t1)*s_H2
h_O2=h_f_O2+h_O2_T-H0_O2
h_O2_T=enthalpy(O2, t=T1)
H0_O2=enthalpy(O2, t=T0)
S_O2=ENTROPY(O2, T=T1, P=P1)
G_O2=h_O2-(t1)*s_O2
DELTA_g_bar_LHV=G_water_LHV-G_H2-0.5*G_O2
V_OCV=-DELTA_g_bar_LHV/(n_e*F)

DV_act=A*ln(i/i_0)
A=R*T1/(ALPHA*n_e*F)
DV_ohm=10*i*(delta_ca*A_ca*exp(B_ca/T1)+delta_an*A_an*exp(B_an/T1)+delta_el*A_el*exp(B_el/T1)+delta_in*A_in*exp(B_in/T1))
V_act_curr=V_OCV-A*ln((i+i_n)/i_0)
Voltage_ac_curr_ohm=V_OCV-A*ln((i+i_n)/i_0)-DV_ohm
```

The **Code 13.5** is used in a parametric table by changing current density, and the results are plotted in Fig. 13.15 as it can be seen at higher current density the ohmic polarization gets bigger.

4.4.4 Concentration polarization

Concentration of oxygen and hydrogen may change during the reactions due to consumption or being fed with other gases such as nitrogen, or carbon dioxide. Oxygen may be used from air that reduces the partial pressure of oxygen on the cathode. Hydrogen consumes continuously and loses pressure as it flows into the anode channels due to the pressure drop and consumption. The concentration of hydrogen reduces, and as a result, the voltage reduces. The concentration loss is usually calculated by the following equation. Although there are different equations for different fuel cells.

$$\Delta V_{conc} = \frac{RT}{n_e F} \left| ln\left(1 - \frac{i}{i_L}\right) \right| \tag{13.31}$$

In which i_L is the limiting current. And the final voltage by considering all of the losses would be as Eq. (13.32):

$$V = V_{rev} - A ln\left(\frac{i + i_n}{i_0}\right) - \Delta V_{ohm} - \Delta V_{conc} \tag{13.32}$$

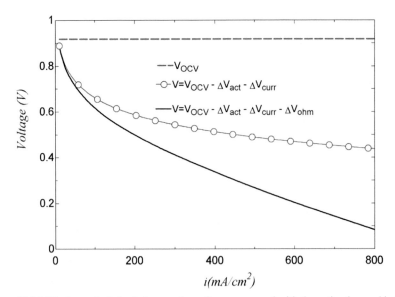

Figure 13.15 The impact of ohmic loss on the voltage compared with the activation and internal current losses with respect to the current density variation.

In other words, the DC power produced by a single cell of the fuel cell is:

$$P_{DC} = VI = ViA_{cell} \tag{13.33}$$

In which A_{cell} is the effective area of the cell. The power loss due to voltage polarization is also as below:

$$P_{loss} = \left(A\ln\left(\frac{i+i_n}{i_0}\right) - \Delta V_{ohm} - \Delta V_{conc}\right)iA_{cell} \tag{13.34}$$

The specific DC power and specific power loss can also be defined as below:

$$p_{DC} = \frac{P_{DC}}{z}, p_{loss} = \frac{P_{loss}}{z} \tag{13.35}$$

In which z is the fuel consumption in $\frac{mol}{s}$.

Example 13.8. Add the voltage drop due to concentration polarization to the fuel cell described in Example 13.7 and plot the final voltage for the limiting current of 800 $\frac{mA}{cm^2}$. Plot the results along with the results presented in Fig. 13.15.

Solution. The concentration loss would be as below:

$$\Delta V_{conc} = \frac{RT}{n_e F}\left|\ln\left(1 - \frac{i}{i_L}\right)\right| = \frac{8.3145 \times 1273.15}{2 \times 96485}\left|\ln\left(1 - \frac{i}{800}\right)\right|$$

By adding the following simple Code 13.6 to the Code 13.5, the problem can be solved, and the results would be plotted by using a parametric table when current density changes from 10 to 800 $\frac{mA}{cm^2}$. To avoid singularity in the *ln* function, current density is changed up to 799 $\frac{mA}{cm^2}$.

Code 13.6. The code added to the Code 13.4 to consider the impact of concentration loss

```
i_L=800
DV_conc=abs(R*T1/(n_e*F)*ln(1-i/i_L))
Voltage_ac_curr_ohm_conc=V_OCV-A*ln((i+i_n)/i_0)-DV_ohm-DV_conc
```

The results are plotted in Fig. 13.16. It shows that concentration polarization for bigger current densities is more significant.

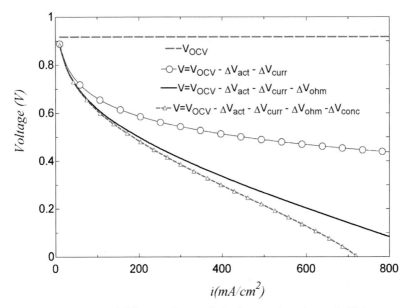

Figure 13.16 The impact of different voltage polarizations on the voltage of a high temperature fuel cell fed with hydrogen and pure oxygen.

4.5 Energy balance of a fuel cell

Depending on the fuel type, the composition, rate of fuel consumption, and the rate of reactions the operating temperature of the fuel cell may change. Usually, it is vital to write an energy balance for the fuel cell to determine the operating temperature of the fuel cell. In addition, not all of the fuel entering the anode is consumed; some of it may leave the fuel cell without taking part in the reactions. To include this in the calculations, the fuel utilization factor is usually introduced. The energy balance for consuming z mole of fuel (hydrogen) per second in an acid electrolyte fuel cell can be written as below:

$$- \Delta \bar{h}_f = p_{DC} + q_{out} + p_{loss} \tag{13.36}$$

Or

$$-z\Delta \bar{h}_f = P_{DC} + zT\Delta s + P_{loss}$$

$$\Delta s = \left(s_{H_2O} - s_{H_2} - 0.5 s_{O_2} \right) + \frac{R}{2} \ln \left(\frac{P_{O2} P_{H2}{}^2}{P_{H_2O}{}^2} \right) \tag{13.37}$$

In which, Δs is the change in specific entropy for an acid electrolyte hydrogen fuel cell.

In many fuel cell simulations, the operating temperature of the cell should be guessed to start the simulations. To correct the first guess, the new temperature (T_n) must be calculated according to the energy balance Eq. (13.37), and by using the new temperature, the second trail can go on. Trial and error continues until the difference between two consecutive temperatures calculated is small enough (ε) for the fuel cell. Hence:

$$T_n = \frac{-z\Delta\bar{h}_f - P_{DC} - P_{loss}}{z\Delta s} \tag{13.38}$$

$$|T_n - T_{n-1}| \leq \varepsilon$$

4.6 Fuel utilization and equilibrium constant

Fuel utilization refers to the molar fraction of the fuel, which is consumed in the electrochemical reaction. For a low-temperature fuel cell, it can be defined as below:

$$U_f = \frac{\dot{n}_{H_2,consummed}}{\dot{n}_{H_2,in}} = \frac{\dot{n}_{H_2,in} - \dot{n}_{H_2,out}}{\dot{n}_{H_2,in}} = 1 - \frac{\dot{n}_{H_2,out}}{\dot{n}_{H_2,in}} \tag{13.39}$$

For high-temperature fuel cells such as MCFC or SOFC, the CO can be used in a shifting reaction to generate hydrogen. In other words, CO is used as a fuel. The water gas shift reaction is as below:

$$CO + H_2O \rightarrow H_2 + CO_2 \tag{13.40}$$

And the fuel utilization is defined as below:

$$U_f = \frac{\dot{n}_{H_2,consummed}}{\dot{n}_{H_2,in} + \dot{n}_{CO,in}} \tag{13.41}$$

The equilibrium constant of the water gas shift reaction can be determined as below:

$$K_{shifting} = \frac{\dot{n}_{CO_2,out} \times \dot{n}_{H_2,out}}{\dot{n}_{CO,out} \times \dot{n}_{H_2O,out}} \tag{13.42}$$

Or

$$K_{shifting} = \exp\left(\frac{\Delta g_{shifting}}{RT}\right) \tag{13.43}$$

$$\Delta g_{shifting} = \Delta h_{shifting} - T\Delta s_{shifting}$$

$$\Delta h_{shifting} = h_{H_2} + h_{CO_2} - h_{CO} - h_{H_2O}$$

$$\Delta s_{shifting} = s_{H_2} + s_{CO_2} - s_{CO} - s_{H_2O}$$

In which $h = h_{formation} + h(T,P) - h(T_0,P_0)$ and $s = s(T,P)$.

Example 13.9.   An SOFC is using CH_4 as the fuel; therefore, it needs to have an internal reformer, in which steam water reacts with CH_4 and generate CO and H_2O. The CO generated in the reforming is then converted to H_2 and CO_2 in the water gas shift process. The reforming and shifting processes are as below:

$$CH_4 + H_2O \rightarrow 3H_2 + CO$$

$$CO + H_2O \rightarrow H_2 + CO_2$$

And the overall process is as below:

$$H_2 + \frac{1}{2}O_2 \rightarrow H_2O$$

In these processes x mole of CH_4, y mole of CO and z mole of H_2 are consumed per second.

a. Write the molar balance for each component on the anode, cathode, and the overall reaction.
b. Write the equations, which can be used to determine the partial pressure of each species, and the voltage and current. Assume the inlet molar rate of each spices are known. In addition, the fuel utilization is known.

Solution.   (a) The molar balance should be written between the inlet molar rate of each species; its consumption and the output molar rate. By rewriting the chemical reactions as below:

$$CH_4 + H_2O \rightarrow 3H_2 + CO \left(x \frac{mole}{s} \right)$$

$$CO + H_2O \rightarrow H_2 + CO_2 \left(y \frac{mole}{s} \right)$$

$$H_2 + \frac{1}{2}O_2 \rightarrow H_2O \left(z \frac{mole}{s} \right)$$

The molar balance would be as below for the anode:

$$\dot{n}_{in,CH_4} = \dot{n}_{out,CH_4} + x \rightarrow \dot{n}_{out,CH_4} = \dot{n}_{in,CH_4} - x$$

$$\dot{n}_{in,H_2O} = \dot{n}_{out,H_2O} + x + y - z \rightarrow \dot{n}_{out,H_2O} = \dot{n}_{in,H_2O} - x - y + z$$

$$\dot{n}_{in,H_2} = \dot{n}_{out,H_2} - 3x - y + z \rightarrow \dot{n}_{out,H_2} = \dot{n}_{in,H_2} + 3x + y - z$$

$$\dot{n}_{in,CO} = \dot{n}_{out,CO} - x + y \rightarrow \dot{n}_{out,CO} = \dot{n}_{in,CO} + x - y$$

$$\dot{n}_{in,CO_2} = \dot{n}_{out,CO_2} + y \rightarrow \dot{n}_{out,CO_2} = \dot{n}_{in,CO_2} + y$$

$$\dot{n}_{total,out} = \dot{n}_{total,in} + 2x$$

And for the cathode:

$$\dot{n}_{in,O_2} = \dot{n}_{out,O_2} + 0.5z \rightarrow \dot{n}_{out,O_2} = \dot{n}_{in,O_2} - 0.5z$$

Since nitrogen enters with air, and it is considered as a neutral gas, then:

$$\dot{n}_{in,N_2} = \dot{n}_{out,N_2}$$

(b) to determine the partial pressures of the components, the output molar rate of each component must be known. To do this, the x, y, and z must be calculated. For this purpose, three equations are necessary.

The equilibrium constant equations can be written for the reforming and shifting reactions as below:

$$K_{reforming} = \frac{P_{CO,out} \times P_{H_2,out}^3}{P_{CH_4,out} \times P_{H_2O,out}} = \frac{\dfrac{\dot{n}_{CO,out}}{\dot{n}_{total}} P_{cell} \times \left(\dfrac{\dot{n}_{H_2,out}}{\dot{n}_{total}} P_{cell}\right)^3}{\dfrac{\dot{n}_{CH_4,out}}{\dot{n}_{total}} P_{cell} \times \dfrac{\dot{n}_{H_2O,out}}{\dot{n}_{total}} P_{cell}}$$

$$= \frac{\dot{n}_{CO,out} \times \dot{n}_{H_2,out}^3}{\dot{n}_{CH_4,out} \times \dot{n}_{H_2O,out}} \left(\frac{P_{cell}}{\dot{n}_{total}}\right)^2$$

Hence, the first equation would be:

$$K_{reforming} = \frac{\dot{n}_{CO,out} \times \dot{n}_{H_2,out}^3}{\dot{n}_{CH_4,out} \times \dot{n}_{H_2O,out}} \left(\frac{P_{cell}}{\dot{n}_{total}}\right)^2$$

In which:

$$K_{reforming} = \exp\left(\frac{\Delta g_{reforming}}{RT}\right)$$

By substituting the molar output of the species, the equilibrium constant would be as below:

$$K_{reforming} = \frac{\left(\dot{n}_{in,CO} + x - y\right) \times \left(\dot{n}_{in,H_2} + 3x + y - z\right)^3}{\left(\dot{n}_{in,CH_4} - x\right) \times \left(\dot{n}_{in,H_2O} - x - y + z\right)} \left(\frac{P_{cell}}{\dot{n}_{total,in} + 2x}\right)^2$$

For the water gas shift process:

$$K_{shifting} = \frac{P_{CO_2,out} \times P_{H_2,out}}{P_{CO,out} \times P_{H_2O,out}} = \frac{\dfrac{\dot{n}_{CO_2,out}}{\dot{n}_{total}}P_{cell} \times \dfrac{\dot{n}_{H_2,out}}{\dot{n}_{total}}P_{cell}}{\dfrac{\dot{n}_{CO,out}}{\dot{n}_{total}}P_{cell} \times \dfrac{\dot{n}_{H_2O,out}}{\dot{n}_{total}}P_{cell}} = \frac{\dot{n}_{CO_2,out} \times \dot{n}_{H_2,out}}{\dot{n}_{CO,out} \times \dot{n}_{H_2O,out}}$$

$$= \frac{\left(\dot{n}_{in,CO_2} + y\right) \times \left(\dot{n}_{in,H_2} + 3x + y - z\right)}{\left(\dot{n}_{in,CO} + x - y\right) \times \left(\dot{n}_{in,H_2O} - x - y + z\right)}$$

Hence, the second equation would be:

$$K_{shifting} = \frac{\left(\dot{n}_{in,CO_2} + y\right) \times \left(\dot{n}_{in,H_2} - 3x - y + z\right)}{\left(\dot{n}_{in,CO} + x - y\right) \times \left(\dot{n}_{in,H_2O} - x - y + z\right)}$$

In which:

$$K_{shifting} = \exp\left(\frac{\Delta g_{shifting}}{RT}\right)$$

And the third equation would be:

$$U_f = \frac{z}{\dot{n}_{H_2,in} + \dot{n}_{CO,in} + 3x + y}$$

By solving the simultaneous equations for U_f, $K_{shifting}$, and $K_{reforming}$, the x, y, and z would be determined. When they are determined calculating the partial pressure would be easy as well as voltage calculation. The partial pressure of each component is as below:

$$P_{speciec} = \frac{\dot{n}_{species,out}}{\dot{n}_{total,out}}P_{cell}$$

By calculating the Nernst equation and the voltage polarizations, the power generated by a single cell can be determined.

5. Problems

1. Write the Nernst equation for the cell reactions in the fourth column of Table 13.2.
2. How the reaction rate can be improved?
3. How the ohmic losses can be reduced?
4. Plot the voltage of an acid electrolyte hydrogen fuel cell with operating temperature of $1000°C$ vs. the current density variation from $10 \frac{mA}{cm^2}$ to $1000 \frac{mA}{cm^2}$ (a) as a reversible fuel cell and (b) by considering the impact of activation loss on the reversible voltage. The exchange current density is $10 \frac{mA}{cm^2}$. In addition, the electron transfer coefficient is 0.5, and pure hydrogen and oxygen are fed into the anode and cathode. The water in the products appears as steam. The formation enthalpy of steam water is $-241,830 \frac{kJ}{kmol}$.
5. Determine a) the $\Delta \bar{g}_f$ for the cell reaction of $CO + \frac{1}{2}O_2 \rightarrow CO_2$ that may happen in an MCFC with temperature of $800°C$ and pressure of 1 bar. Assume the species as ideal gas (b) if the temperature is supposed to change from $700°C$ to $900°C$, how the $\Delta \bar{g}_f$ will change? Plot the changes. (c) keep the temperature constant at $800°C$ but change the pressure from 1 to 5 bars and plot the changes of $\Delta \bar{g}_f$.
6. Determine the OCV for the cell reaction of $CH_4 + 2O_2 \rightarrow 2H_2O + CO_2$ for the temperature range of $1100-1300$ K.
7. Determine the maximum thermal efficiency of an SOFC with the cell reaction of $CH_4 + 2O_2 \rightarrow 2H_2O + CO_2$ by assuming water as steam in the products for the temperature range of $1100-1300$K.
8. Determine (a) the voltage change of a hydrogen fuel cell if the cell pressure is changed, while the molar fraction of the reactants and products and the operating temperature stay the same, (b) the voltage change for the PEMFC with operating temperature of $80°C$, (c) the voltage change for the PAFC with operating temperature of $180°C$, and (d) the voltage change for the SOFC with operating temperature of $950°C$,
9. In a fuel cell operating with pure oxygen, it is decided to use air instead of oxygen. What would be the change in voltage for (a) SOFC with operating temperature of $1000°C$ and (b) PEMFC with operating temperature of $80°C$. The other factors remain constant.
10. Plot the voltage of a PEMFC with operating temperature of $70°C$ vs. the current density variation from 0.1 $5 \frac{mA}{cm^2}$ to $800 \frac{mA}{cm^2}$; (a) as a reversible fuel cell and (b) by considering the impact of activation loss on the reversible voltage. The exchange current density is $0.15 \frac{mA}{cm^2}$. In addition, the electron transfer coefficient is 0.5, and pure hydrogen and oxygen are fed into the anode and cathode. The water in the products appears as steam. The formation enthalpy of steam water is $-242,000 \frac{kJ}{kmol}$.
11. Plot the voltage losses of the Example 13.8 versus current density. Which one has more impact on the voltage?

References

[1] J. Larminie, A. Dicks, Fuel Cell Systems Explained, second ed., John Wiley & Sons Ltd, 2003.
[2] Energy and Environmental Analysis, Prepared for Environmental Protection Agency, Combined Heat and Power Partnership Program, Washington DC.

[3] H.I. Onovwiona, V.I. Ugursal, Residential cogeneration systems: review of the current technology, Renew. Sustain. Energy Rev. 10 (2006) 389–431.

[4] J.H. Hirschenhofer, D.B. Stauffer, R.R. Engleman, and M.G. Klett, Fuel Cell Handbook, U.S. Department of Energy, Office of Fossil Energy, Federal Energy Technology Center.

[5] https://www.energy.gov/eere/fuelcells/types-fuel-cells.

[6] M. Ebrahimi, I. Moradpoor, Combined solid oxide fuel cell, micro-gas turbine and organic rankine cycle for power generation (SOFC-MGT-ORC), Energy Convers. Manag. 116 (2016) 120–133.

[7] A. Volkan Akkaya, Electrochemical model for performance analysis of a tubular SOFC, Int. J. Energy Res. 31 (2007) 79–98.

Further reading

[1] https://americanhistory.si.edu/fuelcells/origins/origins.htm#img1.

Thermoelectric generator

14

Chapter outline

Thermoelectric generator (TEG) is an energy converter device that converts the temperature difference to electricity. TEG like the PV panel and piezoelectric is a direct converter. PV panels by using the *photovoltaic effect* convert light into electric current. Piezoelectric materials convert mechanical stress to electric signals by using the *piezoelectric effect*. Power generators, such as gas turbines, convert the chemical energy of the fuel to mechanical rotation firstly, and then this mechanical rotation is converted to electricity by using an electric generator, so they are indirect converter. TEG also converts the temperature difference to electricity directly by using the *Seebeck effect*. A TEG produces electric current when temperature difference is applied to two legs of the thermocouple. TEGs have p-type (positive charge) and n-type (negative charge) elements (also called legs) connected by a metal interconnector, in which the electrons migrate through and as a result, generate electric current in the opposite direction [1].

TEG has long lifespan without any moving part; therefore, its operation and maintenance costs are too low. They create no noise or vibrations. They can use low- to high-temperature waste heat, which exist in many applications such as exhaust and stack of thermal power plants, different fuel cells, exhaust of car engines, exhaust of heaters and furnaces, etc. They are silent and very environment friendly. They can be coupled with renewable energies and technologies as well. For example, they can be used to cool the hot absorber plate of the thermal photovoltaic panels and while increasing the PV electrical efficiency, generate more electricity by utilizing the temperature difference. Despite all of these advantages, their efficiency is still too low and cost is high. Researchers, scientists, and manufacturers are seeking high-temperature operating thermoelectric materials to achieve higher efficiency. In addition, low cost materials to be used in TEGs and improving figure-of-merit (ZT) in the TEGs are

Power Generation Technologies. https://doi.org/10.1016/B978-0-323-95370-2.00015-6
Copyright © 2023 Elsevier Inc. All rights reserved.

two important challenges, which need further research and investigations. The heat source and heat sink may require pump or fan to circulate the working fluid around the TEG to improve the heat transfer. This means electricity consumption and also a source for maintenance jobs. Since the electrical efficiency of TEG is still low, reducing electricity consumption of the auxiliary systems is vital. TEG has the potential to generate electricity from a fraction of 1 W to even some MW. Fig. 14.1 shows a Peltier module thermoelectric power generator element with dimensions of 40 × 40 mm covered with ceramic. The ampere and open circuit voltage generated for various temperature differences are reported beside the picture.

In the present chapter, the classification of TEG materials would be discussed firstly, then the thermodynamic and governing equations would be presented. The governing equations are fully discussed by presenting applied solve examples with their corresponding EES codes. The EES codes can be further extended by the reader to investigate the impact of different parameters.

1. Classifications of TEG

TEGs can be classified according to the application, type, wattage, temperature, and material used [2].

By application, TEGs can be divided into energy harvesting, direct power/electricity generation, heat recovery, and cogeneration.

Energy harvesting is collecting small amount of energy from different sources such as sun, wind, ocean waves, body motion, moving objects such as cars, salinity gradient, electromagnetic radiation, etc. to power wireless devices such as sensors,

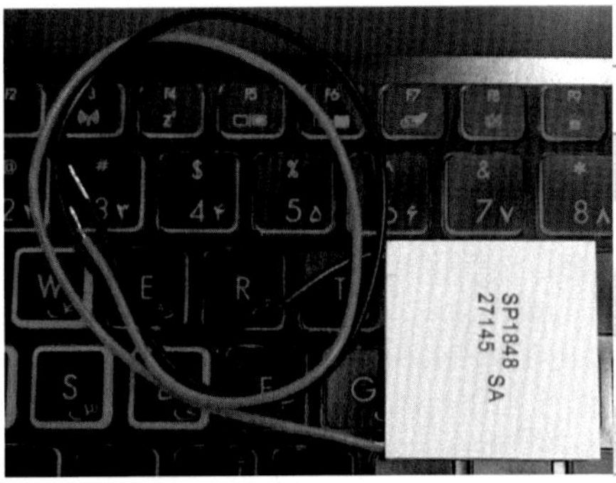

	TEG specifications		
	ΔT (°C)	$V_{OCV}(V)$	$I(mA)$
	20	0.97	225
	40	1.8	368
	60	2.4	469
	80	3.6	558
	100	4.8	669

Figure 14.1 SP1848-27145 thermoelectric generator high-temperature-generation element 40 × 40 mm (personal gallery).

wearable electronics. In addition the energy harvesting devices can power or charge cellphones, portable laptops, radio communication equipment, batteryless watches, etc.

The direct power/electricity generation also refers to the energy conversion equipment or technologies, which directly convert the available energy to power/electricity. Technologies such as PV panels, piezoelectric, and TEG are direct electricity generators because they directly convert sunshine (light), mechanical stress, and temperature difference to electricity. Technologies such as wind turbine and hydro turbines are direct power generator (not electricity). Thermal power plants are indirect power generator because they use a working fluid that must be pressurized first (converting mechanical or electrical power to potential energy), then heated or combusted (converting chemical energy to thermal energy), and then expanded (converting potential and thermal energy to kinetic energy) to turn a shaft and have power (converting kinetic energy to rotation and power); the power is then converted to electricity by using an electric generator.

The cogeneration term refers to the technologies in which several products such as heat and electricity are produced simultaneously by using a single source of energy. For example, TEG can be used in the stack of a boiler or a furnace. The boiler/furnace produces hot steam/gas, while the exhaust heat is utilized to generate electricity by using the TEG.

Heat recovery technology refers to recover the waste heat from a process to increase the fuel energy utilization factor. For example, recovering heat from the exhaust gas and cooling liquid of a reciprocating engine by TEG is a very popular idea, and many research studies are devoted to this application especially for providing electricity to the electrical devices on cars. Today, cars have many electrical devices such as electrical windows, lights, alarms, cameras, radio, entertainments, cellphone charge, cooler, heater, etc. and using these devices mean fuel consumption, while electricity generation from the waste heat means less fuel consumption and less emissions.

TEGs are also classified according to their electrical capacity. They may be classified as low power for capacities smaller than 10 W, medium power for capacities from 1 to 10 kW, and high power for capacities more than 10 kW.

The heat source temperature is also used to categorize TEGs into low temperature for temperature smaller than 80°C, medium temperature for the temperature range from 80°C to 500°C, and high temperature for temperature bigger than 500°C.

Mostly used materials in the TEGs include bismuth telluride (Bi-Te), lead telluride (Pb-Te), and silicon-germanium (Si-Ge) with efficiency range of 4%−6%. The TEG efficiency is still lower than solar PV (15%−25%) and solid-state piezoelectric materials (10%−15%). It is clear that increasing efficiency by using new and cheap materials is a big challenge and open field for the researchers in the world [1].

Thermoelectric materials can be used as an electricity generator or as a cooler. Unlike TEG, thermoelectric coolers (TECs) receive electricity and produce a temperature difference, a cold side and a hot side, the cold side is used as a cooler, while the hot side can be used as a heater.

2. Thermodynamics of TEG

Thermodynamics of TEG is founded on the thermoelectric effects. Thermoelectric effects are reversible phenomena of converting temperature difference to electricity and vice versa.

The thermoelectric effect depends on the physical properties of the thermoelectric material including Seebeck coefficient, electrical conductivity, thermal conductivity, and figure-of-merit. The main thermoelectric phenomena that occur in the thermoelectric devices are Seebeck effect, *Peltier effect*, *Thomson effect*, and *Joule effect*.

Peltier effect occurs in the opposite direction of Seebeck effect. Seebeck effect is the conversion of temperature difference to electricity, while Peltier effect is the conversion of electricity to the temperature difference. The device that uses Peltier effect is called TEC, while the device using Seebeck effect is called TEG.

Thomson effect is the releasing or absorbing heat by a conductive material in which electrical current is flowing in it, while there is a temperature difference between the two ends of the material.

And the Joule effect or *Joule heating* is the generation of heat in a conductive material due to electric current in it.

To simulate the TEGs mathematically and thermodynamically, two semiconductor materials of A and B are attached at two cold and hot junctions as shown in Fig. 14.2. The Seebeck voltage of the circuit shown in Fig. 14.2 is as Eq. (14.1):

$$V = (\alpha_A - \alpha_B) \cdot (T_H - T_C) = \alpha_{AB} \Delta T \tag{14.1}$$

In which $\alpha_A \left(\frac{V}{K} \right)$ and $\alpha_B \left(\frac{V}{K} \right)$ are the Seebeck coefficients of the semiconductors A and B. The magnitude of α_{AB} depends of the operating temperature and other physical properties including thermal and electrical conductivities. The sign (positive or

Figure 14.2 Schematic of a thermocouple made of two semiconductors of A and B for the Seebeck effect in an open circuit.

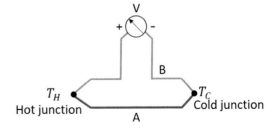

negative) of the thermoelectric depends on the charge carriers (electrons with negative charge, e^-, or holes with positive charge of h^+). If the electric current is conducted by holes the Seebeck coefficient sign would be positive, and if it is conducted with electrons the sign would be negative. The total Seebeck coefficient of the thermocouple (also called *thermopower*) is also related to the temperature gradient and the electric field (E) as below:

$$\alpha_{AB} = \frac{E}{\Delta T} \qquad (14.2)$$

According to the Thomson and Joule effects for a conductance with electrical resistivity of $\rho(\Omega.m)$, electrical conductivity of $\sigma\left(\frac{Siemens}{m}\right)$ in which $\rho\sigma = 1$, current density of $i\left(\frac{A}{m^2}\right)$, the heat which is released or absorbed would be:

$$\dot{Q}\left(\frac{W}{m^3}\right) = \rho i^2 - \mu_{AB} \cdot i \cdot \nabla T \qquad (14.3)$$

The term ρi^2 is the Joule heat and the second term $\mu_{AB} \cdot i \cdot \nabla T$ is the Thomson heat. In addition, the relation between the Thomson coefficient and Seebeck coefficient is as $\mu_{AB} = T\frac{d\alpha_{AB}}{dT}$, and the term ∇T is the temperature gradient along the conductor, and μ_{AB} is the Thomson coefficient in $\frac{V}{K}$.

The sign of the Thomson coefficient is positive when current flows from the cold side to hot side, and heat is absorbed through it, and it is negative when the current is flowing from the hot side to the cold side and it releases heat.

Although the Seebeck, Peltier, and Thomson effects are reversible processes, but generating electricity by the TEG is an irreversible process. According to the second law of thermodynamics, heat generation in the interconnectors due to the Joule effect is an irreversible process. In addition, the heat transfer between the heat source and cold sink is irreversible. If there was no irreversibility, the TEG will reach its maximum efficiency that is the Carnot efficiency of a heat engine operating between T_H and T_C.

The TEG device as demonstrated in Fig. 14.3 comprises of thermocouples. The number of thermocouples can be one or more, and each thermocouple has two legs, one N-type leg, and the other P-type leg. These legs are doped semiconductor charge carriers having excess electrons or holes and the free electrons flow to the holes as they receive enough energy. The P-type semiconductor has a positive Seebeck coefficient with positive holes of h^+, but the N-type's Seebeck coefficient is negative with excess free electrons e^-. The two legs are connected in series electrically, but thermally they are parallel. The interconnect between the two legs may be copper strip or other conductors. As the electrons flow from the N-type leg to the P-type leg through the external circuit, a differential voltage would be generated in the external circuit having an electric load with resistance of R_L. The electric current by convention flows from the P-type with positive charge to the N-type with negative charge, but electrons flow in

Figure 14.3 A thermocouple with two
semiconductor legs of P-type and N-type.

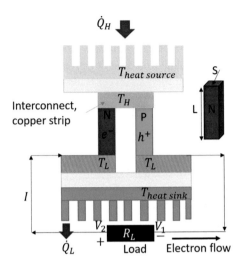

the opposite direction. According to Eq. (14.1), by using bigger temperature difference
or utilizing materials with larger Seebeck coefficient, the electric voltage will increase.
Since the legs are electrically connected in series but thermally parallel, the electrical
resistance and thermal conductance of the TEG with one thermocouple would be:

$$R_1 = \frac{\rho_P L_P}{S_P} + \frac{\rho_N L_N}{S_N}$$

$$\hspace{8cm} (14.4)$$

$$K_1 = \frac{k_P S_P}{L_P} + \frac{k_N S_N}{L_N}$$

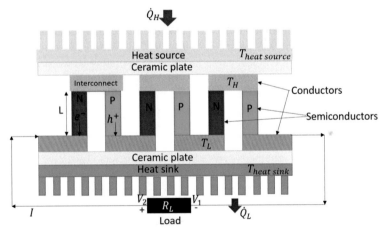

Figure 14.4 A thermoelectric generator (TEG) with n thermocouples, covered with ceramic and
attached to the heat source and heat sink fins.

And for a TEG with n thermocouples such as Fig. 14.4 the total resistance and conductance of the thermocouples would be:

$$R = n\left(\frac{\rho_P L_P}{S_P} + \frac{\rho_N L_N}{S_N}\right)$$
$$K = n\left(\frac{k_P S_P}{L_P} + \frac{k_N S_N}{L_N}\right)$$

(14.5)

In which $R_P = \frac{\rho_P L_P}{S_P}$ and $R_N = \frac{\rho_N L_N}{S_N}$ are the electrical resistivity of each leg, $S(m^2)$ is the cross-section area of each leg, and $L(m)$ is the length of each leg. In addition, $k\left(\frac{W}{m \cdot K}\right)$ is the thermal conductivity and $K\left(\frac{W}{K}\right)$ is the thermal conductance. For a specific case in which both P-type and N-type legs are completely similar, Eq. (14.5) can be simplified as Eq. (14.6):

$$R = 2n\left(\frac{\rho L}{S}\right) = nR_1$$
$$K = 2n\left(\frac{kS}{L}\right) = nK_1$$

(14.6)

The voltage at the TEG terminals is as below:

$$V_2 - V_1 = n(I \cdot R_1 - \alpha_{PN}\Delta T) = n \cdot I \cdot R_1 - V_{Seebeck} \tag{14.7}$$

In which, $\alpha_{PN} = \alpha_P - \alpha_N$, and $V_{TEG} = V_2 - V_1$ and the input electrical current would be:

$$I = \frac{V_{Seebeck}}{R_1 + R_L} = \frac{\alpha_{PN}\Delta T}{R_1 + R_L} \tag{14.8}$$

The electricity consumed by the load would be:

$$P = V_{TEG}I = n\left(R_1 \cdot I^2 - \alpha_{PN} \cdot I \cdot \Delta T\right) \tag{14.9}$$

And the electricity produced by the TEG and delivered to the load is:

$$P_{TEG} = -V_{TEG}I = n\left(\alpha_{PN} \cdot I \cdot \Delta T - R_1 \cdot I^2\right) \tag{14.10}$$

By taking differential from Eq. (14.10) with respect to the current I, the conditions to get the maximum power would be determined as below:

$$\frac{dP}{dI} = n(2R_1 \cdot I - \alpha_{PN} \cdot \Delta T) = 0 \rightarrow$$

$$I = \frac{\alpha_{PN} \cdot \Delta T}{2R_1} \rightarrow \tag{14.11}$$

$$P_{max} = n\left(\frac{\alpha_{PN} \cdot \Delta T}{2R_1}\right)^2 \tag{14.12}$$

Comparing Eq. (14.11) with Eq. (14.8) results in that for the maximum power output $R_L = R$ or $\frac{R_L}{R} = 1$.

The heat absorbed at the hot junction and the heat rejected at the cold junction depends on the Peltier heat, the heat conduction, and the Joule heat as Eq. (14.3).

$$\dot{Q}_H = n\left[\alpha_{PN} \cdot T_H \cdot I - \frac{R_1 \cdot I^2}{2} + K_1 \cdot (T_H - T_C)\right]$$

$$\dot{Q}_C = n\left[\alpha_{PN} \cdot T_C \cdot I + \frac{R_1 \cdot I^2}{2} + K_1 \cdot (T_H - T_C)\right] \tag{14.13a}$$

The thermal efficiency (first law efficiency) of the TEG can be determined as below:

$$\eta_{TEG} = 1 - \frac{\dot{Q}_C}{\dot{Q}_H} = \frac{P_{TEG}}{\dot{Q}_H} \tag{14.13b}$$

It is clear that the maximum efficiency of a TEG under reversible conditions is the efficiency of a Carnot cycle operating between T_H and T_L as below:

$$\eta_{Carnot} = 1 - \frac{T_C}{T_H} \tag{14.14}$$

And the second law efficiency (exergy efficiency) would be:

$$\eta_{II,TEG} = \frac{\dot{X}_{recovered}}{\dot{X}_{supplied}} = 1 - \frac{\dot{X}_{destroyed}}{\dot{X}_{supplied}} = \frac{P_{TEG}}{\dot{Q}_H\left(1 - \frac{T_C}{T_H}\right)} \tag{14.15}$$

And the exergy destruction would be:

$$\dot{X}_{destroyed} = \dot{Q}_H\left(1 - \frac{T_C}{T_H}\right) - P_{TEG} \tag{14.16}$$

Or

$$\dot{X}_{destroyed} = T_0 \dot{S}_{gen} = T_0 \left(\frac{\dot{Q}_C}{T_C} - \frac{\dot{Q}_H}{T_H} \right) \tag{14.17}$$

Example 14.1. For a TEG module, the heat rate to the hot junction at temperature of 100°C is 100 W, and it rejects 95 W of heat to the cold side at temperature of 25°C. Determine (a) the thermal efficiency and the corresponding Carnot efficiency, (b) the exergy destruction, and (c) the exergy efficiency of the TEG module.

Solution. (a) The Carnot cycle efficiency can be calculated as below:

$$\eta_{Carnot} = \left(1 - \frac{T_C}{T_H} \right) \times 100 = \left(1 - \frac{25 + 273.15}{100 + 273.15} \right) \times 100$$

$$\eta_{Carnot,TEG} = 20.1\%$$

The thermal efficiency of the TEG can also be calculated as below:

$$\eta_{TEG} = \left(1 - \frac{\dot{Q}_C}{\dot{Q}_H} \right) \times 100 = \left(1 - \frac{95}{100} \right) \times 100$$

$$\eta_{TEG} = 5\%$$

As it can be seen, the difference between reversible and real thermal efficiencies is very considerable. It shows that the recovered exergy from the TEG is very low, and there are great opportunities to recover more exergy.

(b) The exergy potential of the TEG can be determined as below:

$$\dot{X}_Q = \dot{Q}_H \left(1 - \frac{T_C}{T_H} \right) = 100 \left(1 - \frac{25 + 273.15}{100 + 273.15} \right)$$

$$\dot{X}_Q = 20.1 W$$

The exergy recovered in the TEG is actually equal to the electricity generated by the TEG and is:

$$P_{TEG} = \dot{Q}_H - \dot{Q}_C = 100 - 95$$

$$\dot{X}_P = P_{TEG} = 5W$$

Hence, the exergy destruction would be:

$$\dot{X}_{destroyed} = \dot{X}_Q - \dot{X}_P = 20.1 - 5$$

$$\dot{X}_{destroyed} = 15.1 W$$

The exergy destruction can also be calculated as below:

$$\dot{X}_{destroyed} = T_0 \dot{S}_{gen} = T_0 \left(\frac{\dot{Q}_C}{T_C} - \frac{\dot{Q}_H}{T_H} \right)$$

$$= (25 + 273.15) \left(\frac{95}{25 + 273.15} - \frac{100}{100 + 273.15} \right)$$

$$\dot{X}_{destroyed} = 15.1 W$$

(c) The exergy efficiency of the TEG can be calculated as below:

$$\eta_{II,TEG} = 1 - \frac{X_{destroyed}}{X_{supplied}} = 1 - \frac{15.1}{20.1}$$

$$\eta_{II,TEG} = 24.88\%$$

Or, the second law efficiency can also be calculated by the following equation:

$$\eta_{II,TEG} = \frac{\eta_{TEG}}{\eta_{Carnot}} = \frac{\eta_{TEG}}{\eta_{Carnot}} = \frac{5}{20.1}$$

$$\eta_{II,TEG} = 24.88\%$$

According to the exergy definition, if the magnitudes of the $\dot{Q}_H$ and $\dot{Q}_L$ are kept the same, the best way to decrease exergy destruction (or increasing exergy efficiency) is to have higher T_H and lower T_L. It should be noted that the material used in the TEG should withstand the higher temperature at the source side as well.

The code written for the Example 14.1 is given below (**Code 14.1**):

Code 14.1. The code written for the Example 14.1 in the EES

```
T_L=298.15
T_H=373.15
Q_L=95
Q_H=100
T0=298.15
E_TEG=Q_H-Q_L
eta_carnot=1-(T_L)/(T_H)
eta_TEG=1-Q_L/Q_H
S_gen=-(Q_H/(T_H)-Q_L/(T_L))
X_Q=Q_H*(1-(T0)/(T_H))
X_detruction1=-(Q_H/(T_H)-Q_L/(T_L))*(T0)
X_detruction2=X_Q-E_TEG
eta_II_1=1-X_detruction2/X_Q
eta_II_2=eta_TEG/eta_carnot
```

Example 14.2. The physical properties of one leg of a TEG unit made of bismuth telluride (TGM199-1.4-2) are given in Table 14.1 [3]:

The hot side of the TEG is heated by constant temperature of 200°C, and the cold side is in touch with 30°C water. Determine:

(a) the electricity current under maximum power output condition (I_{mp})
(b) the voltage under maximum electricity output condition (V_{mp})
(c) the maximum electricity generated by the TEG (P_{max})
(d) the thermal efficiency under maximum electricity output condition (η_{mp})
(e) the exergy efficiency under maximum electricity output condition

Solution.
(a) the maximum output power happens if $R_1 = R_L$; hence, the electricity current generated due to this temperature difference would be:

Table 14.1 The physical properties of one leg of the TEG unit [3].

Parameter	Value	Unit
Number of thermocouples, n	199	–
Element cross-section area (S)	1.96	mm^2
Length of element (L)	2	mm
Dimensions (Width × Length × Thickness)	40×40 × 4.4	mm
Seebeck coefficient of one leg (α)	162.856	μV.K^{-1}
Electrical resistivity (ρ)	1.024×10^{-5}	Ω.m
Thermal conductivity (k)	1.5	W.m^{-1}.K^{-1}

$$I_{mp} = \frac{\alpha_{PN} \cdot \Delta T}{2R_1}$$

In which, $\alpha_{PN} = \alpha_P - \alpha_N$, since the materials of the semiconductors are the same for both legs of the thermocouple hence $\alpha_P = -\alpha_N$ and:

$$\alpha_{PN} = 2\alpha_P = 2 \times 162.856 \times 10^{-6}\frac{V}{K}$$

And $\Delta T = T_H - T_C = 473.15 - 303.15 = 170K$, and the total electrical resistance and thermal conductance of one thermocouple would be:

$$R_1 = \frac{\rho_P L_P}{S_P} + \frac{\rho_N L_N}{S_N} = 2\frac{\rho L}{S} = 2\frac{1.024 \times 10^{-5} \times 2 \times 10^{-3}}{1.96 \times 10^{-6}} = 0.0209\Omega$$

$$K_1 = \frac{k_P S_P}{L_P} + \frac{k_N S_N}{L_N} = 2\frac{kS}{L} = 2\frac{1.5 \times 1.96 \times 10^{-6}}{2 \times 10^{-3}} = 0.000294\frac{W}{K}$$

Therefore:

$$I_{mp} = \frac{2 \times 162.856 \times 10^{-6} \times 170}{2 \times 0.0209} = 1.325A$$

(b) the voltage under the maximum power condition would be:

$$V_{mp} = n(R_1 \cdot I_{mp} - \alpha_{PN} \cdot \Delta T)$$
$$= 199 \times (0.0209 \times 1.325 - 2 \times 162.856 \times 10^{-6} \times 170) = -5.509V$$

(c) the maximum electricity would be calculated by multiplying minus voltage of TEG by the electric current under maximum power condition as below:

$$P_{max} = -V_{mp}I_{mp} = -(-5.509) \times 1.325 = 7.3W$$

Or:

$$P_{max} = n(\alpha_{PN} \cdot I_{mp} \cdot \Delta T - R_1 \cdot I_{mp}^2)$$
$$= 199 \times (2 \times 162.856 \times 10^{-6} \times 1.325 \times 170 - 0.0209 \times 1.325^2)$$
$$= 7.3W$$

(d) to determine the thermal efficiency, it is required to calculate the heat input and heat rejection as below:

$$\dot{Q}_H = n \left[\alpha_{PN} \cdot T_H \cdot I_{mp} - \frac{R_1 \cdot I_{mp}^2}{2} + K_1 \cdot (T_H - T_C) \right]$$

$$= 199 \times \left(2 \times 162.856 \times 10^{-6} \times 473.15 \times 1.325 - \frac{0.0209 \times 1.325^2}{2} \right.$$

$$\left. + 0.000294 \times (473.15 - 303.15) \right)$$

$$= 136.4W$$

$$\dot{Q}_C = n \left[\alpha_{PN} \cdot T_C \cdot I_{mp} + \frac{R_1 \cdot I_{mp}^2}{2} + K_1 \cdot (T_H - T_C) \right]$$

$$= 199 \times \left(2 \times 162.856 \times 10^{-6} \times 303.15 \times 1.325 + \frac{0.0209 \times 1.325^2}{2} \right.$$

$$\left. + 0.000294 \times (473.15 - 303.15) \right)$$

$$= 129.1W$$

To validate the results, the energy balance equation can be written as below:

$$P_{TEG} = \dot{Q}_H - \dot{Q}_c = 136.4 - 129.1 = 7.3W$$

Which is exactly the same as the result in part c.
The thermal efficiency under maximum power condition would be:

$$\eta_{mp} = \left(1 - \frac{\dot{Q}_C}{\dot{Q}_H} \right) \times 100 = \left(1 - \frac{129.1}{136.4} \right) \times 100 = 5.35\%$$

(e) the exergy efficiency would be as below:

$$\eta_{II,TEG} = \frac{\eta_{TEG}}{\eta_{Carnot}} = \frac{\eta_{TEG}}{1 - \dfrac{T_C}{T_H}} = \frac{0.0535}{1 - \dfrac{303.15}{473.15}} = 14.89\%$$

The code used to solve the Example 14.2 is presented in Code 14.2.

Code 14.2. The code used to evaluate the performance of the TEG described in Example 14.2

```
n=199
S=1.96*10^(-6)
L=2*10^(-3)
Alpha_P=162.856*10^(-6)
Alpha_N=-Alpha_P
Alpha_PN=Alpha_P-Alpha_N
k_conductivity=1.5
R_L=2*RO*L/S
RO=1.024*10^(-5)
R_1=2*RO*L/S
T_H=473.15
T_C=303.15
K_1=2*k_conductivity*S/L
Delta_T=T_H-T_C
I=Alpha_PN*Delta_T/(R_1+R_L)

P_TGE=n*(Alpha_PN*Delta_T*I-R_1*I^2)
V_TEG=P_TGE/I
Q_H=n*(Alpha_PN*T_H*I-(R_1)*I^2/2+K_1*Delta_T)
Q_C=n*(Alpha_PN*T_C*I+(R_1)*I^2/2+K_1*Delta_T)
DQ=Q_H-Q_C
eta_TEG=P_TGE/Q_H*100
X_supplied=Q_H*(1-T_C/T_H)
ETA_II=P_TGE/X_supplied*100
ETA_2=eta_TEG/(1-T_C/T_H)
```

Example 14.3. The physical properties of one leg of a TEG unit made of bismuth telluride (TG12-4) are given in Table 14.2 [3]:

The hot side of the TEG is heated by constant temperature of 170°C, and the cold side is in touch with 50°C water. Determine:

(a) the electricity output vs. $\frac{R_L}{R_1}$ when it changes from 0 to 4.

(b) the thermal efficiency vs. $\frac{R_L}{R_1}$ when it changes from 0 to 4.

(c) the exergy efficiency versus $\frac{R_L}{R_1}$ when it changes from 0 to 4.

Table 14.2 The physical properties of a one leg of the TEG unit [3].

Parameter	Value	Unit
Number of thermocouples, n	127	—
Element cross section area (S)	1	mm^2
Length of element (L)	1.17	mm
Dimensions (Width × Length × Thickness)	30×30 × 3.4	mm
Seebeck coefficient of one leg (α)	210.769	$\mu V.K^{-1}$
Electrical resistivity (ρ)	1.638×10^{-5}	$\Omega.m$
Thermal conductivity (k)	1.5	$W.m^{-1}.K^{-1}$

Solution. By changing the electricity current from this equation:

$$I = \frac{\alpha_{PN} \cdot \Delta T}{2R_1}$$

To the following equation:

$$I = \frac{\alpha_{PN} \cdot \Delta T}{R_L + R_1}$$

And using the new physical data given in Table 14.2, in the code written for the Example 14.2, the new code would be as Code 14.3:

Code 14.3. The code written for the parametric analysis of the TEG given in Table 14.2

```
n=127
S=1.0*10^(-6)
L=1.17*10^(-3)
Alpha_P=210.769*10^(-6)
Alpha_N=-Alpha_P
Alpha_PN=Alpha_P-Alpha_N
k_conductivity=1.5
RO=1.638*10^(-5)
R_1=2*RO*L/S
R_LtoR_1=1
R_LtoR_1=R_L/R_1
T_H=443.15
T_C=323.15
K_1=2*k_conductivity*S/L
DELTA_T=T_H-T_C

I=Alpha_PN*DELTA_T/(R_1+R_L)
V_TEG=-n*(R_1*I-Alpha_PN*DELTA_T)
P_TGE=n*(Alpha_PN*Delta_T*I-R_1*I^2)
Q_H=n*(Alpha_PN*T_H*I-(R_1)*I^2/2+K_1*Delta_T)
Q_C=n*(Alpha_PN*T_C*I+(R_1)*I^2/2+K_1*Delta_T)
DQ=Q_H-Q_C
eta_TEG=P_TGE/Q_H*100
X_supplied=Q_H*(1-T_C/T_H)
ETA_II=P_TGE/X_supplied*100
ETA_2=eta_TEG/(1-T_C/T_H)
```

By using a parametric table and changing (R_LtoR_1) from 0 to 4, the thermal efficiency, exergy efficiency, and power generate by the TEG would be calculated and plotted as Fig. 14.5.

As Fig. 14.5 shows, the electricity output is maximum at $\frac{R_L}{R_1} = 1$ and is equal to 2.12 W, but the thermal and exergy efficiencies are maximum at $\frac{R_L}{R_1} = 1.3$ in which $\eta_{TEG,max} = 4.015\%$ and $\eta_{II,max} = 14.83\%$.

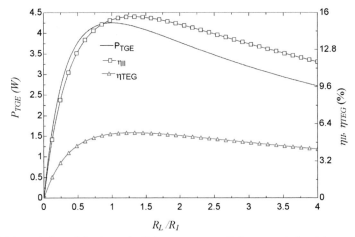

Figure 14.5 Variation of the thermal efficiency, exergy efficiency, and electricity output by the thermoelectric generator (TEG) versus the resistance ratio of $\frac{R_L}{R_1}$.

2.1 Maximum efficiency and dimensionless parameters

The thermal efficiency of a TEG unit can be determined according to some dimensionless parameters by manipulating the equations presented in the previous section. For this purpose, by substituting Eq. (14.8) into 14.10:

$$P_{TEG} = n\left(\alpha_{PN} \cdot \frac{\alpha_{PN}\Delta T}{R_1 + R_L} \cdot \Delta T - R_1 \cdot \left(\frac{\alpha_{PN}\Delta T}{R_1 + R_L}\right)^2\right)$$

$$= \frac{n\alpha_{PN}^2 \Delta T^2}{R_1 + R_L} \cdot \left(1 - \frac{R_1}{R_1 + R_L}\right) = \frac{n\alpha_{PN}^2 (T_H - T_C)^2}{R_1\left(1 + \frac{R_L}{R_1}\right)} \cdot \left(\frac{\frac{R_L}{R_1}}{\left(1 + \frac{R_L}{R_1}\right)}\right)$$

Hence:

$$P_{TEG} = \frac{n\alpha_{PN}^2 \Delta T^2}{R_1\left(1 + \frac{R_L}{R_1}\right)} \cdot \left(\frac{\frac{R_L}{R_1}}{\left(1 + \frac{R_L}{R_1}\right)}\right) \tag{14.18}$$

And by substituting Eq. (14.8) into the 14.13:

$$\dot{Q}_H = n\left[\alpha_{PN} \cdot T_H \cdot \frac{\alpha_{PN} \cdot \Delta T}{R_L + R_1} \frac{R_1}{2} \cdot \left(\frac{\alpha_{PN} \cdot \Delta T}{R_L + R_1}\right)^2 + K_1 \cdot (T_H - T_C)\right]$$

$$= n\left[\alpha_{PN} \cdot T_H \cdot \frac{\alpha_{PN} \cdot \Delta T}{R_1\left(1 + \frac{R_L}{R_1}\right)} - \frac{R_1}{2} \cdot \left(\frac{\alpha_{PN} \cdot \Delta T}{R_1\left(1 + \frac{R_L}{R_1}\right)}\right)^2 + K_1 \cdot \Delta T\right]$$

$$= n\Delta T \times T_H \frac{\alpha_{PN}^2}{R_1\left(1 + \frac{R_L}{R_1}\right)}\left[1 - \frac{R_1}{2}\frac{1}{R_1\left(1 + \frac{R_L}{R_1}\right)}\frac{\Delta T}{T_H} + \frac{K_1 R_1}{\alpha_{PN}^2}\left(1 + \frac{R_L}{R_1}\right)\frac{1}{T_H}\right]$$

Hence:

$$\dot{Q}_H = n\Delta T \times T_H \frac{\alpha_{PN}^2}{R_1\left(1 + \frac{R_L}{R_1}\right)}\left[1 - \frac{R_1}{2}\frac{1}{R_1\left(1 + \frac{R_L}{R_1}\right)}\frac{\Delta T}{T_H} + \frac{K_1 R_1}{\alpha_{PN}^2}\left(1 + \frac{R_L}{R_1}\right)\frac{1}{T_H}\right]$$

$$(14.19)$$

By substituting Eq. 14.18 and 14.19 in the thermal efficiency equation(Eq. 14.13):

$$\eta_{TEG} = \frac{P_{TEG}}{\dot{Q}_H}$$

$$= \frac{\dfrac{n\alpha_{PN}^2 \Delta T^2}{R_1\left(1 + \frac{R_L}{R_1}\right)} \cdot \left(\dfrac{\frac{R_L}{R_1}}{\left(1 + \frac{R_L}{R_1}\right)}\right)}{n\Delta T \times T_H \dfrac{\alpha_{PN}^2}{R_1\left(1 + \frac{R_L}{R_1}\right)}\left[1 - \dfrac{R_1}{2}\dfrac{1}{R_1\left(1 + \frac{R_L}{R_1}\right)}\dfrac{\Delta T}{T_H} + \dfrac{K_1 R_1}{\alpha_{PN}^2}\left(1 + \frac{R_L}{R_1}\right)\dfrac{1}{T_H}\right]}$$

$$= \frac{\dfrac{\Delta T}{T_H}\dfrac{R_L}{R_1}}{\left(1 + \dfrac{R_L}{R_1}\right) - \dfrac{1}{2}\dfrac{\Delta T}{T_H} + \dfrac{K_1 R_1}{\alpha_{PN}^2}\left(1 + \dfrac{R_L}{R_1}\right)^2 \dfrac{1}{T_H}}$$

$$= \frac{\left(1 - \dfrac{T_C}{T_H}\right)\dfrac{R_L}{R_1}}{\left(1 + \dfrac{R_L}{R_1}\right) - \dfrac{1}{2}\left(1 - \dfrac{T_C}{T_H}\right) + \dfrac{K_1 R_1}{\alpha_{PN}^2 T_C}\left(1 + \dfrac{R_L}{R_1}\right)^2 \dfrac{T_C}{T_H}}$$

$$= \frac{\left(1 - \dfrac{T_C}{T_H}\right)\dfrac{R_L}{R_1}}{\left(1 + \dfrac{R_L}{R_1}\right) - \dfrac{1}{2}\left(1 - \dfrac{T_C}{T_H}\right) + \dfrac{1}{ZT_C}\left(1 + \dfrac{R_L}{R_1}\right)^2 \dfrac{T_C}{T_H}}$$

Hence:

$$\eta_{TEG} = \frac{\left(1 - \frac{T_C}{T_H}\right)\frac{R_L}{R_1}}{\left(1 + \frac{R_L}{R_1}\right) - \frac{1}{2}\left(1 - \frac{T_C}{T_H}\right) + \frac{1}{ZT_C}\left(1 + \frac{R_L}{R_1}\right)^2 \frac{T_C}{T_H}} = f\left(\frac{R_L}{R_1}, \frac{T_C}{T_H}, ZT_C\right)$$

$$(14.20)$$

Eq. (14.20) shows that the TEG efficiency is a function of three dimensionless groups of resistance ratio $R_r = \frac{R_L}{R_1}$, temperature ratio $T_r = \frac{T_C}{T_H}$, and ZT_C as the figure-of-merit (FOM) in which $Z = \frac{\alpha_{PN}^2}{K_1 R_1}$. These three dimensionless parameters are very important for designing TEG systems.

Example 14.4. Prove that the maximum thermal efficiency of a TEG happens at:

$$R_r = \left[\frac{1}{2}\left(1 + \frac{1}{T_r}\right)ZT_C + 1\right]^{\frac{1}{2}} \tag{14.21}$$

if the temperature ratio and FOM are constant. Calculate the R_r for the Example 14.3 and validate your answer with the results of the previous example.

Solution. By taking the derivative from Eq. (14.20) with respect to $R_r = \frac{R_L}{R_1}$ and setting the derivative equal to zero we will have:

$$\frac{d\eta_{TEG}}{dR_r} = \frac{(1 - T_r)\left[\frac{1}{2} + \frac{T_r}{2} + \frac{T_r}{ZT_C}(1 - R_r^2)\right]}{\left[(1 + R_r) - \frac{1}{2}(1 - T_r) + \frac{1}{ZT_C}(1 + R_r)^2 T_r\right]^2} = 0 \rightarrow R_r$$

$$= \left[\frac{1}{2}\left(1 + \frac{1}{T_r}\right)ZT_C + 1\right]^{\frac{1}{2}}$$

For the previous example:

$$T_r = \frac{T_C}{T_H} = \frac{323.15}{443.15} = 0.7292$$

$$Z = \frac{\alpha_{PN}^2}{K_1 R_1} = \frac{0.0004215^2}{0.002564 \times 0.03833} = 0.001808\frac{1}{K}$$

$$R_r = \left[\frac{1}{2}\left(1 + \frac{1}{T_r}\right)ZT_C + 1\right]^{\frac{1}{2}} = \left[\frac{1}{2}\left(1 + \frac{1}{0.7292}\right)0.001808 \times 323.15 + 1\right]^{\frac{1}{2}}$$

$$= 1.3$$

Which is exactly equal to the result of the Example 14.3.

By substituting Eq. (14.21) into Eq. (14.20) the maximum efficiency would be determined as below:

$$\eta_{max} = (1 - T_r) \frac{\sqrt{1 + ZT_m} - 1}{\sqrt{1 + ZT_m} - T_r} \tag{14.22}$$

In which $T_m = \frac{T_H + T_C}{2}$ is the mean temperature [4].

2.2 Maximum current and voltage

The maximum current happens at the short circuit, where the load resistance is zero. Hence for $R_L = 0$ the maximum current would be:

$$I_{max} = I_{SC} = \frac{\alpha_{PN} \cdot \Delta T}{R_1} \tag{14.23}$$

And the maximum voltage happens for open circuit voltage condition, that means current must be zero. Hence, by using Eq. (14.7):

$$V_{max} = V_{OCV} = n\alpha_{PN} \cdot \Delta T \tag{14.24}$$

Care must be taken not to confuse the definition of different terms; to avoid confusion a summary is given below:

P_{max} is achieved at $R_r = 1$, and the corresponding thermal efficiency, voltage, and ampere are shown by η_{mp}, V_{mp}, and I_{mp}, respectively.

η_{max} happens at $R_r = \left[\frac{1}{2}\left(1 + \frac{1}{T_r}\right)ZT_C + 1\right]^{\frac{1}{2}} = \sqrt{1 + ZT_m}$ if the temperature ratio and FOM are constants and $T_m = \frac{T_H + T_C}{2}$ is the mean temperature.

I_{max} happens under short circuit condition ($R_L = 0$).

V_{max} happens under open circuit voltage condition ($I = 0$).

Example 14.5. Determine the maximum voltage and maximum ampere for the TEG presented in Table 14.1:

Solution. The maximum current happens under short-circuit condition, that means $R_L = 0$ or:

$$I_{max} = \frac{\alpha_{PN} \cdot \Delta T}{R_1}$$

In which $\alpha_{PN} = \alpha_P - \alpha_N$, since the material of the semiconductor are the same for both legs of the thermocouple hence $\alpha_P = -\alpha_N$ and:

$$\alpha_{PN} = 2\alpha_P = 2 \times 162.856 \times 10^{-6} \frac{V}{K}$$

And $\Delta T = T_H - T_C = 473.15 - 303.15 = 170K$, and the total electrical resistance of one thermocouple would be:

$$R_1 = \frac{\rho_P L_P}{S_P} + \frac{\rho_N L_N}{S_N} = 2\frac{\rho L}{S} = 2\frac{1.024 \times 10^{-5} \times 2 \times 10^{-3}}{1.96 \times 10^{-6}} = 0.0209\Omega$$

Therefore:

$$I_{max} = \frac{2 \times 162.856 \times 10^{-6} \times 170}{0.0209} = 2.65A$$

The maximum voltage happens under open circuit voltage, that means $I = 0$, or:

$$V_{max} = n(\alpha_{PN} \cdot \Delta T) = 199 \times \left(2 \times 162.856 \times 10^{-6} \times 170\right) = 11.02V$$

2.3 Effective and normalized parameters

The maximum values including V_{max}, I_{max}, P_{max}, and η_{max} are important characteristics of TEG products that the manufacturers usually provide them in the catalog of their products. Hence, knowing them and working with them is very essential. These parameters are also used to determine the effective physical parameters such as effective resistivity, effective Seebeck coefficient, effective conductivity, and effective figure-of-merit as below:

$$\rho^* = G\frac{P_{max}}{I_{max}^2} \tag{14.25}$$

$$\alpha^* = \frac{4P_{max}}{nI_{max}\Delta T} \tag{14.26}$$

$$k^* = \frac{\alpha^{*2}}{\rho^* Z^*} \tag{14.27}$$

$$Z^* = \frac{2}{T_C\left(1 + \dfrac{1}{T_r}\right)}\left[\left(\frac{1 + T_r\dfrac{\eta_{max}}{\eta_{Carnot}}}{1 - \dfrac{\eta_{max}}{\eta_{Carnot}}}\right)^2 - 1\right] \tag{14.28}$$

$$Z^* = \frac{4\dfrac{T_r}{T_C}}{\eta_{Carnot}\left(\dfrac{1}{\eta_{mp}} + \dfrac{1}{2}\right) - 2} \tag{14.29}$$

In which $G = \frac{S}{L}$ is the geometry factor of the TEG, and $\eta_{Carnot} = 1 - T_r$.

The normalized characteristics of a TEG are obtained by dividing the parameter to its maximum value. Hence, the normalized power output, normalized voltage, normalized ampere, and normalized efficiency are determined as below:

$$P_n = \frac{P}{P_{max}} = \frac{4R_r}{(R_r + 1)^2} \tag{14.30}$$

$$V_n = \frac{V}{V_{max}} = \frac{R_r}{R_r + 1} \tag{14.31}$$

$$I_n = \frac{I}{I_{max}} = \frac{1}{R_r + 1} \tag{14.32}$$

$$\eta_n = \frac{\eta}{\eta_{max}} = \frac{R_r(\sqrt{1 + ZT_m} + T_r)}{\left[1 + R_r - \dfrac{1}{2}(1 - T_r) + \dfrac{(R_r + 1)^2(1 + T_r)}{2ZT_m}\right](\sqrt{1 + ZT_m} - 1)} \tag{14.33}$$

Example 14.6. Plot the variation of the normalized parameters given in Eqs. (14.30−14.33) vs. R_r when it changes from 0 to 4 for the TEG given in the Example 14.3.

Solution. Since the problem is a parametric analysis, Eqs. (14.30−14.33) are coded in the EES, and the data from Table 14.2 are used in the code. The Code 14.4 is given below:

Code 14.4. The code used to calculate the normalized parameters for the Example 14.6

```
n=127
S=1.0*10^(-6)
L=1.17*10^(-3)
Alpha_P=210.769*10^(-6)
Alpha_N=-Alpha_P
Alpha_PN=Alpha_P-Alpha_N
k_conductivity=1.5
RO=1.638*10^(-5)
R_1=2*RO*L/S
"R_r=1"
T_H=443.15
T_C=323.15
K_1=2*k_conductivity*S/L
T_r=T_C/T_H
Z=Alpha_PN^2/K_1/R_1
T_m=(T_H+T_C)/2

RR=((1/2+1/T_r/2)*Z*T_C+1)^0.5
P_n=4*R_r/(1+R_r)^2
V_n=R_r/(1+R_r)
I_n=1/(1+R_r)
eta_n=R_r*((1+Z*T_m)^0.5+T_r)/((1+R_r-0.5*(1-
T_r)+(1+R_r)^2*(1+T_r)/(2*Z*T_m))*((1+Z*T_m)^0.5-1))
```

By using a parametric table and running the code for the given range of resistance ratio, Fig. 14.6 would be obtained. As it can be seen, the maximum ampere happens when $R_r = 0$, and the maximum power and maximum efficiency do not happen at the same R_r. In addition, the normalized ampere and normalized voltage cross each other at $R_r = 1$ where the normalized power is maximized.

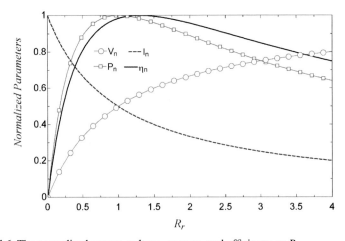

Figure 14.6 The normalized power, voltage, ampere, and efficiency vs. R_r.

3. Problems

1 .What is the maximum operating temperature of the commercialized TEG in the market?

2 .What are the newest materials used in the TEG modules?

3 .For a TEG module the heat rate to the hot side at temperature of 220°C is 100 W and it rejects 95 W of heat to the cold side at temperature of 30°C. Determine (a) the thermal efficiency and the corresponding Carnot efficiency, (b) the exergy destruction, and (c) the exergy efficiency of the TEG module.

4 .The physical properties of one leg of a TEG unit made of bismuth telluride (HZ-2) are given in Table 14.3 [3]:

The hot side of the TEG is heated by constant temperature of 230°C, and the cold side is in touch with 30°C water. Determine:

(a) the electricity current under maximum power output condition (I_{mp}).

(b) the voltage under maximum power output condition (V_{mp}).

(c) the maximum electricity generated by the TEG (P_{max}).

(d) the thermal efficiency under maximum power output condition (η_{mp}).

(e) the exergy efficiency under maximum power output condition.

5 .The physical properties of one leg of a TEG unit made of bismuth telluride (TGM127-1.4-2.5) are given in Table 14.1 [3]:

The hot side of the TEG is heated by constant temperature of 170°C, and the cold side is in touch with 50°C water. Determine:

(a) the electricity output versus $\frac{R_L}{R_1}$ when it changes from 0 to 4.

(b) the thermal efficiency versus $\frac{R_L}{R_1}$ when it changes from 0 to 4.

(c) the exergy efficiency versus $\frac{R_L}{R_1}$ when it changes from 0 to 4.

6 .Prove that the maximum efficiency of a TEG unit can be determined by using the following equation:

$$\eta_{max} = (1 - T_r) \frac{\sqrt{1 + ZT_m} - 1}{\sqrt{1 + ZT_m} - T_r}$$

7 .Determine the maximum efficiency, voltage and maximum ampere for the TEG presented in Table 14.3.

Table 14.3 The physical properties of a one leg of the TEG unit [3].

Parameter	Value	Unit
Number of thermocouples, n	97	—
Element cross-section area (S)	2.1	mm^2
Length of element (L)	2.87	mm
Dimensions(Width $\times$ Length $\times$ Thickness)	30$\times$30 $\times$ 4.5	mm
Seebeck coefficient of one leg (α)	167.526	$\mu V.K^{-1}$
Electrical resistivity (ρ)	1.532×10^{-5}	$\Omega.m$
Thermal conductivity (k)	1.6	$W.m^{-1}.K^{-1}$

Table 14.4 The physical properties of a one leg of the thermoelectric generator (TEG) unit [3].

Parameter	Value	Unit
Number of thermocouples, n	127	—
Element cross-section area (S)	1.96	mm^2
Length of element (L)	2.5	mm
Dimensions(Width $\times$ Length $\times$ Thickness)	$40\times40\times4.4$	mm
Seebeck coefficient of one leg (α)	171.981	$\mu V.K^{-1}$
Electrical resistivity (ρ)	0.9672×10^{-5}	$\Omega.m$
Thermal conductivity (k)	1.7	$W.m^{-1}.K^{-1}$

8 .Determine the maximum efficiency, voltage, and maximum ampere for the TEG presented in Table 14.4.

9 .Plot the variation of the normalized parameters given in Eqs. (14.30−14.33) versus R_r when it changes from 0 to 4 for the TEG given in Table 14.3.

10 .Plot the variation of the normalized parameters given in Eqs. (14.30−14.33) versus R_r when it changes from 0 to 4 for the TEG given in Table 14.4.

11 .Suppose that you have several TEG modules in hand, how should you connect them to get higher voltage?

12 .Suppose that you have several TEG modules in hand, how should you connect them to get bigger ampere?

References

[1] D. Enescu, Thermoelectric Energy Harvesting: Basic Principles and Applications, DOI: 10.5772/intechopen.83495 , https://www.intechopen.com/chapters/65239.

[2] https://www.marketsandmarkets.com/Market-Reports/thermoelectric-generators-market-91553904.html.

[3] A. Elarusi, N. Illendula, H. Fagehi, Performance Prediction of Commercial Thermoelectric Generator Modules Using the Effective Material Properties, Western Michigan University, 2014.

[4] C. L. Hapenciuc, T. Borca-Tasciuc, and I. N. Mihailescu, The relationship between the thermoelectric generator efficiency and the device engineering figure of merit Zd,eng. The maximum efficiency ηmax, AIP Adv. 7, 045007 (2017); https://aip.scitation.org/doi/10.1063/1.4979328.

Cogeneration cycles

15

Chapter outline

Cogeneration or *multigeneration (MG)* is the generation of several forms of products and energy, by combining different energy conversion technologies. To this point, in the present book, different power generators including steam turbines, gas turbines, micro-steam, and micro-gas turbines, reciprocating engines, solar power generators, wind turbines, hydro turbines, fuel cells, and thermoelectric generators (TEGs) are presented. Combining these technologies to reach higher efficiency and producing the energy demands of the customer are the main topics of this chapter. The power generators can also be integrated with different cooling, heating [1], water desalination systems, hydrogen generation, dryers, energy storage systems, etc. to improve the overall efficiency, reduce greenhouse gases, and make economic profits.

In the cogeneration, one of the power generators is the heart of the cycle, and the recovered heat can be used for heating, process heat, cooling, hydrogen generation, desalinated water production, more power generation, etc. as the secondary products. The connection between the prime power generator and the secondary subsystems is usually created by some heat recovery or heat exchanger units.

Multigeneration cycles are getting popular due to their numerous economic, environmental, thermal efficiency, and passive defense advantages.

Since most of the MG cycles reduce fuel consumption by recovering waste heat of the power generator, they reduce the energy bills, which generally are required to be paid for. The secondary units added to the power generators usually have short payback period and make considerable economic income and profits.

Power Generation Technologies. https://doi.org/10.1016/B978-0-323-95370-2.00014-4
Copyright © 2023 Elsevier Inc. All rights reserved.

The overall energy efficiency of the cogeneration cycles is considerably higher than the electrical efficiency of the prime power generator alone. The electrical efficiency of most of the power generators is usually less than 50%, while the overall efficiency of an MG cycle can reach 90% or even more.

The design of MG cycles requires extensive knowledge about the operation principles, main and auxiliary components and equipment, technical specifications, limitations, advantages, thermodynamic characteristics, operation and maintenance jobs, energy and exergy performances, economics, environmental impact, etc. of all types of power generators and the secondary sub-systems.

MG cycles are mostly among the *distributed electricity generator* (DG) technologies, which produce power and other products at the place of demand. DGs omit the electricity transmission losses; they are more safe than the central big power plants against terrorist attacks, wars, and natural disasters.

Correspondingly, the designer should have comprehensive knowledge about the secondary sub-systems, which are integrated with the prime power generator for designing applicable MG cycles.

In addition, the designer must be equipped with the energy, exergy, environmental, and economic evaluation techniques to investigate the impact of design parameters and performing sensitivity and parametric analyses.

In the present chapter, several MG cycles, which are designed and evaluated from different viewpoints, would be introduced as examples. Their operation principles would be discussed, and the main results would be presented. The purpose of this chapter is to make the readers familiar with the MG cycles and giving them the ideas how to combine different technologies to get higher overall efficiency, make less pollution but more economic profits.

1. Cogeneration examples

In the following, several cogeneration or MG cycles with their operational principles are introduced, and some of the main results of the multicriteria evaluations are presented. Many more MG cycles can be found in the published literature [1].

1.1 Combined gas turbine with absorption chiller

Gas turbines have a high-temperature exhaust that can be used for different purposes such as running a steam turbine, generating steam for processes, generating steam for absorption chillers, etc. Another opportunity in the gas turbine is the heat rejection due to using intercooler in the compression process [2]. It can also be utilized to produce hot water or steam as well.

Fig. 15.1 shows the schematic of an MG cycle, which is also called CCHP that stands for *Combined Cooling, Heating, and Power*. The cycle is combined with an absorption chiller to produce cooling load, and the additional heat can be used to produce steam for process or heating. The gas turbine (GT) was originally manufactured by GE and the GT model is PG6561B [3]. The GT has an axial compressor with 17 stages and

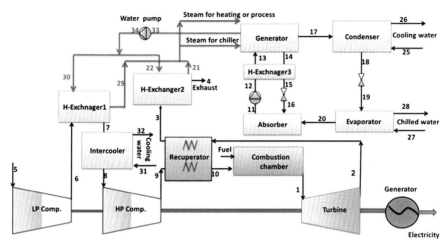

Figure 15.1 A combined cooling heating and power based on a gas turbine [2].

pressure ratio of 11.8. Turbine is also axial with three stages. The GT is simple cycle with no recuperator and single shaft. Fuel type is natural gas, with LHV at lean and rich conditions of, 45,706 kJ.kg^{-1} and 45,742 kJ.kg^{-1}, respectively. The technical design data of the GT for the summer and winter case input data at ambient temperature of 48 °C are presented in Table 15.1 [3].

In order to increase the cycle efficiency, the compression process is assumed to be double stage with an intercooler. Additionally, a recuperator is utilized at the exhaust of turbine to preheat the compressed air before entering the combustion chamber. Although the turbine has used a recuperator and intercooler, but the fuel energy

Table 15.1 Main specifications of the GT model PG6561B [3].

	Unit	Design data			
Parameter		Summer	Summer	Winter	Winter
Ambient temperature	°C	48	43	37	5
Relative humidity	%	65	65	65	100
Atmospheric pressure (average)	bar (absolute)	0.997	0.997	0.997	0.997
Rated output power	MW	35.5	-	-	-
Max output power	MW	-	-	-	47
Output at turbine shaft	MW	32.107	33.383	34.905	42.725
Gear losses	kW	420	421	423	431
Generator output	MW	30.980	32.230	33.720	41.370
Heat rate	kJ.kWh^{-1}	12,110	11,930	11,750	11,170
Air flow	kg.s^{-1}	118.4	121.6	125.4	143.8
Exhaust flow	kg.s^{-1}	122.5	125.8	129.7	149.1
Exhaust temperature	°C	555	551	546	523

utilization can be more improved by recovering heat from the hot compressed air after the first stage of compression and exhaust gases leaving the recuperator.

The heat Exchanger 1 is used before the intercooler to recover the compression heat and at the same time cooling the compressed air. This unit is designed to increase the temperature of the saturated liquid water returning from the absorption chiller and converting it to superheated steam to be used in the chiller. Similar process has occurred in the Exchanger 2 while recovering heat from the exhaust gases.

The Water-LiBr absorption chiller is utilized to consume the generated steam for cooling purpose. The main processes of the chiller are as below:

At the beginning, the superheated steam enters the generator and evaporates the refrigerant (water), and then the refrigerant vapor enters the condenser and is condensed. After that the pure condensed refrigerant (water) pours down over the evaporator's tubes and absorbs the required latent heat for evaporation from the returned line of the chilled water. The returned line of chilled water gets chilled and can be used for cooling purposes. After that the refrigerant vapor enters the absorber to be absorbed by the strong absorber solution which is poured over the cooling water tubes inside the absorber. Finally the strong absorber solution comes from the generator, which was heated in the first step and its heat is used for preheating (in the H-Exchanger 3) of the strong refrigerant solution which is pumped from the absorber to the generator. The strong absorber solution while passing through the Exchanger 3 becomes cooler before entering the absorber.

The cycle can reach the overall energy and exergy efficiencies of about 56% and 69%. Electricity production is 42.8 MW and 6.6% and 9.4% of the fuel energy can be recovered from the compression heat and exhaust gases. This heat recovery can be used to produce 24.5 tons/hr of superheated steam at temperature of 134.8 $^{\circ}C$ and pressure of 167.4 kPa. The chiller can produce 7.8 MW of cooling by using the superheated steam and its coefficient of performance is 0.68. The emission reduction for carbon dioxide is about 17%.

1.2   Combined MGT and desalination

Drinking water and clean air are the most precious material on earth. Combined cycles, which can produce drinking water and at the same time reduce air pollution, deserve to be studied and investigated in detail. For example, Ebrahimi and et al. [4] proposed a cycle (Fig. 15.2) based on a micro gas turbine (MGT), which was coupled with a small multistage flashing (MSF) desalination system. The exhaust heat of the MGT is recovered in a brine heater (BH), which is actually a heat exchanger. The MSF has two main sections of heat recovery and heat rejection. The heat recovery comprises of five flashing stages, while the heat rejection section has two stages.

The operation description of the proposed cycle is as below:

The feed brine at the inlet of each stage flashes due to installing an orifice at the inlet gate of the stage. The distilled water is produced by condensing the flashed steam. While the feed brine flows through the condenser in the heat recovery section it is preheated by the flashed steam and produces distilled water, it leaves the condenser tubes at point 4. Then, it enters the brine heater (*BH*) unit, and is heated to the top brine temperature

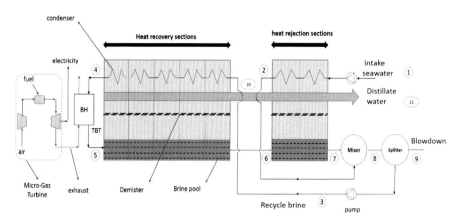

Figure 15.2 The combined cycle of a micro gas turbine and a small multistage flashing desalination system [4].

($TBT = 100\ °C$) at point 5. The heated feed brine flows through the first orifice and flashes as it enters the first stage of the MSF. Steam will be generated due to flashing, and it moves up due to lower density and distills over the condenser tubes. The condensation heat is transferred to the intake seawater flowing in the condenser tubes and thus preheating it. A portion of the remaining liquid flashes again as it enters the next stages. This process continues through the stages consecutively, and the remaining water leaves the heat recovery section and enters the heat rejection section at point 6. Inside this section, water flashes again and the vapor moves up to preheat the intake seawater from point 1 to point 2. While the preheated intake seawater leaves the heat rejection section, the remaining brine feed also leaves the heat recovery section, and these two streams get mixed in the mixer. After the mixing chamber, a splitter is installed. The output of the mixing chamber enters the splitter, and a portion of the flow goes to the blowdown at point 9 to keep the concentration at acceptable level, while the remaining is pumped back to the heat recovery section as the recycle brine at point 3 to complete the desalination cycle [4]. The main conclusions of the proposed cycle include [4]:

- A 30 kW MGT of Capstone can provide the input heat to BH of the small-MSF, which is supposed to provide the drinking water for 32 people in a residential building.
- According to the design parameters and the calculations the footprint area required to install the small-MSF is only 0.25 m^2.
- The project can make 250, 000 $US\$$ of net present value (NPV) profit during the project lifespan and the payback period of the project is less about 2 years.
- The cycle is environment friendly and can reduce CO_2 production for about 39 tons/year.

Fuel cells have great potential for MG cycles. Many cycles proposed by researchers can be found in the published investigations. Fuel cells exist in different types, diverse operating temperature, and various capacities. In the following, some examples of the cogeneration or MG cycles based on polymer exchange membrane fuel cell (PEMFC), solid oxide fuel cell (SOFC), molten carbonate fuel cell (MCFC), thermoelectric cooler (TEC) and thermoelectric generator (TEG) are introduced.

1.3 PEMFC integrated with TEG, TEC, and adsorption chiller

PEMFCs having operating temperature of 60 °C to 80 °C are popular for residential, electric car, and flight applications such as drones. In the following, two examples of MG cycles based on PEMFC are introduced [5,6].

Fig. 15.3 shows a PEMFC, which is fed with hydrogen and compressed air to produce electricity, heat, and liquid water. A jacket of cooling water is circulated around the PEMFC that produces hot water with temperature of 78 °C. Electricity is stored in a battery, and liquid water is added to the inlet cold water at point 10. The 78 °C hot water is used in a highly effective plate heat exchanger (PHE) to produce domestic hot water with temperature of 60 °C and store it in a heat storage tank. The hot water after passing through the PHE returns to the PEMFC jacket with a temperature of 71 °C. A portion of the electricity, which is stored in the battery, is used in a thermoelectric cooling system that can be a small refrigerator on a car for example. The remaining electricity can be used for other electricity demands [5].

The overall efficiency of the MG cycle in Fig. 15.3 has reached 76.94%, and a fuel energy saving ratio is 43.25%. The exergy efficiency is 53.86%. Furthermore, the carbon dioxide production has reduced about 2.58 $kg.hr^{-1}$. The overall weight of the proposed cycle is estimated less than 100 kg [5].

In another MG cycle [6], the PEMFC is integrated with TEG and adsorption chiller (Fig. 15.4). In addition, there are two tanks, hot and cold temperature damping tanks (HTDT and CTDT) to stabilize the cold and hot temperature fluctuations due to periodic operation of the adsorption chiller [6].

The electricity generated due to electrochemical reactions can be directly used or stored, but the heat should be recovered from the jacketing water of the fuel cell. An effective PHE is utilized for the heat recovery, and a pump circulates the jacketing water. The returning water from the HTDT enters the PHE at point 7 and leaves the PHE at a proper and steady temperature at point 8. The hot water enters a silica gel-water bed adsorption chiller to regenerate the silica-gel and vaporizing the water content in the adsorption bed. The temperature of the heating source leaving the chiller is time dependent at point 13; it enters an HTDT to limit its temperature fluctuations. In

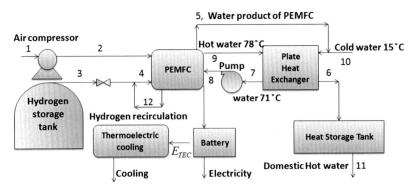

Figure 15.3 Schematic diagram of the PEMFC-TEC-based micro-CCHP system [5].

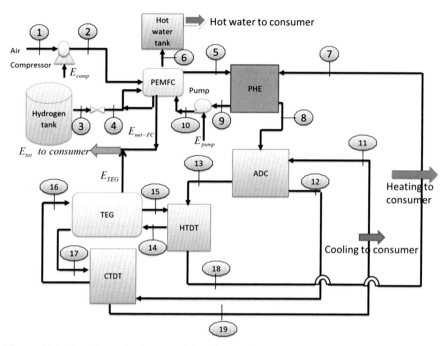

Figure 15.4 The schematic diagram of the micro-CCHP proposed in this research [6].

addition, a branch is taken from the HTDT to be used as the heat source of the TEG modules to generate more electricity. The chilled water temperature produced by the adsorption chiller at point 12 is time dependent too; to stabilize the chilled water temperature, it enters a CTDT. To improve the heat transfer in the TEG, the heat sink side is cooled with the cold water taken from CTDT. Another branch is also taken from the CTDT to provide cooling for the consumer before returning to the chiller inlet at point 11. The hot water produced due to the electrochemical reactions is directly stored in a storage tank after point 6 [6].

The results of the research [6] showed that for normal operating conditions, the overall energy/exergy efficiencies are 75% and 53%. The fuel energy saving ratio is 39%, and production of CO_2 has reduced by 16.9 tons/year. In addition, the proposed cycle can compete with conventional systems with the same outputs for hydrogen price of about 2 US$ [6].

1.4 SOFC integrated with micro gas/steam turbine

SOFC with high operating temperature provides great potential for multigeneration cycles. It can be used to generate more power, heating, cooling, desalinated water, hydrogen, etc. Fig. 15.5 shows an example in which the SOFC is coupled with an MGT and an MST for generation of more power [7].

As the schematic of the cycle shows, air and methane are compressed by using compressors before entering the recuperator and steam reformer. Then, the compressed air

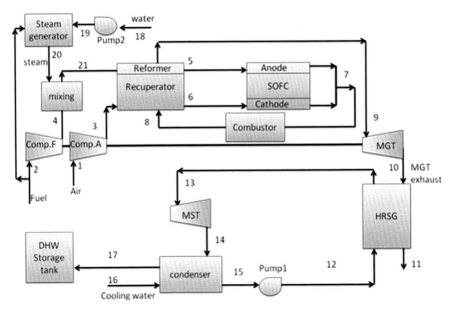

Figure 15.5 Schematic of the SOFC-MGT-organic Rankine cycle (ORC) cycle [7].

is heated in the recuperator, while fuel is mixed with steam in a mixer before entering the reformer. Methane takes part in the steam reforming reaction to produce hydrogen and carbon monoxide. The carbon monoxide can be transformed to carbon dioxide when it takes part in the water gas shift reaction. Since in the water gas shift process, hydrogen is generated, the carbon monoxide acts like a fuel for the SOFC. Oxygen ions ($O^=$) are the charge carrier in this type of fuel cell that flow from cathode to the anode to form water when it is combined with hydrogen. In return, electrons flow from anode to the cathode through the external circuit to produce electrical current in the opposite direction. Another byproduct of the SOFC is the hot stack gases of the SOFC. The stack gases of the SOFC first enter the combustor to burn the remaining fuel in the stack gases. Then, the combustor output is used to deliver heat for the reformer and recuperator to preheat anode and cathode inlet gases. The MGT is installed after the reformer to generate electricity and also deliver the compression power demanded by the air and fuel compressors [7].

The MGT exhaust is used in a heat recovery steam generator (HRSG) unit to produce saturated or superheated steam for the organic Rankine cycle (ORC) that operates based on an MST. The working fluid in this cycle is assumed to be R123 because it is dry refrigerant, nontoxic, nonflammable, and noncorrosive and does not need to be superheated. The critical pressure and temperature of R123 are 36.68 bar and 183.7°C; as a result, it can be used for low-quality heat sources, which is the main characteristic of the ORC technology [7]. In other words, this cycle could generate power in three steps of SOFS, MGT, and MST according to the quality of the heat source available.

The exhaust of the MST is then condensed in a condenser to be pumped again into the HRSG. The condensing heat can also be recovered to increase cycle efficiency for producing hot water. The hot water can be stored in a storage tank if it exceeds the domestic hot water (DHW) demand. The results show that the overall efficiency of the proposed can reach more than 65% [7].

1.5 MCFC integrated with stirling engine, kalina cycle, and chiller

MCFC is another high-temperature fuel cell that can be used for cogeneration or MG cycles. It operates with different fuels by using internal/external reformer. Fig. 15.6 shows an example of MG cycles based on MCFC. It integrates a Stirling engine with, a Kalina cycle, and an absorption chiller [8].

In this cycle, fuel is compressed to the point 2 and then gets mixed with steam, and the mixture enters the internal steam reformer at point 3 where the reforming and water gas shifting reactions occur inside the fuel cell to produce hydrogen from the methane. On the other side, the compressed air enters the cathode at point 1, and the overall electrochemical reaction occurs. The MCFC exhaust with a temperature of about 600°C–700°C enters a heat exchanger to heat the working fluid of the Stirling engine, which is assumed to be air. The MCFC exhaust, which is mostly water vapor, enters the absorption chiller at point 5 with a temperature of about 250°C. In the Stirling engine, the air flows between the two temperatures; the maximum temperature (hot cylinder) and the minimum temperature (cold cylinder). The hot air after leaving the Stirling engine enters the evaporator of the Kalina cycle at point 8, where its heat is transferred to the ammonia–water mixture and reaches to the minimum temperature of the Stirling engine at point 9. After heating the ammonia–water mixture, a two-phase flow enters the separator at point 10. In the separator, steam, which has a higher

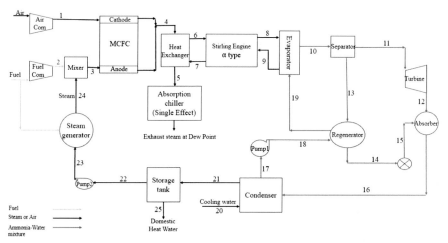

Figure 15.6 Schematic of the MG cycle comprises of MCFC, stirling, kalina cycle and absorption chiller [8].

ammonia fraction (point 11), enters the turbine. The saturated liquid at point 13 leaves the separator and enters the regenerator, where it preheats the ammonia—water mixture at point 18—19 before the evaporator. At the exit of the regenerator at point 14, the saturated liquid expands through a throttle valve and enters the absorber at point 15, where it is mixed with the turbine output. The output of the absorber is then condensed and pumped to the regenerator for preheating before entering the evaporator. On the other side, the heat from the ammonia—water mixture in the condenser is transferred to the cooling water entering. Part of the heated water can be used to produce hot water for domestic demands (DHW), and the rest can be pumped into the steam generator [8]. Care must be taken that the steam generator must be fed with demineralized water to avoid fouling and scale forming.

1.6 Photovoltaic coupled with thermoelectric generator

Photovoltaic systems need sunshine to generate electricity, when sun sets there will be no solar electricity. To solve this problem, solar systems are oversized to store some electricity in the battery banks for the night time. Batteries are still expensive and their lifespan is short. Hence, reducing the battery size by combining the solar system with other power generation technologies such as TEG or wind turbines can extend the power generation at night without the need for oversizing the PV system and using battery bank. An example of such attempts can be seen in Figs. 15.7 and 15.8 in which the

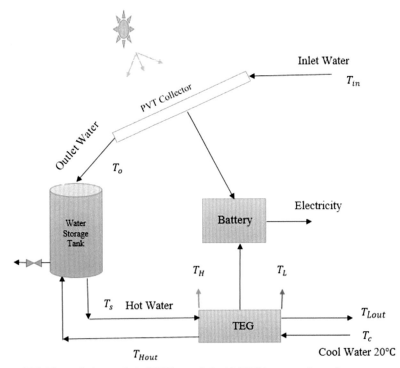

Figure 15.7 Thermal photovoltaic (PVT) coupled with TEG in a once through water system [9].

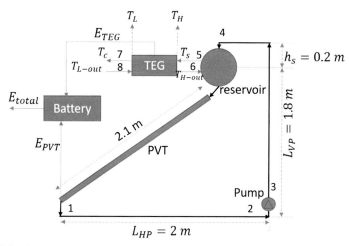

Figure 15.8 Thermal photovoltaic (PVT) coupled with TEG in a water circulating system [9].

thermal photovoltaic (PVT) is coupled with TEG. Fig. 15.7 is a once through system and the cooling water is stored in the storage tank after passing through the PVT. The cycle in Fig. 15.8 is a circulating system and the stored water is pumped through the PVT and the TEG system. In these systems, the hot water stored during the day time is used as the heat source to generate electricity during at night [9].

The analyses showed that the cycle in Fig. 15.7 is more effective from the energy, exergy, environmentally, and economic viewpoints. Moreover, both cycles have positive net present value (NPV) for interest rates of less than 10%. By calculating the internal rate of return, it was revealed that the first cycle is less risky for investment because its NPV is zero at interest rate of 15% [9].

1.7 Solar heater combined with TEG and wind turbine

Solar heat can be used for power generation in different ways. It can be used in the PV panels and concentrated solar power (CSP) plant such as those presented in Chapter 10. However, there are other creative ways to use the solar heat in the MG cycles. For example, Fig. 15.9 shows a solar system, which is combined with TEG, solar chimney, and wind turbine. In addition to generating electricity, the final hot air is used for drying agriculture products such as fruits and vegetables.

The solar collector comprises a double-glazed glass (point 1) and a copper absorber plate (point 2) that faces the sun. The side walls and back cover of the solar collector are insulated. With the purpose of increasing the air temperature and heat gain, four side mirrors are hinged to the collector casing perimeter (point 9) for concentrating solar radiation on the collector. The absorber temperature increases due to absorbing solar radiation, and heat is transferred to the air flow inside the duct. As the air inside the duct gets hot, it accelerates and ascends due to reducing density and buoyancy effect. The solar chimney is utilized to increase the air velocity for rotating the wind turbine. As the air arises through the solar chimney (point 6), its velocity increases, and at the

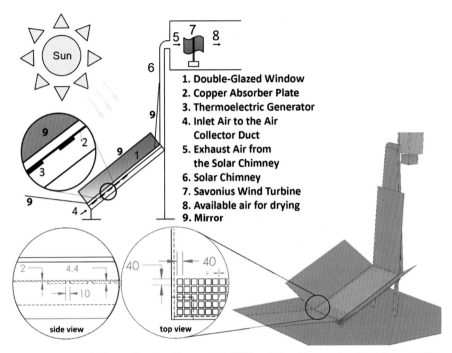

Figure 15.9 Combining solar air collector with TEG, wind turbine, and solar chimney [10].

exit point, it strikes the turbine blades and rotates the turbine (point 7). Meanwhile, TEGs (point 3) are installed under the absorber plate. The absorber plate is used as a heat source, and airflow within the duct acts as a heat sink. The TEG converts the temperature difference between the absorber plate and incoming air to electricity. The airflow at point 8 after the turbine is still warm and can be used for drying or heating purposes or to be stored in a thermal energy storage system for the time when heat is needed [10]. The simulation results proved that the collector efficiency, TEG efficiency, and power coefficient of the wind turbine in a certain hour are 61%, 4.54%, and 19.56%. The average air velocity outlet from the chimney is 2.32 m/s that generates 2.17 W when striking the wind turbine. Furthermore, the economic evaluations showed that NPV and payback are approximately 19,707 $ and 1.34 years. Finally, the cycle has the potential to reduce carbon dioxide production by 5111 kg per year [10].

1.8 Stirling engine, integrated with MST, and adsorption chiller

Stirling engines are external combustion reciprocating engines, which are capable of using different gaseous, liquid, and solid fossil fuels or the renewable energies such as solar energy, and geothermal heating. According to the waste heat temperature of these engines, they can be integrated with technologies such as organic Rankine cycles to generate power as the bottoming cycle. Fig. 15.10 shows an MG cycle based on an α-type Stirling engine [11]. The system comprises of some pumps, evaporators, gearboxes, generators,

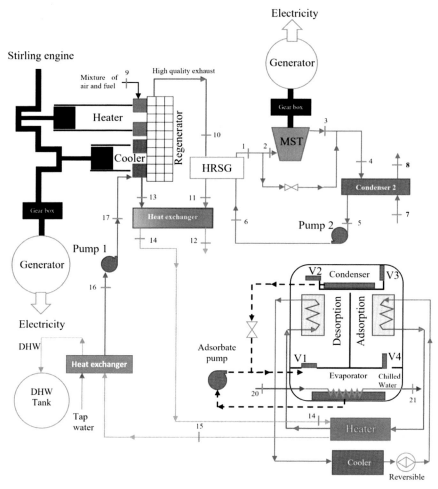

Figure 15.10 Combining stirling engine, with micro steam turbine (MST) and adsorption chiller [11].

condensers, heat exchanger, Stirling engine, heat recovery steam generator (HRSG), MST, and adsorption chiller. A description of the process is given below.

According to the cycle, methane is combusted with air, and the required input heat for the hot cylinder of the Stirling engine is provided (point 9). The hot exhaust enters an HRSG at point 10. As a result, R123, which is the working fluid of the ORC, becomes saturated steam and gets prepared to enter the MST at point 1. Since R123 is a dry working fluid, it becomes superheated as it expands through the MST.

The exhaust at point 11 still has energy to be extracted for running an adsorption chiller; therefore, it is passed through a heat exchanger to provide hot water for the chiller. The hot water enters the chiller at point 14 and leaves it at point 15. At point

15, it enters a heat exchanger to give up its energy to provide hot water for domestic uses and then enters the cold cylinder of the Stirling engine at point 17.

The results of simulation of the cycle in Fig. 15.10 revealed that the overall efficiency, fuel energy saving ratio with respect to the conventional systems, and the exergy efficiency are 76.5%, 42.13%, and 52.27%. Carbon dioxide reduction due to higher efficiency and fuel saving with respect to the conventional systems with the same amount of output energies is 28.87% [11].

1.9 MGT-ORC and steam ejector refrigerator

MGT exhaust temperature may reach 300 $^{\circ}C$ in case of using a recuperator. This temperature is still high and usually is wasted, but it can be recovered by using other low temperature operating technologies such as MST and ejector compression refrigeration cycles.

Fig. 15.11 shows such cycle [12] in which the MGT is coupled with the MST and an ejector refrigeration cycle, which can be operated in both summer and winter. In fact, it is a trigeneration or CCHP cycle that produces cooling, heating, and power simultaneously. In this cycle, the MGT and MST produce electricity, the evaporator produces cooling load, and the condenser and heat exchanger provide the heating load and domestic hot water (DHW).

The results showed the energy saving of 37% and 24% for summer and winter operating modes. In addition, the overall efficiency is about 78% in summer [12].

1.10 Internal combustion engine, integrated with absorption chiller and land fill gas

Internal combustion engines are the most matured heat engines and are available in different capacities. Although they have more maintenance jobs, but there are enough

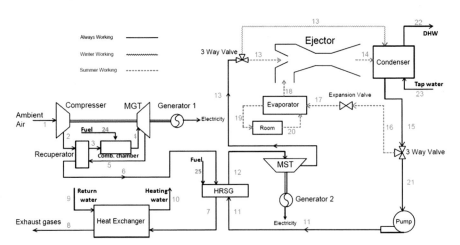

Figure 15.11 MG cycle for CCHP applications by combining micro gas turbine (MGT), micro steam turbine (MST), and ejector [12].

skilled repairmen in the world for these engines. Their start up is fast. They have high-quality exhaust, high-temperature cooling liquid from the engine jacketing, and lube oil cooler. These potentials make them an appropriate candidate for cogeneration and MG cycles.

Ref. [13] proposed a trigeneration cycle operating with land fill gas (LFG). The system consists of three main sub-systems: LFG collection system, LFG processing system, and the CCHP system itself. The CCHP system is based on a reciprocating internal combustion engine. The schematic of the system is redrawn in Fig. 15.12. According to the cycle, the LFG is collected through some pipes, which are connected to a blower. The blower is an induced draft fan that sucks the LFG and then the particulate pollution are filtered by the filteration system. The scrubber also cleans the LFG from the chlorine, sulfur, and ammonia, and the water vapor is condensed, separated, and drained in the cooler. The extra LFG will be flared in the flare line [1].

2. Problems

1 Describe how the cycle presented in Fig. 15.13 operates?
2 Explain how the schematic cycle presented in Fig. 15.14 works?
3 Define how the cycle illustrated in Fig. 15.15 operates?
4 Explain the operation principals of the cycle presented in Fig. 15.16?
5 Describe how the schematic diagram presented in Fig. 15.17 works?

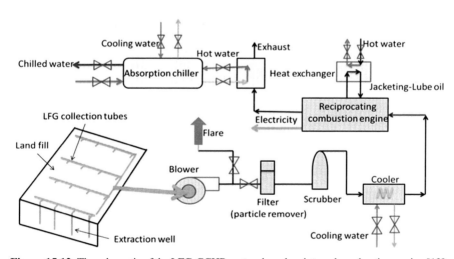

Figure 15.12 The schematic of the LFG-CCHP system based on internal combustion engine [13].

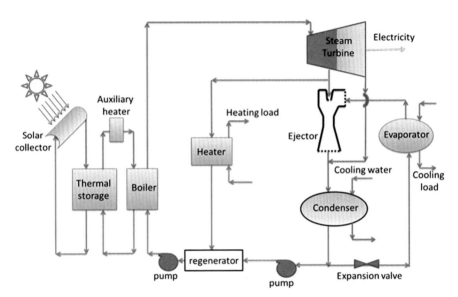

Figure 15.13 Ejector cooling ORC-CCHP working with solar energy [1,14].

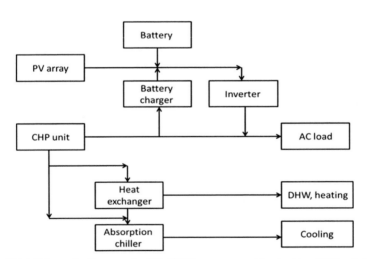

Figure 15.14 Schematic diagram of a PV-CCHP system combined with a CHP unit [1,15].

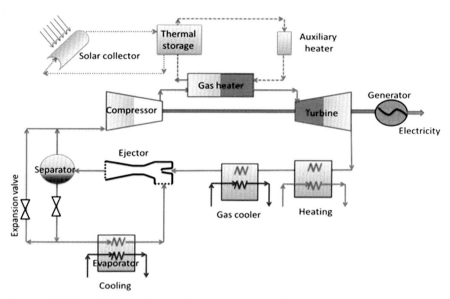

Figure 15.15 The CCHP cycle comprises of micro gas turbine (MGT), solar collector, and ejector refrigeration unit [1,16].

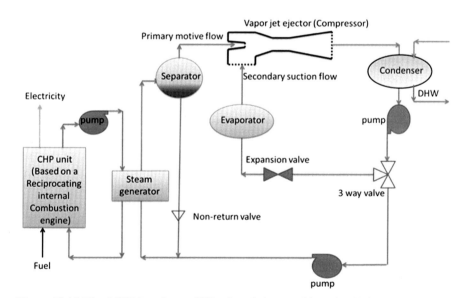

Figure 15.16 The CCHP based on a CHP unit and ejector refrigeration [1,17].

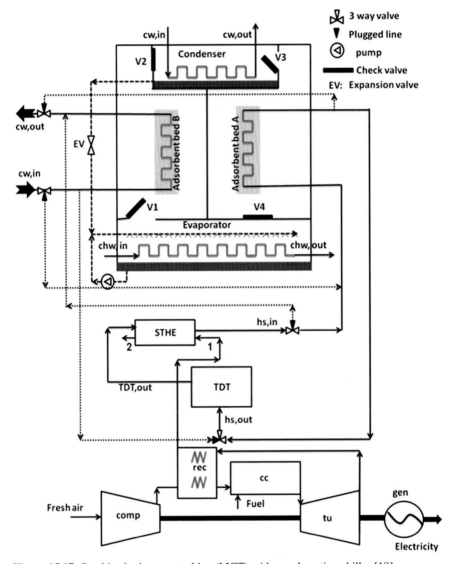

Figure 15.17 Combined micro gas turbine (MGT) with an adsorption chiller [18].

References

[1] M. Ebrahimi, A. Keshavarz, Combined Cooling, Heating and Power, Decision-Making, Design and Optimization, first ed., Elsevier, 2014.

[2] M. Ebrahimi, S. Majidi, Exergy-energy-environ evaluation of combined cooling heating and power system based on a double stage compression regenerative gas turbine in large scales, Energy Conversion and Management 150 (2017) 122–133.

[3] Gas turbine data sheet, GE energy products Germany GmbH, Model GT Frame 6 PG6561B. https://www.ge.com/.

[4] M. Ebrahimi, B. Naghali, M. Aryanfar, Thermoeconomic and environmental evaluation of a combined heat, power, and distilled water system of a small residential building with water demand strategy, Energy Conversion and Management 258 (2022) 115498.

[5] M. Ebrahimi, E. Derakhshan, Design and evaluation of a micro combined cooling, heating, and power system based on polymer exchange membrane fuel cell and thermoelectric cooler, Energy Conversion and Management 171 (2018) 507−517.

[6] M. Ebrahimi, E. Derakhshan, Thermo-environ-economic evaluation of a trigeneration system based on thermoelectric generator, two-bed adsorption chiller, and polymer exchange membrane fuel cell, Energy Conversion and Management 180 (2019) 269−280.

[7] M. Ebrahimi, I. Moradpoor, Combined solid oxide fuel cell, micro-gas turbine and organic rankine cycle for power generation (SOFC-MGT-ORC), Energy Conversion and Management 116 (2016) 120−133.

[8] I. Moradpoor, M. Ebrahimi, Thermo-environ analyses of a novel trigeneration cycle based on clean technologies of molten carbonate fuel cell, Stirling engine and Kalina cycle, Energy 185 (2019) 1005−1016.

[9] M. Ebrahimi, M. Amin Ahmadi, E. Khalife, Multi-criteria evaluation, and dynamic modeling of combining thermal photovoltaic and thermoelectric generators to extend electricity generation at night, Journal of Cleaner Production 344 (2022) 131107.

[10] K. Kazemi, M. Ebrahimi, M. Lahonian, A. Babamiri, Micro-scale heat and electricity generation by a hybrid solar collector-chimney, thermoelectric, and wind turbine, Sustainable Energy Technologies and Assessments 53 (2022) 102394.

[11] A. babamiri, M. Gharib, M. Ebrahimi, Multi-criteria evaluation of a novel micro-trigeneration cycle based on α-type Stirling engine, organic Rankine cycle, and adsorption chiller, Energy Conversion and Management 253 (2022) 115162.

[12] M. Ebrahimi, K. Ahookhosh, Integrated energy-exergy optimization of a novel micro-CCHP cycle based on MGT-ORC and steam ejector refrigerator, Applied Thermal Engineering 102 (2016) 1206−1218.

[13] X. Hao, H. Yang, G. Zhang, Trigeneration: a new way for landfill gas utilization and its feasibility in Hong Kong, Energy Policy 36 (2008) 3662−3673.

[14] J. Wang, Y. Dai, L. Gao, S. Ma, A new combined cooling, heating and power system driven by solar energy renewable, Energy 34 (2009) 2780−2788.

[15] A. Nosrat, J.M. Pearce, Dispatch strategy and model for hybrid photovoltaic and trigeneration power systems, Applied Energy 88 (2011) 3270−3276.

[16] J. Wang, P. Zhao, X. Niu, Y. Dai, Parametric analysis of a new combined cooling, heating and power system with transcritical CO2 driven by solar energy, Applied Energy 94 (2012) 58−64.

[17] J. Godefroy, R. Boukhanouf, S. Riffat, Design, testing and mathematical modelling of A small-scale chp and cooling system (small CHP-ejector trigeneration), Applied Thermal Engineering 27 (2007) 68−77.

[18] M. Ebrahimi, M. Soleimanpour, Design and evaluation of combined cooling, heating and power using micro gas turbine, adsorption chiller and a thermal damping tank in micro scale, Applied Thermal Engineering 127C (2017) 1063−1076.

Environmental impacts of power plants

16

Chapter outline

Power generation has serious impacts on the environment, including air pollution, water pollution, thermal pollution, global warming, climate change, destruction of natural landscapes, and solid waste dumping. The emission from fossil fuel combustion is the major cause of urban air pollution [1,2]. If extensive life cycle assessments or cradle to grave evaluations were conducted, it would be revealed that not only the fossil fuel−based power generation systems but also renewable energy−based power generation technologies have harmful impacts on the environment. Human has founded the new civilization on energy, especially heat and electricity. Without energy, the planet would be dark and cold. Generating power without negative impact on the environment (air, water, and soil) is impossible, but it can be minimized by optimizations in different steps from the cradle to grave. Or it can be compensated by using technologies with a very short energy payback period.

Combusting fossil fuels is the main source of greenhouse gases (GHGs). Fig. 16.1 shows the carbon dioxide emissions by different fuels. It shows that coal, oil, and gas are the most CO_2 generators in the world.

In addition, when comparing different sectors involved in GHG generation (Fig. 16.2), it shows that 73.2% of the GHG is generated due to using energy in three main sectors of industries (24.2%), buildings (17.5%), and transport (16.2%) in 2016. Agriculture, forest, and land use are also responsible for 18.4% of the total GHG emissions. According to the pie chart in Fig. 16.2, it is clear that the potential to reduce GHG production in the energy systems is high.

The trend of equivalent[1] CO_2 production in different sectors from 1990 to 2018 is shown in Fig. 16.3. It shows that electricity and heat are the main sources of CO_2 emissions globally. The vertical axis shows the equivalent CO_2 production in billion

---

[1] "This means non-CO_2 gases are weighted by the amount of warming they cause over 1100-year timescale" [3].

Power Generation Technologies. https://doi.org/10.1016/B978-0-323-95370-2.00016-8
Copyright © 2023 Elsevier Inc. All rights reserved.

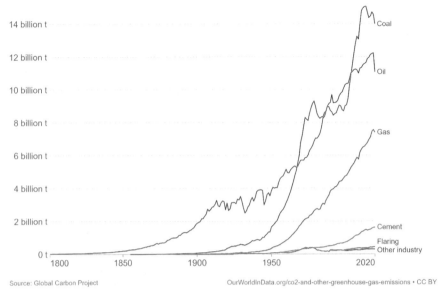

Figure 16.1 Global CO_2 emissions by fuel type [3].

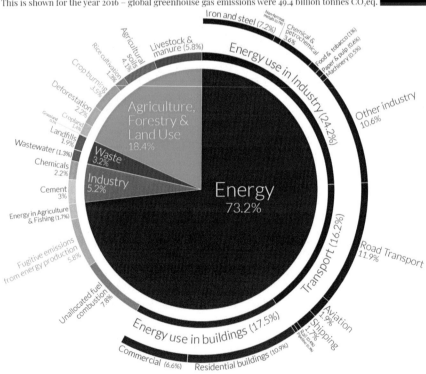

Figure 16.2 The pie chart of global greenhouse gas (GHG) emissions by sector in 2016 [3].

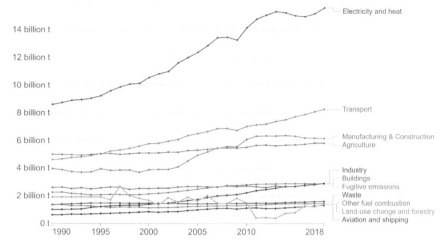

Greenhouse gas emissions by sector, World

Emissions are measured in carbon dioxide equivalents (CO2eq). This means non-CO2 gases are weighted by the amount of warming they cause over a 100-year timescale.

Source: CAIT Climate Data Explorer via Climate Watch   OurWorldInData.org/co2-and-other-greenhouse-gas-emissions • CC BY
Note: Greenhouse gases are weighted by their global warming potential value (GWP100). GWP100 measures the relative warming impact of one molecule of a greenhouse gas, relative to carbon dioxide, over 100 years.

Figure 16.3 The trend of equivalent CO_2 emission by different sectors from 1990 to 2018 [3].

(10^9) tons. After electricity and heat, transport produces the biggest amount of equivalent CO_2, and it is still increasing. The trend of electricity, heat, and transport is highly increasing in comparison with other sectors such as construction, agriculture, and waste. In order to slow down the growth rate of global carbon dioxide production in electricity and heat sectors, special attention should be paid to the development of renewable technologies and heat recovery from the conventional systems. As discussed in the Chapter 15, heat recovery is very essential for multi-generation (MG) cycles in which, the recovered heat could be used for the production of electricity, power, desalinated water, heating, cooling, drying, heat required processes, hydrogen, etc., to reduce carbon dioxide generation by reducing fossil fuel consumption.

1. Environmental evaluation criteria

For the purpose of specifying environmental evaluation criteria to compare the power generation technologies the published researches, standards, and technical reports must be reviewed. In the following two popular criteria, which are usually used are presented. One is the life cycle assessment (LCA) and the other is the energy payback (EP), which both of them can be used to compare different power generation technologies.

1.1   *Energy payback period*

The definition of energy payback period (EP) is similar to the economical definition for the payback period of an investment. In economics, it is said that *"it takes money to make money"* in the energy systems we can say *"it takes energy to save energy"* [4]. The definition of EP implies that how long does it take for an energy system to recover the energy and the related emissions and GHG that went into making the energy system from the first place. For example, for a wind turbine, the EP would be the number of years that it must operate to generate an amount of renewable energy, which is equivalent to the amount of energy consumed to manufacture the wind turbines from the first place (from mining the row material required to build the wind turbine) to the time when they start generating renewable energy. Hence:

$$EP = \frac{E_{man}}{E_{gen}} \qquad (16.1)$$

In which:

E_{man}: is the amount of energy consumed for manufacturing of the energy system from the moment of decision to extract the necessary materials in the manufacturing until the moment when it starts to produce energy.

E_{gen}: is the amount of yearly energy saved or renewable energy generated.

Fig. 16.4 shows the EP of different PV technologies. It shows that PV systems can generate pollution-free and GHG-free electricity for 26–29 years if the total lifespan could be 30 years.

Fig. 16.4 shows that the EP for the systems using current multicrystalline-silicon PV modules is about 4 years, for the systems using current thin-film modules, it is 3 years, for the PV systems using anticipated multicrystalline modules, it is 2 years,

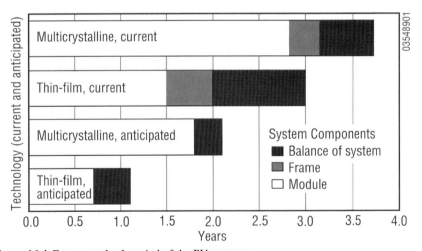

Figure 16.4 Energy payback period of the PV systems.
(Credit of US department of energy office of energy efficiency and renewable energy) [4].

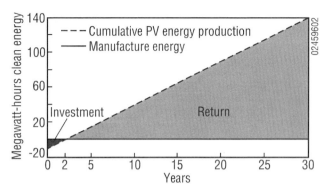

Figure 16.5 Return versus the investment energy for a rooftop photovoltaic (PV) system providing half of the electricity demand by an American household consuming 830 kWh/month.
(Credit of U.S. Department of Energy, office of energy efficiency and renewable energy) [4].

and for the anticipated thin-film modules, it is only 1 year. It means 87%—97% of the total electricity generated by the PV systems is pollution free, GHG free without depletion of energy resource [4].

Fig. 16.5 shows the cumulative PV energy production, and manufacturing energy of a rooftop PV system that meets half of the electrical demand for a US household that consumes an average energy of 830 kWh per month. The EP is assumed 2 years. It shows that during the 28 years of pollution-free operation, it eliminates 0.5 ton of SO_2, 0.33 ton of NO_x, and 100 tons of CO_2 [4].

1.2 Life cycle assessment

LCA has be defined as following [5]:

> *LCA studies the environmental aspects and potential impacts throughout a product's life cycle (i.e. cradle-to-grave) from raw materials acquisition through production, use and disposal. The general categories of environmental impacts needing consideration include resource use, human health, and ecological consequences.*

The National Renewable Energy Laboratory (NREL) in 2012 led the Life Cycle Assessment (LCA) Harmonization Project. In this project, results of 400 published research studies of PV systems including crystalline silicon (c-Si) (mono-crystalline and multicrystalline) and thin film (TF) (amorphous silicon (a-Si), cadmium telluride (CdTe), and copper indium gallium diselenide (CIGS)) were reviewed and screened. Among them, 17 studies passed the screening. In these 17 studies, 46 life cycle GHG emissions for the PV technologies were reported [6].

The results show that if the life cycle of a PV electricity generator and a coal-fired thermal power plant could be divided into three stages of upstream processes, operational processes, and downstream processes, the PV system consumes 60%—70% of the total energy during the upstream processes, 21%—26% of the total energy during

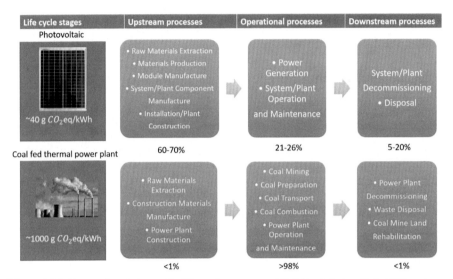

Figure 16.6 Comparison between the life cycle assessment (LCA) of photovoltaic (PV) systems and coal fed thermal power plants, regenerated from Ref. [6].
(Credit of U.S. Department of Energy, office of energy efficiency and renewable energy).

the operational processes, and 5%−20% of the total energy during the downstream process. For a coal-fired thermal power plant, these values are about 1% in the upstream, 98% during operational processes, and about 1% in the downstream processes. It shows that for the PV panels, the biggest potential for improvement is in the upstream processes, while for the coal-fired thermal power plants, the major opportunity for improvements is in the operational processes. The actions for each stage are shown in Fig. 16.6. This figure also shows that the equivalent carbon dioxide generation due to generating each kWh of solar electricity is only 40 g, while it is 1000 g for coal-fired power plants [6].

LCA for equivalent GHG emissions in $\frac{gCO_2}{kWh}$ for other power generation technologies such as wind turbines, nuclear power plants, coal-fired power plants, concentrating solar power plant, and PV technologies are also reported by NREL and presented in Fig. 16.7. As it can be seen, the mean value for the crystalline silicon is 40 $\frac{gCO_2eq}{kWh}$; for wind turbine, it is about 10 $\frac{gCO_2eq}{kWh}$, while it is 1000 $\frac{gCO_2eq}{kWh}$ for the coal-fired power plants. According to the results of LCA, the renewable technologies are highly clean in comparison with fossil fuel−based technologies.

2. Hydrogen-fired power generators

Most of fossil fuel−fired power generators such as steam turbines, gas turbines, MGT, MST, high-temperature fuel cells such as SOFC, and MCFC can use hydrogen or a mixture of hydrogen with other conventional fuels to reduce emissions. Today,

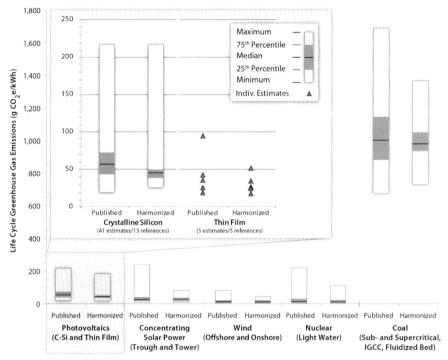

Figure 16.7 Comparing the life cycle greenhouse gas (GHG) emissions for some power generators [6].
(Credit of U.S. department of energy office of energy efficiency and renewable energy).

most of the leading manufacturers of the gas turbine power generators such as GE and Siemens are proposing or planning to have products that can be operated with hydrogen volume fractions from 5% to 100% [7,8].

Burning hydrogen instead of natural gas in power plants such as gas turbines has some challenges. Since there are differences between the combustion of hydrogen with other fuels such as natural gas, its impact on all of the gas turbine components and systems must be considered. In addition, in a power plant with some hydrogen-fired turbines, the balance of the plant must be revised as well. Changes may be needed in the fuel accessories and the bottoming cycle components in the cogeneration or multigeneration cycles [7]. Hydrogen, which can be used in the power generators, is categorized according to the production process, and they are usually classified by color as below [7]:

Gray (or black) hydrogen is produced by gasification of coal or reforming of natural gas without carbon capture [7].

Blue hydrogen is produced by steam reforming of methane with carbon capture and storage [7].

Green hydrogen is produced by electrolysis of water using renewable power such as photovoltaic panels or wind turbines [7].

Pink (red) hydrogen is the hydrogen produced by electrolysis of water using nuclear power.

Turquoise hydrogen can be produced by pyrolysis of methane, which produces hydrogen and solid carbon as a by-product [7].

White hydrogen is produced by gasification or other process using 100% biomass as a feedstock [7].

3.   Net zero emission by 2050 (NZE2050)

United nation (UN) has defined the NZE2050 as below [8]:

> *Put simply, net zero means cutting greenhouse gas emissions to as close to zero as possible, with any remaining emissions re-absorbed from the atmosphere, by oceans and forests for instance.*

Scientific researches shows that in order to prevent the harmful effects of global warming and protect the livability of the planet, the increase in temperature should be limited to 1.5 $°C$ above the preindustrial temperature. Until now, the temperature has increased by 1.1 $°C$ compared to the 1800s, and GHGs are produced more and more every day. In order to prevent the increase in temperature by more than 1.5 $°C$, the amount of pollutants should be reduced by 45% by 2030 and reach zero by 2050 (Paris Agreement) [8].

China, the United States, and European Union are the main three GHG emitters, and they are responsible for more than 46% of the total pollution [8].

The main three emission producers with seven other countries of India, Russia Federation, Japan, Brazil, Indonesia, Iran, and Canada are responsible for 68% of the emission production.

These 10 emission producers need to take the most important steps toward reducing emission.

Global warming is the most important challenge of human kind and need fast and precise actions. We need to take care about what we produce, consume and move about.

Energy sector, especially heat and electricity, is the most important sector that needs improvement, and it should provide energy from renewable energies such as solar, wind, hydro, etc. In addition, the fossil fuel—based technologies need to be replaced with hydrogen-fired technologies specially those using green or blue hydrogen. Heat recovery and multigeneration cycles also play an important role for reducing GHG.

Example 16.1. A Siemens gas turbine product, SGT5-9000HL has the following specifications reported in Table 16.1:

By recovering the exhaust heat and using it for more power generation in a bottoming cycle, fuel would be saved. Due to fuel saving, the production of carbon dioxide by the bottoming cycle would be avoided. Assume CO_2 production due to burning natural gas in the gas turbine is $0.2 \frac{kg\,of\,CO_2}{kWh\,of\,natural\,gas}$, and the gas turbine could be operated for

Table 16.1 the SGT5-9000HL specifications [9].

Parameter	Magnitude
Power output	593 MW
Frequency	50 Hz
Gross efficiency	43%
Turbine speed	3000 rpm
Pressure ratio	24:1
Exhaust mass flow	$1050 \frac{kg}{s}$
Exhaust temperature	$670\ °C$

8000 h per year. Assume the combustion efficiency and the effectiveness of the heat recovery to be 100%. Assume the combustion gases as pure air.

Determine:

a. The magnitude of the carbon dioxide emitted to the environment by this turbine in tons per year.
b. How much carbon dioxide would be avoided due to the heat recovery in 1 year, if the exhaust temperature could be cooled to $120\ °C$ by the heat recovery steam generator of the bottom cycle?

Solution. **a.** According to the definition of thermal efficiency and considering the combustion efficiency of 100% the input heat rate would be:

$$\eta_{th} = \frac{P_{GT}}{\dot{Q}_{in}} \rightarrow \dot{Q}_{in} = \frac{P_{GT}}{\eta_{th}} = \frac{593 \times 10^3}{0.43} = 1.379 \times 10^6 kW$$

To determine the energy consumed by the simple cycle GT, the input heat rate must be multiplied by the operational hours per year:

$$Q_{yearly} = \dot{Q}_{in} \times H = 1.379 \times 10^6 kW \times 8000 \frac{h}{year} = 1.103 \times 10^{10} \frac{kWh}{year}$$

And the generated carbon dioxide would be determined by multiplying the emission index of natural gas by the consumed energy as below:

$$E_{CO_2} = Q_{yearly}\left(\frac{kWh}{year}\right) \times i_{CO_2}\frac{kg\,of\,CO_2}{kWh} = 1.103 \times 10^{10} \times 0.2$$

$$= 2.207 \times 10^{10}\frac{kg\,of\,CO_2}{year}$$

$$= 2.207 \times 10^6 \frac{tons\,of\,CO_2}{year}$$

That means, each year this GT will emit 2.207×10^6 tons of CO_2 into the environment.

b. The heat recovery avoid CO_2 emissions in the bottoming cycle. The magnitude of the recovered heat can be estimated by assuming the exhaust gases as pure air as below:

$$T_{ex} = 670°C \rightarrow h_{ex} = 981.6 \frac{kJ}{kg}$$

$$T_{out} = 170°C \rightarrow h_{out} = 394.3 \frac{kJ}{kg}$$

The recovered heat rate would be:

$$\dot{Q}_{rec} = \dot{m}_{ex}(h_{ex} - h_{out}) = 1050 \times (981.6 - 394.3) = 616597 kW$$

And the yearly energy recovered would be:

$$Q_{rec,yearly} = \dot{Q}_{rec} \times H = 616597 \times 8000 = 4.933 \times 10^9 \frac{kWh}{year}$$

And the avoided emission would be:

$$E_{CO_2} = Q_{rec,yearly} \left(\frac{kWh}{year} \right) \times i_{CO_2} \frac{kg\,of\,CO_2}{kWh} = 4.933 \times 10^9 \times 0.2$$
$$= 9.866 \times 10^8 \frac{kg\,of\,CO_2}{year} = 9.866 \times 10^5 \frac{tons\,of\,CO_2}{year}$$

Which means by using the heat recovery and bottoming cycle, approximately one million tons of carbon dioxide would be avoided yearly. The problem is also coded in the EES, which is presented in Code 16.1:

Code 16.1. The EES code for the Example 16.1

```
ETA=0.43
Power=593*10^3                          "kw"
ETA=Power/Q_input
H=8000                        "Hours per year"
Q_yearly=Q_input*H            "kWh of natural gas"
i_CO2=0.2                     " kg of CO2 per kWh of natural gas"
E_CO2=Q_yearly*i_co2/1000     "ton of CO2 in one year"
m_ex=1050 "kg/s"
T_ex=670
h_ex=enthalpy(air, T=T_ex)
T_out=120
h_out=enthalpy(air, T=T_out)
Q_rec=m_ex*(h_ex-h_out)
Q_rec_yearly=Q_rec*H
E_avoided_CO2=Q_rec_yearly*i_CO2/1000
```

Example 16.2. In the previous example, suppose the thermal efficiency of the bottoming cycle is 45%. If it was supposed to generate the same amount of power generated by the bottoming cycle by PR using a coal fired power plant, how much carbon dioxide would be generated during the life cycle of the plant. Assume the plant could operate for 35 years. Assume the effectiveness of the heat recovery system to be 100%.

Solution. The thermal efficiency of the bottoming cycle would be as below:

$$\eta_{th} = \frac{P_{bottming}}{\dot{Q}_{rec}} \rightarrow 0.45 = \frac{P_{bottming}}{616597} \rightarrow P_{bottming} = 277469kW$$

According to Fig. 16.6 for each kWh of electricity generated by a coal-fired power plant 1 kg of equivalent carbon dioxide would be produced during the life cycle of the plant. Hence:

$$LC_{CO_2,coal} = 1 \frac{kgofeqCO_2}{kWh}$$

Hence, the electrical energy generated for 8000 h of operation during 35 years of its life cycle would be:

$$P_{LC} = P_{bottming} \times H \times Y = 277469kW \times 8000 \frac{h}{year} \times 35 years$$

$$= 7.769 \times 10^{10} kWh$$

And the emitted equivalent carbon dioxide would be:

$$E_{LC} = P_{LC} \times LC_{CO_2,coal} = 7.769 \times 10^{10} kWh \times 1 \frac{kgofeqCO_2}{kWh}$$

$$= 7.769 \times 10^{10} kgofeqCO_2 = 7.769 \times 10^7 tonsofeqCO_2$$

That means if the heat would not be recovered and we had to establish a coal-fired power plant to generate the same amount of power as the bottoming cycle, 7.769×10^7 tons of extra carbon dioxide would be emitted to the environment during the 35 years. This problem clearly shows the potential of heat recovery to reduce GHG emissions. The problem is also coded in EES and presented below:

Code 16.2. The EES code prepared for the Example 16.2

```
H=8000                          "Hours per year"
m_ex=1050 "kg/s"
T_ex=670
h_ex=enthalpy(air, T=T_ex)
T_out=120
h_out=enthalpy(air, T=T_out)
Q_rec=m_ex*(h_ex-h_out)
Y=35            "life cycle"
eta_bottoming=0.45
eta_bottoming=P_bottoming/Q_rec
LC_CO2_coal=1    "kg of CO2  per kWh"
P_LC=P_bottoming*H*Y
E_LC=P_LC*LC_CO2_coal
```

Example 16.3. A photovoltaic power plant produces 100 MWh of electricity during 30 years of operation. How much carbon dioxide has been emitted to the environment in the upstream processes?

Solution. According to the NREL reports (Fig. 16.6), the PV systems emit 60%−70% of the emissions during the upstream processes. Hence, by estimating the life cycle emissions, the upstream emissions would be estimated as below:

$$E_{total} = Energy_{LC} \times i_{CO2,eq} = 100MWh \times 40\frac{gCO_2eq}{kWh} = 4000kgCO_2eq$$

$$= 4tonsCO_2eq$$

The upstream processes produce 60%−70% of this value; hence:

$$E_{upstream} = (0.6\ to\ 0.7)E_{total} = 2.4\ to\ 2.8\ tonsCO_2eq$$

Therefore, the upstream processes would produce 2.4−2.8 tons of equivalent carbon dioxide during 30 years of operation.

Example 16.4. How much equivalent GHG would be generated if a coal-fired power plant was used to produce the 100 MWh of electricity during 30 years?

Solution. According to the NREL reports (Fig. 16.6), the LC emission index of coal-fired power plants is $1000\frac{gCO_2eq}{kWh}$; hence:

$$E_{total} = (energy producedinLC) \times i_{CO2,eq} = 100MWh \times 1000\frac{gCO_2eq}{kWh}$$
$$= 100000kgCO_2eq = 100tonsCO_2eq$$

Comparing the Examples 16.3 and 16.4 proves the critical role of solar systems in reducing GHG emissions.

4. Problems

1. What is carbon capturing? How does it help to make environment clean?
2. Propose a combined cycle which is able to produce gray and green hydrogen at the same time. Draw the schematic diagram of the power plant.
3. Two Siemens gas turbine products, have the following specifications reported in Table 16.2:

By recovering the exhaust heat and using it for more power generation in a bottoming cycle, fuel would be saved. Due to fuel saving, the production of carbon dioxide by the bottoming cycle would be avoided. Assume CO_2 production due to burning natural gas in the gas turbine is $0.2\frac{kgofCO_2}{kWhofnaturalgas}$, and the gas turbine could be operated for 7800 h per year. Assume the combustion efficiency and the effectiveness of the heat recovery to be 100%. Assume the combustion gases as pure air.

Table 16.2 the SGT6-9000HL and SGT5-8000HL specifications [9].

	SGT6-9000HL	**SGT5-8000HL**
Power output	440 MW	450 MW
Frequency	60 Hz	50 Hz
Gross efficiency	43.2%	41.2%
Turbine speed	3600 rpm	3000 rpm
Pressure ratio	24:1	21:1
Exhaust mass flow	$725 \frac{kg}{s}$	$935 \frac{kg}{s}$
Exhaust temperature	$670\,°C$	$630\,°C$

Determine the followings for both GTs and compare the results:

a. The magnitude of the carbon dioxide emitted to the environment by the turbines in tons per year.

b. How much carbon dioxide would be avoided due to the heat recovery in 1 year, if the exhaust temperature could be cooled to 110 °C by the heat recovery steam generator of the bottoming cycle?

4. In the previous problem suppose the thermal efficiency of the bottoming cycle is 40%. If it was supposed to generate the same amount of power as the bottoming cycle by using a photovoltaic power plant, how much carbon dioxide would be generated during the life cycle of the plant. Assume the plant could operate for 30 years.

5. A photovoltaic power plant produces 1000 MWh of electricity during 30 years of operation. How much carbon dioxide would be emitted to the environment in the downstream and operational processes?

6. Make a literature review about the LC emissions emitted by a micro gas turbine, how much does it emit emissions per kWh of electricity?

7. Make a literature review about the global LC emissions of gas turbines per kWh of electricity production.

References

[1] https://www.eea.europa.eu/help/glossary/eea-glossary/environmental-impact-of-energy.
[2] https://www.eia.gov/energyexplained/electricity/electricity-and-the-environment.php.
[3] https://ourworldindata.org.
[4] https://www.nrel.gov/docs/fy04osti/35489.pdf.
[5] W. Klopffer, B. Grahl, Life Cycle Assessment (LCA), Wiley-VCH Verlag GmbH & Co. KGaA, 2014, pp. 1−2.
[6] https://www.nrel.gov/docs/fy13osti/56487.pdf.
[7] https://www.ge.com/gas-power/future-of-energy/hydrogen-fueled-gas-turbines.
[8] https://www.un.org/en/climatechange/net-zero-coalition.
[9] https://www.siemens-energy.com/global/en/offerings/power-generation/gas-turbines/sgt-600.html.

Index

United States
Publisher Services